PROGRESS IN HYPERTENSION
Volume 2

PROGRESS IN HYPERTENSION
Volume 2

ANTIHYPERTENSIVE DRUGS TODAY

Editors
H. Saito [1], M. Minami [2] and S.H. Parvez [3]

[1]*First Department of Pharmacology, Hokkaido University School of Medicine, Sapporo 060, Japan*

[2]*Department of Pharmacology, Faculty of Pharmaceutical Sciences, Higashi-Nippon-Gakuen University, Ishikari - Tobetsu 061-02, Hokkaido, Japan*

[3]*Neuropharmacology Unit, University of Paris XI, Bât. 440, 91405 Orsay, France*

///VSP///
Utrecht, The Netherlands
Tokyo, Japan
1992

VSP BV
P.O. Box 346
3700 Zeist
The Netherlands

First published in 1992

ISBN 90-6764-140-5

CIP-GEGEVENS KONINKLIJKE BIBLIOTHEEK, DEN HAAG

Antihypertensive

Antihypertensive drugs today / ed. H. Saito, M. Minami and
S.H. Parvez. – Utrecht : VSP. – (Progress in hypertension
; vol. 2)
Met index.
ISBN 90-6764-140-5 geb.
Trefw.: bloeddruk / geneesmiddelen bij hoge bloeddruk.

Printed by Koninklijke Wöhrmann BV, The Netherlands

This book is dedicated to

Professor Tsuneyoshi Tanabe

as a homage to our Master and Teacher for his:
Outstanding Leadership
Brilliant Scientific Achievements and
Continued Support to his students

Preface

The investigations in the field of "antihypertensive drugs" have been greatly enriched during the past decade. Many new compounds with clinically useful characteristics have appeared in research and clinical practice. It's important that each class of antihypertensive drug is useful clinically and that new compounds are still being sought in each class of antihypertensive agents to improve their efficacy, potency, duration of action and adverse reactions.

Recently, the primary treatment of hypertension is targeted at the pathophysiology of the individual patient. Namely, the defects in normal physiology in clinical cases are the most important focus for the drug treatment. Therefore, the appropriate selection of antihypertensive drugs is the strategy for the underlying physiological derangement. Since the pathophysiology of hypertension has made great advances during the past few years, we can explain more accurately the underlying physiological derangement in individual patients with hypertension. Therefore, the development of better drugs with different new mechanisms have to be designed for antihypertensive drug therapy.

Now, we would like to look toward the complications of hypertension. Statistics in Japan indicate that hypertensive and borderline hypertensive account for the majority of the aged population; in males beyond the seventh decade of life, more then one-third are hypertensive and the other one-third are borderline hypertensive; the definite risk factor for stroke is hypertension. Moreover, it has been realized that the disturbance of microcirculation in the brain due to the complication of stroke might be the cause of vascular dementia. Therefore, primary prevention of hypertension in Japan can be targeted at over two-thirds of the population. Since senile diseases such as cerebrovascular dementia are increasing despite the recent gradual reduction in the incidence of stroke, suitable antihypertensive drug therapy is the most important prevention for the complications of hypertension, such as stroke and stroke-related cerebral disorders.

Next, it is necessary for us to look toward the great advances in pathophysiology of hypertension. During the past few years, there is significant evidence indicating that endothelin, endothelium derived relaxing factor (EDRF) and biogenic amines especially $5HT_2$ receptor agonists are involved in the regulation, development and maintenance of cardiovascular hypertension in both central and peripheral sites. Moreover, it has been reported that ouabain (g-strophanthin) might be the endogenous digitalis-like substance. This endogenous digitalis-like substance produces a decrease in Na^+, K^+-ATPase activity. Since intracellular Ca^{2+} concentration may rise, it can increase arterial tone, total peripheral resistance and finally onset of hypertension. These advances in pathophysiology of hypertension have been the driving force for the development of new antihypertensive drugs.

The second volume of "**PROGRESS IN HYPERTENSION**" summarizes the recent advances in basic and clinical research dedicated to new antihypertensive drugs and related interactions. Each chapter reviews up-to-date information on specific fields along with experimental findings. The editors expect that this volume will be useful to those interested in the development of better antihypertensive drugs as well as to researchers in clinical and basic pharmacology and cardiovascular medicine.

We are extremely grateful to the members of Department of Pharmacology, Higashi- Nippon-Gakuen University for their participation and help in the completion of this book. We also thank Dr. Groesbeek, V.S.P. Press for his constant support.

The Editors

Contents

Preface by The Editors vii

The pharmacological background of the antihypertensive activity and side effects of urapidil
VAN ZWIETEN, P. A. (Amsterdam, The Netherlands) 1

Antihypertensive and natriuretic effects of bunazosin on stroke-prone spontaneously hypertensive rats.
MINAMI, M., HAMAUE, N., TOGASHI, H., YOSHIOKA, M., MATSUMOTO, M., SAITO, H. and PARVEZ, S. H. (Ishikari, Sapporo, Japan and Paris, France). 15

α_1-Adrenoceptor antagonists : Studies on the sympathoinhibitory activity in anesthetized rats
TOGASHI, H., YOSHIOKA, M., TOCHIHARA, M., MATSUMOTO, M., MINAMI, M. and SAITO, H. (Sapporo and Ishikari, Japan) 31

Primary prevention with metoprolol in hypertensive patients
WIKSTRAND, J. (Gothenburg, Sweden) 49

Bopindolol : Preclinical study and clinical correlations
ISHIBASHI, T. and IMAI, S. (Niigata, Japan) 73

Newly developed beta-blockers : Basic and clinical investigations
SHIMAMOTO, K. and IIMURA, O. (Sapporo, Japan) 93

Clinical aspects of oral angiotensin converting enzyme inhibitors.
BURSZTYN, M. and GAVRAS, H. (Jerusalem, Israel and Boston, Mass., USA) 107

Molecular pharmacology of intracellular Ca^{2+} antagonists.
HIDAKA, H., ISHIKAWA, T. and CHIJIWA, T. (Nagoya and Tokushima, Japan) 131

Clinical aspects of calcium antagonists
NOMURA, A. and YASUDA, H. (Sapporo, Japan) 145

Pharmacology of central dopaminergic modulation in hypertension.
VAN DEN BUUSE, M. and DE JONG, W. (Strasbourg, France) 157

Serotonin-2 receptor antagonist : Ketanserin, focus on the sympathoinhibitory action.
MATSUMOTO, M., YOSHIOKA, M., TOGASHI, H., MINAMI, M. and SAITO, H. (Sapporo and Ishikari, Japan) 189

Basic and clinical pharmacology of fenoldopam and quinpirole.
VAN DER NIEPEN, P. and DUPONT, A. G. (Brussels, Belgium) 207

NTS neuronal mechanisms and clonidine action.
SUGA, T. and SUZUKI, M. (Saitama, Japan) 233

Centrally acting antihypertensive alpha-2 adrenoceptor agonists
SCHOLTYSIK, G. (Berne, Switzerland) 257

Indapamide as an antihypertensive agent.
TODA, N. (Ohtsu, Japan) 269

Pharmacology of a new diuretic S-8666
IMAI, M., SHIMIZU, T. and NAKAMURA, M. (Tochigi and Osaka, Japan) 291

Budralazine : A direct-acting vasodilator
YOSHIOKA, M., MINAMI, M., PARVEZ, H. S. and SAITO, H. (Sapporo, Ishikari, Japan and Paris, France) 309

Possible role of endothlium-derived nitric oxide in the regulation of blood pressure.
SAKUMA, I., KOBAYASHI, T., SATO, M., YASUDA, H., KATO, M., TOGASHI, H., YOSHIOKA, M. and SAITO, H. (Sapporo, Japan) 319

Immunological factors in genetic hypertension
BATAILLARD, A., TOURAINE, J. L. and SASSARD, J. (Lyon, France) 331

Pharmacokinetic properties of antihypertensive drugs.
MIYAZAKI, K., ISEKI, K. and NOMURA, A. (Sapporo, Japan) 355

Author's index 367

Subject index 369

Progress in Hypertension, Vol. 2, pp. 1-13.
Saito *et al.* (Eds)

THE PHARMACOLOGICAL BACKGROUND OF THE ANTIHYPERTENSIVE ACTIVITY AND SIDE EFFECTS OF URAPIDIL

P.A. VAN ZWIETEN

Departments of Pharmacotherapy and Cardiology, Academic Medical Center, University of Amsterdam, Meibergdreef 15, 1105 AZ Amsterdam. The Netherlands.

SUMMARY

Urapidil, a phenylpiperazine-substituted derivative of uracil, is a selective α_1-adrenoceptor antagonist and therefore lowers blood pressure and peripheral resistance. Apart from its α_1-adrenoceptor blocking potency, which is somewhat less selective than that of prazosin, urapidil appeared to display substantial central hypotensive activity.

Initially, this central mechanism was assumed to be the same as that of clonidine, i.e. agonistic activity towards central α_2-adrenoceptors. This view, however, appeared to be incorrect for the following reasons: (1) in radioligand binding studies urapidil did not show any affinity for α_2-adrenoceptors; (2) the central hypotensive activity of urapidil, injected into cat's vertebral artery remained uninfluenced by the α_2-adrenoceptor blocking agents yohimbine and rauwolscine, which are known to suppress the central hypotensive effect of clonidine. The central hypotensive effect of urapidil so far remains unknown in detail, although a few (but not all) experimental arguments would plead for the 5-HT_{1A}-agonistic potency of urapidil as an explanation of its central activity. A presynaptic α_2-agonistic activity can only be demonstrated in certain models but not in the classical pithed rat preparation. It is unlikely to play a relevant role in urapidil's hypotensive activity, also because of the poor affinity of the drug for α_2-adrenoceptors.

Finally, a weak β_1-adrenoceptor activity of urapidil can be demonstrated, but this mechanism is unlikely to significantly contribute to the drug's hypotensive activity.

INTRODUCTION

Urapidil (Fig.1), chemically derived from uracil, has been introduced into the treatment of hypertensive disease about a decade ago and in a few European countries it has been registered as an antihypertensive drug. Its efficacy and safety appear to be well established now, but its definite position with respect to that of other antihypertensives remains to be established. Older pharmacological data (Schoetensack *et al.*, 1977; 1983) suggested that urapidil reduces blood pressure predominantly as a result of vasodilatation and a fall in total peripheral resistance. In a later stage the drug was characterized more precisely as a selective antagonist with respect to vascular postsynaptic α_1-adrenoceptors. As such urapidil resembles prazosin, the prototype of selective α_1-adrenoceptor antagonists.

In addition it has been submitted that urapidil would stimulate presynaptic α_2-adrenoceptors and hence reduce the release of endogenous noradrenaline (Eltze, 1979), similarly as observed for clonidine. It was also recognized that urapidil displays central hypotensive activity apart from its peripheral effects. Initially, the mechanism of this central effect was poorly understood and for that matter believed to be similar to that of clonidine, the prototype of centrally acting antihypotensives (Schoetensack *et al.*, 1977).

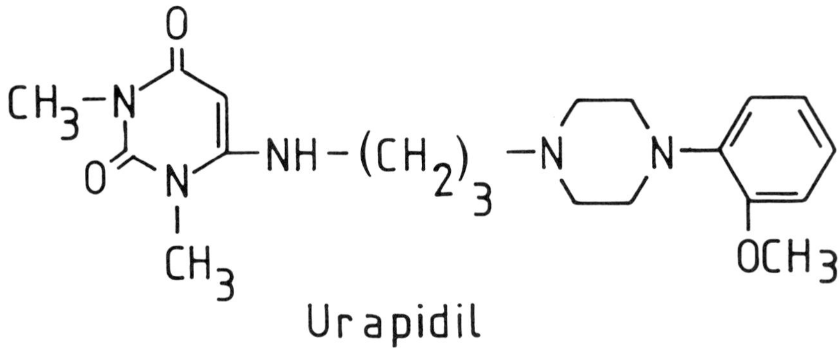

Figure 1. Chemical structure of urapidil, a derivative of uracil. Note that the molecule does not contain chiral centers and hence is not composed of different enantiomers.

Accordingly, urapidil was (incorrectly) claimed to combine the pharmacodynamic properties of both prazosin and clonidine. This view proved to be erroneous when in a later stage it was recognized that urapidil's central hypotensive mechanism is greatly different from that of clonidine (Van Zwieten *et al.*, 1985a). Nevertheless, it appears to be justified that urapidil has combined pharmacodynamic properties, including at least peripheral α_1-adrenoceptor blockade and a central hypotensive component which differs from that of clonidine and related α_2-adrenoceptor agonists. The rather unusual combination of hypotensive mechanism together with the fairly positive clinical judgement of the drug's value in antihypotensive treatment, however, has led to a renaissance of interest in urapidil's pharmacological profile. Accordingly, many

RADIOLIGAND BINDING STUDIES

Radioligand binding studies have been carried out to establish the affinity of urapidil for a number of relevant receptors. So far, all studies have been performed with homogenates of rat brain. The results of the two most important receptor binding studies (Kellar *et al.*, 1984; Van Zwieten *et al.*, 1985a) show satisfactory mutual agreement. An example of a radioligand binding study with urapidil in rat brain membranes is shown in Fig.2.

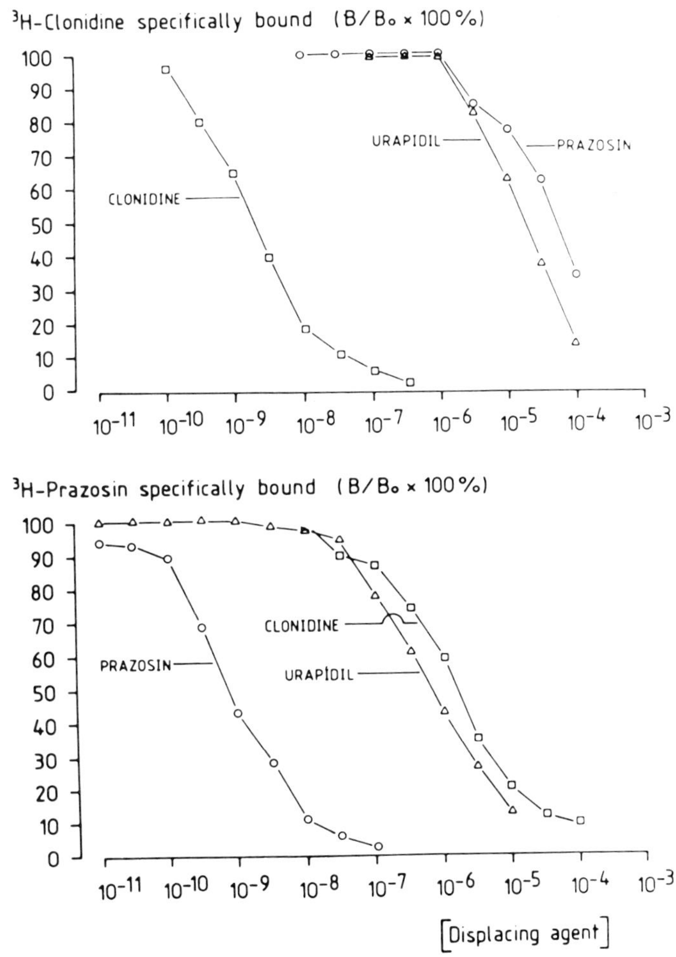

Figure 2. Displacement of [^{3}H] prazosin and [^{3}H] clonidine from specific binding sites in rat brain membranes by urapidil, clonidine and prazosin. Symbols represent means of 4 separate determinations, performed in duplicate. B=fractions of radioligand specifically bound in the presence of competing displayer; B_0=in the absence of competing displayer. (Data from van Zwieten *et al.*,1985a.)

new data have significantly changed the general opinion on urapidil's mode (s) of action. The older literature on this drug has been reviewed extensively by Schoetensack and co-workers (1983). The present paper deals in particular with the more recently acquired experimental data on urapidil and their interpretation.

[^{3}H]-Prazosin was used in this connection as selective ligand for α_1-adrenoceptor binding sites. The substantial affinity of urapidil is reflected by its IC_{50}-value for the displacement of [^{3}H]-prazosin: $IC_{50} = 8 \times 10^{-7}$ M. In similar experiments with [^{3}H]-clonidine as a radioligand with high affinity for α_2-adrenoceptors the IC_{50}-value for the displace-ment by urapidil amounted to 1.4×10^{-5} M (data from van Zwieten *et al.*, 1985a). Accordingly, in radioligand binding experiments on rat brain membranes urapidil displayed obvious selectivity for [^{3}H]-prazosin binding sites over [^{3}H]-clonidine binding sites. This selectivity suggests that urapidil has a stronger affinity for α_1- than for α_2- adrenoceptors. However, urapidil is rather less selective for the α_1- adrenoceptor subtype than prazosin, the prototype of selective antagonists. Nevertheless, it should be realized that urapidil's absolute affinity for α_2-adrenoceptors is even lower than that of prazosin because of the higher potency of prazosin on a molar base.

In recently published radioligand binding studies with various different rat tissues urapidil was shown to possess a modest preference for α_{1A}-over α_{1B}-binding sites. In the same series of experiments (Hanft and Gross, 1989) it was demonstrated that urapidil's congener 5-methylurapidil clearly and optimally distinguishes between α_{1A} and α_{1B}-binding sites and for this reason this compound is considered as a suitable tool to discriminate between the two aforementioned subtypes of the α_1-binding site. The functional relevance of these two distinguishable α_1-binding sites so far remains unclear. Urapidil appears to possess, in addition, some modest affinity for dopaminergic receptors, but no affinity for muscarinic receptors (van Zwieten *et al.*, 1985b). Urapidil's affinity for dopaminergic receptors is lower than that of clonidine(Freissmuth *et al.*, 1984). A modest affinity of urapidil for β-adrenoceptors was found in a cardiac ventricular membrane preparation. A non-selective radioligand was used, hence no conclusions could be drawn with respect to a preference for β_1- or β_2-receptors. Owing to the availability of more selective radioligands for the various subtypes of serotonergic receptors, three different groups of investigators demonstrated that urapidil displays a predominant affinity for 5-HT_{1A}-receptors, whereas its affinity for the 5-HT_{1B}-5-HT_{1C} and 5-HT_2-receptor subtypes is much lower (Fozard *et al.*, 1987; Kolassa *et al.*, 1987; Gross *et al.*, 1987). The selective affinity of urapidil for the 5-HT_{1A}-receptor subtype bears relevance to the presumed mechanism of urapidil's central hypotensive activity.

INTERACTION OF URAPIDIL WITH VASCULAR POSTSYNAPTIC α-ADRENOCEPTORS

In an early stage of urapidil's development it was recognized that its vasodilator activity is predominantly caused by α-adrenoceptor antagonism. Accordingly, α-adrenoceptor antagonism of urapidil had already been detected by Schoetensack *et al.* (1977), using the non-selective α-adrenoceptor antagonist dihydro-ergotamine. In later studies, when selective agonists and antagonists for the α_1- and α_2- adrenoceptor subtypes had become available, it became clear that in functional studies urapidilappears to be a selective antagonist of postjunctional α_1-adrenoceptors in the following animal models: anaesthetized cats or dogs (Schoetensack *et al.*, 1977; Sanders and Jurna, 1985; Zimmerman and Largent, 1983); conscious SHR and normotensive rats (van Zwieten *et al.* , 1985a; Sanders and Jurna, 1985; Verberne and Rand, 1984; Zimmerman 1984; Brody *et al.*, 1984); anaesthetized intact rats (van Zwieten *et al.*, 1985a); pithed rats (van Zwieten *et al.*, 1985a; Bousquet *et al.*, 1983; Sanders *et al.*, 1983); conscious dogs (Sanders and Jurna, 1985; Gillis *et al.*, 1987); isolated rat aorta and caudal artery preparations (van Zwieten *et al.*, 1985a; Verberne and Rand, 1985) and also in healthy human volunteers and hypertensive patients (Leonetti *et al.*, 1986a; Gerber *et al.*, 1985; Renondin *et al.*, 1988; Prichard *et al.*, 1989).

As an example, experiments in pithed rats are shown, where the pressor effect of the selective α_1-adrenoceptor agonist cirazoline is competitively antagonized by urapidil (Fig.3). The value for pA_2, calculated according to Arunlakshana and Schild (1959) amounts to 6.46, whereas in similar experiments with the selective α_2-adrenoceptor stimulant B-HT 920 a pA_2-value of 4.92 was found, thus confirming the

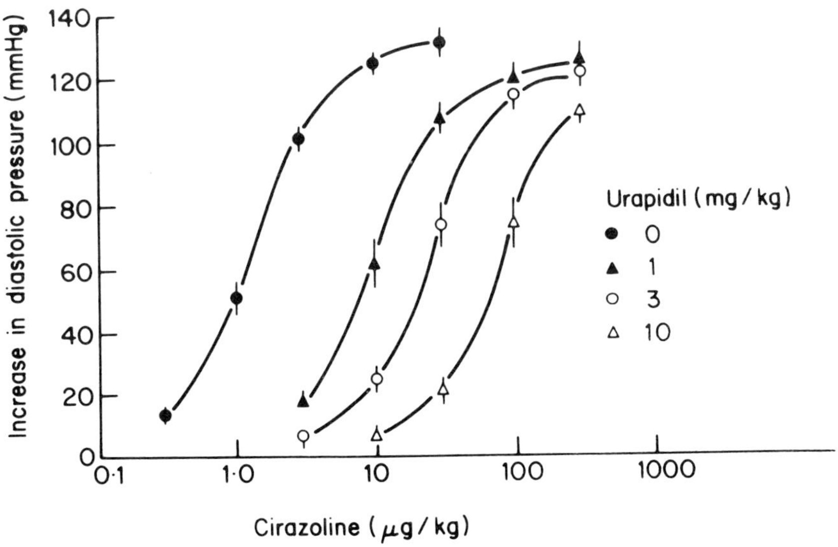

Figure 3. Competitive antagonism of urapidil toward vascular postsynaptic α_1-adrenoceptors in pithed rats. Dose-response curves for the increase in diastolic pressure by intravenous administration of the selective α_1-adrenoceptor antagonist cirazoline in pithed normotensive rats after intravenous pretreatment (15 min previously) with various doses of urapidil. Symbols represent mean value ± S.E.M. (n = 6). (Data from van Zwieten *et al.*,1985a.)

selectivity of urapidil for α_1-over α_2-adrenoceptors, also in functional experiments (van Zwieten *et al.*, 1985a). As already discussed in the paragraph on radioligand binding studies urapidil's selectivity for α_1-adrenoceptors is somewhat less than that of prazosin. However, the extremely weak absolute affinity of urapidil for α_2-adrenoceptors probably rules out any relevant participation of postsynaptic α_2-adrenoceptor blockade of the drug's vasodilator effect.

INTERACTION OF URAPIDIL WITH PRESYNAPTIC α_2-ADRENOCEPTORS

A presynaptic α_2-adrenoceptor agonistic effect of urapidil, accompanied by a reduction of noradrenaline release from the sympathetic nerve endings, had been presumed to exist on the basis of older pharmacological studies (Schoetensack *et al.*, 1977; Eltze 1979), but this effect could not be confirmed by several authors using modern methodology (van Zwieten *et al.*, 1985a). Accordingly, the weak α_2-adrenoceptor antagonistic activity of urapidil does not lead to any significant presynaptic release of noradrenaline (Sanders *et al.*, 1983). For these reasons presynaptic effects of urapidil should be regarded as practically irrelevant. This view is in accordance with the extremely weak absolute affinity of urapidil for α_2-adrenoceptors observed in radioligand binding studies.

INTERACTION OF URAPIDIL WITH β-ADRENOCEPTORS

Modest β-adrenoceptor blockade by urapidil has been described by a few authors (van Zwieten *et al.*, 1985a; Freissmuth *et al.*, 1984; Verberne and Rand, 1985). The stimulation of adenylate cyclase in a guinea-pig ventricular membrane preparation was reduced by urapidil via a competitive mechanism at the level of β-receptors. Both in guinea-pig and rat isolated atria urapidil proved a competitive antagonist of the chronotropic response to isoprenaline. Urapidil also blunted the tachycardiac response to noradrenaline in pithed rats. Pharmacological analysis with selective antagonists has revealed that urapidil is a selective antagonist of β_1- but not β_2-adrenoceptors. In animal studies the β_1-antagonistic effect of urapidil is associated with a modest degree of intrinsic sympathetic activity (ISA) (van Zwieten *et al.*, 1985a). Accordingly, as far as β-receptors, are concerned urapidil should be characterized as a modest β_1-selective adrenoceptor blocking agent with moderate ISA. It is doubtful, however, whether β-receptor blockade plays a relevant role upon oral application of urapidil in humans in therapeutic doses. Jamieson *et al.* (1986) established the influence of urapidil on the rise in heart rate induced by isoprenaline in normal human volunteers and they found no inhibition, even when urapidil doses were applied that caused similar plasma concentrations to those found during the treatment of hypertension. In similar experiments, Renondin *et al.* (1988) did not find any evidence for β-adrenoceptor blockade of urapidil, also in healthy volunteers. The influence of urapidil on endogenous β-adrenoceptor stimulation, induced by physical exercise, was also studied. Neither intravenously given urapidil in normal subjects (Belz *et al.*, 1985) nor urapidil administered orally to hypertensive patients (for 4 weeks, 30 mg b.i.d.) (Leonetti *et al.*, 1986b) depressed exercise-induced tachycardia, in contrast to the classical β_1-blocker metoprolol (Leonetti *et al.*, 1986b). It therefore seems unlikely that the antihypertensive effect of urapidil is associated with β_1-adrenoceptor blockade. For a review of this subject see Prichard *et al.* (1989).

CENTRAL HYPOTENSIVE ACTIVITY

Earlier studies as reviewed by Schoetensack (1983) had already suggested the existence of a central component in the hypotensive activity of urapidil, although the evidence from these experiments was not conclusive (Freissmuth *et al.*, 1984). Several later studies largely confirmed that urapidil possesses considerable central hypotensive activity in various models. The drug was injected by different techniques into ventricular cavities or into the vertebral artery (Wambach *et al.*, 1986; van Zwieten *et al.*, 1985a; Bousquet *et al.*, 1983). Comparing the hypotensive effect of

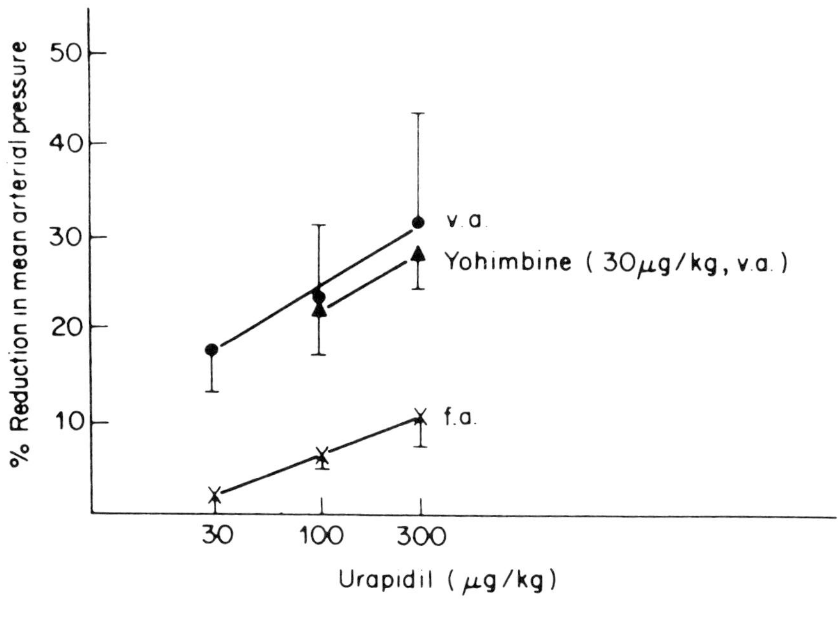

Figure 4. Demonstration of the central hypotensive activity of urapidil. After infusion into the left vertebral artery, including perfusion of the brain stem with the drug, the hypotensive effect is much stronger than after systemic administration through a femoral artery. The central hypotensive effect of urapidil is not reduced by prior treatment with the α_2-adrenoceptor antagonist yohimbine. Dose-response curves for the reduction in mean arterial pressure of chloralose-anesthetized cats by urapidil after infusion through the vertebral artery (v.a.) or through the femoral artery (f.a.) and after infusion through the v.a. 15 min after v.a. infusion of yohimbine (30 μ/kg). Symbols represent mean values ± S.E.M. (n = 4 to 5). $^*p < 0.05$. (Data from van Zwieten *et al.*, 1985a.)

urapidil when injected into various brain regions it could be demonstrated that the strongest hypotensive activity was observed upon injection into the cat's hind-brain region via the IVth ventricle (Gillis *et al.*, 1987). In this respect there was a significant difference with respect to clonidine, which is known to be more active when injected into several other brain areas, in particular the pontomedullary region in the brain stem (van Zwieten *et al.*, 1975). Furthermore, the clonidine-induced central hypotensive effect, which is known to be mediated by central α_2-adrenoceptors, remained unchanged after pretreatment with urapidil (Gillis *et al.*, 1987). This finding suggests that the blockade of central α_1-adrenoceptors by urapidil and possibly other components of urapidil's pharmacodynamic spectre do not interfere with the hypotensive action caused by the activation of central α_2-adrenoceptors by clonidine.

A further discrepancy of urapidil's effect with respect to the central hypotensive activity of clonidine was demonstrated by studying the influence of yohimbine on the centrally initiated depressor response to urapidil, injected into the cat's vertebral artery. This classical technique which has been employed, for instance, to demonstrate the central hypotensive activity of clonidine, guanfacine and α-methyl-DOPA (van Zwieten *et al.*, 1975; 1984; 1985a) did indeed reveal a potent centrally induced depressor effect of urapidil (see Fig.4). Furthermore, in contrast to the effect of clonidine the central hypotensive activity of urapidil could not be antagonized by the selective α_2-adrenoceptor antagonist yohimbine (injected into the vertebral artery prior to urapidil). This finding which proves that the central hypotensive activity of urapidil is not mediated by central α_2-adrenoceptors could be confirmed by others in a different experimental model (Gillis *et al.*, 1987). Furthermore, the hypotensive effect of urapidil, induced upon injection of the drug into the left vertebral artery of cats (van Zwieten *et al.*, 1985a; 1985b) could not be suppressed by the blockade with selective, appropriate antagonists of the following receptors: histamine (both H_1 and H_2), dopamine DA_2, serotonin (5-HT_2), muscarine (M_1), α-adrenoceptor and opioid-receptors. These findings clearly indicate that the central hypotensive effect of urapidil is different from that of clonidine, no central α_2-adrenoceptors being involved. This discrepancy between the central effects of the two drugs is accentuated by the different brain regions where their effects are initiated. Whereas clonidine and other centrally acting α_2-adrenoceptor agonists are assumed to act in the ponto-medullary region, more in particular in the region disclosed by the nucleus tractus solitarii, the vasomotor centre and the vagal nucleus, the central hypotensive action is believed to be located at the B_1/B_3-regions (raphe neurons) of the ventral surface of the medulla (Gillis *et al.*, 1987; 1989).

Until recently it was assumed that none of the better known receptor types appeared to be playing a role in the central hypotensive activity of urapidil. However, recent experimental findings strongly suggest that serotonergic neurons in the brain and 5-HT_{1A}-receptors may play an important role in the central hypotensive effect of urapidil.

INTERACTION OF URAPIDIL WITH CENTRAL 5-HT1A-RECEPTORS UNDERLYING ITS CENTRAL HYPOTENSIVE ACTIVITY

Various investigations, to be discussed below in some more detail, have supported a role for central serotonergic neurons and receptors in the regulation of blood pressure. A central hypotensive action, at least partly caused by an interaction with such neurons and receptors, has been postulated for methysergide (Kuhn *et al.*, 1980) and also for α-methyl-DOPA (Head *et al.*, 1983), although the latter drug is known to interact predominantly with central α_2-adrenoceptors (van Zwieten *et al.*, 1984).

The possibility that urapidil might exert its central hypotensive activity via central 5-HT_{1A}-receptors and serotonergic neurons was based upon radioligand binding studies by Fozard and Mir (1987) and Gross *et al.* (1987). These investigators

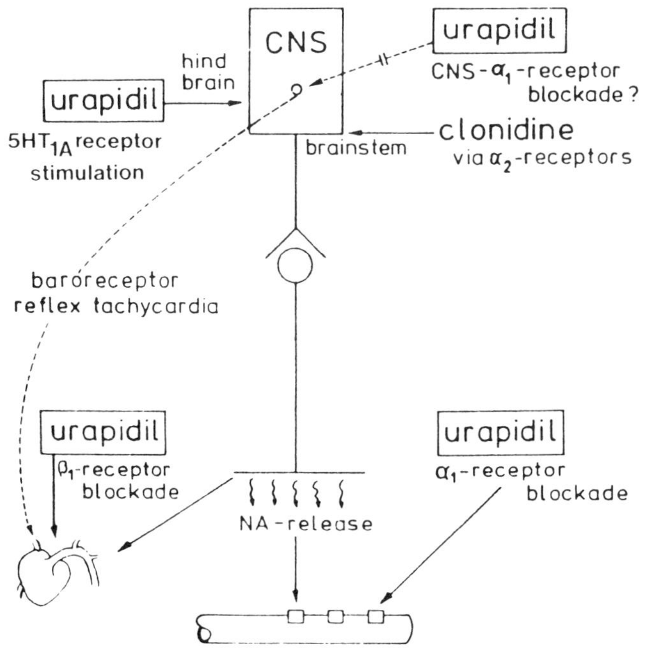

Figure 5. Mechanisms underlying the hemodynamic changes induced by urapidil. (1) Blockade of vascular postsynaptic α_1-adrenoceptors cause dilatation of the precapillary arterioles, a reduction in total peripheral resistance and thus a decrease in blood pressure. (2) A central hypotensive effect probably triggred by the stimulation of $5HT_{1A}$-receptors, initiated probably at the level of the hind-brain. Accordingly, peripheral sympathetic tone is reduced, causing vasodilatation. (3) Baroreflex receptor- mediated tachycardia is blunted, probably at a central level, possibly involving blockade of central α_1-adrenoceptors. (4) Urapidil shows modest antagonistic activity on cardiac β_1-adrenoceptors. This mechanism may offer a moderate contribution to hypotensive activity and to the impairment of reflex tachycardia. CNS = central nervous system; NA = noradrenaline.

demonstrated that with respect to serotonergic receptors urapidil has a selective affinity for 5-HT_{1A}-receptors, which is approximately ten-fold higher than that for α_1-adrenoceptors.

In the study by Gross *et al.* (1987) it was observed that 3 analogs of urapidil with different substituents at the 5-position of the uracil moiety showed a 16-to 200-fold higher affinity for 5-HT_{1A} binding sites than displayed by unsubstituted urapidil. There existed a significant correlation between the affinities of these various compounds for 5-HT_{1A}-receptors and their central hypotensive activity (Kolassa *et al.*, 1986; Gillis *et al.*, 1988). From a functional point of view urapidil appeared to act as a partial agonist in the in vitro model of the electrically stimulated guinea-pig ileum (Fozard and Mir, 1987).

A further argument supporting the contribution of 5-HT_{1A}-receptors in the central nervous system is offered by the central hypotensive activity of the pharmacological tool substance 8-hydroxy-2-(di-n-propylamino)tetralin (8-OH-DPAT), which is known to be a highly selective agonist for 5-HT_{1A}-receptors (Fozard *et al.*, 1987). The central hypotensive activity, which was demonstrated in different animal species and models, is antagonized by 5-HT_{1A}-receptor antagonists like spiroxatrine, spiperone, methiothepine and (-) pindolol (McCall *et al.*, 1987; Fozard *et al.*, 1987; Dabiré *et al.*, 1987; Doods *et al.*, 1988).

Furthermore it was shown that after full α_1-adrenoceptor blockade with prazosin, an additional hypotensive and sympathoinhibitory effect could still be induced with 5-HT_{1A}-agonistic drugs.

The central hypotensive activity of urapidil and of 8-OH-DPAT, induced by injection of the drugs into the left vertebral artery of the cat, was inhibited competitively by the 5-HT_{1A}-antagonist spiroxatrine (Dabiré *et al.*, 1987).

Accordingly, several experimental arguments support the presumption that the acute central hypotensive effect of urapidil is induced by the stimulation of 5-HT_{1A}-receptors in the brain. It seems likely that the fall in blood pressure is brought about by a depression of peripheral sympathetic nervous activity, which is triggered by the stimulation of central 5-HT_{1A}-receptors (Kolassa *et al.*, 1989). This sympathoinhibitory action initiated in the central nervous system also readily explains why urapidil does not cause reflex tachycardia, in spite of its potent vasodilator activity.

Gillis *et al.* (1988) have attempted to localize the target of urapidil in the central nervous system and for that matter the position of the 5-HT_{1A}-receptors. Experiments with topically applied urapidil have suggested that the central hypotensive action of urapidil is initiated at the level of the chemosensitive zone of the ventral medullary surface (Schlaefke's area) and at the nucleus lateralis reticularis. Based on these findings it has been hypothesized, that urapidil inhibits the activity of serotonergic neurons located in the B_1/B_3 regions of the ventral medulla. The neurons in the B_1/B_3 region are known to be excitatory to the preganglionic sympathetic neurons in the intermediolateral cell column of the spinal cord. The stimulation of 5-HT_{1A}-receptors will hence lead to a reduced activity of the aforementioned preganglionic sympathetic neurons, and concomitantly to a fall in blood pressure.

PHARMACOLOGICAL BACKGROUND OF ADVERSE REACTIONS TO URAPIDIL

Vast clinical experience with many thousands of patients treated with urapidil indicates that the drug is usually well tolerated. Dizziness, headache, fatigue and palpitations are the most common adverse effects to urapidil. These side-effects are usually mild and transient and generally do not require discontinuation of treatment (for review see Langtry *et al.*, 1989). These adverse reactions are those to be anticipated globally for vasodilator drugs. The impression is obtained that urapidil causes less severe orthostatic hypotension than provoked by the classical α_1-adrenoceptor blocking agent prazosin. So far there is no documented evidence that urapidil would cause a first dose hypotension, as currently experienced with prazosin. In spite of the vasodilator effect of urapidil reflex tachycardia is usually not experienced. Fatigue, although a rather unspecific complaint, may reflect the existence of a central mechanism of urapidil, although there is no evidence for this view.

CONCLUSION

A vast body of pharmacological evidence indicates that urapidil is a multifactorial or hybrid drug with at least two modes of action which contribute to its antihypertensive activity.

There is no doubt whatsoever that urapidil is a selective α_1-blocker, thus pharmacodynamically resembling prazosin or doxazosin. The α_1-adrenoceptor antagonistic potency of the drug is probably the major basis of its vasodilator and antihypertensive mechanism. α_2-Adrenoceptor agonism and β_1-adrenoceptor blockade, although demonstrable at high doses of urapidil are probably irrelevant.

However, the additional mechanism of urapidil is now well documented by means of various methods and in different animal species. It seems very likely that central 5-HT_{1A}-receptors and neurons at the B_1/B_3 levels of the ventral medulla are involved, where urapidil is presumed to act as a 5-HT_{1A}-receptor agonist. It remains difficult to judge to which degree this central mechanism contributes to the antihypertensive activity of urapidil in patients. Indirect arguments, however, strongly suggest that peripheral α_1-adrenoceptor blockade cannot account for all of the effects of urapidil on blood pressure and heart rate that an additional central mechanism will thus play a part. It seems of interest to note that the various mechanisms of urapidil (see Fig.5 for a schematic representation) are contained in one single molecule, since urapidil has no enantiomers.

The adverse reactions caused by urapidil are explained on the basis of α_1-adrenoceptor blockade and vasodilatation, with a possible contribution of a so far unknown central mechanism.

REFERENCES

Arunlakshana, O. and Schild, H. O. (1959). Some quantitative uses of drug antagonists. *Br. J. Pharmacol.* **1,** 48-58.

Bousquet, P., Decker, N., Feldman, J. and Schwartz, J. (1983). Etude des effects cardiovasculaires de l'urapidil. *J. Pharmacol.* **14,** 465-472.

Belz, G. G., Matthews, J.H., Graf, D. and Stern, H. (1985). Dynamic responses to intravenous urapidil and dihydralazine in normal subjects. *Clin. Pharmacol. Ther.*

Brody, M. J., Webb, R. L., Mangiapane, M. L., Porter, J. P., Bonham, A. C. and Trapani, A. J. (1984). Comparative central and peripheral antihypertensive mechanisms of urapidil and prazosin. *Am. J. Med.* **77** (suppl 4A), 74-80.

Dabiré, H., Cherqui, C., Fournier, B. and Schmitt, H. (1987). Comparison of effects of some 5-HT_{1A}-agonists on blood pressure and heart rate of normotensive anaesthetized rats. *Eur. J. Pharmacol.* **140,** 259-266.

Doods, H. N., Boddeke, H.W.G.M., Kalkman, H. O., Hoyer, D., Mathy, M-J. and van Zwieten, P. A. (1988). Central $5HT_{1A}$-receptors and the mechanism of the central hypotensive effect of (+)8-OH-DPAT, DP-5-CT, R 28935 and urapidil. *J . Cardiovasc. Pharmacol.* **11,** 432-437.

Eltze, M. (1979). Investigations on the mode of action of a new antihypertensive drug, urapidil, in the isolated vas deferens. *Eur. J. Pharmacol.* **59,** 1-6.

Fozard, J. R. and Mir, A. K. (1987). Are 5HT-receptors involved in the antihypertensive effects of urapidil ? *Br. J. Pharmacol.* **90,** 24p.

Fozard, J. R., Mir, A. K. and Middlemiss, D. N. (1987). Cardiovascular response to 8-hydroxy-2-(di-n-propylamino) tetralin (8-OH-DPAT) in the rat: site of action and pharmacological analysis. *J. Cardiovasc. Parmacol.* **9,** 328-347.

Freissmuth, M., Tuisi, E., Steuer, G. and Schuetz, W. (1984). Interaction of the antihypertensive drug urapidil with cardiac ß-adrenoceptors in vitro. *Eur. J. Pharmacol.* **104,** 169-172.

Gerber, A., Weidmann, P., Marone, C., Uehlinger, D. and Riesen, W. (1985). Cardiovascular and metabolic profile during intervention with urapidil in humans. *Hypertension* **7,** 963-971.

Gillis, R. A., Dretchen, K. L., Namath, I., Anastasi, N., Dia., Souza, J., Hill, K. and Browne, R. K. (1987). Hypotensive effect of urapidil: CNS-site and relative contribution. *J. Cardiovasc. Pharmacol.* **9,** 103-109.

Gillis, R. A., Kellar, K. J., Quest, J. A., Namath, I. J., Martino-Barrows, A., Hill, K., Gatti, P. J. and Dretchen, K. (1988). Experimental studies on the neurocardiovascular effects of urapidil. *Drugs* **35** (suppl 6), 20-33.

Gross, G., Hanft, G. and Kolassa, N. (1987). Urapidil and some analogues with hypotensive properties show high affinities for 5-hydroxytryptamine binding sites of the $5HT_{1A}$ -subtype and for α_1-adrenoceptor binding sites. *Naunyn-Schmiedeberg's Arch. Pharmacol.* **336,** 597-601.

Hanft, G. and Gross, G. (1989). Subclassification of α_1-adrenoceptor recognition sites by urapidil derivatives and other selective antagonists. *Br. J. Pharmacol.* **97,** 691-700.

Head, G. A., Korner, P. I., Lewis, S.L. and Badoer, E. (1983). Contribution of noradrenergic and serotonergic neurons to the circulatory effects of centrally acting clonidine and α-methyldopa in rabbits. *J. Cardiovasc. Pharmacol.* **5,** 945-953.

Jamieson, M. J., Jackson, S.H.D., Patel, S. S., Shepherd, A.M.M. and Galbraith, H. (1986). The assessment of the beta-blocking activity of urapidil: a new method. *Eur. J. Clin. Pharmacol.* **31,** 149-154.

Kellar, K. J., Quest, J. A., Spera, A. C., Buller, A., Confort, A., Dias Souza, J. and Gillis, R. A. (1984). Comparative effects of urapidil, prazosin and clonidine on ligand binding to central nervous system receptors, arterial pressure and heart rate in experimental animals. *Am. J. Med.* **44** (suppl 4A), 87-95.

Kolassa, N., Bischler, P. and Sanders, K. H. (1986). Hypotensive effects of urapidil analogues in cats: peripheral versus central site of action (abstr). 11th Sci. Meet Int. Soc. Hypertension, Heidelberg, 0157.

Kolassa, N., Bischler, P., Biller, K. D. and Sanders, K. H. (1987). Inhibition of the hypotensive effect of urapidil and 8-OH-DPAT by the 5HT-receptor antagonist spiroxatrine in cats. 10th

IUPHAR-Congress, Sydney.

Kolassa, N., Beller, K. D. and Sanders, K. H. (1989). Involvement of brain $5HT_{1A}$-receptors in the hypotensive response to urapidil. *Am. J. Cardiol.* **64**, 7D-10D.

Kuhn, D. M., Wolf, W. A. and Lovenberg, W. (1980). Review of the role of central serotonergic neuronal systems in blood pressure regulation. *Hypertension* **2**, 243-255.

Langtry, H. D., Mammen, G. J. and Sorkin, E. M. (1989). Urapidil. A review of its pharmacodynamic and pharmacokinetic properties, and therapeutic potential in the treatment of hypertension. *Drugs* **38**, 900-940.

Leonetti, G., Mazzola, C., Boni, S., Guffabti, E., Meani, A. and Zanchetti, A. (1986a). Comparison of the antihypertensive effect of urapidil and metoprolol in hypertension. *Eur. J. Clin. P harmacol.* **30**, 637-640.

Leonetti, G., Terzoli, L., Rupoli, L., Gradnik, R. and Zanchetti, A. (1986b). Effects of intravenous urapidil on blood pressure, renal plasma flow and responsiveness to vasoconstrictor agents in hypertensive patients. In: *Treatment of hypertension with urapidil: preclinical and clinical update* (Ed) Amery, A., Roy. Soc. Med. Serv., London, pp. 11-17, ICSS 101.

McCall, R. B., Patel, B. N. and Harris, L. T. (1987). Effects of $serotonin_2$ receptor agonists and antagonists on blood pressure, heart rate and sympathetic nerve activity. *J. Pharmacol. Exp. Ther.* **242**, 1152-1159.

Prichard, B.N.C., Tomlinson, B. and Renondin, J. C. (1989). Urapidil, a multiple action alpha-blocking drug. *Am. J. Cardiol.* **64**, 11D-15D.

Renondin, J. C., Tomlinson, B., Graham, B. R. and Prichard, B.N.C. (1988). Urapidil, α- and β-adrenoceptor blocking activity in healthy volunteers. *Br. J. Clin. Pharmacol.* 216p.

Sanders, K.H., Kilian, U., Kolassa, N. and Schoetensack, W. (1983). Influence of urapidil on pre- and postsynaptic α-adrenoceptors in pithed rats. *Naunyn-Schmiedeberg's Arch. Parmacol.* **324** (suppl), R15.

Sanders, K. H. and Jurna, I. (1985). Effects of urapidil, clonidine, prazosine and propranolol on autonomic nerve activity, blood pressure and heart rate in anaesthetized rats and cats. *Eur. J. Pharmacol.* **110**, 181-190.

Schoetensack, W., Bischler, P., Dittman, E. and Steinijans, V. W. (1977). Tierexperimentelle Untersuchungen ueber den Einfluss des Antihypertensivums Urapidil auf den Kreislauf und die Kreislaufregulation. *Arzneimittelforschung* **27**, 1908-1919.

Schoetensack, W., Bruckschen, E.G. and Zech, K. (1983). Urapidil. In *New Drugs Annual: Cardiovascular drugs.* (Ed) Scriabine, A., Raven Press, New York, pp. 19-48.

Verberne, A.J.M. and Rand, M. J. (1985). Effect of urapidil on β-adrenoceptors of rat atria. *Eur. J. Pharmacol.* **108**, 193-196.

Wambach, G., Godehardt, E., Lang, R., Heitz, W., Meurer, K. A. and Kaufmann, W. (1986). Pharmacodynamics of urapidil in essential hypertension and in chronic renal failure. In: *Treatment of hypertension with urapidil. Preclinical and clinical update* (Ed) Amery, A., Roy. Soc. Med. Serv., London, pp. 63-69.

Zimmerman, B. G. and Largent, R. D. (1983). Relationship of alpha-receptor types to hypotension and renal vasodilatation caused by alpha-blockers in conscious dogs. *Hypertension* (suppl 1), 170-174.

Zimmerman, B. G. (1984). Determination of the role of sympathetic and renin-angiotensin systems in Goldblatt-hypertension with urapidil and captopril. *J. Hypertension* **2**, 485-491.

Van Zwieten, P. A. (1975). Antihypertensive drugs with a central action. *Progr. Pharmacol.* **1**, 1-63.

Van Zwieten, P. A., Thoolen, M.J.M.C. and Timmermans, P.B.M.W.M. (1984). The hypotensive activity and side-effects of methyldopa, clonidine and guanfacine. *Hypertension* **6** (suppl 11), 28-33.

Van Zwieten, P. A., de Jonge, A., Wilffert, B., Timmermans, P.B.M.W.M., Beckeringh, J.J. and Thoolen, M.J.M.C. (1985a). Cardiovascular effects and interaction with adrenoceptors of urapidil. *Arch. Int. Pharmacodyn.* **276**, 180-201.

Van Zwieten, P. A., Mathy, M-J. and Thoolen, M.J.M.C. (1985b). Deviating central hypotensive activity of urapidil in the cat. *J. Pharm. Pharmacol.* **37**, 810-811.

Progress in Hypertension, Vol. 2, pp. 15-30.
Saito *et al.* (Eds)

ANTIHYPERTENSIVE AND NATRIURETIC EFFECTS OF BUNAZOSIN ON STROKE-PRONE SPONTANEOUSLY HYPERTENSIVE RATS

MASARU MINAMI, NAOYA HAMAUE, HIROKO TOGASHI,* MITSUHIRO YOSHIOKA,* MACHIKO MATSUMOTO,* HIDEYA SAITO* and HASAN S. PARVEZ†

Department of Pharmacology, Faculty of Pharmaceutical Sciences, Higashi-Nippon-Gakuen University, Ishikari-Tobetsu, Hokkaido 061-02; *1st Department of Pharmacology, Hokkaido University School of Medicine, Sapporo 060, Japan; and †Neuropharmacology Unit, University of Paris XI, 91405 France.

SUMMARY

Bunazosin is a potent and highly selective postsynaptic α_1- adrenoceptor antagonist, which was recently developed in Japan. Its pA_2 (6.09) value for α_1- blocking potency is almost equipotent to prazosin ($pA_2 = 6.03$), while, its α_2-blocking activity ($pA_2 = 8.63$) is weak, similar to that of prazosin ($pA_2 = 8.54$).

In addition to an α_1-adrenoceptor blocking activity, a possible calcium antagonistic action of bunazosin has been indicated by potassium-depolarized cat aortic strips and dog coronary arterial strips. This hypothesis was, however, not confirmed by an electrophysiological analysis of calcium currents in the cell membrane. In normotensive rats, bunazosin did not reduce blood pressure. However, bunazosin produced significant fall in blood pressure of SHR without marked increase in heart rate. The hypotensive effects of oral bunazosin appeared 2-3 hours after administration of doses > 0.1mg/kg and the effect lasted 6-9 hours in various hypertensive animal models. This hypotensive effect was proven to be mainly due to bunazosin's peripheral effect and was not mediated by a central action. Bunazosin did not increase urinary excretion of catecholamines in SHRSP. In these rats, bunazosin increased urinary sodium excretion, while prazosin decreased sodium excretion. In a multicenter open trial involving 184 patients with essential hypertension, bunazosin (3-9mg, t.i.d.) revealed an effective hypotensive action for up to 12 weeks. There was a low rate of occurrence of tachycardia and the overall rate of side effects was low. In controlled comparative clinical studies of bunazosin and prazosin, bunazosin does not seem to increase circulating blood volume or dilate capacitance vessels, in contrast to prazosin. This lack of dilatation may be the reason for less orthostatic hypotension in patients treated with bunazosin. Bunazosin has been introduced and registered for the treatment of hypertensive disease in Japan since 1986.

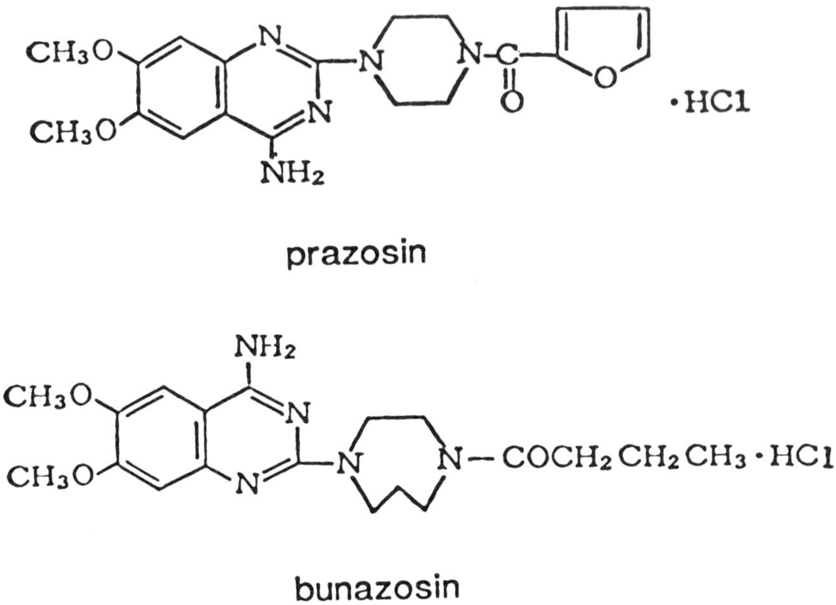

Figure 1. Chemical structures of prazosin and bunazosin.

INTRODUCTION

A newly synthesized alpha-1 adrenergic blocking agent, bunazosin [4-amino-2-(4-butanoyl-hexahydro-1H-1, 4-diazepin-1-yl)-6,7-dimethoxy quinazoline hydrochloride], is a selective alpha-1 antagonist (Shoji *et al.*, 1980;1981). As shown in Fig.1, the chemical structure of bunazosin (E-643) resembles that of prazosin, a postsynaptic α_1 adrenoceptor antagonist (Cambridge *et al.*, 1977). Using spontaneously hypertensive rats (SHR) and DOCA-salt hypertensive rats, Igarashi *et al.* (1976) reported that bunazosin (E-643) reduced blood pressure in a dose-related manner. Kawasaki *et al.* (1981) demonstrated that bunazosin caused a significant reduction in the systolic and diastolic blood pressure of patients with essential hypertension. Thus, bunazosin has an antihypertensive effect both on animal models of hypertension and patients with essential hypertension. No reports are available in the literature concerning the diuretic activity in stroke-prone spontaneously hypertensive rats (SHRSP) treated with bunazosin. The present investigation was designed to study the effects of bunazosin on blood pressure, urinary electrolyte excretion and urinary volume in SHRSP. In addition, we determined plasma catecholamine and serotonin concentrations and sympathetic efferent nerve discharge.

METHODS

Animals

Fifteen- to twenty-week-old male Wistar-Kyoto rats (WKY), SHR and stroke-prone spontaneously hypertensive rats (SHRSP) (Okamoto *et al.*, 1974) were used. The experiment was done in a light-dark alternation cycle. Rats were fed a regular diet (Clea Japan, Inc., CE-2) mixed with bunazosin or prazosin. The estimated dose of both groups was 10 mg/kg/day. These bunazosin- and prazosin-treated rats received this mixed diet for 4 weeks. The SHRSP control group received the same rat diet without either drugs. Thus, food intake, water drinking, body weight and urinary volume were measured in three groups of SHRSP. The tail cuff method was used for non-invasive measurement of blood pressure and heart rate prior to and during the pretreatment period. Rats were anesthetized with urethane (500 mg/kg, i.p.) and alpha-chloralose (50 mg/kg, i.p.) after the administration of bunazosin. Blood was withdrawn through a cannula inserted in a femoral artery.

Assay of humoral factors

Urinary sodium and potassium concentrations were determined by flame photometry using lithium as the internal standard. Urinary creatinine concentrations were measured by routine autoanalyzer methods. Plasma and urinary catecholamine (NE and E) contents were determined using a high performance liquid chromatographic method (Minami *et al.*, 1983). Plasma and urinary aldosterone were measured by RIA.

Sympathetic nerve recordings

Each rat was anesthetized with urethane (500mg/kg, i.p.) and alpha chloralose (50mg/kg, i.p.). The trachea was cannulated and attached to an artificial respirator (Harvard apparatus, USA, model 683). The left femoral vein was cannulated for drug administration. The renal and adrenal nerves were exposed retroperitoneally through a flank incision under a binocular microscope. Efferent nerve discharges from the proximal cut end of the nerves were recorded with bipolar platinum wire electrodes in warm liquid paraffin (Togashi, 1983; Yoshioka *et al.*, 1987). Nerve activity was amplified and displayed on an oscilloscope (Iwatsu, Japan, MS 5100A) and a thermal array recorder (Nihon Kohden, Japan, WS-682G). The mean nerve activity per second was counted by a real-time data analyzer (Nihon Kohden, Japan, ATAC-450).

Statistics

Data were recorded as mean ± SE. Comparison of more than two groups was performed by analysis of variance (ANOVA) followed by multiple comparison tests (modified *t*-test according to Bonferroni).

RESULTS

Changes in food intake, water drinking and body weight

Body weight gradually increased in all three groups (Table 1). No significant difference was demonstrated among the three groups in respect to food and water intake. Since there was no significant difference in food intake among the groups, it was assumed that sodium intake did not differ.

Effects of bunazosin on blood pressure and heart rate
As shown in Table 1, the blood pressure of control-SHRSP increased progressively from 189.3 ± 5.5 mmHg during the pretreatment period to 203.5 ± 6.5 mmHg 4 weeks later. The blood pressure of bunazosin-SHRSP was 188.6 ± 5.5 mmHg during the pretreatment period and then decreased to 181.4 ± 7.2 mmHg 4 weeks after treatment with bunazosin. A significant difference in blood pressure was noted between the bunazosin-SHRSP and the control-SHRSP ($p < 0.05$) 4 weeks after bunazosin administration. A similar significant difference was observed between the prazosin and non-drug groups after 4 weeks of prazosin treatment. There was no significant difference in heart rate among the three groups.

Effects of oral bunazosin or prazosin on urinary volume, urinary sodium and urinary potassium excretion
As shown in Fig. 2, there was a significant increase in urinary sodium excretion 4 weeks after bunazosin administration compared with that of SHRSP in the non-drug control group and prazosin group. In regard to urinary volume and urinary creatinine and potassium excretion, significant differences were noted between the bunazosin-treated SHRSP and prazosin-treated SHRSP.

Table 1. Body weight, food intake, systolic blood pressure (SBP) and heart rate (HR) after oral bunazosin or prazosin administration in stroke-prone spontaneously hypertensive rats (SHRSP).

	Initial	1 week	4 weeks
Body weight (g)			
non-drug (12)	269.2 ± 12.0	290.0 ± 13.5**	293.8 ± 14.8**
bunazosin (14)	271.9 ± 8.4	291.6 ± 8.2**	303.4 ± 9.2**
prazosin (6)	271.7 ± 18.8	319.7 ± 19.6**	310.5 ± 25.2**
Food intake (g/day)			
non-drug (12)	17.0 ± 0.7	18.5 ± 0.5	13.9 ± 0.8*
bunazosin (14)	17.1 ± 0.5	18.9 ± 0.4*	15.4 ± 0.5**
prazosin (6)	19.3 ± 0.7	20.4 ± 0.6	13.9 ± 1.3**
SBP (mmHg)			
non-drug (12)	189.3 ± 5.2	198.8 ± 8.0	203.5 ± 6.5**
bunazosin (14)	188.6 ± 5.5	186.1 ± 6.7	181.4 ± 7.2#
prazosin (6)	191.0 ± 7.1	191.2 ± 11.0	153.8 ± 6.1##**§
HR (bpm)			
non-drug (12)	498.2 ± 13.4	508.9 ± 18.3	522.4 ± 17.3
bunazosin (14)	485.4 ± 11.1	497.0 ± 14.5	506.9 ± 13.5
prazosin (6)	471.7 ± 27.4	445.5 ± 24.6	500.0 ± 17.7

mean ± SE, (*n*), #$p < 0.05$, ##$p < 0.01$ *vs* non-drug control group, §$p < 0.05$ bunazosin group *vs* prazosin group, *$p < 0.05$, **$p < 0.01$ *vs* initial value.

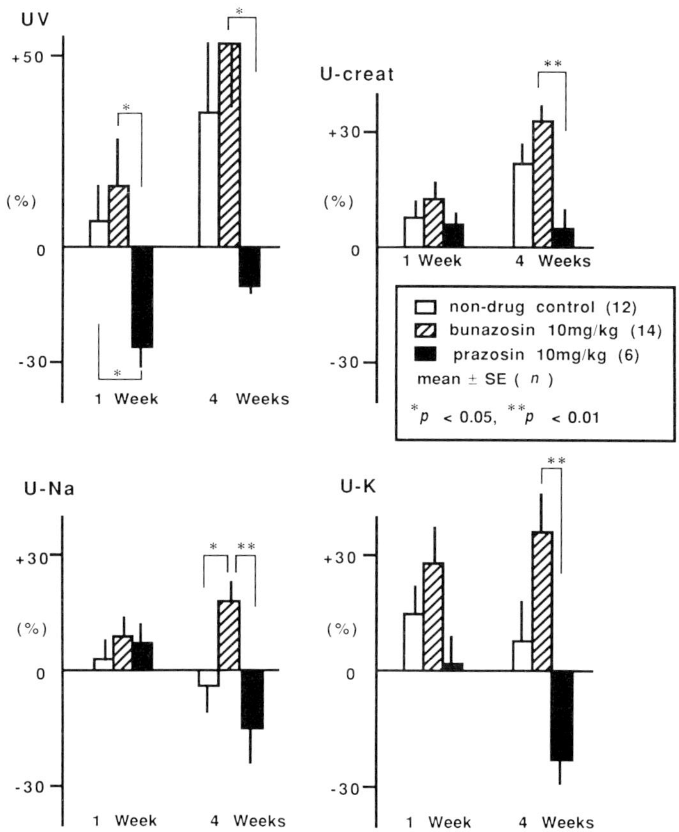

Figure 2. Changes in urinary volume (UV), urinary creatinine (U-creat), urinary sodium (U-Na) and urinary potassium (U-K) excretion after oral administration of bunazosin or prazosin in SHRSP. Data shows percent changes from initial value, 1 week and 4 weeks after administration of drugs.

Effects of bunazosin on urinary concentrations of norepinephrine (NE), epinephrine (E) and aldosterone

As shown in Table 2, the urinary catecholamine content 4 weeks after drug treatment did not differ significantly among the three groups. There were slight decreases in urinary aldosterone excretion in animals treated with bunazosin and prazosin compared with that of animals in the non-drug group.

Effect of bunazosin on plasma catecholamine concentrations and plasma serotonin concentration

There was no significant difference in plasma catecholamines 4 weeks after oral administration among the three groups (Table 3). Plasma serotonin concentrations in bunazosin-treated animals and prazosin-treated animals were higher than that of animals in the non-drug group. There was, however, no significant difference among the three groups.

Effect of bunazosin on sympathetic nerve activity
Intravenous administration of bunazosin (10 µg/kg) did not produce significant changes in the sympathetic renal nerve activity of normotensive Wistar rats (Table 4). Bunazosin also did not produce any significant effects on either sympathetic renal nerve activity (Fig.3) or adrenal nerve activity (Fig.5) in SHR. The changes induced by exogenously administered NE (blood pressure rise, tachycardia and the changes in sympathetic nerve activity) were antagonized by pretreatment with bunazosin in SHR (Fig.4). These findings suggest that a therapeutic dose of bunazosin works as a peripherally acting α_1 antagonist.

DISCUSSION

This study demonstrated that bunazosin has both antihypertensive and natriuretic effects on SHRSP. It is generally accepted that alpha-adrenoceptor blocking drugs induce a decrease in blood pressure and tachycardia through a reflex mechanism. Mulvihil-Wilson *et al.* (1979) studied the effect of phenoxybenzamine and prazosin

Table 2. Urinary norepinephrine (NE), epinephrine (E) and aldosterone excretion 4 weeks after oral administration of bunazosin or prazosin in SHRSP.

	NE	E	aldosterone
non-drug control	56.2 ± 3.8 (*n* = 12)	3.5 ± 0.8 (*n* = 5)	54.3 ± 11.6 (*n* = 5)
bunazosin (10 mg/kg)	53.4 ± 2.6 (*n* = 14)	6.9 ± 2.1 (*n* = 6)	36.7 ± 1.8 (*n* = 6)
prazosin (10 mg/kg)	54.7 ± 3.7 (*n* = 6)	4.5 ± 0.9 (*n* = 6)	38.9 ± 4.8 (*n* = 6)

mean ± S.E., urinary norepinephrine and epinephrine excretion; ng/mg.creatinine/day, urinary aldosterone excretion; ng/day

Table 3. Plasma norepinephrine (NE), epinephrine (E) and serotonin (5-HT) concentrations 4 weeks after oral administration of bunazosin or prazosin in SHRSP.

	NE (pg/ml)	E (pg/ml)	5-HT (ng/ml)
nondrug control (*n* = 5)	186.2 ± 18.2	42.8 ± 11.0	3.2 ± 0.5
bunazosin (10mg/kg) (*n* = 6)	188.0 ± 18.8	34.0 ± 4.9	6.4 ± 2.3
prazosin (10mg/kg) (*n* = 6)	198.7 ± 33.5	27.7 ± 3.3	4.3 ± 1.3

mean ± S.E., SHRSP= stroke-prone spontaneously hypertensive rat

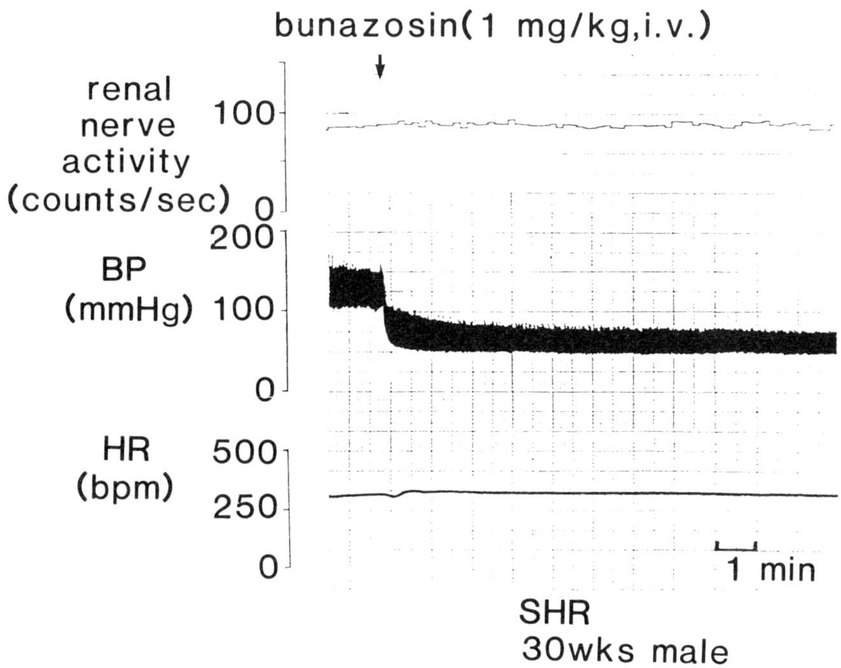

Figure 3. Effect of bunazosin (E-643) on renal nerve activity, blood pressure (BP) and heart rate (HR) in spontaneously hypertensive rats (SHR).

Table 4. Effect of bunazosin on renal nerve activity (RNA) in normotensive rats

%RNA	PT	10 min.	20 min.	30 min.
bunazosin ($n = 5$)	100%	82.4 ± 28.9	88.0 ± 22.7	88.6 ± 47.8
saline ($n = 5$)	100%	101.4 ± 7.3	102.0 ± 6.9	103.7 ± 5.1

mean ± S.D., PT = pretreatment.

on blood pressure, heart rate and plasma catecholamine concentrations in patients with essential hypertension. Both drugs reduced blood pressure and significantly increased heart rates and plasma catecholamine levels. The cardiac stimulation effect and increased plasma catecholamine concentration appear to be accounted for by the effects on presynaptic receptors of alpha-adrenoceptor blocking drugs (Roach and Cavero, 1979). Bunazosin, a newly developed alpha-1 adrenergic blocking drug, did not appear to have a cardiac stimulation effect (Figs.3, 4 and 5), nor did it increase the plasma catecholamine concentrations (Table 3) through a reflex mechanism.

There are many reports about salt and water retention caused by antihypertensive drugs. Perera (1954) found that reserpine produced transient sodium retention. Guanethidine and methyldopa also caused salt and water retention in hypertensive patients (Smith, 1965; Hansen, 1968) and in normotensive subjects (Weil and Chidsey, 1968).

Treatment with certain vasodilators such as hydralazine (Taylor *et al.*, 1952),

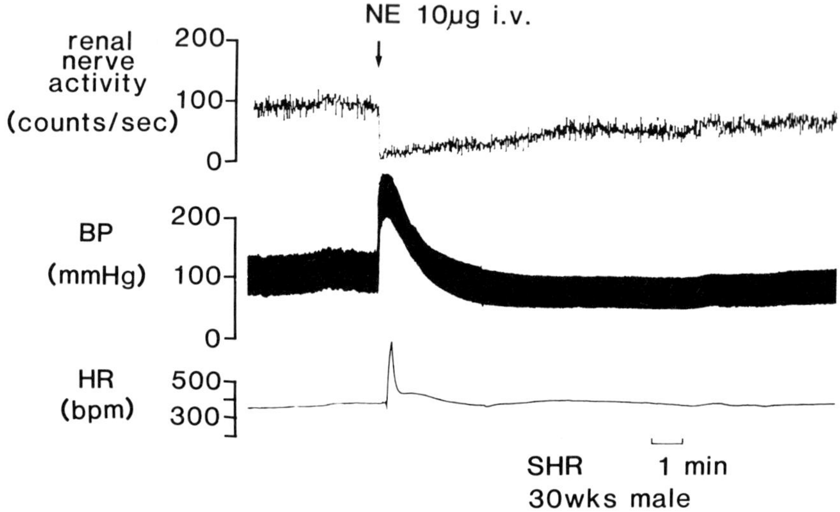

Figure 4. Effects of exogenously administered norepinephrine (NE) on blood pressure (BP), heart rate (HR) and renal nerve activity in spontaneously hypertensive rats (SHR).

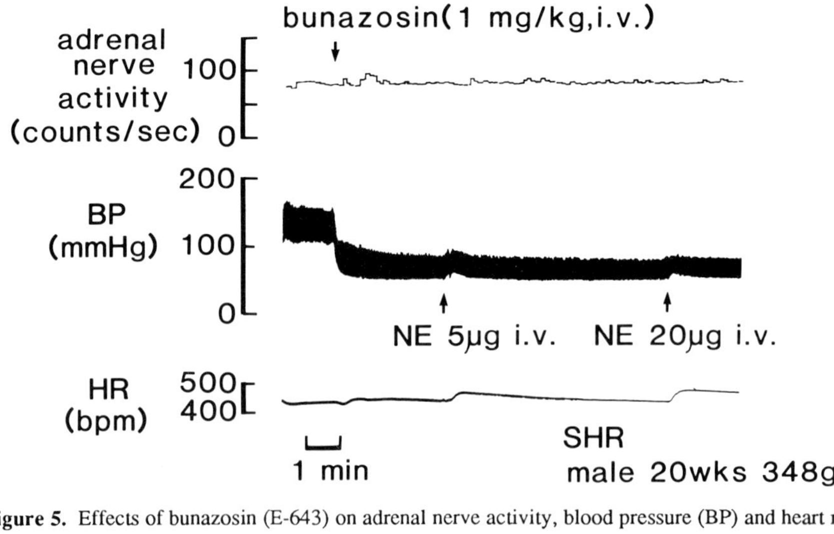

Figure 5. Effects of bunazosin (E-643) on adrenal nerve activity, blood pressure (BP) and heart rate (HR) in spontaneously hypertensive rats (SHR). After bunazosin administration, NE did not produce BP rise, increased HR or decreased sympathetic nerve activity in SHR.

diazoxide (Finnerty *et al.*, 1970) and minoxidil (Dustan, 1983) was also associated with fluid retention. Generally, a drug-induced vasodilation produced reflex tachycardia, increased cardiac output and an increase in sympathetic vasomotor outflow in the kidney (Dustan *et al.*, 1953). Increased plasma renin activity resulted in reflex adrenergic stimulation. These factors could have evoked sodium reabsorption. It was demonstrated that bunazosin did not produce significant changes in sympathetic renal nerve activity in Wistar rats and SHR. It is possible that bunazosin does not have a significant hypotensive effect via the central nervous system. Hypertensive patients often have decreased renal blood flow and GFR and, thus, diminished sodium excretion. Arterial blood pressure is also an important factor in sodium excretion. Orthostatic hypotension often presents with oliguria during the day and diuresis at night. The fluid retention that occurs with a vasodilator most likely results from a combination of blood pressure decrease, increase in renal sympathetic activity and increased renin release. Alpha-adrenergic activity might decrease sodium excretion due to increased sodium reabsorption triggered by renal nerve stimulation (DiBona *et al.*, 1977). This enhanced reabsorption was inhibited by phenoxybenzamine (DiBona *et al.*, 1977). Thus, it seems possible that alpha-adrenergic antagonism might facilitate sodium excretion. Lessening alpha-adrenergic tone (Gill *et al.*, 1964) as well as decreasing renin release through beta-adrenergic inhibition may facilitate sodium excretion.

In conclusion, bunazosin caused a significant decrease in arterial pressure, but no increase in sympathetic renal nerve activity. The presence of diuretic action without reflex stimulation may account for the lack of pressor effect that usually accompanies fluid retention. These findings suggest that bunazosin is a peripherally acting α_1 antagonist with natriuretic effect.

MINI-REVIEW OF BUNAZOSIN (EXPERIMENTAL DATA)

Mechanism of Action

The hemodynamic effects of bunazosin are due to its competitive blockade of vascular post-synaptic alpha-1 adrenergic receptors; its pA_2 value is 6.09, almost equipotent to prazosin ($pA_2 = 6.03$) (Shoji *et al.*, 1981). While its alpha-2 blocking activity is weak, ($pA_2 = 8.63$) similar to that of prazosin ($pA_2 = 8.54$) (Shoji *et al.*, 1981). This alpha-1 selectivity was confirmed by the binding of ^{3}H-bunazosin to rat brain and peripheral organs (Hata *et al.*, 1982). Prazosin as well as bunazosin inhibited ^{3}H-bunazosin binding more than yohimbine and clonidine. There was a thousand fold difference in the k1 values of these substances (Hata *et al.*, 1982). These facts strongly support that bunazosin is a potent and highly selective post-synaptic α-blocker. The vasodilatation produced by bunazosin is seen in both the capacitance and resistance vessels which reduce cardiac preload and afterload. Bunazosin does not significantly increase heart rate, plasma renin activity or norepinephrine concentration (Minami *et al.*,1984). Hemodynamic studies confirmed that bunazosin induced a fall in blood pressure without a concomitant significant change in cardiac output, heart rate, renal blood flow or glomerular filtration rate (Sakanashi *et al.*, 1986). There is no measurable negative chronotropic effect.

Blood pressure

In normotensive rats, bunazosin did not affect the systemic blood pressure (Igarashi *et al.*, 1976; 1977; Minami *et al.*, 1984). However, in spontaneously hypertensive rats, intravenous bunazosin produced an immediate fall in systolic blood pressure (Igarashi *et al.*, 1976; 1977). In stroke-prone spontaneously hypertensive rats, oral bunazosin also prevented the progressive development of hypertension over a 4-week period (Minami *et al.*, 1984;1985). These results suggest that bunazosin has an antihypertensive activity in these hypertensive models. The hypotensive effects of oral bunazosin appeared after 2-3 hours and lasted 6-9 hours in various types of hypertensive animal models including SHR, DOCA-salt hypertension rats and renal hypertensive dogs (Sugimoto and Ebihara, 1985). In anesthetized open-chest dogs (Sakanashi *et al.*, 1986) and cats (Hoshi and Fujino, 1980), intravenous bunazosin produced a significant fall in systolic blood pressure. This hypotensive effect was shown to be mainly due to bunazosin's peripheral effect, and was not mediated by a central action, as bunazosin changed neither water drinking behavior nor sympathetic renal nerve discharge in rats (Minami *et al.*, 1985). And also, bunazosin is water soluble and lipid insoluble.

Heart rate

In conscious normotensive rats, bunazosin did not reduce blood pressure but markedly increased the heart rate (Igarashi *et al.*, 1976; 1977; Minami *et al.*, 1984). On the other hand, bunazosin reduced the blood pressure of SHR without a marked increase in heart rate (Igarashi *et al.*, 1976; 1977; Minami *et al.*, 1984; 1985). The reasons for the differences in heart rate responses to bunazosin in different types of rats have not yet been clarified. The transient increase in heart rate observed after bunazosin administration in anesthetized open-chest dogs (Sakanashi *et al.*, 1986) can be attributed to a sympathetic reflex due to hypotension induced by bunazosin. In these experiments, the hypotensive effect lasted after the heart rate returned to the control level (Sakanashi *et al.*, 1986).

Cardiac hypertrophy

Ventricular hypertrophy is an adaptive response of the heart to hemodynamic overload (Swynghedauw *et al.*, 1982). Tamai *et al.*(1989) showed a significant reduction in left ventricular hypertrophy of guinea-pigs during treatment with bunazosin. Ikegaya *et al.* (1988) reported that bunazosin at a dose of 10mg/kg inhibited the progression of cardiomyopathy in the Syrian hamster of inbred strain BIO 14.6.

Cardioprotection in ischemic heart

Phentolamine and bunazosin suppressed hypoxia-induced release of CK and ATP metabolites, rise in resting tension and increase in tissue calcium content of rabbits (Takeo *et al.*, 1988). Premedication with bunazosin prevented increase in mitochondrial calcium content. Furthermore, bunazosin prevented both mitochondrial dysfunction and reperfusion arrythmias in canine hearts (Hanaki *et al.*, 1988). One possible mechanism for the beneficial effect on oxygen deficiency-induced impairment of the myocardium might be attributed to their alpha-blocking action.

Arrythmias

In the reserpinized guinea-pig ventricular myocardium, bunazosin has an antiarrhythmic effect, especially in the arrhythmia induced by abnormal automaticity or triggered activity, probably through the inhibition of fast Na^+ and slow Ca^{2+} currents independently of blockade of cardiac alpha adrenoceptors (Hasegawa *et al.*, 1988). Seven dogs (29%) of the reperfusion group developed reperfusion arrhythmias. In the bunazosin group, however, none developed arrhythmias (Sugiyama *et al.*, 1987).

Blood flow

Sakanashi *et al.* (1986) measured, simultaneously and continuously, the vertebral, coronary, renal and aortic blood flows in anesthetized dogs. They showed that the flow of blood to important organs is well maintained even during the hypotension induced by intravenous bunazosin. In infarcted rats with bunazosin treatment, the mean aortic pressure and total vascular resistance index were reduced, the left ventricular end-diastolic pressure was modestly lessened, and the cardiac index was maintained (Itagaki *et al.*, 1989).

Humoral factors

In normotensive rats, oral bunazosin had no significant effect on the plasma levels of epinephrine, norepinephrine, serotonin, vasopressin and angiotensin II (Minami *et al.*, 1984).

Absorption, Metabolism, Excretion

According to Yamato *et al.* (1982), bunazosin is rapidly absorbed after oral administration in rats and rabbits. The plasma ^{14}C-bunazosin levels in rabbits increase monophasically and reach peaks 1-2 hours after oral treatment. In rats, the radioactivity levels increased biphasically and reached the first peak in 1 hour and second peak 3-4 hours after treatment. Thereafter the concentration of bunazosin declines according to first order kinetics with a half-life of 1.5 hours. The half-life of bunazosin's hemodynamic effect appears to be considerably longer than that of the plasma concentration. During the initial period of therapy, plasma concentrations correlate well with antihypertensive effects. There was no induction of hepatic drug-metabolizing enzymes in the rats and no abnormal accumulation in rat tissue (Yamato *et al.*, 1985).

Metabolite

After oral administration, approximately 10% and 24% of bunazosin was excreted in the urine of rats and rabbits, respectively (Yamato *et al.*, 1982). Highest tissue levels of ^{14}C-bunazosin and its metabolites were found in the liver of rats and the kidney of rabbits. Unchanged bunazosin and its metabolites were excreted in the urine, bile and feces of rats and rabbits. The major metabolites of bunazosin were the O- and N-glucuronides of 7-O-desmethyl bunazosin in rats and the N-glucuronide of bunazosin in rabbits. The blood concentration-time curve is apparently monoexponential, despite the enterohepatic circulation of bunazosin and its metabolites.

Urinary Na excretion
Bunazosin and prazosin did not increase the urinary excretion of catecholamines during the antihypertensive treatment of SHRSP. In these rats, bunazosin increased urinary Na excretion, while prazosin decreased Na excretion (Minami *et al.*, 1985). This difference may be important because Na and water retention are disadvantages of prazosin and vasodilators.

Toxicology
In acute toxicity study in beagles, bunazosin at 100-500 mg/kg, p.o. caused myosis, conjunctival congestion, ptosis and lacrimation. Table 5 shows the LD_{50} of bunazosin in mice and rats. In a chronic toxicity study, rats received bunazosin at 1-1000 mg/kg for 5 weeks to 12 months showed ptosis. Examination of blood and urine did not reveal any significant changes except for diuresis and saluresis at 30-100 mg/kg, p.o.. Bunazosin at 30-100 mg/kg, p.o. increased the weight of liver, kidney, spleen and adrenal glands in female rats. In an electronmicroscopic study, no significant changes were observed. In reproduction studies of the rat, bunazosin at 25 mg/kg, p.o. and more tended to decrease the pregnancy rate and at 100 mg/kg, p.o. was embryotoxic and inhibited body weight gain in newborn rats. Bunazosin had no mutagenic activity.

Table 5. Acute toxicity of bunazosin, LD_{50} (mg/kg)

Route of Administration	Sex	Mouse	Rat
Oral	male	1,201	980
	female	1,250	1,280
Subcutaneous	male	730	430
	female	630	365
Intravenous	male	67	41-51
	female	80	50

DATA OF CLINICAL STUDY

Blood pressure and heart rate
Patients with essential hypertension significantly reduced blood pressure. The systolic and diastolic blood pressures showed decreased from 172 ± 31/100 ± 12 to 151 ± 28/89 ± 14 mmHg in the supine position and from 158 ± 22/ 101 ± 11 to 138 ± 28/ 89 ± 5 mmHg in the sitting position (Kawasaki *et al.*,1981). Heart rate tended to increase from 67 ± 10 to 69 ± 10 beats/ min in the supine position and from 68 ± 10 to 73 ± 9 beats/min in the sitting position. Tsukiyama and Otsuka (1984) also reported that blood pressure was significantly reduced from 170 ± 3/105 ± 3 to 150 ± 7/92 ± 3 and heart rate was not significantly changed.

Hemodynamic study, renal circulation
Tsukiyama and Otsuka(1984) studied the hemodynamic effects of bunazosin at

6mg/day given 3 times daily for 6 weeks in 7 patients with essential hypertension. Cardiac index and total peripheral resistance index tended to decline from 4.12 ± 0.35 to 4.08 ± 0.25 $l/min/m^2$ and from 2,557 ± 206 to 2,230 ± 161 $dynes.S/cm^5/m^2$, respectively. Sakai and Ueda (1985) investigated the acute hemodynamic effects of a single dose of 1 mg p.o. bunazosin in 7 patients with congestive heart failure. They observed reductions of pulmonary end-diastolic pressure, pulmonary vascular resistance, systemic vascular resistance, systolic and diastolic blood pressures. On the other hand, the cardiac index increased from 2.12 ± 0.20 to 2.82 ± 0.28 $l/min/m^2$. These data indicate that bunazosin reduces both preload and afterload. In eight patients with chronic heart failure, after the administration of bunazosin (1-4 mg/day, p.o.), renal blood flow were significantly increased accompanied by a significant increase in urine volume (Matsuzaki *et al.*, 1989). In six patients with essential hypertension, oral bunazosin (6 mg/day) for 4 to 12 weeks also significantly increased effective renal blood flow and creatinine clearance by 34% and 37%, respectively (Hirata *et al.*, 1989). An increase in renal vascular resistance has been found to be sympathetically mediated (Hollenberg *et al.*, 1975). Therefore, bunazosin might be expected to improve the altered renal circulation.

Pharmacokinetics

Bunazosin is well absorbed from the intestinal tract, subjected to hepatic metabolism and the metabolites are excreted into the feces via the bile (Yamato *et al.*, 1985). Urinary excretion of unchanged bunazosin in healthy volunteers after ingestion of 3 mg was as low as 0.7% in the first 24 hours. In healthy volunteers, the plasma concentration of bunazosin reached a peak level in 1 hour and decreased over 8 hours with a half-life of 1.53 hours (Yamato *et al.*, 1985). The plasma concentration of bunazosin in the morning had a tendency to be greater than that in the evening (Kumagai *et al.*, 1989). This might be one of the possible mechanisms for explaining that a marked hypotension caused by alpha-1 blockers is preventable by the evening dosage of the drug. Nakashima *et al.* (1983) reported that the plasma concentration of bunazosin reached its peak value of about 20ng/ml 1 hour after oral administration of 2 mg. The elimination half-life was about 2 hours in healthy volunteers and those with borderline hypertension. The correlation of bunazosin plasma concentrations and hypotensive effect was fairly good. The pharmacokinetic parameters of 6 patients with severe congestive heart failure were as follows: plasma peak concentration level was 36.2 ng/ml, time of the peak level was 2.33 hours and elimination half life of beta phase was 2.31 hours (Ito *et al.*, 1983). Other pharmacokinetic characteristics obtained in animal studies pertaining to clinical bunazosin pharmacokinetics are : transfer of a small percentage of bunazosin through the placenta, excretion of a small amount in milk and high plasma protein binding (94% in humans and 71% in rats) (Yamato *et al.*, 1985). There was no induction of hepatic drug-metabolizing enzymes in rats (Yamato *et al.*, 1985).

Humoral Factors, lipid metabolism

During continuous administration of bunazosin, the reduction of blood pressure remained unchanged, the heart rate showed a tendency to decline, and the plasma levels of epinephrine, norepinephrine, dopamine-beta-hydroxylase activity, renin activity and aldosterone were not significantly changed. The antihypertensive effect of bunazosin was more prominent in patients with high plasma catecholamine and dopamine-beta-hydroxylase activity and less in patients with high plasma renin and aldosterone levels (Kawasaki *et al.*, 1981). After 12 weeks of bunazosin (1 to 3mg t.i.d.) treatment, significantly lowered apoprotein B and apoprotein B/ A-I ratio were observed in contrast to no changes in the propranolol group (10 to 40mg t.i.d.)(Sasaki et al., 1989). Yoshino *et al.*(1989) also demonstrated the beneficial effect of bunazosin on the lipid metabolism of hypertensive diabetic patients. Moreover, bunazosin did not affect glucose metabolism (Yoshino *et al.*, 1989; Takeda *et al.*, 1989).

REFERENCES

Cambridge, D., Davey, M. J. and Massingham, R. (1977). Prazosin, a selective antagonist of postsynaptic alpha-adrenoceptors. *Br. J. Pharmacol.* **59,** 514 - 515.

DiBona, G. F., Zambraski, E. J., Aguilera, A.H. and Kaloyanaides, G. J. (1977). Neurogenic control of renal tubular sodium reabsorption in the dog: a brief review of preliminary report concerning possible humoral medication. *Circ. Res.* **40** (suppl I), 127 -130.

Dustan, H. P.(1983). Causes of inadequate responses to antihypertensive drugs, volume factors. *Hypertension* **5** (Suppl III) , 26 - 30.

Dustan, H. P., Corcoran, A. C., Taylor, R. D. and Page, I. H. (1953). Renal vascular responses to intravenous 1-hydrazinophthalazine in relation to therapeutic responses to chronic oral dosage (abstract). *Clin. Res.* **1,** 6.

Finnerty, F.A.Jr., Davidov, M., Mroczek, W. H. and Gavrilobich, L. (1970). Influence of extracellular fluid on response to antihypertensive drugs. *Circ. Res.* **26** and **37** (suppl 1),71 - 82.

Gill, J.R.Jr., Mason, D. T. and Bartter, F. C. (1964). Adrenergic nervous system in sodium metabolism: Effects of guanethidine and sodium retaining steroids in normal man. *J. Clin. Invest.* **43,** 177- 184.

Hanaki, Y., Saito, H., Sugiyama, S. and Ozawa, T. (1988). Effect of the α1-blocker bunazosin on reperfusion-induced mitochondrial dysfunction in canine hearts. *Arzneim. Forsch./ Drug Res.* **38** (I)1, 11 - 13.

Hansen, J. (1968). Alpha-methyl-dopa (Aldomet) in the treatment of hypertension. *Acta Med. Scand.* **183,** 323 - 327.

Hasegawa, J., Hirai, S., Saitoh, M., Kotake, H. and Mashiba, H. (1988). Antiarrhythmic effects of alpha-adrenoceptor antagonists in guinea pig ventricular myocardium. *J. Am. Coll. Cardiol.* **12,** 1590-1598.

Hata, F., Kondo, E., Kondo, S., Kagawa, K. and Ishida, H. (1982). Characteristics of [^{3}H] E-643-binding to alphaadrenoceptors. *Japan J. Pharmacol.* **32,** 181 - 187.

Hirata, Y., Fukui, K., Dan, K., Matsuoka, H., Sugimoto, T. and Ishii, M. (1989). Renal and hormonal efects of alpha1-adrenoceptor blockade by bunazosin in essential hypertension. *Eur. J. Clin. Pharmacol.* **36,** 575 - 578.

Hollenberg, N. K., Adams, D. F., Solomon, H., Chenitz, W. R., Burger, R. B., Abrams, H. L. and Merril, J. P. (1975). Renal vascular tone in essential and secondary hypertension. *Medicine* **54,** 29 - 44.

Hoshi, K. and Fujino, S. (1980). Effects of alpha blockers on blood pressure and on the Ca-contracture of cat aortic strips. *Japan J. Pharmacol.* **30,** 427 - 435.

Igarashi, T., Nakajima, Y. and Ohtaka, S. (1976). Comparison of effects of alpha-blocking agents on blood pressure and heart rate in unanesthetized spontaneously hypertensive (SHR) and normotensive rats. *Clin. Exp. Pharmacol. Physiol.* (suppl) **3,** 89 - 92.

Igarashi, T., Nakajima, Y. and Ohtaka, S. (1977). Antihypertensive effect of combined treatment with α- and β -adrenergic blockers in the spontaneously hypertensive rats. *Jpn. Circ. J.* **41,** 903- 911.

Ikegaya, T., Nishiyama, T., Kobayashi, A. and Yamazaki, N. (1988). Role of α_1-adrenergic receptors and the effect of bunazosin on the histopathology of cardiomyopathic syrian hamsters of strain BIO 14.6. *Japan. Circ. J.* **52,** 181 - 187.

Itagaki, T., Toma, Y., Umemoto, S., Yamakawa, K., Yoshinaga, T., Fukuta, S. and Kusukawa, R. (1989). Hemodynamic benefits of long-term bunazosin therapy in rats with myocardial infarction and heart failure. *Japan. Circ. J.* **53,** 313 - 318.

Ito, K., Kohya, T., Nishijima, H., Minami, M., Tsujinaga, H. and Yasuda, H. (1983). Acute effect of new alpha1-receptor blocking drug, E-643 in severe congestive heart failure (in Japanese). *Kokyuu to Junkan* **31**(10), 75 - 79.

Kawasaki, T., Uezono, K., Abe, I., Nakamuta, S., Ueno, M., Kawazoe, N .and Omae, T. (1981). Antihypertensive effect of E-643, a new alpha-adorenergic blocking agent. *Eur. J. Clin. Phrmacol.* **20,** 399 - 405.

Kumagai, Y., Fujimura, A., Nakashima, H., Shitoh, T., Sugimoto, K., Oribe, H. and Ebihara, A. (1989). Chronopharmacological study of bunazosin hydrochloride in healthy subjects. *Intern. J. Clin. Pharm. Res.* **9**(5), 335 - 340.

Matsuzaki, M., Kohno, M., Matsuda, Y., Ogawa, H., Ozaki, M., Toma, Y., Fujii, H. and Kusukawa, R. (1989). The effects of vasodilators on hemodynamics and renal circulation in stable chronic heart failure. *Japan. Circ. J.* **53,** 463 - 464.

Minami, M.,Yasuda, H., Yamazaki, N., Kojima, S., Nishijima, H., Matsumura, N., Togashi H., Koike, Y. and Saito, H. (1983). Plasma norepinephrine concentration and plasma dopamine-beta-hydroxylase activity in patients with congestive heart failure. *Circulation* **67,** 1324 - 1329.

Minami, M., Togashi H., Sano, M., Saito, I., Nomura, A., Saito, H., Shimamura, K. and Sakurai, H. (1984). Effects of E-643 on sympathetic nerve activity and water drinking behavior in rats [English abstract]. *Kiso-to-Rinsho* **18,** 2849-2858.

Minami, M., Togashi H., Sano, M., Yoshioka, M., Saito, I., Morii, K., Tochihara, M., Hirogami, M., Nomura, A. and Saito, H. (1985). Effects of new antihypertensive agent, E-432, on blood pressure, urinary sodium excretion and urinary norepinephrine excretion rate in stroke-prone spontaneously hypertensive rats.[English abstract]. *Hokkaido-Igaku-Zasshi* **60,** 536 - 543.

Mulvihl-Wilson, J., Graham, R. M., Pettinger, W. A., Mucklerog, C., Anderson, S., Gaffney, F. A. and Blomqvist, C. G. (1979). Comparative effects of prazosin and phenoxybenzamine on arterial blood pressure, heart rate and plasma catecholamine in essential hypertension. *J. Cardiovasc. Pharmacol.* (Suppl) **1,** 1 - 7.

Nakashima, M., Ooguchi, S.and Takiguchi, Y. (1983). Hypotensive and cardiac effects of E-643, an alpha blocker, on healthy volunteers. *Rinsho To Kenkyu* **60,** 1393 - 1398.

Okamoto, K.,Yamori, T. and Nagaoka, A. (1974). Establishment of the stroke-prone SHR. *Circ. Res.* (Suppl 1) **34-35,** 143-153.

Perera, G. A. (1954). Rauwolfia serpentina in hypertensive vascular disease. *Proc. Soc. Exp. Biol. Med.* **86,** 453 - 456.

Roach, A. G. and Cavero, I., (1979). Comparison of cardiovascular actions of dihydralazine, phentolamine, and prazosin in spontaneously hypertensive rats. *J. Cardiovasc. Pharmacol.* **1,** 19.

Sakai, M. and Ueda, K. (1985). Acute hemodynamic effects of Ea-0643 in aged patients with congestive heart failure [English abstract]. *ICU-CCU* **9,** 675- 681.

Sakanashi, M., Noguchi, K., Kato, T., Miyamoto, Y., Nakasone, J. and Takeo, S. (1986). Acute effects of bunazosin on aortic, vertebral, coronary and renal blood flows of anesthetized dogs. *Arzneim. Forsch./Drug Res.* **36**(II), 1202 - 1205.

Sasaki, T., Koga, S., Uzawa, H. and Arakawa, K. (1989). Comparative effects of bunazosin and propranolol on serum lipids and apoproteins in patients with essential hypertension. *Intern. J. Pharmacol. Ther. Toxicol.* **27**(8), 402 - 407.

Shoji, T., Daiku, Y. and Igarashi, T. (1980). Alpha-adrenoceptor blocking properties of a new antihypertensive agent,2-[4-(n-butyryl)-homopiperazine-lyl]-4-amino6,7-dimethoxy-quinazoline (E-

643). *Japan. J. Pharmacol.* **30,** 763 - 772.

Shoji, T.(1981). Comparison of pre- and postsynaptic alpha-adorenoceptor blocking effect of E-643 in the isolated vas deferens of the rat. *Japan. J. Pharmacol.* **31,** 361 - 368.

Smith, A. J. (1965). Fluid retention produced by Guanethidine; changes in body exchangeable sodium, blood volume and creatinine clearance. *Circulation* **31,** 490 - 496.

Sugimoto, K. and Ebihara, A.(1985). First Japanese alpha -1 blocker, bunazosin. *Curr. Ther.* **3,** 1741- 1744.

Sugiyama, S. and Ozawa, T. (1987). Biochemical basis for reperfusion arrhythmias. *J. Mol. Cell. Cardiol.* **19** (suppl. V), 67 - 75.

Swynghedauw, B. and Delcayer, C. (1982). Biology of cardiac overload. *Pathobiol. Ann.* **12,** 137-183.

Takeda, N., Iwai, T., Tanamura, A. and Nagano, M. (1989). Effects of bunazosin on plasma glucose levels in hypertensive diabetic patients. *Clin. Ther.* **11**(5), 568 - 571.

Takeo, S., Tanonaka, K., Matsumoto, M., Miyake, K. and Minematsu, R. (1988). Cardioprotective action of alpha-blocking agents, phentolamine and bunazosin, on hypoxic and reoxygenated myocardium. *J. Pharmacol. Exp. Ther.* **246,** 674 - 681.

Tamai, J., Hori, M., Kagiya, T., Iwakuma, K., Iwai, K., Kitabatake, A., Watanabe, Y., Yoshida, H., Inoue, M. and Kamada, T. (1989). Role of α1-adrenoceptor activity in progression of cardiac hypertrophy in guinea-pig hearts with pressure overload. *Cardiovasc. Res.* **23,** 315 - 322.

Taylor, R. D., Dustan, H. P., Corcoran, A. and Page, I. H. (1952). Evaluation of 1-hydrazinophthalazine (apresoline) in treatment of hypertensive disease. *Arch. Intern. Med.* **90,** 734 - 749.

Togashi, H. (1983). Central and peripheral effects of clonidine on the adrenal medullary function in spontaneously hypertensive rats. *J. Pharmacol. Exp. Ther.* **225,** 191 - 197.

Tsukiyama, H. and Otsuka, K. (1984). Hemodynamic effects of the α, β blocker arotinol and the α blockers bunazosin and prazosin in essential hypertension [English abstract]. *Jpn. J. Clin. Pharmacol. Ther.* **15,** 341 - 352.

Weil, J. V. and Chidsey, C. A. (1968). Plasma volume expansion resulting from interference with adrenergic function in normal man. *Circulation* **37,** 54 - 61.

Yamato, C., Takahashi, T., Kinoshita, K. and Fujita, T. (1982). The metabolic fate of 4-amino-2-(4-butanoyl-hexahydro-1H, 4-diazepin-i-yl)-6,7-dimethoxy-quinazoline HCl. *Xenobiotica* **12,** 549-559, 1982.

Yamato, C., Takahashi, T..and Tadano, K. (1985). The metabolic fate of bunazosin hydrochloride(2). Studies on the transfer to fetus and milk, plasma protein binding, tissue accumulation and effect on drug metabolizing enzymes. *Pharmacometrics* **30,** 257 - 265.

Yoshino, G., Kazumi, T., Okada, K., Nakata, Y., Matsuura, M., Narutaki, K., Yoshida, A., Ishihara, K., Tsutou, A., Chuma, J., Sakagami, M., Yokono, K. and Baba, S. (1989). The effect of bunazosin, a new alpha-1 blocker, on plasma lipids in diabetic subjects with essential hypertension. *Curr. Ther. Res.* **45**(4), 677 - 683.

Yoshioka, M., Matsumoto, M., Togashi, H., Minami, M. and Saito, H. (1987). Central sympathoinhibitory action of ketanserin in rats. *J. Pharmacol. Exp. Ther.* **243,** 1174-1178

Progress in Hypertension, Vol. 2, pp. 31-47.
Saito *et al.* (Eds)

α_1-ADRENOCEPTOR ANTAGONISTS: STUDIES ON THE SYMPATHOINHIBITORY ACTIVITY IN ANESTHETIZED RATS

HIROKO TOGASHI, MITSUHIRO YOSHIOKA, MASAHIRO TOCHIHARA, MACHIKO MATSUMOTO, MASARU MINAMI* and HIDEYA SAITO.

First Department of Pharmacology, Hokkaido University School of Medicine, Sapporo 060, Japan; *Department of Pharmacology, Faculty of Pharmaceutical Sciences, Higashi-Nippon-Gakuen University, Ishikari-Tobetsu, Hokkaido 061-21, Japan.

INTRODUCTION

The most important hemodynamic change in established hypertension is characterized by the increased peripheral vascular resistance. Vasodilators, including α-adrenoceptor blocking drugs which can decrease the peripheral vascular tone through the blockade of sympathetic vasoconstriction, are therefore theoretically suitable for the treatment of patients with high blood pressure. However, the clinical use of classical α-adrenoceptor antagonists such as phentolamine and phenoxybenzamine has been limited because of unacceptable side effects due to their lack of selectivity for α-adrenoceptor subtypes. The subsequent development of the selective antagonists for the postsynaptic α_1- adrenoceptors made it possible to treat hypertensive patients with these drugs.

Prazosin is a prototype of the selective postsynaptic α_1-adrenoceptor antagonists, which has been used as an antihypertensive agent (Cambridge *et al.*, 1977; Cavero and Roach, 1980). Several new α-adrenoceptor antagonists with (so called "hybrid drugs") or without the additional pharmacological property have been developed (van Zwieten, 1988). Beneficial effects of α_1-adrenoceptor antagonists in the treatment of essential hypertension have been reported. Because they cause few metabolic disturbances, these selective α_1-adrenoceptor antagonists make it possible to treat various types in hypertensive patients: those with excessive catecholamines as in pheochromocytoma; those with contraindications to diuretics such as in gout and diabetes and those with contraindications to β-adrenoceptor antagonists such as in asthma or cardiac failure (Barbieri *et al.*, 1980; Leren *et al.*, 1980; Warren and Dollery, 1988; Yoshino *et al.* 1989). In addition, these drugs have been found to be effective in the treatment of voiding dysfunction.

Pharmacological investigations using autoradiographic methods showed that α_1-

EVALUATION OF EFFECTS OF α_1-ADRENOCEPTOR ANTAGONISTS ON SYMPATHETIC NERVE ACTIVITY IN RATS

Efferent discharges of the postganglionic renal and cardiac sympathetic nerves and/or the preganglionic adrenal sympathetic nerves were recorded in α-chloralose and urethane anesthetized rats using electrophysiological techniques as previously reported (Togashi *et al.*, 1990). Genetically hypertensive animals, stroke-prone spontaneously hypertensive rats (SHRSP), were used in some experiments. The acute effects of α_1-adrenoceptor antagonists on sympathetic nerve activity were compared with that of α_2-adrenoceptor agonist, clonidine, which possesses the most potent central sympathoinhibitory activity (van Zwieten, 1975; Togashi, 1983). In some experiments, the repetitive effect of the drugs was also evaluated.

Prazosin

Quinazoline derivative, prazosin, is a prototype of the selective postsynaptic α_1-adrenoceptor antagonist. The chemical structure of prazosin is shown in Fig. 1. Binding studies have shown that prazosin is approximately 20,000 times more selective for α_1-than for α_2-receptor sites (Kellar *et al.*, 1984). Prazosin produced a decrease in sympathetic nerve activity in normotensive and hypertensive rats and in buffer nerve denervated cats (McCall and Humphrey, 1981; Huchet *et al.* 1983;

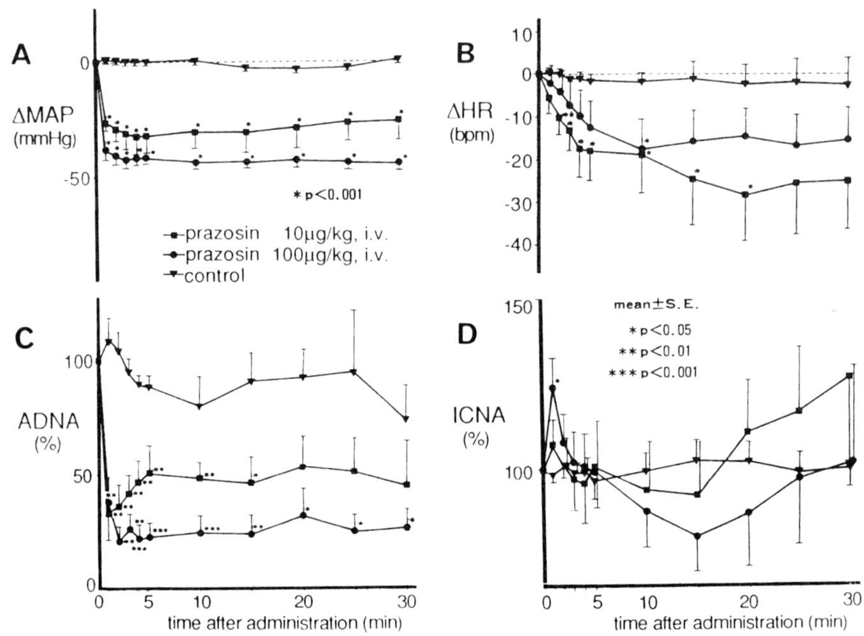

Figure 2. Time course changes in mean arterial blood pressure (A), heart rate (B), aortic depressor nerve activity (C) and inferior cardiac nerve activity (D) after systemic *i.v.* injection of prazosin in anesthetized rats. Five to 6 animals were used for each group.

adrenoceptors are distributed in the central nervous system (Greengrass and Brenner, 1979; Hata *et al*.,. 1982; Palacios *et al*., 1987; Wilcox *et al*., 1990). The medulla oblongata, an important region for cardiovascular control, is included in this distribution (Andén *et al*., 1978; Huchet *et al*., 1981; Kim *et al*., 1982; Kellar *et al*., 1984; Sinha *et al*., 1984). Although the functional role of central α_1-adrenoceptors is not fully understood, there is evidence that the blockade of α_1-adrenoceptors in the medulla oblongata acts to lower blood pressure (Bhargaba *et al*., 1972; McCall and Humphrey, 1981; Kim *et al*., 1982; Kubo *et al*., 1987). There is no doubt that the depressor effect of α_1-adrenoceptor blocking agents, when administered systemically, is predominantly of peripheral origin. However, the fact that these drugs caused a more moderate degree of reflex tachycardia than those expected from their potent vasodilating effects may depend upon the central component suppressing the sympathetic reflex activation triggered by vasodilatation. Actually, prazosin penetrates the blood-brain barrier. Systemically administered prazosin decreased efferent sympathetic nerve activity in cats and normotensive and hypertensive rats (Huchet *et al*., 1983; Ramage, 1986). Furthermore, undesirable side effects which depend on central action such as dry mouth, lassitude and vivid dreams have also been reported (van Zwieten *et al*., 1978; Warren and Dollery, 1988). However, little is known concerning the central component of hypotensive activity of α_1-adrenoceptor antagonists in spite of wide range use of the α_1-adrenoceptor antagonists in the clinical applications mentioned above. There are no systematic comparative studies that evaluate the sympathoinhibitory potency of α_1-adrenoceptor antagonists.

The purpose of this chapter is to review the sympathoinhibitory activity of novel α_1-adrenoceptor blocking agents and their contribution to hypotensive activity. The influence of α_1-adrenoceptor antagonists on sympathetic nerve activity was determined by direct recordings of nerve activity. Understanding the sympathetic effects of α_1- adrenoceptor antagonists should serve as a basis for the development of antihypertensive drugs that act via central or peripheral catecholamine receptors.

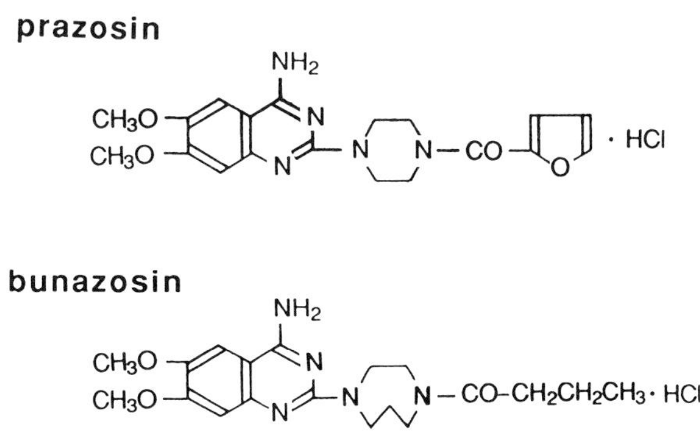

Figure 1. Chemical structures of prazosin and bunazosin.

Ramage, 1986). Since it has been reported that prazosin penetrates the blood-brain barrier, prazosin- induced sympathetic inhibition is thought to be of central origin. In clinical trials, undesirable side effects based on its central actions supported this possibility. On the other hand, conflicting results that the hypotensive effects of prazosin appear to be opposed by sympathetic stimulation in hypertensive patients have been reported (Izzo *et al.*, 1981). Concerning the central sympathoinhibitory action of prazosin, Ito *et al.* (1988) reported recently that it was not mediated by the direct blockade of central α_1-adrenoceptors.

In our study using direct nerve recordings, prazosin (10 and 100μg/kg) produced a dose dependent reduction of mean arterial pressure and aortic depressor nerve activity (Fig. 2). The same dose of prazosin failed to decrease inferior cardiac sympathetic nerve activity, although it produced a significant decrease in heart rate (Figs. 2 & 3). Preganglionic adrenal nerve activity increased significantly at a dose of 100μg/kg (Fig. 4). This might be responsible for the attenuated response in heart rate observed in 100μg/kg administered group. Thus, it is unlikely that prasozin acts as a potent sympathoinhibitory drug when administered systemically, although the possibility that it might have a weak central sympathoinhibitory activity that is masked by reflex sympathetic stimulation via the baroreceptor afferent nerves cannot be ruled out.

Coinciding with clinical observation, prazosin did not produce any significant increase in atrial pressure (Table 1). This is based on its pharmacological property that prazosin blocks α_1-adrenoceptor located on both arterioles and capacitance vessels. This hemodynamic property might partly explain the anti-tachycardiac effect of prazosin.

Bunazosin

Synthetized quainzoline derivative, bunazosin, is a selective α_1-adrenoceptor antagonist (Shoji *et al.*, 1980; Shoji, 1981; Hata *et al.*, 1982). It has been reported that bunazosin produced a significant decrease in systolic and diastolic blood pressure both in rats (Igarashi *et al.*, 1976 & 1977) and patients with essential hypertension (Kawasaki *et al.*, 1981). This drug has recently been put on the market in Japan. The chemical structure of bunazosin ([4-amino-2-(4-butanoyl-hexahydro-1H-1,4-diazepin-1-yl)-6,7- dimethoxy quinazoline hydrochloride]) is shown in Fig. 1. The

Table 1. Changes in central venous pressure (CVP) after intravenous administration of prazosin or bunazosin in normotensive rats.

% CVP	pre	5 min	10 min	20 min	30 min
control (*n* = 6)	100	103.36±9.93	102.60±10.78	108.83±15.32	102.43±12.99
prazosin					
10 μg/kg (*n* = 6)	100	103.90±3.61	102.96±5.18	105.88±8.68	108.61±14.39
100 μg/kg (*n* = 6)	100	98.71±4.28	100.06±4.15	100.76±5.07	102.83±4.47
bunazosin					
10 μg/kg (*n* = 7)	100	97.94±8.61	98.50±10.53	106.47±15.44	105.98±13.46
100 μg/kg (*n* = 6)	100	96.26±5.01	96.40±9.81	94.56±11.53	97.48±11.25

Values are expressed as mean ±S.E.

antagonism of the contractile effects of norepinephrine by bunazosin is almost equipotent to that of prazosin (Shoji, 1981). The acute effects of bunazosin on cardiac and renal sympathetic nerve activity, blood pressure and heart rate were evaluated (Yoshioka *et al.*, 1989). Intravenous administration of bunazosin (10µg/kg) produced a decrease in inferior cardiac nerve activity associated with a decrease in blood pressure and heart rate (Fig. 5A & 5B). On the other hand, the same dose of bunazosin did not produce any significant decrease in renal sympathetic nerve activity (Fig. 5C), which is thought to be responsible for sympathetic influences at the systemic arterioles. Moreover, bunazosin (10, 100 and 1000µg/kg) induced dose-dependent increases in preganglionic adrenal nerve activity that correlated negatively with hypotensive activity as evaluated by the response-time area (Fig. 4). The fact that bunazosin decreased blood pressure without significant changes in renal nerve activity indicates that sympathoinhibitory activity does not contribute to bunazosin-induced hypotensive activity. In other words, its depressor effect when administered systemically is predominantly of peripheral origin. On the contrary, it was of note that bunazosin decreased inferior cardiac nerve activity at a low dose (10µg/kg). Moreover, bunazosin did not produce any significant changes in right atrial pressure (Table 1). These might explain the finding that bunazosin is less effective in cardiac stimulation (Igarashi *et al.*, 1976; Chiba *et al.*, 1981). As shown in Table 1, bunazosin did not produce any significant increase in central venous pressure. This hemodynamic property might also contribute to the anti-tachycardiac effect of bunazosin.

In our previous study (Minami *et al.*, 1985), repetitive administration of bunazosin did not trigger any reflex tachycardia in spite of its significant antihypertensive activity in the hypertensive animal model. This finding led us to suggest that the inhibitory

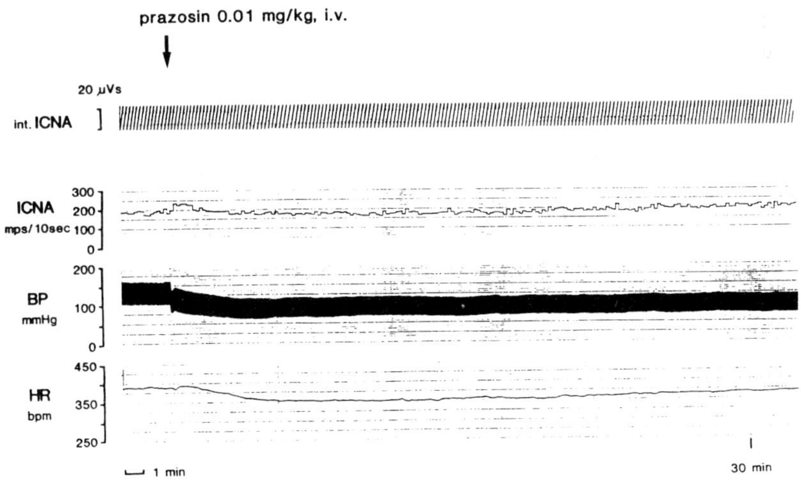

Figure 3. Effects of systemic *i.v.* injection of prazosin on inferior cardiac nerve activity (ICNA) in anesthetized rats. Traces from top to bottom are integrated (int.) ICNA, counts of impulses, blood pressure (BP) and heart rate (HR).

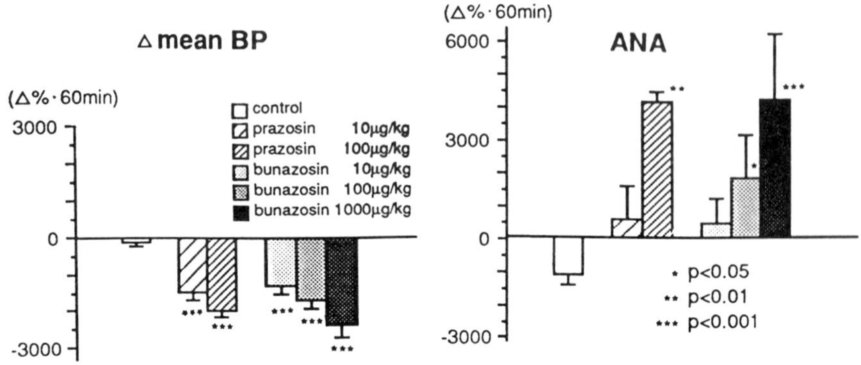

Figure 4. Comparison of the effects of prazosin and bunazosin on mean blood pressure and adrenal nerve activity (ANA) in anesthetized rats. Values (mean±S.E. of 4 to 8 rats) are expressed as the response area (Δ% X60min) which were calculated from the response (per cent change of control value) for 60 min after administration.

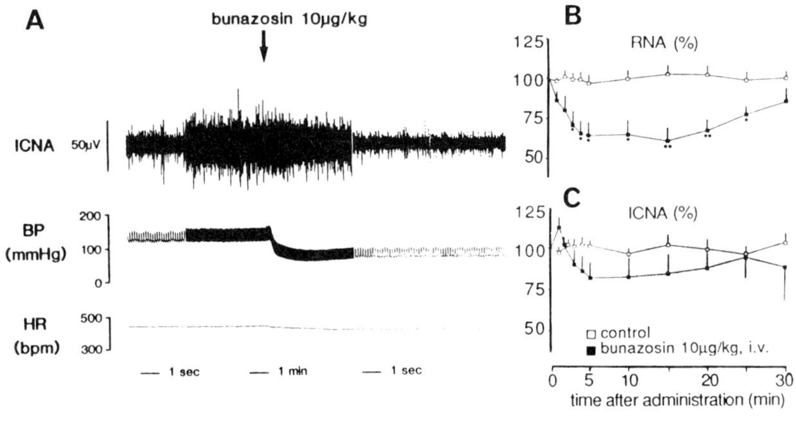

Figure 5. Effects of systemic *i.v.* injection of bunazosin in anesthetized rats. (A) Specimen recordings; Traces from top to bottom are actual inferior cardiac nerve activity (ICNA), blood pressure (BP) and heart rate (HR). (B) and (C) Time course changes in ICNA and renal nerve activity (RNA) after bunazosin administration (10μg/kg, *i.v.*), respectively.

action of bunazosin on the cardiac sympathetic nerves might be responsible for the lack of compensatory cardiac stimulation, even in long-term treatment.

SGB-1534

A newly synthetized α_1-antagonist, SGB-1534, 3-{2-[4-(o-methoxyphenyl)-1-

piperazinyl] ethyl}-2,4(1H,3H)-quinazolinedione monohydrochloride, is a phenylpiperazine derivative (Fig. 6A). This compound is also one of the quinazoline derivatives like prazosin and bunazosin. Chemically, SGB-1534 has a quinazolinedione structure and produces a decrease in blood pressure in normotensive and hypertensive animals (Nabata *et al.*, 1985; Sakai *et al.*, 1985), similar to that of serotonergic (5-HT_2) receptor antagonist, ketanserin (van Nueten *et al.*, 1981). However, it has been demonstrated that SGB-1534 does not influence the serotonin-induced pressor response, but abolishes the norepinephrine and methoxamine-induced responses (Fig. 6B). This indicates that it is unlikely that serotonergic receptor subtypes are involved in the hypotensive mechanisms of SGB-

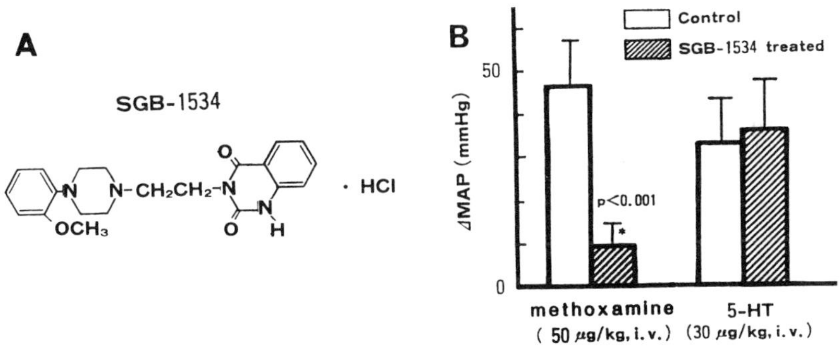

Figure 6. Chemical structure of SGB-1534 (A) and antagonism of methoxamine- or 5-HT-induced pressor response by SGB-1534 (B) in anesthetized rats.

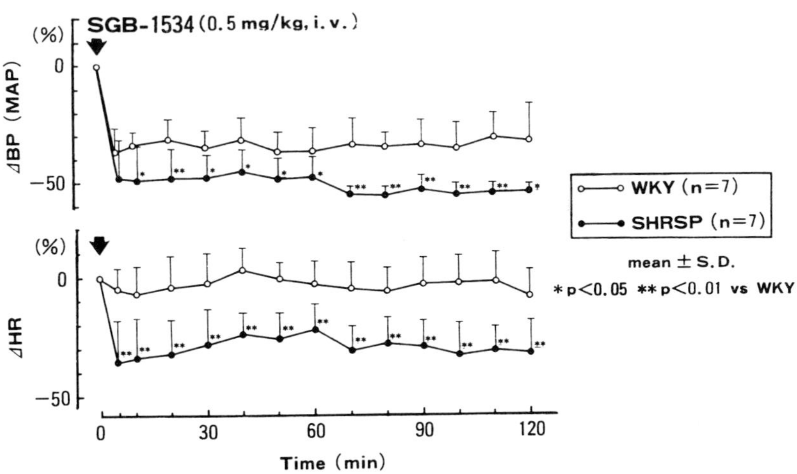

Figure 7. Time course changes in mean arterial blood pressure (MAP) and heart rate (HR) after systemic *i.v.* injection of SGB-1534 in anesthetized WKY and SHRSP.

1534. (Matsumoto *et al.*, 1988).

Intravenous administration of SGB-1534 (50 and 500 µg/kg) decreased arterial blood pressure in both normotensive and hypertensive (SHRSP) rats. The response to SGB-1543 was significantly marked in SHRSP (Fig. 7). Heart rates also decreased in SHRSP but not in normotensive animals in spite of a significant decrease in blood pressure. In order to elucidate the influence of SGB-1534 on sympathetic function, preganglionic adrenal sympathetic nerve activity was recorded. The lack of significant changes of adrenal nerve activity was noted in normotensive rats treated with SGB-1534 (Matsumoto *et al.*, 1988). This indicates that SGB-1534 lacks potent sympatho-inhibitory activity. However, SGB-1534 treated buffer nerve intact animals showed little or no change in heart rate and adrenal nerve activity, in spite of a marked decrease in blood pressure. Thus, it is possibile that SGB-1534 has a weak central sympathoinhibitory action which might overcomes reflex cardiac and sympathetic acceleration.

The different potency of SGB-1534 in normotensive and hypertensive rats indicates that α_1-adrenaoceptor mediated mechanisms, which play a fundamental role in maintaining the peripheral vascular tone, are a possible pathophysiological factor in SHRSP.

SM-2470

SM-2470 is a quianzoline derivative that is a selective α_1-adrenoceptor antagonist. This is not currently on the market. The chemical structure of SM-2470 is 4-amino-2-[4-(bicyclo (2,2,2) oct-2-ene-5-carbonyl) -1- piperazinyl] -6,7- dimethoxy-quinazoline hydrochloride (Fig 8A). α_1-adrenoceptor blocking activity of SM-2470 is as potent as that of prazosin when tested in rabbit and rat aortic strips and rat vas deferens preparation (Takayanagi *et al.*, 1985).

Concerning the sympathoinhibitory effect of SM-2470, Yoshioka *et al.*, (1988) have reported in anesthetized normotensive rats. Intravenous administration of SM-2470 (10, 30 and 100µg/kg) produced a decrease in mean arterial blood pressure in a

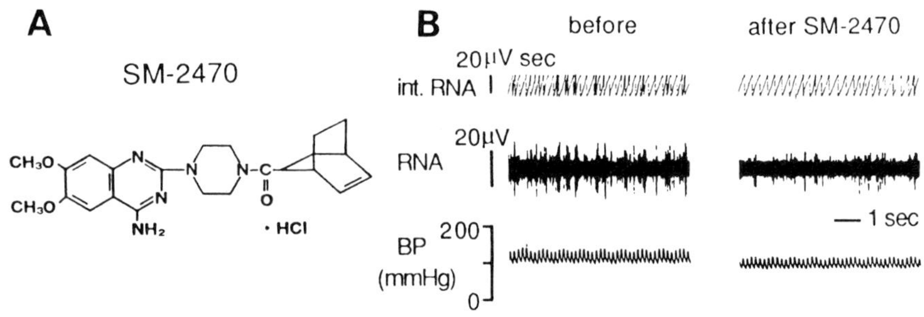

Figure 8. Chemical structure of SM-2470 (A) and effects of systemic *i.v.* injection of SM-2470 in anesthetized rats (B); Traces from top to bottom are integrated renal nerve activity (RNA), actual RNA and blood pressure (BP).

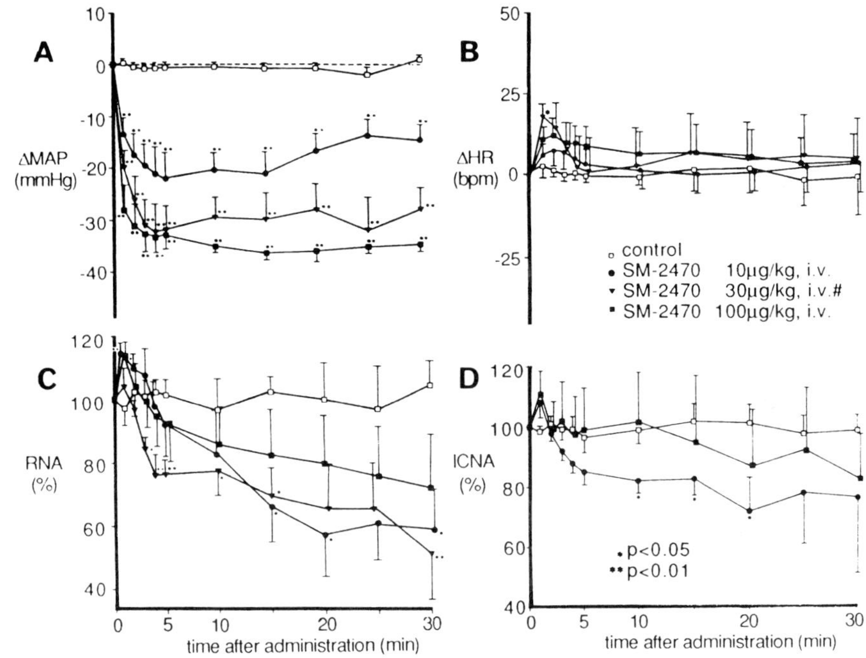

Figure 9. Time course changes in (A) mean arterial blood pressure (MAP), (B) heart rate (HR), (C) renal nerve activity (RNA) and (D) inferior cardiac nerve activity (ICNA) after SM-2470 administration in anesthetized rats. Five to 6 animals were used for each group.

dose dependent manner without any significant change in heart rate (Fig. 9A & 9B). Renal and cardiac sympathetic nerve activity decreased by a low dose of SM-2470 (Fig. 9C & 9D). Preganglionic adrenal nerve activity was also decreased by a low dose of SM-2470. These findings indicate that SM-2470 possesses a central sympatho-inhibitory activity. An equi-hypotensive dose of SM-2470, however, demonstrated less sympathoinhibitory activity than clonidine, centrally acting antihypertensive drug . Moreover, with an equi-sympathoinhibitory dose, SM-2470 exerted a more marked decrease in mean blood pressure than clonidine.

In chronic study where SM-2470 (3 and 10mg/kg/day) was administered orally over a period of a month, antihypertensive effect and bradycardia were observed in SHRSP. In addition, the age-related increase in urinary epinephrine excretion which was observed in age-matched SHRSP was significantly abolished by SM-2470 (Fig. 10). This finding is thought to reflect that SM-2470 decreased adrenal sympathetic nerve activity, as shown in acute study using normotensive rats. In SHRSP, the enhancement of the adreno-medullary function associated with the development of hypertension is considered to be a possible pathophysiological factor. Therefore, the SM-2470-induced suppression of the adreno-medullary function might play a role in its anti-hypertensive and anti-tachycardiac effects in this hypertensive animal model. It is further suggested that the central sympathoinhibitory activity may in part contribute as a central hypotensive component, even in long-term treatment.

YM-617

YM-617, {R-(-)-5-[2-[(2-(o-ethoxyphenoxy)ethyl]amino)propyl]-2-methoxybenzene-sulfonamide hydrochloride}, is a phenethylamine derivative and a new structural type of α_1-adrenoceptor antagonists (Fig. 11A). YM-617 is the R-(-)-enatiomer of YM-12617 which possesses blocking activity 2-18 times more potent than prazosin as evaluated by the pKi values of binding assay (Honda *et al.*, 1985). YM-617 is known to be a more potent antagonist for α_1-adrenoceptors than YM-12617 (Honda and Nakagawa, 1986).

The effect of YM-617 on sympathetic nerve activity was evaluated electrophysiologically in normotensive rats by Yoshioka *et al.* (1990). Intravenous administration of YM-617 (2, 10, and 50 μg/kg) produced a dose-dependent hypotension accompanied with bradycardia and renal sympathetic nerve activity (Fig. 12A-C). Cardiac sympathetic nerve activity (Fig. 11B & 12D) and preganglionic adrenal nerve activity were also decreased by this α_1-antagonist. When compared with a centrally acting antihypertensive drug, clonidine, an equi-hypotensive dose of YM-617 produced significantly less potent sympathoinhibitory activity (Fig. 13). On the other hand, the activity of the aortic depressor nerves, which contain mainly afferent nerves from baroreceptors in the aortic arch (Sapru and Krieger, 1977) was decreased significantly

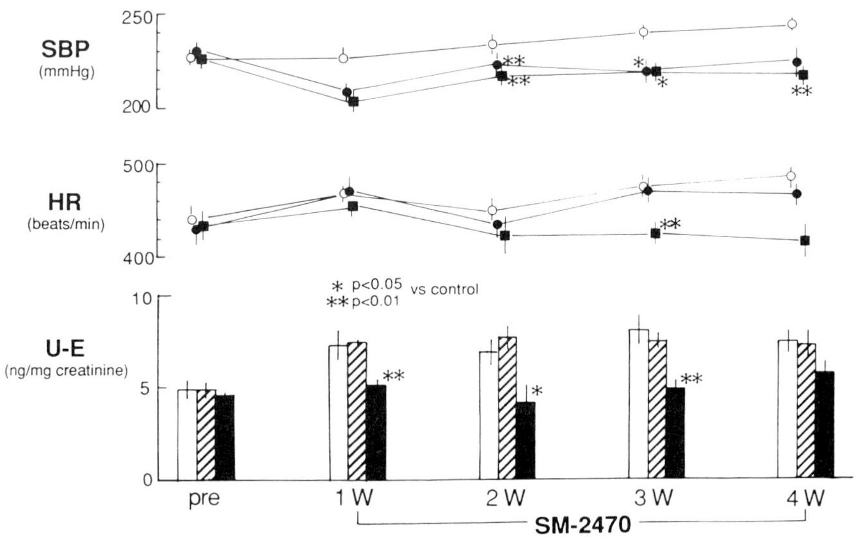

Figure 10. Effects of the chronic administration of SM-2470 on systolic blood pressure (SBP), heart rate (HR) and urinary epinephrine excretion (U-E) in SHRSP. SM-2470 was administered orally via the drug-mixed food for 4 weeks. SBP and HR were measured using the tail-cuff method. U-E was determined by HPLC-ECD methods. Six animals were used for each group. Data are expressed as mean±S.E.

after YM-617 administration. This indicaties that the sympathoinhibitory action of YM-617 is not mediated by the baroreceptor afferent nerve. The sympathoinhibitory activity of YM-617 is thought to contribute to its anti-tachycardiac or bradycardiac effect. Moreover, a component which is mediated via central sympathoinhibition might be involved in the hypotensive activity of YM-617. Of note is the fact that this

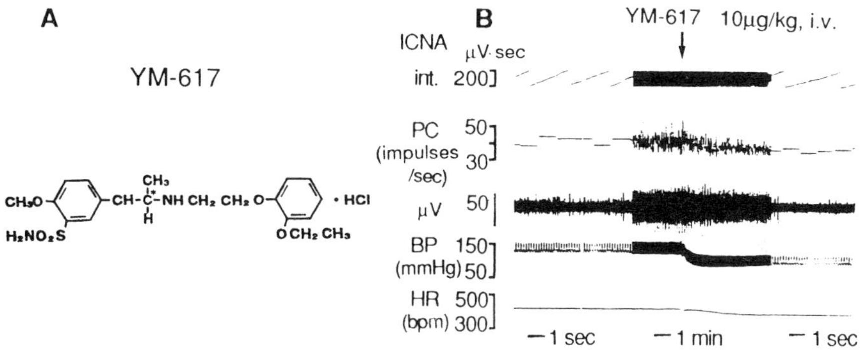

Figure 11. Chemical structure of YM-617 (A) and effects of systemic *i.v.* injection of YM-617 (10μg/kg) in anesthetized rats.(B); Traces from top to bottom are integrated inferior cardiac nerve activity (ICNA), counts of impulses (PC) and actual ICNA, blood pressure (BP) and heart rate (HR).

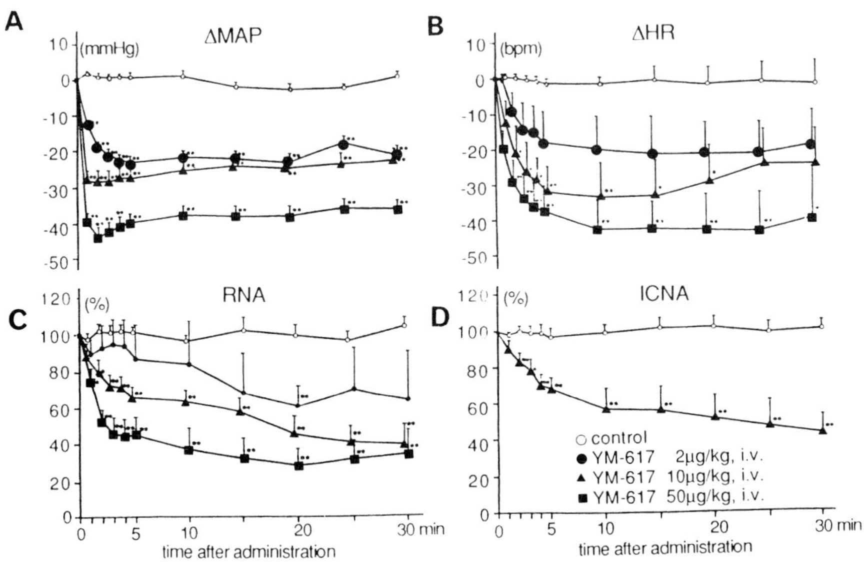

Figure 12. Time course changes in (A) mean arterial blood pressure (MAP), (B) heart rate (HR), (C) renal nerve activity (RNA) and (D) inferior cardiac nerve activity (ICNA) after YM-617 administration in anesthetized rats. Six to twelve animals were used for each group.

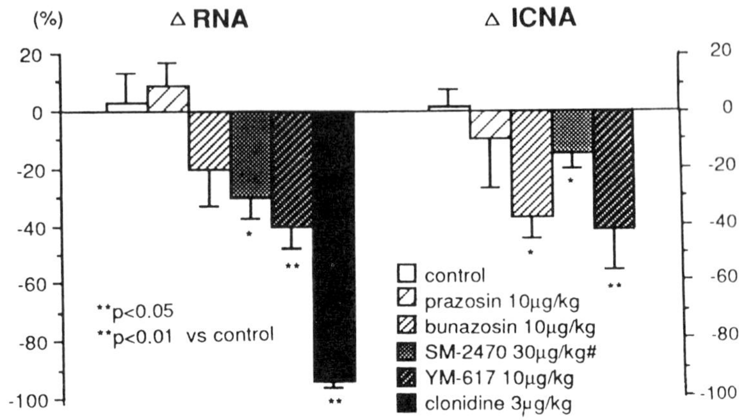

Figure 13. Comparison of sympathoinhibitory effects of α_1-adrenoceptor antagonists in anesthetized rats. Equi-hypotensive doses of α_1-adrenoceptor antagonists which produced a decrease in mean blood pressure by around 30 mmHg were administered intravenously. The response (per cent change from the initial value) 20 min after administration were compared with that in the control group. # Ten µg/kg of SM-2470 was administered in the ICNA experiment. Six to 10 animals were used for each group. Values are expressed as mean±S.E.

structurally new α_1-antagonist demonstrated a significant sympathoinhibitory action at the low dose of 2 µg/kg as compared with quinazoline derivatives such as prazosin and bunazosin.

THE COMPARISON OF SYMPATHOINHIBITORY ACTIVITY OF α_1-ADRENOCEPTOR ANTAGONISTS

The α_1-adrenoceptor antagonists, of which prazosin is a prototype, were electrophysiologically evaluated for their sympathoinhibitory activity in anesthetized rats. Generally, α_1-adrenoceptor antagonists are classified as vasodilators. It is known that vasodilators produce a decrease in blood pressure accompanied with tachycardia (Brunner *et al.*, 1967; Murphy *et al.*, 1983). In another series of experiments, vasodilators such as hydralazine (Fig. 14) and potassium channel opener, pinacidil, produced reflex increases in heart rate and sympathoexcitatory responses along with hypotension in anesthetized rats (Yoshioka *et al.*, 1986 & 1990). On the other hand, α_1-adrenoceptor antagonists demonstrated no changes or decreases in heart rate and sympathetic nerve activity, in spite of expected reflex tachycardia resulting from a marked hypotension. Our results suggest that these drugs might possess sympathoinhibitory activity.

A comparative study with a centrally acting antihypertensive drug, clonidine, showed that the sympathoinhibitory actions of all α_1-adrenoceptor antagonists evaluated were less potent than clonidine when compared in equi-hypotensive doses (Fig. 13). Among the α_1-antagonists used, YM-617 possessed the most potent sympathoinhibitory activity. More specifically, YM-617 showed a significant decrease in renal sympathetic nerve activity at the low dose of 2μg/kg (Fig. 12C). Heart rate was also reduced by YM-617, accompanied with a significant decrease in inferior cardiac nerve activity (Fig. 12B & C). SM-2470 also caused significant sympathoinhibitory activity. Ten μg/kg of SM-2470 decreased not only the postganglionic renal and inferior cardiac sympathetic nerve actvity but also the preganglionic adrenal sympathetic nerve activity as well. However, dose response of sympathoinhibitory activity to SM-2470 were not observed. Namely, it was noted that high doses of SM-2470 attenuated the sympathoinhibitory action observed in the low dose group. This seems to result from the reflex sympathetic stimulation overcoming sympathetic inhibition caused by SM-2470. Bunazosin also decreased heart rate significantly. The effects of bunazosin on sympathetic nerve activity were specific to the inferior cardiac nerve. Renal and adrenal nerve activity were not decreased significantly, but were, in fact, increased. The reason why bunazosin produced this discrepancy among the sympathetic nerves is an unsolved problem. This might be a positive characteristic and provide a logical explanation for bunazosin's anti-tachycardiac effect. Prazosin and SGB-1347, on the other hand, did not reduce sympathetic nerve activity under our experimental conditions. Therefore, these drugs seem unlikely to be potent sympathoinhibitory drugs, although there is a possibility that they have a weak central sympathoinhibitory action which overcomes reflex cardiac and sympathetic acceleration.

The structure-sympathoinhibitory activity relationships which have been reported concerning the sedative effects and hypotensive activity of α_2-agonists (Timmermans *et al.*, 1980) have not been fully evaluated in the α_1-antagonists. Recently, Tsuchihashi and Nagatomo (1989) determined α_1-blocking activity of α_1-adrenoceptor antagonists using radioligand binding methods. They reported that the side chains in the drugs with quinazoline or quinazolinedione residues are important

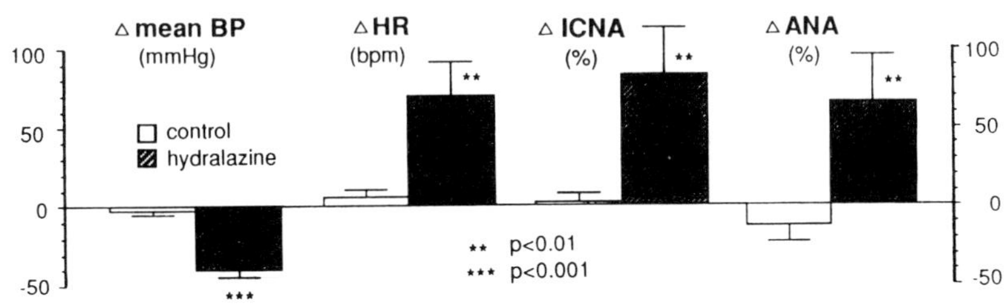

Figure 14. Changes in mean arterial blood pressure (mean BP), heart rate (HR), inferior cardiac nerve activity (ICNA) and adrenal nerve activity (ANA) 20min after systemic *i.v.* injection of hydralazine (5mg/kg) in anesthetized rats.

for the antagonistic potencies of α_1-adrenoceptors. Moreover, the contribution of the effect on sympathetic function on hypotensive activity has not yet been determined. The central sympathoinhibitory activity, however, is probably responsible for the anti-tachycardiac effects through blunting the baroreceptor reflex activity in response to vasodilation.

CONCLUSION

This chapter reviewed the electrophysiological studies on novel α_1-adrenoceptor antagonists. It has been demonstrated that they have anti-tachycardiac or bradycardic activity, in spite of their peripheral vasodilatation. Direct evaluation using electrophysiological nerve recordings pointed to the possibility that these α_1-adrenaoceptor antagonists might possess some sympathoinhibitory activity. However, this sympatho-inhibitory activity was less potent than that caused by clonidine, which is the most potent central sympathoinhibitory α_2-agonist.

Investigations of patients with different stages of hypertension or congestive heart failure demonstrate that α_1-adrenoceptor antagonists provide a useful therapeutic approach in the treatment of cardiovascular disease (Miller *et al.*, 1977; Warren and Dollery, 1988). Both clinical experience and animal studies have shown that reflex tachycardia is not a major problem with these drugs, unlike other vasodilators. The fact that they produced less reflex tachycardia than expected from their peripheral vasodilatation might result from their sympathoinhibitory action. Namely, the blockade of central α_1-adrenoceptors results in blunting the baroreceptor reflex activity in response to vasodilation.

Concerning the role of the central α_2-adrenoceptors, it is thought that presynaptically located α_2-adrenoceptors mediate sedation and postsynaptically located α_2-adrenoceptors are responsible for central hypotensive and sympathoinhibitory activity. However, the functional role of the central α_1-adrenoceptors is not yet fully understood. It is also known that the balance of effects can alter with the dose. Therefore, it is important to understand the complex response which arise from the balance of central and periphral alpha-adrenoceptor antagonism in the α_1-antagonists.

Whether central sympathoinhibitory activity is responsible for the antihypertensive activity after α_1-adrenoceptor antagonists is uncertain. Studies of the long-term effects of α_1-adrenoceptor antagonists on the sympathetic function remain to be understood. As shown in the chronic study of SM-2470 using the hypertensive animal model, the sympathoinhibitory activity of the drug seemed to contribute to the anti-tachycardiac effect during long-term administration.

The ideal antihypertensive drug would be free of any central component that limits its clinical use. However, it is also true that the central sympathoinhibitory activity of the α_1-adrenoceptor antagonists induce little unwanted reflex cardiac acceleration. Based on the above considerations, the concept of combinng the peripheral α_1-adrenoceptor blockade and the activity of central sympathoinhibition seems to be logical. Therefore, understanding the complex response which arises from the balance of central and peripheral α-adrenoceptor antagonism along with additional pharmacological properties should lead to an effective treatment of hypertension. An α_1-antagonist with a mild sympathoinhibitory activity is theoretically beneficial because the blockade of α_1-adrenoceptors can eliminate peripheral vasoconstriction and the central sympatho-inhibitory activity overcomes reflex tachycardia in response to vasodilation via the blockade of the peripheral α_1-receptors.

REFERENCES

Andén, N.E., Gomes, C., Persson, B. and Trolin, G. (1978). R28935 and prazosin: effects on central and peripheral alpha-adrenoceptor activity and on blood pressure. *Naunyn-Schmiedeberg's Arch. Pharmacol.* **302,** 299-306.

Barbieri, C., Caldara, R., Ferrari, C., Dal B., G. A., Paracchi, A., Romussi, M. and Curtarelli, G. (1980). Metabolic effects of parazosin. *Clin. Pharmacol. Ther.* **27,** 313-316.

Bhargaba, K. P., Mishra, N. and Tangri, K. K. (1972). An analysis of central adrenoseptors for control of cardiovascular function. *Br. J. Pharmacol.* **45,** 596-602.

Brunner, H., Hedwall, P. R. and Meier, M. (1967). Influence of adrenergic beta-receptor blockade on the acute cardiovascular effects of hydralazine. *Br. J. Pharmacol.* **30,** 123-133.

Cambridge, D., Davey, M. J. and Massingham, R. (1977). Prazosin, a selective antagonist of post-synaptic α-adrenoceptors. *Br. J. Pharmacol.* **59,** 514p-515p.

Cavero, I. and Roach, A. G. (1980). The pharmacology of prazosin, a novel antihypertensive agent. *Life Sciences* **27,** 1525-1540.

Chiba, S., Furukawa, Y. and Watanabe, H. (1981). Chronotropic and inotropic effects of three kinds of alpha-adrenergic blockers on the isolated dog atria. *Jpn. Heart J.* **22**, 637-641.

Greengrass, P. and Brenner, R.(1979). Binding characteristics of ^{3}H-prazosin to rat brain α-adrenergic receptors. *Eur. J. Pharmacol.* **55,** 323-326.

Hata, F., Kondo, E., Kondo, S., Kagawa, K. and Ishida, H. (1982). Characteristics of (^{3}H)E-643-binding to alpha adrenoceptors. *Japan. J. Pharmacol.* **32,** 181-187.

Honda, K., Miyata-Ozawa, A. and Takenaka, T. (1985) α_1-Adrenoceptor subtype mediating contraction of smooth muscle in the lower urinary tract and prostate of rabbits. *Naunyn-Schmiedeberg's Arch. Pharmacol.* **330,** 16-21.

Honda, K. and Nakagawa, C. (1986). Alpha-1 adrenoceptor antagonist effects of the optical isomers of YM-12617 in rabbit lower urinary tract and prostate. *J. Pharmacol. Exp. Ther.* **239,** 512-516.

Huchet, A. M., Chelly, J. and Schmitt, H. (1981). Role of α_1-and α_2- adrenoceptors in the modulation of the baroreflex vagal bradycardia. *Eur. J. Pharmacol.* **71,** 455-461.

Huchet, A. M., Huguet, F., Ostermann, G., Bakri-Logeais, F, Schmitt, H. and Narcisse, G. (1983). Central α_1-adrenoceptors and cardiovascular control in normotensive and spontaneously hypertensive rats. *Eur. J. Pharmacol.* **95,** 207-213.

Igarashi, T., Nakajima, Y. and Ohtake, S. (1976). Comparison of effects of alpha-blocking agents on blood pressure and heart rate in unanethetized spontaneously hypertensive (SHR) and normotensive rats. *Clin. Exp. Pharmacol. Physiol.* **3** (suppl), 89-92.

Igarashi, T., Nakajima, Y. and Ohtake, S. (1977). Antihypertensive effect of combined treatment with α- and β-adrenergic blockers in the spontaneously hypertensive rats. *Clin. Jpn. Circ. J.* **4,** 903-911.

Ito, T., Hey, J.A. and Koss, C. (1988). Studies on the mechanism of prazosin induced sympatho-inhibition. *Eur. J. Pharmacol.* **158,** 225-231.

Izzo, J. L., Horwitz, D., and Keiser, H. R. (1981). Physiologic mechanisms opposing the hemodynamic effects of prazosin. *Clin. Pharmacol. Ther.* **29,** 7-11.

Kawasaki, T., Uezono, K., Abe, I., Nakamuta, S., Ueno, M., Kawazoe, N. and Omae, T. (1981). Antihypertensive effect of E-643, a new alpha-adrenergic blocking agent. *Eur. J. Clin. Pharmacol.* **20,** 399-405.

Kellar, K. J., Quest, J. A., Spera, A. C., Buller, A., Conforti, A., Souza, J. D. and Gillis, R. A. (1984). Comparative effects of urapidil, prazosin, and clonidine on ligand binding to central nervous system receptors, arterial pressure, and heart rate in experimental animals. *Am. J. Med.* 87-95.

Kim, Y. I., Paik, Y. H., Kang, S. S. and Kim, J. H. (1982). Effects of α-adrenoceptor antagonists administered intraventricularly on central hypotensive action of clonidine and on central hypertensive action of methoxamine in rabbits. *Arch. Int. Pharmacodyn.* **257,** 66-76.

Kubo, T., Kihara, M., Hata, H. and Misu, Y. (1987). Cardiovascular effects in rats of alpha1 and alpha2 adrenergic agents injected into the nucleus tractus solitarii. *Naunyn-Schmiedeberg's Arch. Pharmacol.* **335,** 274-277.

Leren, P., Foss, P. O., Helgeland, A., Hjermann, I., Holme, I. and Lund-Larsen, P. G. (1980). Effect of propranolol and prazosin on blood lipids. *Lancet* 4-6.

Matsumoto, M., Yoshioka, M., Togashi, H., Minami, M., Nagao, A. and Saito, H. (1988). Effects of SGB-1534, a new synthesized phenylpiperazine derivative, on blood pressure and heart rate in stroke-prone spontaneously hypertensive rats. *Hokkaido J. Med. Sci.* **63,** 36-43.

McCall, R. B. and Humphrey, S. J. (1981). Evidence for a central depressor action of postsynaptic α_1- adrenergic receptor antagonists. *J. Autonomic Nerv. Sys.* **3,** 9-23.

Miller, R. R., Awan, N. A., Maxwell, K. S. and Mason, D. T. (1977). Sustained reduction of cardiac impedance and preload in congestive heart failure with the antihypertensive vasodilator prazosin. *New. Eng. J. Med.* **297,** 303-307.

Minami, M., Togashi, H., Sano, M., Yoshioka, M., Saito, I., Morii, K., Tochihara, M., Hirogami, M., Nomura, A. and Saito, H. (1985). Effects of new antihypertensive agents, 2-[4-(n-butyryl) homopiperazine-1-yl]-4-amino 6,7-dimothoxy-quinazoline (E-643), on blood pressure, urinary sodium excretion and urinary norepinephrine excretion rate in stroke-prone spontaneously hypertensive rats[in Japanese]. *Hokkaido J. Med. Sci.* **60,** 536-543.

Murphy, M. B., Scriven, A., Brown, M. J., Causon, R. and Dollery, C. (1983) The effects of nifedipine and hydralazine induced hypotension on sympathetic activity. *Eur. J. Clin. Pharmacol.* **23,** 479-482.

Nabata, H., Aono, R., J.,Ishizuka, N. and Sakai, K. (1985). Antihypertensive effects of a novel phenylpiperazine derivative, SGB-1534, on several hypertensive models of rats. *Arch. int. Pharmacodyn.* **277,** 104-118.

Palacios, J.M., Hoyer, D. and Cortes, R. (1987). α_1-adrenoceptors in the mammalian brain: similar pharmacology but different distribution in the rodents and primates. *Brain Res.* **419,** 65-75.

Ramage, A.G. (1986). Evidence for a central sympathoinhibitory action of prazosin and indoramin. *Eur. J. Pharmacol.* **121,** 83-89.

Sakai, K.,Kuromaru, O., Akima, M. and Yoshizawa, M. (1985). Cardiovascular effects of a novel antihypertensive agent, SGB-1534, in miniature pigs. *Arch. int. Pharmacodyn.* **277,** 119-125.

Sapru, H.N. and Krieger, A.J. (1977). Carotid and aortic chemoreceptor function in anesthetized rats and cats. *J. Appl. Physiol.* **42,** 344-348.

Shoji, T. (1981). Comparsion of pre- and postsynaptic α-adrenoceptor blocking effects of E-643 in the isolated vas deferens of the rat. *Japan. J. Pharmacol.* **31,** 361-368.

Shoji, T., Daiku, Y. and Igarashi, T. (1980). α-adrenoceptor blocking properties of a new antihypertensive agent, 2-[4-(n-butyryl)-homopiperazine-1-yl]-4-amino-6,7-dimethoxyquinazoline (E-643). *Japan. J. Pharmacol.* **30,** 763-772.

Sinha, J. N., Schramm, D. K., Gurtu, S., Pant, K. K. and Bhargaba, K. P. (1984). Nucleus locus coeruleus: evidence for α_1-adrenoceptor mediated hypotension in the cat. *Naunyn-Schmiedeberg's Arch. Pharmacol.* **326,** 193-197.

Takayanagi, I., Konno, F., Arai, H., Chimoto, K. and Kitada, T. (1985). Newly synthesized α_1-adrenoceptor blockers (SM911 and SM2470) and characterization of α-adrenoceptors in rat aortic strips, rat vas deferens preparations and rabbit aortic strips. *Gen. Pharmacol.,* **16,** 613 - 615.

Timmermans, P.B.M.W.M., Hoefke, W., Stähle, H. and van Zwieten, P. A. (1980). Structure-activity relationships in clonidine-like imidazolidines and related compounds. In: *Progress in pharmacology* (Vol. 3), Timmermans, P.B.M.W.M., Hoefke, W., Stähle, H. and van Zwieten, P.A. (Eds.). Gustav Fischer Verlag, Stuttgart and New York, pp.1-104.

Togashi, H. (1983). Central and peropheral effects of clonidine on the adrenal medullary function in spontaneously hypertensive rats. *J. Pharmacol. Exp. Ther.* **225,** 191-197.

Togashi, H., Yoshioka, M., Tochihara, M., Matsumoto, M. and Saito, H. (1990). Differential effects of hemorrhage on adrenal and renal nerve activity in anesthetized rats. *Am. J. Physiol.* **259,** H1134-H1141.

Tsuchihashi, H. and Nagatomo, T. (1989). Alpha-blocking potencies of antihypertensive agents (Prazosin, Terzosin, Bunazosin, SGB-1534 and Ketanserin) having with quinazoline or quinazolinedione as assessed by radioligand binding assay methods in rat brain. *J. Pharmacobio-Dyn.* **12,** 170-174.

Van Nueten, J. M., Janssen, P.A.J., van Beek, J., Xhonneux, R., Verbeuren, T. J. and Vanhoutte, P.M. (1981). Vascular effects of ketanserin (R 41 468), a novel antagonist of 5-HT_2 serotonergic receptors. *J. Pharmacol. Exp. Ther.* **218,** 217-230.

Van Zwieten, P. A. (1975). Antihypertensive drugs with a central action. In: Progress in pharmacology (Vol. 1), van Zwieten, P.A. (Ed.). Gustav Fischer Verlag, Stuttgart, pp.1-63.

Van Zwieten, P. A. (1988). Basic pharmacology of alpha-adrenoceptor antagonists and hybrid drugs. *J. Hypertension* **6** (suppl 2), S3-S11.

Van Zwieten, P. A., Lam, E., Timmermans, P.B.M.W.M. (1978). The interaction between prazosin and clonidine. *Clin. Sci.* **55** (suppl 4), 259s-261s.

Warren, J. B. and Dollery, C. T. (1988). The usefulness of alpha-adrenoceptor blockade in the treatment of hypertension. *J. Hypertension* **6** (suppl 2), S43-S49.

Wilcox, B. J., Raskind, M. A., Ko, G. N., Baskin, D. G., Paccualy, M. and Dorsa, D. M. (1990). Localization of 3H-prazosin binding sites in the supraoptic and paraventricular nuclei of the human hypothalamus. *Neuroendocrinology* **51,** 315-319.

Yoshino, G., Kazumi, T., Okada, K., Nakata, Y., Matsuura, M., Narutaki, K., Yoshida, A., Ishihara, K., Tsutou, A., Chuma, J., Sakagami, M., Yokono, K. and Baba, S. (1989). The effect of bunazosin, a new alpha1-blocker, on plasma lipids in diabetic subjects with essential hypertension. *Current Therapeutic Research* **45,** 677.

Yoshioka, M., Matsumoto, M., Togashi, H., Abe, M., Ikeda, T., Morii, K., Minami, M. and Saito, H. (1988). Effect of SM-2470, a newly synthetized α_1-adrenoceptor antagonist, on sympathetic nerve activity in anesthetized rats. *Arch. int. Pharmacodyn. Ther.* **293,** 162-174.

Yoshioka, M., Togashi, H., Abe, M., Ikeda, T., Matsumoto, M. and Saito, H. (1990). Central sympathoinhibitory action of a new type of alpha-1 adrenoceptor antagonist, YM-617, in rats. *J. Pharmacol. Exp. Ther.* **253,** 427-431.

Yoshioka, M., Togashi, H., Matsumoto, M., Minami, M. and Saito, H. (1989). Effect of bunazosin on cardiac sympathetic nerve activity in anesthetized rats. *Clin. and Exper. Hyper. -Theory and Practice* **A11,** 137-147.

Yoshioka, M., Togashi, H., Minami, M. and Saito, H. (1986). Effects of hydralazine on adrenal and cardiac sympathetic nerve activity in anesthetized rats. *Res. Comm. Chem. Pathol. Pharmacol.* **54,** 313-320.

Yoshioka, M., Togashi, H., Tochihara, M , Saito, Y., Matsumoto, M. and Saito, H. (1990). Effects of potassium channel opener, pinacidil, on symparhetic nerve activity in the rat. *Pharmacodyn. Ther. (Life Sci. Adv.)* **9,** 87-92.

Progress in Hypertension, Vol. 2, pp. 49-72.
Saito *et al.* (Eds)

PRIMARY PREVENTION WITH METOPROLOL IN HYPERTENSIVE PATIENTS

JOHN WIKSTRAND

Wallenberg Laboratory for Cardiovascular Research, Sahlgrenska Hospital, Gothenburg University, S-413 45 Gothenburg, Sweden.

INTRODUCTION

In patients with mild to moderate hypertension, the risk of dying from coronary events is greater than the risk of dying from stroke. In addition, data show that the risk of having a nonfatal myocardial infarction is much greater than that of having a nonfatal stroke, and that there are more nonfatal events than fatal ones (Kannel, 1977; Kannel *et al.*, 1985; MacMahon *et al.*, 1986; Wikstrand, 1986).

During the 1970s and early 1980s, several widely cited studies of thiazide diuretics as first-line treatment for mild to moderate hypertension proved conclusively that this treatment prevents or postpones stroke.

However, the effect of diuretics in such studies on coronary events was not encouraging. Therefore, much would be gained if an alternative to the use of thiazide diuretics as initial therapy could reduce not only stroke but also coronary events. Findings from two studies published in the middle and late 1970s-one, the first randomized secondary preventive study of post-myocardial infarction patients, the other, a nonrandomized study of hypertensives-indicated beneficial effects of β-blocker treatment on total mortality and also on coronary events (Wilhelmsson *et al.*, 1974; Berglund *et al.*, 1978).

In part because of these findings, four large-scale, randomized, primary preventive studies were set up in the middle and late 1970s to compare the efficacy of diuretics and β-blockers in reducing total mortality and other complications, such as coronary events and stroke, in hypertensive patients (Medical Research Council Working Party, 1985; The IPPPSH Collaborative Group, 1984; Wilhelmsen *et al.*, 1987; Wikstrand *et al.*, 1988a; Tuomilehto *et al.*, 1989b; Olsson *et al.*, 1991; Wikstrand *et al.*, 1991a). The MAPHY study was one of these four studies and aimed at investigating whether metoprolol, a relatively lipophilic and β_1-selective β-blocker lowered the risk for coronary events better than thiazide diuretics. This chapter summarizes the mortality and morbidity results from the MAPHY study, present supporting evidence and discusses clinical implications. Furthermore, comments on the design of the HAPPHY study are given in an appendix to the present chapter.

PATIENTS AND METHODS

Study Population

The MAPHY study included 3234 patients, 1609 randomized to metoprolol and 1625 to thiazide diuretics. Patients were recruited from 66 clinical centers in 11 countries (see Acknowledgments). Patients were included only if they gave consent after full explanation of the nature and purpose of the investigation.

Both previously treated patients and those with newly detected and untreated hypertension were eligible for inclusion in the study. All were male outpatients aged 40 to 64 years. The patients' diastolic blood pressure in the sitting position was 100 mm Hg or greater and less than 130 mm Hg at randomization, calculated as the mean of two readings taken two weeks apart and including two readings on each occasion. Since this was a study of primary prevention, previous myocardial infarction, angina pectoris, and stroke were exclusion criteria. Additional exclusion criteria were secondary hypertension, second- and third-degree atrioventricular block, cardiac failure, obstructive lung disease not well controlled by β_2 -stimulants, diabetes mellitus (defined as two separate positive dipstick tests for glucosuria and a fasting blood glucose level > 6.8 mm/L (> 123 mg/dL), gout, and serious diseases such as malignant neoplasms or severe alcoholism.

Study Design

This was an international, multicenter, stratified, randomized, open, controlled study of primary prevention in parallel groups preceded, for previously treated patients, by a four- to eight-week run-in period without treatment. The open study design can be discussed. The advantage of this design is that it mirrors everyday clinical practice. The stratification and randomization procedures together with the results of baseline characteristics, the blind recording of blood pressure, the blind evaluation of hard end-points, and the completeness of follow-up make it unlikely that there is any significant bias stemming from the study design. The vast majority of patient-years came from previously untreated patients, found partly by screening procedures.

Before randomization, patients were stratified according to risk of coronary heart disease into nine risk groups depending on age (40 to 49, 50 to 59, or > 60 yrs), first systolic blood pressure recorded < 190 or > 190 mmHg), cholesterol level (< 7.0, 7.0 to 8.6, or > 8.7 mmol/L [< 270, 270 to 330, or > 330 mg/dL]), and smoking habits (never smoked, ex-smoker, 1 to 14 g/d, or > 15 g/d [one cigarette = 1 g]). An ex-smoker was defined as a smoker who had stopped at least one month previously.

After stratification patients were randomized to treatment with 200 mg/d of metoprolol or a thiazide diuretic (50 mg/d of hydrochlorothiazide or 5 mg/d of bendroflumethiazide). These dosages were doubled, or additional drugs (hydralazine, spironolactone, or others, but not blockers or thiazide diuretics) were given, if necessary, to reach the treatments goal of diastolic blood pressure lower than 95 mm Hg. To modernize the therapy, the maximum daily dosages of the baseline drugs were reduced to 200 mg/d of metoprolol or 50 mg/d of hydrochlorothiazide (5 mg/d of bendroflumethiazide) in 1981.

According to protocol criteria stated in an addendum to the protocol from 1982, a patient in the diuretic group could receive β-blockade if the physician considered this to be indicated for clinical reasons (i.e., in some patients suffering a nonfatal myocardial infarction during the study).

Methods
Blood pressure recordings were taken after five minutes' rest twice at an interval of 15 minutes. Blood pressure was measured blindly (Hawksley random-zero sphygmomanometer or London School of Hygiene blind manometer) in the right arm with the patient in the sitting position.

The following laboratory studies were performed: serum cholesterol, serum potassium, serum uric acid, serum creatinine, dipstick test for glucosuria and albuminuria, and a 12-lead resting electrocardiogram. Body weight and height were also measured.

At the last follow-up visit, all investigators were asked to specify all drugs and dosages used.

MAIN END-POINTS AND CLASSIFICATION OF END-POINTS

The main end-points, defined in advance, were total mortality, sudden cardiac death, the pooled incidence of fatal and nonfatal coronary heart disease and stroke. A classification of all reported fatal and nonfatal acute events was made by the Independent End-Point Committee. This classification, the Minnesota coding of ECGs (see below) and data quality were audited by the Independent Data Audit Committee. All cases were judged without any knowledge of actual treatment or to which treatment the patients had originally been randomized.

Cause-Specific Mortality
At the end of the study, 3085 patients were confirmed to be alive and 148 were dead. One patient was unavailable for follow-up (randomized to diuretic). A decision regarding cause of death was made by the Independent End-Point Committee for each of the 148 deaths.

The cause of death was defined by using all available information, such as hospital records, physicians' reports, police reports, death certificates, and necropsy reports. If the death occurred within 28 days of the onset of an event, that event was classified as the fatal event; otherwise the event was classified as nonfatal, and a subsequent event, if fatal, was coded as the cause of death.

First the death was classified as cardiovascular or noncardiovascular and then the cause-specific mortality was defined within each of these two subgroups. For a fatal myocardial infarction to be recorded two out of the three classical criteria had to be fulfilled (see below). Sudden cardiac death was defined as a death occurring within 24 hours after onset of symptoms with no other obvious cause of death identifiable (Olsson *et al.*, 1991). For a diagnosis of fatal stroke to be recorded, unequivocal signs of focal or global neurological deficit with sudden onset judged to be vascular in origin were to be present.

Nonfatal Myocardial Infarctions
Acute Myocardial Infarction For a diagnosis of an acute myocardial infarction to be

confirmed, at least two of the following three criteria were to be fulfilled: central chest pain of more than 15 minutes' duration, transient elevation of enzymes indicating myocardial necrosis or typical ECG changes. If a myocardial infarction was suspected but only one criterion was fulfilled, the condition was recorded as a possible myocardial infarction.

Silent Myocardial Infarction A 12-lead resting ECG was recorded at randomization

Table 1. Clinical characteristics of patients at entry and at last follow-up visit (n = 3234). From Wikstrand *et al.*, 1988a.

	At entry Randomized to Metoprolol	Diuretics	At last follow-up Randomized to Metoprolol	Diuretics
Systolic blood pressure mm Hg	166.9±17	166.8±17	142.4±17	142.7±16
Diastolic blood pressure mm Hg	107.6±6	107.5±6	88.7±87	89.5±8*
Heart rate, beats/min	78.2±10	77.3±10	64.1±10	74.1±11**
Age, years	52.6±7	52.6±6	57.7±7	57.6±7
Body weight, kg	83.2±12	82.7±12	84.7±12	82.6±12**
Body mass index, kg/m2	27.3±3.5	27.1±3.5	27.6±3.7	26.9±3.5**
Serum cholesterol, mmol/L	6.32±1.2	6.29±1.2	6.15±1.2	6.32±1.2**
(mg/dL)	(244±46)	(243±46)	(237±46)	(244±46)**
Serum creatinine, umol/l	93.1±17	93.1±16	97.7±19	96.8±16
(mg/dL)	(1.05±0.19)	(1.05±0.18)	(1.10±0.21)	(1.09±0.18)
Serum potassium, mmol/L (mEq/L)	4.28±0.4	4.27±0.4	4.28±0.4	3.97±0.4**
Serum urate, umol/l	349±76	347±80	364±75	384±82**
(mg/dL)	(5.9±1.3)	(5.8±1.3)	(6.1±1.3)	(6.4±1.4)**
Serum potassium < 3.6 mmol/L (mEq/L), %	5.2	6.3	1.7	13.0**
Serum urate > 450 umol/L (> 7.5 mg/DL), %	9.2	9.0	11.9	18.2**
Glucosuria, %	1.5	0.4	2.3	1.6
Albuminuria, %	4.7	3.9	3.0	2.7
Smokers, %	34	33	23	25
Patient years[2]				
For total mortality			8110	8070
For definite CV disease events			7830	7720
For definite CHD events			7880	7790

Values are mean ± SD. *p = 0.012 and **p = 0.001 refer to between group comparison at last follow-up visit. Abbreviations: CV = cardiovascular; CHD = coronary heart disease. For definition see Methods.

and repeated on a yearly basis and sent to the administrative center at the Sahlgrenska Hospital, Gothenburg University, Gothenburg, Sweden, for Minnesota coding (Rose and Blackburn, 1968). Altogether 16987 ECGs were Minnesota coded, each by two independent technicians, 8575 from patients randomized to metoprolol and 8412 from patients randomized to diuretics. The occurrence of a new code 1:1 or 1:2 in the Minnesota code (definite Q or Q/S pattern) without other clinical signs of myocardial infarction was defined as a definite silent myocardial infarction. Similarly, a new code 1:3 was coded as a possible silent myocardial infarction.

Nonfatal Stroke
For a diagnosis of nonfatal stroke to be recorded, unequivocal signs of focal or global neurological deficit with sudden onset, of a duration greater than 24 hours and which were thought to be vascular in origin were to be present.

STATISTICAL METHODS

All data were analyzed according to the original randomized allocation, even if actual treatment deviated from this (intention-to-treat). With the Gehan-Wilcoxon nonparametric test for survival analysis, the null hypothesis was tested, i.e., that there was no difference between the two treatments in risk (SAS User's Guide, 1983; Friedman *et al.*, 1983). The Gehan-Wilcoxon test does not assume a constant ratio between the hazard functions of the two groups during follow-up, and therefore

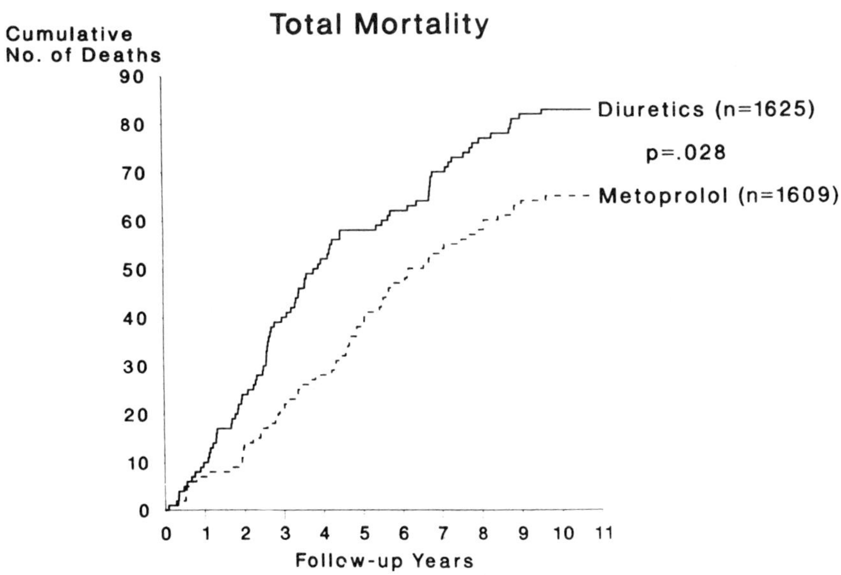

Figure 1. Cumulative numbers for total mortality in the two randomization groups in the MAPHY study. The p-value refers to the difference in risk between the two randomizations during the entire study period. From Wikstrand *et al.*, 1988a.

cannot be used to calculate risk ratios and confidence intervals. Even so, in an attempt to illustrate quantitatively the difference between the two treatment regimens in risk during follow-up, crude risk ratios and confidence intervals for risk ratios for the whole study population have been given at the median follow-up time (4.2 years) and at the end of the study (10.8 years) for all first definite coronary events, using the Fisher statistics (SAS User's Guide, 1983; Friedman *et al.*, 1983). In patients with an event, patient-years have been calculated to the time of the first event, otherwise to the last follow-up date. In a *post hoc* subgroup analysis, data were analyzed according to smoking status at randomization to allow comparisons with data presented from the MRC and IPPPSH reports. Furthermore, data were analyzed according to the occurrence of resting ECG abnormalities at randomization. The ECG abnormalities were defined as any of the codes 1:1-3 (Q- or QS-items); 2:1 (left axis deviation); 4:1-3 (S-T-changes); 5:1-3 (T-wave changes); 7:1 or 7:2 (left or right bundle branch block). Of course any post hoc subgroup analysis must be interpreted with great caution. *P* values below 0.05 (two-sided tests) were considered significant and all *p* values referring to the Gehan-Wilcoxon statistics refer to the total study period until the end of the trial (10.8 years).

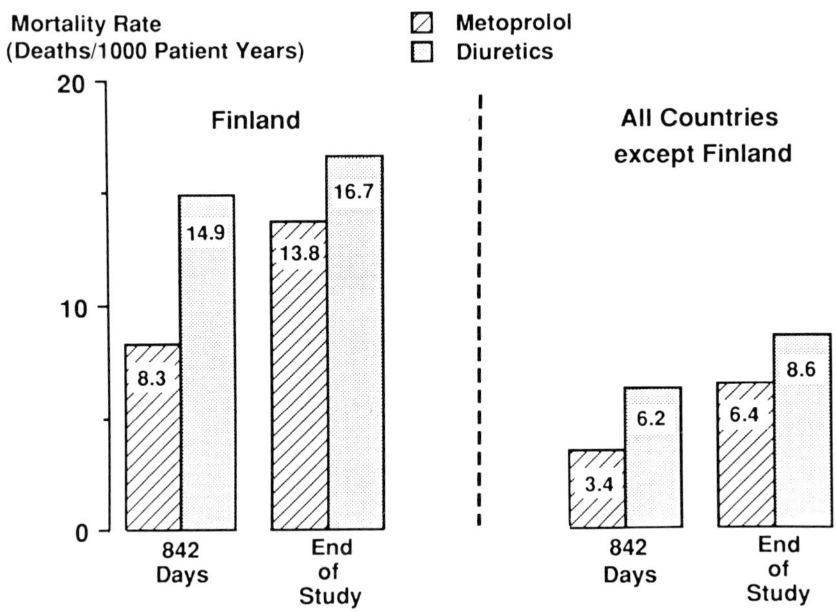

Figure 2. Cumulative mortality rate in patients from Finland and in patients from all other countries except Finland at the shortest common follow-up period for all patients (842 days) and at the end of the MAPHY study. Observe that mortality rate is increasing with increasing time of follow-up and also that the benefit of the beta-blocker treatment in the MAPHY study is independent of the mortality rate, *i.e.* independent of risk. In Finland 21.2 % of the patient-years and 35.1 % of the total number of deaths occurred. The figures for the total study group at the end of the MAPHY study are given in Table 6. From Wikstrand *et al.*, 1991, in press.

RESULTS

Clinical characteristics at randomization and during follow-up are given in Table 1. Blood pressure during follow-up was similar in the two randomization groups: 142/89 mm Hg in the metoprolol group (n = 1609) and 143/90 mm Hg in the diuretic group (n = 1625).

Mortality

Total Mortality Total mortality was significantly lower in patients randomized to metoprolol than in patients randomized to diuretics (p = 0.028) (Fig. 1 and Table 2). Although the mortality rate (deaths per 1000 patient-years) differed substantially between different populations, the percentage difference in mortality between patients randomized to metoprolol and to diuretics indicated that the benefit was spread over different geographic regions (Fig. 2). Furthermore, the figures implied benefits both in centers comparing metoprolol with hydrochlorothiazide (5.0 vs 8.9 deaths per 1000 patient-years at the median follow-up time) and in centers comparing metoprolol with bendroflume-thiazide (4.5 vs 10.2 deaths per 1000 patient-years at the median follow-up time). Finally, the figures implied benefits both in patients with mild hypertension (diastolic blood pressure (≤ 104 mm Hg at entry) (5.7 vs 8.4 deaths per 1000 patient-years) and in those with diastolic blood pressure at or above 105 mm Hg (4.3 vs 10.5 deaths per 1000 patient-years). This type of subgroup analysis must, however, be interpreted with great caution.

Cause-Specific Mortality Significantly lower cardiovascular mortality (p = 0.012),

Table 2 Cause-Specific Mortality. From Wikstrand *et al.*, 1988a.

	Patients Randomized to Metoprolol (n = 1609)	Patients Randomized to Diuretics (n = 1625)	p
Cardiovascular	**42**	**57**	**0.012**
Fatal coronary heart disease (myocardial infarction or sudden death coronary)	36	43	0.048
Sudden death, pulmonary embolism	0	3	-
Fatal stroke	2	9	0.043
Aortic aneurysm	1	2	-
Heart failure	3	0	-
Noncardiovascular	**23**	**26**	-
Cancer	16	18	-
Suicide	3	3	-
Accident	1	4	-
Infection	2	0	-
Other	1	1	-
Total	**65**	**83**	**0.028**

Table 3. All first definite and also definite plus possible coronary heart and cardiovascular disease events. Observe that fatal cases preceded by definite nonfatal cardiovascular events have not been given in the table. For full information on fatal events see Table 2.

	All first Coronary Heart Disease Events					All first Cardiovascular Disease[1] Events				
	Metoprolol			Diuretics		Metoprolol			Diuretics	
	n	rate[2]	*p*[3]	*n*	rate[2]	*n*	rate[2]	*p*[3]	*n*	rate[2]
Fatal										
CHD	29	3.7		35	4.6	29	3.7		34[1]	4.5
Stroke						2	0.3		7	0.9
Other CV[4]	-	-		-	-	1	0.1		4	0.5
Definite Nonfatal										
Acute MI	44	5.7		54	7.1	44	5.7		531	7.0
Silent MI[5]	38	4.9		55	7.2	371	4.8		541	7.1
Acute and Silent MI	82	10.6	0.0034	108	14.3					
Nonfatal Stroke	-	-		-	-	21	2.7		18	2.4
All Definite	111	14.3	0.0010	144	18.9	134	17.3	0.0009	170	22.3
Definite and Possible	157	20.5	0.0015	202	27.0	178	23.3	0.0011	228	30.5

Abbreviations: CHD = coronary heart disease; CV = cardiovascular disease; MI = myocardial infarction. [1]A nonfatal stroke preceded a definite coronary heart disease event in one patient in the metoprolol group and three patients in the diuretic group. [2]Rate = cases/1000 patient-years. [3]The p values refer to the difference in risk during the entire study period. [4]For details see Table 3. [5]New Q/QS items (1:1 or 1:2 according to the Minnesota code, in annual ECGs). Only silent events ($p = 0.016$) given in the table (see Methods section for details). Altogether 57 definite Q/QS codes were coded in the metoprolol group and 73 in the diuretic group ($p = 0.014$).

coronary heart disease mortality ($p = 0.048$), and stroke mortality ($p = 0.043$) were observed in patients randomized to metoprolol (Table 2). Three patients randomized to β-blockade died of severe heart failure several months to years after nonfatal myocardial infarctions. According to protocol criteria these were included as definite coronary events in the analysis of pooled fatal and nonfatal events.

MORTALITY AND MORBIDITY

Definite Coronary Heart Disease Events

Fatal and Nonfatal Coronary Heart Disease The pooled incidence of all first definite coronary events was significantly lower during follow-up in patients on metoprolol than in patients on diuretics (Fig. 3, Tables 3 and 4). Of all first definite coronary events, 25 % were fatal ones, 39 % were definite nonfatal acute myocardial infarctions and 36 % were definite silent myocardial infarctions. There was a statistically significant reduction in acute events (sudden death and fatal and nonfatal acute myocardial infarction, $p = 0.024$) as well as in definite silent myocardial infarctions ($p = 0.016$). In one patient in the metoprolol group and three patients in the diuretic group, the first definite coronary event had been preceded by a nonfatal stroke (Table 3).

Nonfatal Myocardial Infarctions The incidence of all first definite nonfatal myocardial infarctions was significantly lower during follow-up in patients randomized to metoprolol than in patients randomized to diuretics ($p = 0.0034$, 10.6 vs 14.3 cases /1000 patient-years at the end of the trial, Table 3).

Definite Cardiovascular Disease Events

The pooled incidence of all first definite cardiovascular disease events was significantly lower during follow-up in patients randomized to metoprolol than in patients randomized to diuretics (Tables 3 and 4). First stroke events accounted for 16 % of these events, 23 cases in the metoprolol group (2 fatal and 21 nonfatal) and 25 cases in the diuretic group (7 fatal and 18 nonfatal), Fig. 3. The pooled incidence of all first definite cardiovascular disease events was significantly lower during follow-up in patients randomized to metoprolol both in patients with and those without resting ECG abnormalities present at randomization (ECG changes absent: $p = 0.008$; 16.7

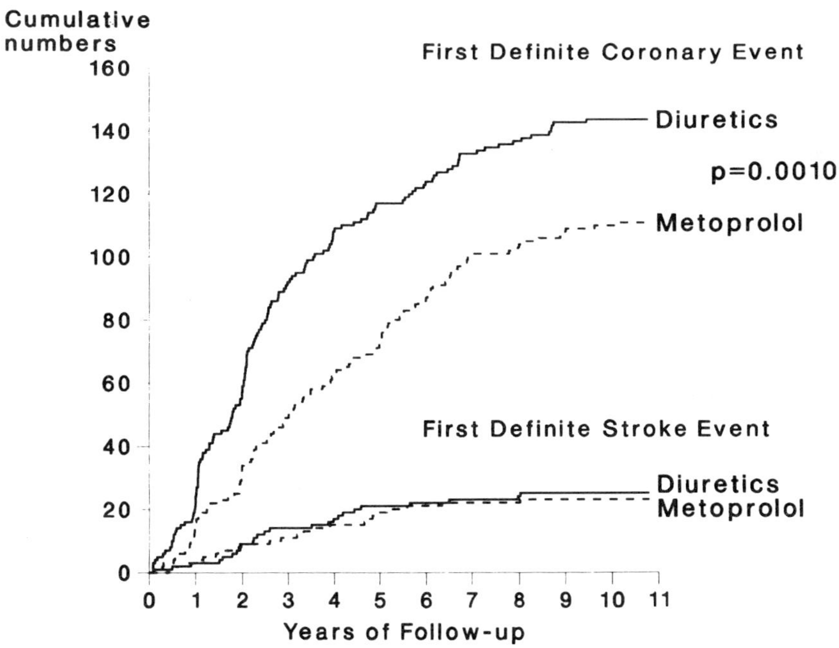

Figure 3. Cumulative numbers of all first definite coronary events and stroke events in the two randomization groups in the MAPHY study. The *p* -value refers to the difference in risk between the two randomization groups during the entire study period. From Wikstrand *et al.*,1991, in press.

vs 21.8 cases /1000 patient-years at the end of the trial. ECG changes present: p = 0.03; 19.4 vs 24.6 cases /1000 patient-years at the end of the trial).

Sub-group Analysis by Smoking Status at Randomization - Definite Events
The pooled incidence of all first definite coronary events and also definite cardiovascular disease events was significantly lower during follow-up in nonsmoking patients randomized to metoprolol than in nonsmoking patients randomized to diuretics (Fig. 4, Table 5).

Total and coronary heart disease mortality was significantly lower in smokers randomized to metoprolol than in smokers randomized to diuretics (Fig. 4). However, in smokers there was no statistically significant difference between the two treatment groups in the pooled incidence of fatal and definite nonfatal coronary or cardiovascular disease events, although the absolute figures were lower in smoking patients randomized to metoprolol than in those randomized to diuretics (Table 5).

Definite and Possible Coronary Heart Disease and Cardiovascular Disease

Events
In the whole study population as well as in the presented subgroups (nonsmokers/ smokers; ECG-changes absent/present), the pooled incidence of all first definite and possible coronary events was significantly lower during follow-up in patients randomized to metoprolol compared with patients randomized to diuretics (Tables 3 and 5).

Noncardiovascular Deaths
There were 23 noncardiovascular deaths in the metoprolol group and 26 in the diuretic group (Table 2). One noncardiovascular death in the metoprolol group and two noncardiovascular deaths in the diuretic group were preceded by definite cardiovascular disease events.

Table 4. Risk ratios (metoprolol/diuretics) and confidence intervals for risk ratios for all first definite coronary heart and cardiovascular disease events at the median follow-up time (4.2 years) and at the end of the trial (10.8 years). For definition of events see Table 3. From Wikstrand *et al.*, 1991, in press.

	Median (4.2 years)		End (10.8 years)	
	Risk Ratio	95 % Confidence Interval	Risk Ratio	95 % Confidence Interval
CHD events	0.58	0.41-0.80	0.76	0.58-0.98
CV events	0.60	0.44-0.81	0.77	0.61-0.98

Abbreviations: CHD = coronary heart disease; CV = cardiovascular.

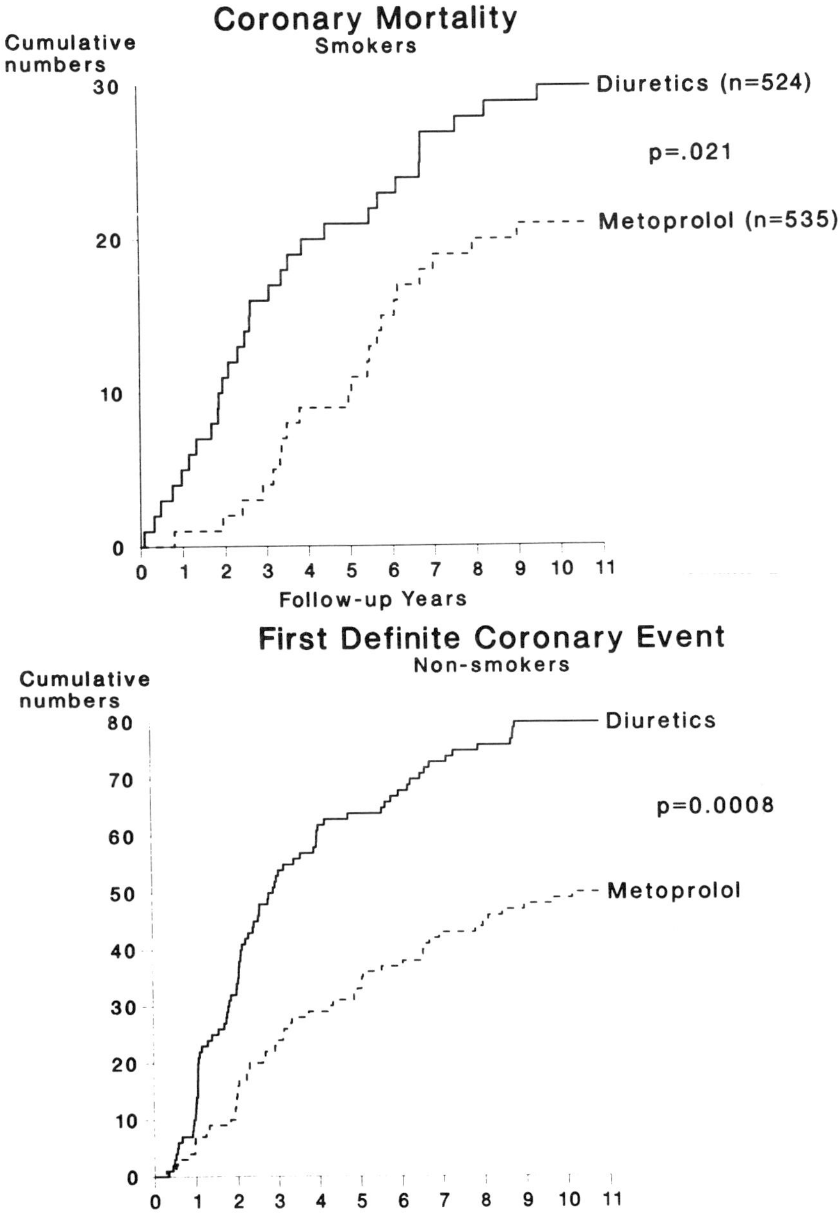

Figure 4. Upper panel: Cumulative numbers for coronary mortality in patients classified as smokers in the two randomization groups in the MAPHY study. From Tuomilehto *et al.*, 1989b. Lower panel: Cumulative numbers of all first definite coronary events in patients classified as non-smokers at randomization in the two randomization groups in the MAPHY study. From Wikstrand *et al.*,1991, in press. The *p* -values refer to the difference in risk between the two randomization groups during the entire study period.

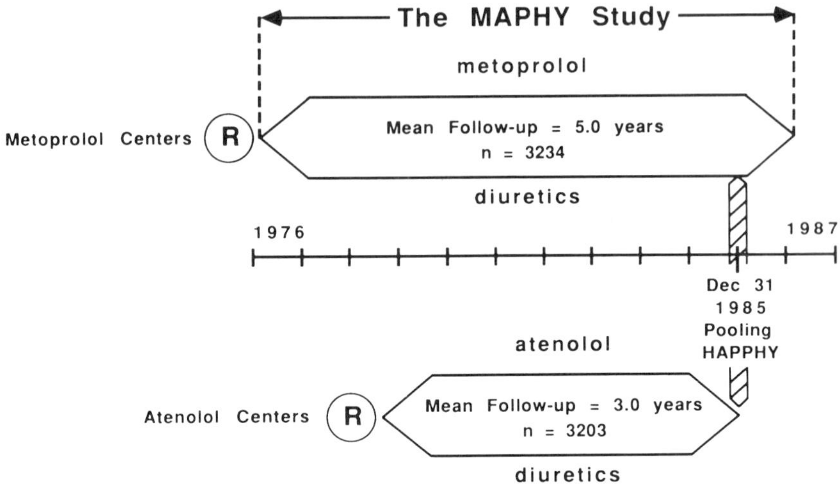

Figure 5. Design of the MAPHY study and the HAPPHY study. For simplicity reasons the single propranolol center has not been illustrated (Information on this center see Table 6). R = randomization. From Wikstrand *et al.*, 1991, in press.

Table 5. All first definite and definite plus possible coronary heart disease and cardiovascular disease events in patients classified as nonsmokers and smokers at randomization. For definition of events see Table 3. From Wikstrand *et al.*, 1991, in press.

	Nonsmokers[1]					Smokers[1]				
	Metoprolol			Diuretics		Metoprolol			Diuretics	
	n	rate[2]	*p*[3]	*n*	rate[2]	*n*	rate[2]	*p*[3]	*n*	rate[2]
Definite Events										
CHD	50	9.7	0.0008	80	16.1	59	23.7	0.09	64	26.8
CV disease	64	12.5	0.001	94	18.9	68	27.5	0.07	76	32.1
Definite and Possible Events										
CHD	84	16.5	0.0072	117	24.0	71	30.0	0.024	85	36.5
CV disease	97	19.2	0.0066	131	26.9	79	32.4	0.019	97	42.0

Abbreviations: CHD = coronary heart disease events; CV = cardiovascular disease events.

[1]For definition of nonsmokers and smokers see Methods section. Information on smoking habits were missing in 72 patients, two of whom had a definite nonfatal cardiovascular event.

[2]Rate = events /1000 patients-years of follow-up.

[3]The P values refer to the difference in risk during the entire study period.

Table 6. Blood pressure inclusion criteria (diastolic), mean follow-up time and mortality rate (total mortality/1000 patient years) in the atenolol arm of the HAPPHY study and in the MRC, MAPHY, IPPPSH and HDFP trials. Data are also given for the single propranolol center in the HAPPHY study. Studies are presented in increasing value of mortality rate in the male control groups.

	Inclusion diastolic BP mmHg	Mean follow-up time years	Mortality rate deaths/1000 patient years Men and women	Men only
Atenolol arm, HAPPHY (men only)	100-130	3.0		
Randomized to diuretics				5.5
Randomized to atenolol				6.9
MRC (men and women)	90-109	4.9		
Randomized to diuretics			6.0	7.5
Randomized to propranolol			5.6	6.7
MAPHY (men only)	100-130	5.0		
Randomized to diuretics				10.3
Randomized to metoprolol				8.0
IPPPSH (men and women)	100-125	4.1		
Randomized to non-blocker[1]			8.8	12.7
Randomized to oxprenolol			8.3	9.3
HDFP (white men and women)	90-above	5.0		
Randomized to referred care[1]			12.0	13.5
Randomized to stepped care			10.9	11.5
Propranolol arm, HAPPHY[2] (men only)	100-130	6.9		
Randomized to diuretics				13.7
Randomized to propranolol				16.1

Abbreviations: HAPPHY = Heart Attack Primary Prevention in Hypertension;
MRC = Medical Research Council Trial;
MAPHY = Metoprolol Atherosclerosis Prevention in Hypertensives;
IPPPSH = Inter-national Prospective Primary Prevention Study in Hypertension 2;
HDFP = Hypertension, Detection and Follow-up Program.
BP = Blood pressure. [1]Mainly diuretics. [2]There were only 9 deaths in this center, 4 on diuretics and 5 on propranolol.

Coronary artery by-pass surgery

There were few cases of coronary artery by-pass surgery performed during the course of the trial, 6 in patients randomized to metoprolol, 8 in patients randomized to diuretics.

Table 7. Pooled results from men in MRC, IPPPSH and MAPHY (studies using relatively lipophilic beta-blockers) regarding total mortality and the pooled incidence of nonfatal and fatal coronary heart disease. From Wikstrand *et al.*,1991b, in press.

Men Only	Number of Events β-blocker	Non-β-blocker[1]		Percentage Difference	95 % Confidence Interval
n	5499	5452			
patient years	25941	25491	*p*-value[2]		
Total mortality	202	247	($p = 0.024$)	-20 %	-32 % to -1 %
CHD (nonfatal + fatal)	262	330	($p = 0.003$)	-22 %	-33 % to -6 %

CHD = coronary heart disease, [1]Mainly diuretics, [2]Fisher's Exact Test.

DISCUSSION

The results from the MAPHY primary prevention study in white men aged 40-64 years with diastolic blood pressure above 100 mm Hg showed that metoprolol given as initial antihypertensive treatment was associated with a lesser risk for death and coronary events as compared with initial diuretic treatment.

Baseline characteristics, blood pressure during follow-up and stroke rates were similar in the two treatment groups. Therefore the difference in outcome is probably due to mechanisms other than blood pressure control.

Comments on differences in mortality rate relating to differences in study-design and recruitment

Since the MAPHY study lacked a placebo group (for reasons see Appendix), the results could be explained by suggesting that the risk in the diuretic group was high and that the high risk in the diuretic group rather than a preventive effect of the beta-blocker explained the observed difference in risk. It is therefore of interest to compare mortality rates (i.e. risk) in the MAPHY study with mortality rates in the other studies performed.

In the MAPHY study, stratified randomization was performed within each center in blocks of four patients. This means that two treatment groups were created that were well balanced as regards classical risk factors for coronary heart disease (Wikstrand *et al.*, 1988a). In addition, randomization by center created a balance for other factors related to cardiovascular risk such as length of follow-up time and geographical background (Miettinen, 1983).

These factors relating to study design and recruitment are of great importance for the mortality rate which varies between different randomized trials in hypertension (Table 6). Mortality rate increases with increasing length of follow-up time (Table 6, Fig. 2). Furthermore, inclusion of patients from geographical regions with a high risk for cardiovascular disease will increase the overall mortality rate in a trial. Twenty per cent of the patients in the MAPHY study were recruited in North Karelia in Finland, a region with one of the highest cardiovascular risks in the world, as is illustrated in

Fig. 2 (Tuomilehto *et al.*, 1989a). Data given in Fig. 2 also illustrate, however, that the benefit of the beta-blocker treatment in the MAPHY study was independent of the level of risk.

There is an apparent difference in mortality in diuretic-treated patients between the MRC trial, the IPPPSH and the MAPHY study (Table 6). This is partly explained by the inclusion of women in the MRC and IPPPSH, but not in MAPHY. In men, however, the mortality rates in the three studies are quite comparable (Table 6). The MRC participants had lower inclusion blood pressure and this is probably one contributing factor for the slightly lower mortality rate in men in the MRC trial compared with men in the other trials (see Table 6). The mortality rate in the comparable group of the Hypertension, Detection and Follow-up Program (HDFP) study (white males) was higher than in the MAPHY study (Table 6) (Hypertension Detection and Follow-up Program Cooperative Group, 1979).

Thus, other trials have shown similar mortality rates in their male control groups so there is no reason to believe that the MAPHY diuretic-treated patients represented an extreme sample of the population under investigation (for comments on HAPPHY, see Appendix). Furthermore, as illustrated in Fig. 2 the data indicate that the difference between the two treatment groups of men in the MAPHY study is independent of the mortality rate, i.e. the risk.

Positive results both in nonsmokers and smokers

Post hoc subgroup analysis from the MRC and IPPPSH showed that the pooled incidence of fatal and nonfatal coronary (MRC) and cardiac (IPPPSH) event rates was significantly lower in nonsmoking men on beta-blockade. The pooled incidence of all first definite coronary events was significantly lower during follow-up in nonsmoking patients randomized to the beta-blocker than in nonsmoking patients randomized to diuretics also in the MAPHY study (Fig. 5). Results in nonsmoking men are accordingly consistent in three randomized clinical trials with relatively lipophilic beta-blockers and furthermore are supported by one observational study (Fletcher *et al.*, 1988).

No effect was seen for cardiovascular events in smoking men randomized to beta-blockade in MRC and IPPPSH. In the MAPHY study, on the other hand, total and coronary mortality (Fig. 4) was significantly lower in smoking men on beta-blockade than in those on diuretics, and the possible role of cardioselectivity and other mechanisms involved have been discussed separately (Tuomilehto *et al.*, 1989b). There was no statistically significant difference in smoking patients in definite nonfatal events, however, indicating that coronary heart disease in the metoprolol treated smoking patients partly is prevented or postponed but also may partly take a milder form. Even so, the mortality risk in the smokers is much higher than in nonsmokers, which emphasizes the importance of measures to help smokers stop smoking.

Mechanisms

There are many possible mechanisms besides the antihypertensive effect by which a beta-blocker may lower cardiovascular risk. These may be listed as cardiac anti-ischemic effects, antifibrillatory effects, antithrombotic and antiatherosclerotic effects. Beta-blockers have been shown in several experimental, clinical and animal studies to influence favourably these and other factors that could be important for the observed prevention of coronary events and thromboembolic complications in the studies performed, (Chobanian, 1983; Rydén *et al.*, 1983; Norris *et al.*, 1984; Spence, 1984; Chobanian *et al.*, 1985; Imperi *et al.*, 1987; Kaplan *et al.*, 1987; Åblad *et al.*, 1987; Beckmann *et al.*, 1988; Berglund *et al.*, 1988; Hartford *et al.*, 1988; Pettersson and Åblad, 1988; Åblad *et al.*, 1988; Östlund-Lindqvist *et al.*, 1988; Strawn *et al.*, 1989; Teger-Nilsson *et al.*, 1990; Thubrikar *et al.*, 1990; Lindén *et al.*, 1990; Åblad *et al.*, 1990; Bondjers *et al.*, 1990; Olsson *et al.*, 1991; Pettersson *et al.*, 1990; Wikstrand, 1990). Furthermore, in individual patients electrolyte disturbances caused by thiazide diuretic treatment might increase the risk for sudden death (Wikstrand, 1986).

A small difference in serum cholesterol was observed between the two randomization groups during follow-up in the MAPHY study, with a significantly lower value on metoprolol than on diuretics, and this might contribute in part to the difference that was observed in coronary events (Wikstrand *et al.*, 1988a; Tuomilehto *et al.*, 1989b). A clinical study also indicates that metoprolol may decrease LDL-binding to arterial proteoglycans and thereby decrease accumulation of LDL in the arterial wall (Lindén *et al.*, 1990). Experimental studies performed in animals have shown a decreased accumulation of LDL-cholesterol in the arterial vessel wall after metoprolol treatment (Strawn *et al.*, 1989; Thubrikar *et al.*, 1990). These findings are further investigated to elucidate responsible mechanisms involved.

While it is tempting to postulate that the pharmacological and clinical actions of beta-adrenergic blocking drugs are due to a "class action" a number of recent observations and several conceptual problems make this hypothesis difficult to defend (Harrison, 1983; Raftery and Carrageta, 1985; Åblad *et al.*, 1987; Streufert *et al.*, 1988; Åblad *et al.*, 1988; MAPHY and the two arms of HAPPHY, 1989; Tuomilehto *et al.*, 1989b; Åblad *et al.*, 1990). Clinical and experimental data indicate that not only sympathetic tone but also vagal tone plays an important role in the maintenance of cardiac electrical stability and for the prevention of ventricular fibrillation (Bigger *et al.*, 1988; Billman and Hoskins, 1989). Other data also indicate that metoprolol increases cardiac vagal tone due to beta1-blockade in the central nervous system and that this action was an important contributor for the prevention of ventricular fibrillation (Åblad *et al.*, 1990). Additional research is needed to clarify the relative clinical importance of certain pharmacological qualities such as lipophilicity and beta$_1$-selectivity for the preventive effect on myocardial infarction and sudden death. However, to date, results from several clinical studies in hypertensives and also results from experimental, clinical and animal studies give supporting evidence and biological explanations for a cardioprotective action of lipophilic beta- blockers. The difference in mortality and coronary events between patients treated with the beta-blockers and patients treated with thiazide diuretics or other non-beta-blocker regimens is probably, as initially postulated, mediated by actions independent of the reduction in blood pressure. Pooled data in men with relatively lipophilic beta-blockers (MRC, IPPPSH and MAPHY) show a statistically significant 20 per cent lower total mortality and a 22 per cent lower incidence of definite coronary events as compared with a non-beta-blocker, mainly diuretic-based schedule (Table 7). This difference in outcome between two first-line drugs in hypertension seems to have important implications for clinical practice because of the high prevalence of hypertension in the community and the increased risk for coronary events in hypertensive males and in elderly patients regardless of sex (MacMahon *et al.*, 1986).

CLINICAL IMPLICATIONS

Therapy choices for initial antihypertensive treatment have become broadened to the four most popular classes of drugs in the latest report from the Joint National Committee on Detection, Evaluation, and Treatment of High Blood Pressure in the US (The 1988 Report of the Joint National Committee on Detection, Evaluation, and Treatment of High Blood Pressure, 1988). Each practitioner must decide which agent is appropriate for an individual patient. Of course, most of these decisions are empiric.

One approach to this "tailoring" of therapy is to give the hypertensive patient at increased risk for a coronary event optimal chances of not suffering sudden death or a fatal or nonfatal acute or silent myocardial infarction, the most common complication to hypertension (see Fig. 3) (Kannel, 1977; Kannel *et al.*, 1985; MacMahon *et al.*, 1986; Wikstrand, 1986). Available evidence summarized in this chapter indicate that metoprolol is superior to thiazide diuretics for this purpose.

The significantly lower incidence of coronary events in the metoprolol group was due to a significantly reduced risk for acute coronary events as well as a reduced risk for silent myocardial infarctions (Table 3). The data from the Framingham study showed that 38 per cent of all first definite coronary events among treated hypertensive men were unrecognized or silent (Kannel *et al.*, 1985) a figure well in accordance with the results from the MAPHY study. Analysis of data from the Framingham study also indicates that silent myocardial infarctions carry a prognosis as grave as symptomatic myocardial infarctions. Therefore it seems clinically important that the given treatment reduced the risk both for silent and symptomatic coronary events. Clinical trials have not yet produced longterm prognostic data on the effects of alpha-blockers, ACE-inhibitors or calcium antagonists on coronary events in hypertensive patients (The 1988 Report of the Joint National Committee on Detection, Evaluation, and Treatment of High Blood Pressure, 1988). Nor are there any trial data to indicate that any of these agents would be generally superior to no treatment or to a cardioselective beta-blocker as regards quality of life (Nugent *et al.*, 1987; Frohlich, 1988; Gengo *et al.*, 1988; Streufert *et al.*, 1988; Dimenäs *et al.*, 1989; Herrick *et al.*, 1989; Dimenäs *et al.*, 1990). Data indicate, however, that cardioselective beta-blockers as well as ACE-inhibitors are superior to a nonselective beta-blocker as regards quality of life (Croog *et al.*, 1986; Dimenäs *et al.*, 1990).

The experience from earlier longterm studies with diuretics and certain β-blockers in hypertensive patients and from studies with calcium antagonists in secondary prevention after myocardial infarction has clearly shown that we cannot take for granted that any antihypertensive agent will prevent coronary events (Veterans Administration cooperative Study Group on Antihypertensive Agents, 1970; Australian Therapeutic trial in Mild Hypertension, 1980; Coope and Warrender, 1986; Lubsen *et al.*, 1987; Yusuf and Furberg, 1987; The Multicenter Diltiazem Postinfarction Trial Research Group, 1988; MAPHY and the two arms of HAPPHY, 1989). Further long-term large-scale randomized trials are therefore needed to define the usefulness of alpha-blockers, ACE inhibitors, and calcium antagonists for this purpose.

CONCLUSION

In conclusion then, the nonpharmacological approach seems to be the most logical choice in low risk patients. In those not responding to such nonpharmacological treatment any of the four most popular classes of drugs could be an excellent choice. An analysis of the longterm cost-effectiveness of various initial monotherapies for the treatment of primary hypertension may be of importance in defining the preferential initial option (Edelson *et al.*, 1990). Comparative data on risk reduction for coronary events will probably never be gathered in low risk patients because of statistical and biological reasons.

Available data support the choice of a beta-blocker as antihypertensive therapy in patients with increased risk for coronary events unless contra-indications limit the choice of these agents. Extrapolation from these results to patients with a low absolute risk for coronary events such as middle-aged white women with uncomplicated hypertension would need further evidence. Lipophilicity and cardioselectivity appear to be an advantage but the clinical relevance of these pharmacological properties for the preventive effect on coronary events has to be further explored.

However, it should always be kept in mind that it is not "enough" to lower blood pressure; the goal of treatment should be to prevent progression of the underlying atherosclerotic disease, to reduce major complications, improve longevity and quality of life (The 1988 Report of the Joint National Committee on Detection, Evaluation, and Treatment of High Blood Pressure, 1988). Furthermore, primary prevention and risk reduction in hypertensives is dependent not only on the choice of the antihypertensive agent but also on improved detection and management of the other cardiovascular risk factors (Samuelsson *et al.*, 1987; The 1988 Report of the Joint National Committee on Detection, Evaluation, and Treatment of High Blood Pressure, 1988; Tuomilehto *et al.*, 1989b).

APPENDIX

METHODOLOGICAL CONSIDERATIONS -MAPHY AND HAPPHY- THE BACKGROUND

The publication of the MAPHY mortality data (Wikstrand *et al.*, 1988a) aroused great interest as well as a number of comments in the literature (Primary Prevention with Metoprolol in Patients with Hypertension, 1988; Kaplan, 1988; Moser and Sheps, 1989; MAPHY and the two arms of HAPPHY, 1989) mainly because the earlier publication of the HAPPHY trial (Wilhelmsen *et al.*, 1987) confused the picture somewhat. Fortunately, the situation can be clarified by reviewing carefully the planning, the execution and the interpretation of the two parts of this investigation (Wikstrand, 1990; Wikstrand *et al.*, 1991, in press b).

Background plans were drawn up in 1975 in the Sahlgrenska Hospital,

Gothenburg, for the design of a primary preventive study comparing beta-blockade with thiazide diuretic treatment in hypertensives, and it was suggested that metoprolol, a newly invented cardioselective beta-blocker, should be used. The study "Beta-blockers vs Diuretics in Hypertension" was aimed at achieving 20 000 patient years with only men eligible. When this study was initiated data already existed showing that hypertensive men with diastolic blood pressure above 100 mm Hg benefited from antihypertensive treatment (Veterans Administration Cooperative Study Group on Antihypertensive Agents, 1970). For ethical reasons, therefore, a placebo control group could not be included since only men with diastolic blood pressure above 100 mm Hg were eligible for inclusion (Wikstrand *et al.*, 1988a).

The original protocol from 1976 stated that patients could be randomly assigned to treatment with metoprolol or a thiazide diuretic. Propranolol was used in only one center (Table 6). Thus, the original study mainly aimed at comparing metoprolol and thiazide diuretics.

THE HAPPHY STUDY - POOLING OF METOPROLOL AND ATENOLOL DATA

In 1978, more than two years after the first patient was randomized in the original study, atenolol had become available in many countries and the protocol for the original study was modified to allow for additional centers that, in a separate part, could randomize patients to either atenolol or diuretics (Fig. 5) (Wikstrand *et al.*, 1988a; Wikstrand *et al.*, 1988b; Tuomilehto *et al.*, 1989b; Wikstrand, in press). This became known as the HAPPHY trial but the two parts were kept completely separate as regards randomization since the HAPPHY protocol stipulated that only one beta-blocker could be used within each center. The reason for adding a new beta-blocker was to increase the number of patients available for enrolment in the study. Results covering 20 000 patient years with the pooled data (the HAPPHY study) were initially analyzed and published (Wilhelmsen *et al.*, 1987), although the mean follow-up time was considerably shorter in the atenolol part as compared with the metoprolol part (Fig. 5, Table 6) (The Steering Committee of the HAPPHY trial, 1989). The MAPHY study deals with the outcome exclusively in the centers that used metoprolol in comparison with thiazide diuretics as baseline drugs.

The HAPPHY study showed no difference in mortality between beta- blocker (metoprolol, atenolol or propranolol) and thiazide diuretic treatment (Wilhelmsen *et al.*, 1987). Data analysed after the main analysis had been performed in the HAPPHY trial, however, showed a positive effect of metoprolol on total mortality, but a nonsignificant negative effect for atenolol thus explaining why the pooled data showed no effect (see below, The Steering Committee of the HAPPHY Trial, 1989). Mortality rates in the atenolol arm of the HAPPHY study, especially the mortality rate in the diuretic control group, were definitely low compared with the other primary preventive studies performed in hypertensive males (Table 6). Since mean follow-up time is an important factor defining mortality rate (see Fig. 2) this is probably one factor explaining the overall lower mortality in the atenolol part of the HAPPHY study than in other studies. Factors besides the classical risk factors for coronary heart disease and length of follow-up time such as geographical background, socioeconomic status or other uncontrolled factors may also be of importance. However, the low mortality rate or the shorter mean follow-up time does not seem to explain why there was no positive effect of atenolol: at similar mortality rate in the diuretic control group in the MAPHY study there was a beneficial effect of metoprolol (see Fig. 2, data at 842 days in all countries except Finland: 3.4 vs 6.2 deaths /1000 patients years). Furthermore, at exactly the same mean follow-up time in the MAPHY study as in the atenolol part of the HAPPHY study (3.0 years, Table 6) there was 22 deaths on metoprolol and 41 deaths on diuretics corresponding to 4.6 vs 8.5 deaths/1000 patients years in MAPHY as compared to 33 deaths on atenolol and 26 deaths on diuretics corresponding to 6.9 vs 5.5 deaths /1000 patients years in the atenolol part of the HAPPHY study. The difference between the two beta-blockers in their effect as compared to diuretics was significant at this point in time, that is at comparable mean follow-up times (p = 0.03, Fisher's exact test, comparison between two binominal distributions). The reason why the atenolol part of the HAPPHY study, in contrast to the other studies in male hypertensives, did not show data in favour of beta-blockade (Table 6) thus remains unanswered and can only be answered through new studies.

THE MAPHY STUDY GROUP

Steering Committee:
Gunnar Olsson, MD, PhD, Department of Medicine, Danderyd Hospital, Karolinska Institute, Stockholm; Jaakko Tuomilehto, MD, PhD, Department of Epidemiology, National Public Health Institute, Helsinki; Ingrid Warnold, PhD (nonvoting secretary), Medical Department, Hässle Cardiovascular Research Laboratories, Mölndal, Sweden; John Wikstrand, MD, PhD (chairman), Wallenberg Laboratory for Cardiovascular Research, Sahlgrenska University Hospital, Gothenburg, Sweden.

Advisory Committee:
Hamish J. Barber, MD, Department of General Practice, University of Glasgow, Scotland; Keith Eliasson, MD, PhD, Department of Internal Medicine, Karolinska Hospital, Karolinska Institute, Stockholm; Birthe Jastrup, MD, Medical Clinic, Odder (Denmark) Sygehus; Nicholas B. Karatzas, MD, D Phil, Hypertension Research Clinic, First Department of Propedeutic Medicine, Laikon Hospital, University of Athens; John Leer, MD, Legecenteret i Florvåg, Norway; Fausto Marchetta, MD, Ospedale S Orsola, Bologna, Italy; Johann Ragnarsson, MD, Medical Department, Reykjavik (Iceland) City Hospital; Michelle Robitaille, MD, M PH, Institut de Cardiologie, Quebec; Liba Valkova, MD, National Public Health Institute, Prague; Harry Wesseling, MD, PhD, Department of Pharmacology and Clinical Pharmacology, University of Groningen, The Netherlands.

Independent End-Point Committee:
Cause-specific mortality: Robert Bergstrand, MD, PhD, Associate Professor and Lars Wilhelmsen, MD, PhD, Associate Professor, Department of Internal Medicine, Eastern Hospital, Gothenburg University, Gothenburg, Sweden. Nonfatal End-Points: John Kjekshus, MD, PhD, Professor, Department of Internal Medicine, Baerum Hospital, Oslo University, Oslo, Norway and Lars Rydén, MD, PhD, Professor, Department of Cardiology, Karolinska Hospital, Karolinska Institute, Stockholm, Sweden. Independent Data Audit Committee: Aram Chobanian, MD, Professor, Cardiovascular Institute, Boston University School of Medicine, Boston, Massachusetts and Göran Berglund, MD, PhD, Professor, Department of Internal Medicine, Malmö Hospital, Lund University, Malmö, Sweden.

Statistical Consultant:
Anders Odén, PhD, Department of Mathematics, Gothenburg University, Gothenburg, Sweden.

Data System Manager:
Johnny Lindqvist, Computing Center, Gothenburg University, Gothenburg, Sweden.

Investigators:
Canada (8.5 % of patient-years): J. George Fodor, MD, PhD, S. Paul Handa, MD, Gordon D. Hardacre, MD, N. Michelle Robitaille, MD, MPH, Robin S. Shearer, MD, Robin C. Wollam, MD; Czechoslovakia (3.5 % of patient-years): Frantisek Boudik, MD, Zdenek Hejl, MD, Zbynek Pisa, MD, Jiri Valek, MD, Liba Valkova, MD; Denmark (0.4 % of patient-years): Birthe Jastrup, MD; Finland (21.2 % of patient-years): Matti Honkavaara, MD, Pasi Kuusisto, MD, Martti Mönkkönen, MD, Arto Mönttinen, MD, Matti Pohjola, MD, Jaakko Tuomilehto, MD, PhD, Seija Vinni, MD, Olavi Vänskä, MD; Greece (0.6 % of patient-years): Nicholas B. Karatzas, MD; the Netherlands (2.2 % of patient- years): Bart S. ten Berge, MD, Theo H.J.J.M. Bloem, MD, Johan Frans May, MD, Menso J. Nubé, MD, Jo L.J.Jansen, MD; Iceland (5.9 % of patient-years): Thordur Hardarson, MD, Johann Ragnarsson, MD, Snorri Snorrason, MD; Italy (2.7 % of patient-years): Ettori Ambrosioni, MD, Roberto Fogari, MD, Fausto Marchetta, MD, Anna Pirelli, MD; Norway (1.9 % of patient-years): Svein Z. Bratland, MD, Åge Dahl, MD, Arne Haraldsson, MD, John Leer, MD; Sweden (43.8 % of patient-years): Bernd Atmer, MD, Agneta Brouneus, MD, Birger Brouneus, MD, Arne Bruhn, MD, Jan Candefjord, MD, Gunnar Ekblad, MD, Keith Eliasson, MD, PhD, Åke Ellborg, MD, Dag Elmfeldt, MD, PhD, Sven-Erik Fagerberg, MD, PhD, Lars Fries, MD, Johan Erik Frisell, MD, Karl Gumaelius, MD, Christer Gunnarsson, MD, Raoul Hagströmer, MD, Leo Hassler, MD, Staffan Hjalmers, MD,Kjell Hultsten, MD, Brita Hylander, MD, Anders Häggmark, MD, Bertil Johansen, MD, Jörgen Jonsson, MD, Jan Kjellberg, MD, Lars-Göran Lindgren, MD, Gösta Linné, MD, Janos Luptovics, MD, Per-Olof Magnusson, MD, Lennart Nilsson, MD, Lars-Göran Persson, MD, Åke Sjöstrand, MD, Ted Strömvall, MD, Tunc Tovi, MD, Halina Wajntraub, MD; United Kingdom (9.4 % of patient-years): Barry J. Adams-Strump, MD, Steven Bird, MD, L. Malcolm Campbell, MD, Allan R. Cuthill, MD, John R. Goves, MD, Peter F. Grundy, MD, Keith G. Harding, MD, Robert Harvard Davis, MD, John H. McLauchlan, MD, Chris J.L. Morgan, MD, Alexander L. Muir, MD, Trevor G. Pinker, MD, Sudhi Sudhakar, MD, Stanley H. Taylor, MD, Chris G. Wathen, MD.

REFERENCES

Åblad, B., Abrahamsson, T., Adler, G., Björkman, J.-A., Bjurö, T., Ek, L., Ervik, M., Sjöquist, P.-O., Sutherland, I. and Svensson, L. (1987). Cardiac antiischemic effect of metoprolol: Role of β-blockade within the ischemic region. *J Cardiovasc Pharmacol.* **10** (Suppl 2), S117-S125.

Åblad, B., Björkman, J. A., Gustafsson, D., Hansson, G., Östlund-Lindqvist, A. M. and Pettersson, K. (1988). The role of sympathetic activity in atherogenesis: Effects of beta-blockade. *Am Heart J.* **116** (suppl), 322-327.

Åblad, B., Bjurö,, T., Björkman, J. A., Edström, T. and Olsson, G. (1991). Role of central nervous beta-adrenoceptors in the prevention of ventricular fibrillation through augmentation of cardiac vagal tone (abstract). American College of Cardiology, 40th Annual Scientific Session, March 3-7, 1991, Atlanta, Georgia. *JACC* **17,** 165A.

Australian Therapeutic Trial in Mild Hypertension (1980). A five-year controlled drug trial. *Lancet* **1,** 1261-1267.

Beckmann, M. L., Gerber, J. G., Byyny, R. L., LoVerde, M. and Nies, A. S. (1988). Propranolol increases prostacyclin synthesis in patients with essential hypertension. *Hypertension* **12,** 582-588.

Berglund, G., Wilhelmsen, L., Sannerstedt, R., *et al.* (1978). Coronary heart disease after treatment of hypertension. *Lancet* **1,** 1-5.

Berglund, U., Wallentin, L. and von Schenck, H. (1988). Platelet function and plasma fibrinogen and their relations to gender, smoking habits, obesity and beta-blocker treatment in young survivors of myocardial infarction. *Thromb Haemostas.* **60,** 21-24.

Bigger, J. T., Jr., Kleiger, R. E., Fleiss, J. L., Rolnitzky, L. M., Steinman, R.C., Miller, J. P. and the Multicenter Postinfarction Research Group. (1988). Components of heart rate variability

measured during healing of acute myocardial infarction. *Am J Cardiol.* **61,** 208-215.
Billman, G. E. and Hoskins, R. S. (1989). Time-series analysis of heart rate variability during submaximal exercise. Evidence for reduced cardiac vagal tone in animals susceptible to ventricular fibrillation. *Circulation* **80,** 146- 157.
Bondjers, G., Hansson, G., Olsson, G. and Pettersson, K. (1990). Smoking, catecholamines and their effects on endothelial cell integrity. In: *Tobacco smoking and atherosclerosis. Pathogenesis and cellular mechanisms.* Diana, J.N. (Eds.). *Adv Exp Med Biol.* **273,** 51-59.
Chobanian, A. V. (1983). The influence of hypertension and other hemodynamic factors in atherogenesis. *Prog Cardiovasc Dis.* **26,** 177-196.
Chobanian, A. V., Brecher, P. and Chan, C. (1985). Effects of propranolol on atherogenesis in the cholesterol-fed rabbit. *Circ Res.* **56,** 755-762.
Coope, J. and Warrender, T. S. (1986). Randomized trial of treatment of hypertension in elderly patients in primary care. *Br Med J.* **293,** 1145-1151.
Croog, S. H., Levine, S., Testa, M. A., *et al.* (1986). The effects of antihypertensive therapy on quality of life. *N Engl J Med.* **314,** 1657-1664.
Dimenäs, E. S., Wiklund, I. K., Dahlöf, C. G., Lindvall, K. G., Olofsson, B. K. and De Faire, U. H. (1989). Differences in the subjective well-being and symptoms of normotensives, borderline hypertensives and hypertensives. *J Hypertension* **7,** 885-890.
Dimenäs, E., Ker,r D. and Macdonald, I. (1990). Beta-adrenoceptor blockade and CNS-related subjective symptoms: A randomized, double-blind, placebo-controlled comparison of metoprolol CR/ZOK, atenolol and propranolol LA in healthy subjects. *J Clin Pharmacol.* **30**(suppl), S103-S107.
Edelson, J. T., Weinstein, M. C., Tosteson, A.N.A., Williams, L., Lee, T. H. and Goldman, L.(1990). Long-term cost-effectiveness of various initial monotherapies for mild to moderate hypertension. *JAMA* **263,** 408-413.
Fletcher, A. E., Beevers, D. G., Bulpitt, C. J., *et al.*(1988). Beta-adrenoceptor blockade is associated with increased survival in male but not femalch ypertensive patients: A report from the DHSS Hypertensive Care Computing Project (DHCCP). *J Hum Hypertens.* **2,** 219-227.
Friedman, L., Furberg, C. and DeMets, D.(1983). Fundamentals of Clinical Trials. Littleton, Mass, John Wright-PSG Inc, pp. 174-194.
Frohlich, E.D.(1988). β-blockers and mental performance. *Arch ntern Med.* **148,** 777-778.
Gengo, F. M., Fagan, S. C., De Padova, A., Miller, J. K. and Kinkel, P.R. (1988). The effect of β-blockers on mental performance on older hypertensive patients. *Arch Intern Med* **148,** 779-784.
Harrison, D.C.(1983). Beneficial effects of beta blockers: A class action or individual pharmacologic spectrum? *Circulation* **67** (suppl I), I-77-I-82.
Hartford, M., Wendelhag, I., Berglund, G., Wallentin, I., Ljungman, S. and Wikstrand, J. (1988). Cardiovascular and renal effects of long-term antihypertensive treatment. *JAMA* **259,** 2553-2557.
Herrick, A. L., Waller, P. C., Berkin, K. E., *et al.* (1989). Comparison of enalapril and atenolol in mild to moderate hypertension. *Am J Med.* **86,** 421-426.
Hypertension Detection and Follow-up Program Cooperative Group. (1979). Five-Year Findings of the Hypertension Detection and Follow-up Program. II. Mortality by Race-Sex and Age. *JAMA* **242,** 2572-2577.
Imperi, G. A., Lambert, C. R., Coy, K., Lopez, L. and Pepine, C.J . (1987). Effects of titrated beta blockade (metoprolol) on silent myocardial ischemia in ambulatory patients with coronary artery disease. *Am J Cardiol.* **60,** 519-524.
The IPPPSH Collaborative Group. (1985). Cardiovascular risk and risk factors in a randomized trial of treatment based on the beta-blocker oxprenolol: The International Prospective Primary Prevention Study in Hypertension (IPPPSH). *J Hypertens.* **3,** 379-392.
Kannel, W. B. (1977). Importance of hypertension as a risk factor in cardiovascular disease. In: *Hypertension: physiopathology and treatment.* J. Genest, E. Koiw, O. Kuchel (Eds). McGraw-Hill, New York, pp. 889-910.
Kannel, W. B., Dannenberg, A. L. and Abbot, R. D. (1985). Unrecognized myocardial infarction and hypertension: The Framingham study. *Am Heart J.* **109,** 581-585.

Kaplan, J. R., Manuck, S. B., Adams, M. R. and Clarkson, T.B. (1987). The effects of beta-adrenergic blocking agents on atherosclerosis and its complications. *Eur Heart J.* **8,** 928-944.
Kaplan, N. M. (1988). Critical comments on recent literature. SCRAAPHY about MAPHY from HAPPHY. *Am J Hypertens.* **1,** 428-430.
Lindén, T., Camejo, G., Wiklund, O., Warnold, I., Olofsson, S.-O. and Bondjers, G.(1990). Effect of short term beta-blockade on serum lipid levels and on the interaction of LDL with human arterial proteoglycans. *J Cardiovasc Pharmacol.* **30,** S124-S131.
Lubsen, J., Tijssen, J. G. P. and Simoons, M. L. (1987). Beta-blockers and calcium antagonists in the treatment of prolonged anginal pain at rest: Recent evidence from clinical trials. *Curr. Opinion Cardiol.* **2,** 628-639.
MacMahon, S., Cutler, J. A., Furberg, D. and Payne, G. H. (1986). The effects of drug treatment for hypertension on morbidity and mortality from cardiovascular disease: A review of randomized controlled trials. *Prog Cardiovasc Dis.* **29** (suppl 1), 99-118.
MAPHY and the two arms of HAPPHY. (1989). Letters to the Editor. *JAMA* **262,** 3272-3274.
Medical Research Council Working Party (1985). MRC trial of treatment of mild hypertension: Principal results. *Br Med J.* **291,** 97-104.
Miettinen, O. S. (1983). The need for randomization in the study of intended effects. *Stat Med.* **2,** 267-271.
Moser, M. M. and Sheps, S. (1989). Confusing messages from the newest of the β-blocker/diuretic hypertension trials. The Metoprolol Atherosclerosis Prevention in Hypertensives Trial. *Arch Intern. Med.* **149,** 2174-2175.
The Multicenter Diltiazem Postinfarction Trial Research Group (1988). The effect of diltiazem on mortality and reinfarction after myocardial infarction. *New Engl. J. Med.* **319,** 385-392.
Norris, R. M., Brown, M. A., Clarke, E. D., *et al.* (1984). Prevention of ventricular fibrillation during acute myocardial infarction by intravenous propranolol. *Lancet* **2,** 883-886.
Nugent, L. W., Miola, S. R. and Walker, J.F. (1987). Comparison of enalapril and metoprolol as initial therapy for mild to moderate hypertension. *J Clin Pharmacol.* **27,** 461-467.
Olsson, G., Tuomilehto, J., Berglund, G., Elmfeldt, D., Warnold, I., Barber, H., Elisson, K., Jastrub, B., Karatzas, N., Leer, J., Marchetta, F., Ragnarsson, J., Robitaille, N.M., Valkova, L., Wesseling, H. and Wikstrand, J. (1991). Primary prevention of sudden cardiovascular death in hypertensive patients: Mortality results from the MAPHY study. *Am J Hypertension* **4,** 151-158.
Östlund-Lindqvist, A.-M., Lindqvist, P., Bräutigam, J., Olsson, G., Bondjers, G. and Nordborg, C. (1988). Effect of metoprolol on diet-induced atherosclerosis in rabbits. *Arteriosclerosis* **8,** 40-45.
Pettersson, K. and Åblad, B. (1988). Metoprolol inhibits platelet deposition at arterial bifurcations in rabbits with sympathetic activation. *FASEB* **2**(5), A1580(7456) Abstract.
Pettersson, K., Bejne, B., Björk, H., Strawn, W.B. and Bondjers, G. (1990). Experimental sympathetic activation causes endothelial injury in the rabbit thoracic aorta via β_1-adrenoceptor activation. *Circ Res.* **67,** 1027-1034.
Primary prevention with metoprolol in patients with hypertension. (1988). Letters to the Editor. *JAMA* **260,** 1713-1716.
Raftery, E. B. and Carrageta, M. O. (1985). Hypertension and beta-blockers. Are they all the same? *Int J Cardiol.* **7,** 337-346.
The 1988 Report of the Joint National Committee on Detection, Evaluation, and Treatment of High Blood Pressure. (1988). *Arch Intern Med.* **148,** 1023-1038.
Rose, G. A. and Blackburn, H. (1968). Classification of the electrocardiogram for population studies. In *Cardiovascular Survey Methods*, World Health Organization Monograph Series No 56, Geneva, pp. 137-154.
Rydén, L., Ariniego, R., Arnman, K., *et al.* (1983). A double-blind trial of metoprolol in acute myocardial infarction. Effects on ventriculartachyarrhythmias. *N Engl J Med* **308,** 614-618.
Samuelsson, O., Wilhelmsen, L., Andersson, O. K., Pennert, K. and Berglund G.(1987). Cardiovascular morbidity in relation to change in blood pressure and serum cholesterol levels in treated hypertension. Results from the primary prevention trial in Göteborg, Sweden. *JAMA*

258, 1768-1776.

SAS User's Guide. (1983). Cary, NC, SAS Institute Inc.

Spence, J.D. (1984). Hemodynamic effects of antihypertensive drugs: Possible implications for the prevention of atherosclerosis. *Hypertension* **6,** 163-168.

The Steering Committee of the HAPPHY Trial: MAPHY and the two arms of HAPPHY. (1989). Letter to the Editor. *AMA* **262,** 3273-3274.

Strawn, W. B., Clarkson, T. B., Kaplan, J. R., Bondjers, G., Shively, C. A. and Åblad, B. (1989). Inhibition by beta-adrenergic blockade of psychosocially-induced endothelial injury in the coronary arteries of monkeys. *Circulation* **80**(Suppl II), 435.

Streufert, S., DePadova, A., McGlynn, T., Pogash, R.and Piasecki, M. (1988). Impact of β-blockade on complex cognitive functioning. *Am Heart J.* **116,** 311.

Teger-Nilsson, A.C., Dahlöf, C., Haglund, E., Hedman, C., Olsson, G. and Åblad, B. (1990). Influence of metoprolol CR/ZOK on plasminogen activator inhibitor (PAI-1) in man: a pilot study. *J Clin Pharmacol.* **30**(suppl), S132-S137.

Thubrikar, M.J., Christie, A.M., Cao-Danh, H.C., Holloway, P.W. and Nolan, S. P. (1990). Metoprolol reduces low density lipoprotein uptake in aortic regions prone to atherosclerosis. (Abstract) *FASEB J* **4,** A1151.

Tuomilehto, J., Puska, P., Korhonen, Mustaniemi, H., Vartainen, E., Nissinen, A., Kuulasmaa, K., Niemensivu, H. and Salonen, J. T. (1989a). Trends and determinants of ischaemic heart disease mortality in Finland: with special reference to a possible levelling off in the early 1980s. *Int J Epidemiol.* **18** (Suppl 1), S109-S117.

Tuomilehto, J., Wikstrand, J., Olsson, G., Elmfeldt, D., Warnold ,I., Barber, H., Eliasson, K., Jastrup, B., Karatzas, N., Leer, J., Marchetta ,F., Ragnarsson, J., Robitaille, N.-M., Valkova, L., Wesseling, H. and Berglund, G. (1989b). Decreased coronary heart disease in hypertensive smokers. Mortality results from the MAPHY Study. *Hypertension* **13,** 773-780.

Veterans Administration Cooperative Study Group on Antihypertensive Agents: Effect of treatment on morbidity in hypertension.(1970). II. Results in patients with diastolic blood pressure averaging 90 through 114mm Hg. *JAMA* **213**(7), 1143-1152.

Wikstrand, J.(1986). Initial therapy for mild hypertension. *Pharmacotherapy* **6,** 64-72.

Wikstrand, J., Warnold, I., Olsson, G., Tuomilehto, J., Elmfeldt, D., Berglund, G., on behalf of the Advisory Committee. (1988a). Primary prevention with metoprolol in patients with hypertension. Mortality results from the MAPHY Study. *JAMA* **259,** 1976-1982.

Wikstrand, J., Warnold, I., Olsson, G., Tuomilehto, J., Elmfeld,t D. and Berglund, G.(1988b). Primary prevention with metoprolol in patients with hypertension. Letter to the Editor. In reply. *JAMA* **260,** 1715-1716.

Wikstrand, J. (1990). Primary prevention with β-blockade in patients with hypertension: Review of results and clinical implications. *J Cardiovasc Pharmacol.* **16**(Suppl 5), S64-S75.

Wikstrand, J., Warnold, I., Tuomilehto, J., Olsson, G., Barber, H.J., Eliasson, K., Elmfeld, D., Jastrub, B., Karatzas, N.B., Leer, J., Marchetta, F., Ragnarsson, J., Robitaille, N.M., Valkova, L., Wesseling, H. and Berglund G. (1991a). Metoprolol versus thiazide diuretics in hypertension. Morbidity results from the MAPHY Study. Hypertension, in press.

Wikstrand, J., Berglund, G., Tuomilehto, J. Beta-blockade in the primary prevention of coronary heart disease in hypertensives. Review of results from randomized trials. Submitted for publication, b.

Wilhelmsen, L., Berglund, G., Elmfeldt, D., Fitzsimons, T., Holzgreve, H., Hosie, J., Hörnkvis,t P.-E., Pennert, K., Tuomilehto, J. and Wedel, H. (1987). Beta-blockers versus diuretics in hypertensive men: Main results from the HAPPHY trial. *J Hypertens.* **5,** 561-572.

Wilhelmsson, C., Wilhelmsen, L., Vedin, J. A., *et al.* (1974). Reduction of sudden deaths after myocardial infarction by treatment with alprenolol. *Lancet* **2,** 1157-1160.

Yusuf, S.and Furberg, C. D. (1987). Effects of calcium channel blockers on survival after myocardial infarction. *Cardiovasc. Drug Ther.* **1,** 343-344.

Progress in Hypertension, Vol. 2, pp. 73-92.
Saito *et al.* (Eds)

BOPINDOLOL: PRECLINICAL STUDY AND CLINICAL CORRELATIONS

TAKAHARU ISHIBASHI and SHOICHI IMAI

Department of Pharmacology, Niigata University School of Medicine, Niigata 951, Japan.

INTRODUCTION

Bopindolol (4-[2-benzoyloxy-3-tert-butylaminopropoxy]-2-methyl-indol hydrogen malonate) is an indol β-adrenoceptor antagonist bearing a benzoyl ester residue on the β-carbon atom of the propanolamine side chain (Fig. 1) which is reported to be a potent and specific β-adrenoceptor antagonist possessing partial agonist activity and a long duration of action. However, only a brief report of an abstract form has been published concerning the basic characters of the compound (Berthold *et al.*, 1981). Furthermore, though bopindolol itself is known to be a pro-drug which is rapidly transformed during its absorption and passage through the liver into active main metabolite responsible for the pharmacological effect (Dayer *et al.*, 1985), characteristics of the active metabolite itself has not been examined.

Its first metabolite is 18-502 (hydrolysed bopindolol shown in Fig. 1) and the metabolism to this compound is considered to be common in the rat, dog and human. Though different metabolic processes are recognized in different species, many other intermediate metabolites have no pharmacological activities except for 20-785 (Fig. 1) which is found in the dog and human (Sandoz Basle Report by J. P. Baldeck and F. Kalbser, personal communication). Therefore, we examined the basic characters of bopindolol and its two metabolites (18-502 and 20-785) and tried to correlate the results obtained in animal experiments to those in human trials.

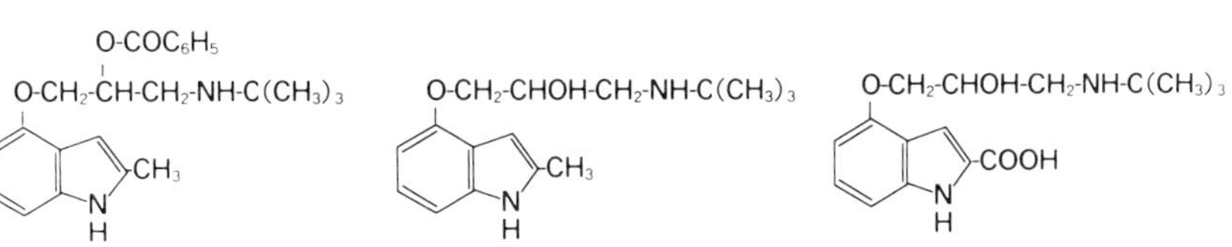

Bopindolol 18 – 502 20 – 785

Figure 1. Chemical structures of bopindolol and its two metabolites.

METHODS AND RESULTS

β-blocking activity in isolated organs

Isolated right atria, left atria and tracheal smooth muscle from guinea-pigs were used to assess the β-blocking activity of bopindolol and its two metabolites. These preparations were suspended in organ baths filled with 10 ml of Krebs-Henseleit solution (32°C) aerated with 95% O_2 and 5% CO_2 (Chin *et al.*, 1989). Bopindolol, 18-502 and 20-785 did not affect basal conditions of the spontaneous beat of right atria, the contractile state of left atria (driven by field stimulation of 1 Hz and 1 msec in duration) and the tension of tracheal smooth muscle. However, the positive chronotropic effect, positive inotropic effect and relaxant effect of isoproterenol in respective preparations were attenuated and dose-response curves for isoproterenol in these preparations were shifted rightwards (Figs. 2-4). Calculated pA_2 values are listed in Table 1 with those of propranolol and pindolol as references. It clearly shows that not only bopindolol but also its metabolite, 18-502, have β-blocking activity more potent than that of propranolol.

The pA_2 value of 18-502, 9.88-9.95, was the highest we have ever experienced. Potency ratio and $β_1$-selectivity of three agents are shown in Fig. 5 which indicate that

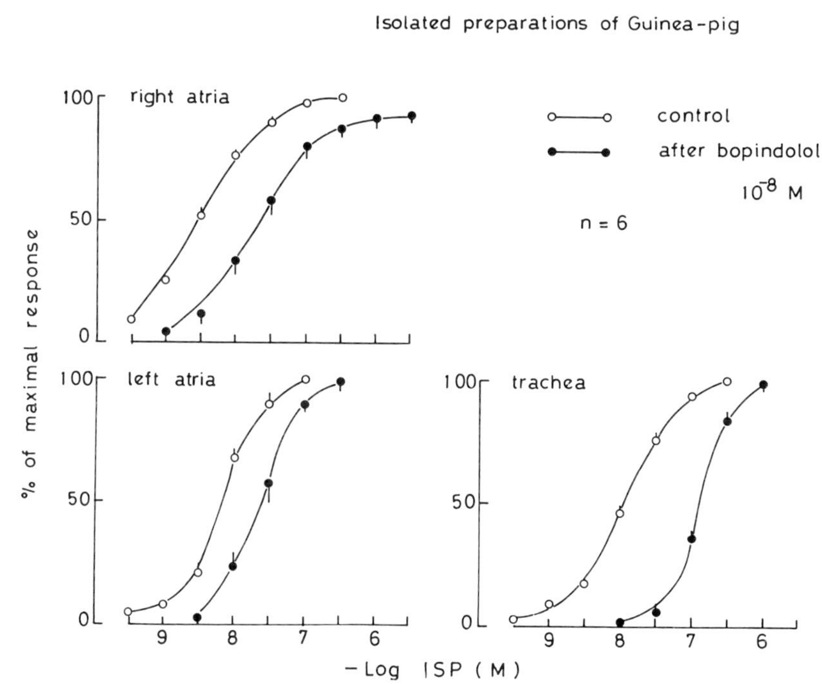

Figure 2. Antagonistic effects of bopindolol against the positive chronotropic, positive inotropic and smooth muscle relaxant effects of isoproterenol (ISP) in the isolated right atria, left atria and tracheal preparations of the guinea-pig. [From Chin *et al.* (1989) with permission.]

bopindolol itself is somewhat β_2-selective, while the most potent metabolite, 18-502, is non-selective. 20-785 is the most β_1-selective among the three. In other words, β_1-selectivity increases as the metabolism of bopindolol progresses.

Membrane stabilizing activity

Isolated papillary muscle of the guinea-pig was superfused with Krebs-Ringer bicarbonate solution (34°C) aerated with 95% O_2 and 5% CO_2. Microelectrode filled with 3M KCl was inserted into the muscle cell to record the action potential induced by field stimulation (1 Hz, 1 msec in duration) (Chin *et al.*, 1989). Bopindolol and its two metabolites had no effect on the resting membrane potential. Significant reductions in the action potential amplitude and the duration were observed by bopindolol and propranolol, whereas only the reduction of duration was observed by 18-502 and pindolol and that only at higher concentration. 20-785 had no effect on every parameter of the action potential examined in this study. The maximal rate of upstroke of action potential (Vmax) was significantly reduced by bopindolol, 18-502, propranolol and pindolol, and the potency order of membrane stabilizing activity was bopindolol >18-502 > propranolol > pindolol (Table 2).

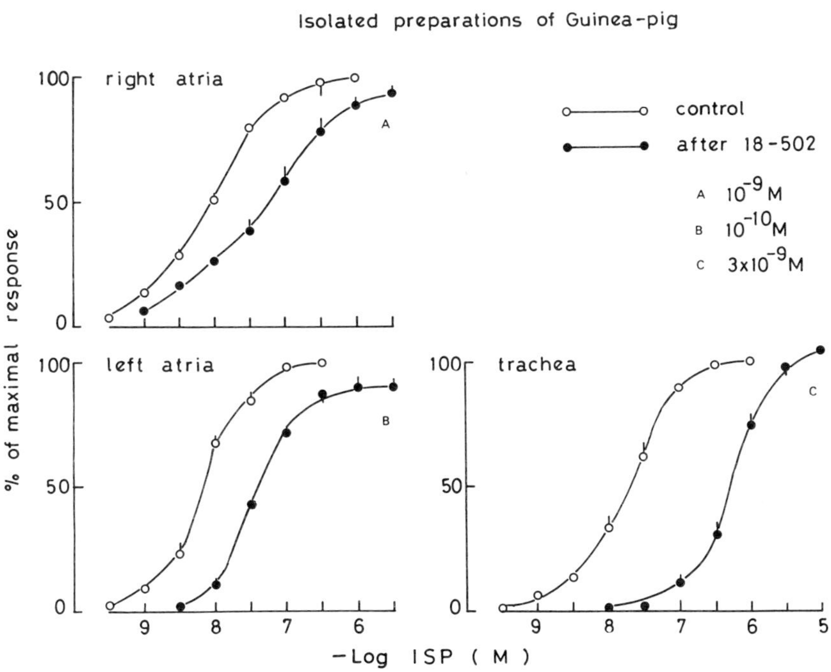

Figure 3. Antagonistic effects of 18-502 against the positive chronotropic, positive inotropic and relaxant effects of isoproterenol (ISP) in the isolated right atria, left atria and tracheal preparation of the guinea-pig. [From Chin *et al.* (1989) with permission.]

β-blocking activity and cardiohemodynamic effect of bopindolol in conscious beagle dogs

Chronically instrumented conscious beagle dogs were used in this study (Ishibashi *et al.*, 1989). Cumulative intravenous administration of bopindolol (3-100 μg/kg) had no effect on the heart rate. On the other hand, pindolol exhibited a marked increase in heart rate which was significantly attenuated by pretreatment of propranolol (3 mg/kg) (Fig. 6). Though bopindolol has no effect on the left ventricular pressure and dp/dt at all, there was a slight increase in cardiac output (the maximum was 109% of the control at 10-100 μg/kg). A slight decrease in blood pressure was also observed by bopindolol at each dose. Total peripheral resistance was not affected by bopindolol until the dose was increased up to 100 μg/kg and the decrease was prevented (though not significantly) by pretreatment of propranolol (3 mg/kg) (Fig. 7). Pindolol showed a marked and dose-dependent decrease in the total peripheral resistance and this was attenuated by propranolol (3 mg/kg) significantly.

Intravenous injection of isoproterenol (0.1 μg/kg) increased the heart rate (about 75 beats/min) and decreased the total peripheral resistance (about 6 mmHg/1/min). The effects of the same dose of isoproterenol were attenuated dose-dependently 20-30 min after each cumulative intravenous administrations of bopindolol and pindolol as shown in Fig. 8. ID_{50}s of bopindolol for the increase in heart rate and the decrease in total peripheral resistance were 9.5 ± 3.3 and 25.6 ± 7.2 μg/kg, respectively, and those of

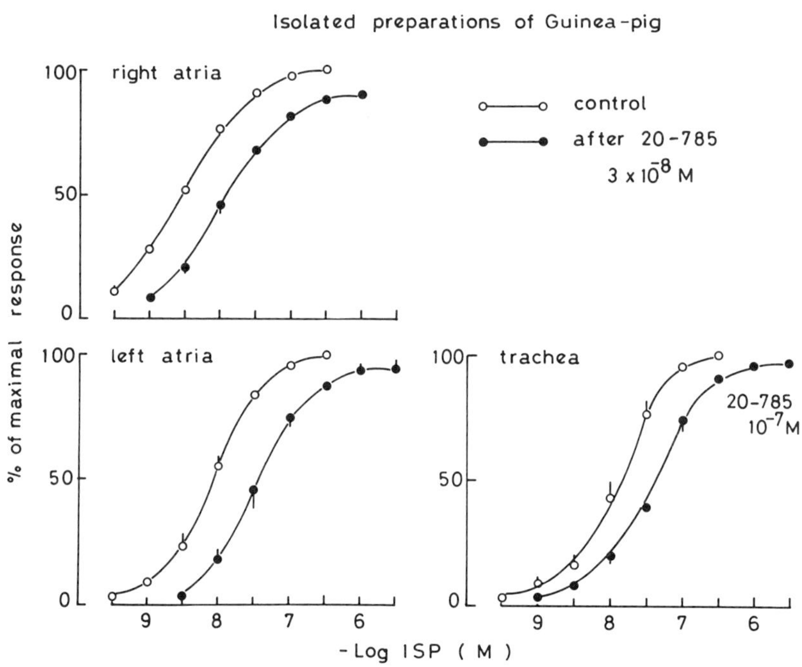

Figure 4. Antagonistic effects of 20-785 against the positive chronotropic, positive inotropic and relaxant effects of isoproterenol (ISP) in the isolated right atria, left atria and tracheal preparation of the guinea-pig.

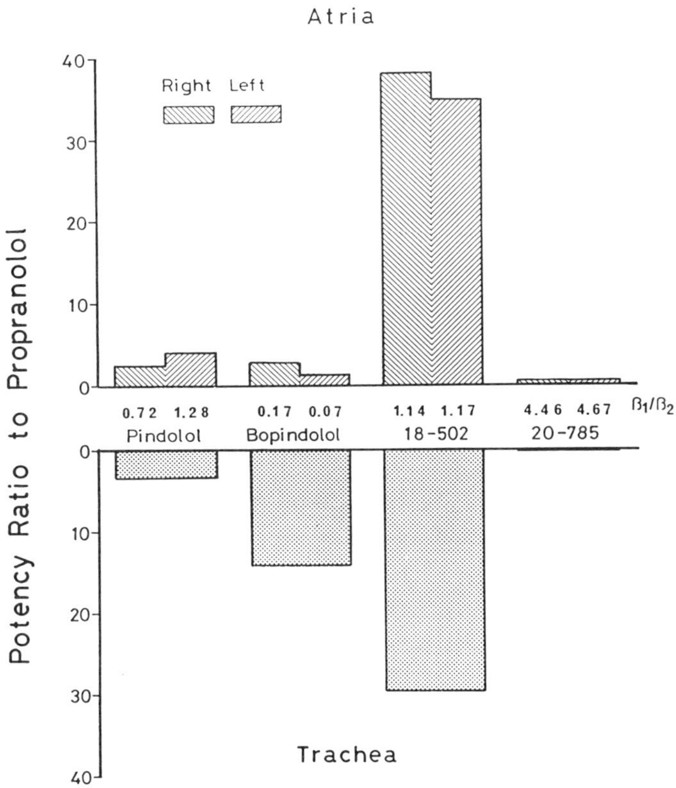

Figure 5. Potency ratio of bopindolol and its two metabolites. Ordinate: Potency ratio to propranolol was calculated from pA_2 values of each agent obtained from respective preparation. Selectivity for β-adrenoceptor subtype was indicated in abscissa (β_1/β_2).

pindolol were 7.2 ± 2.7 and 6.7 ± 2.5 μg/kg, respectively. Therefore, the ratio (ID_{50} for heart rate) / (ID_{50} for total peripheral resistance) of bopindolol was 0.47 ± 0.25 and that of pindolol was 1.29 ± 0.45 indicating that the selectivity of these two agents as β-blocker was not marked.

Partial agonist activity of bopindolol in pithed rat

Male Wistar rat was anesthetized with thiopental (50-60 mg/kg) and was respirated artificially. Both adrenal glands were removed and after injection of atropine sulphate (1 mg/kg), both vagal nerves were cut at the level of cervical region. Destruction of the spinal cord was achieved by the use of a steel rod through left orbita (Gillespie and Muir, 1967; Chin *et al.*, 1989). After this procedure, the blood pressure and the heart rate fell down from 137.5 ± 3.1 to 46.5 ± 1.2 mmHg and from 414.1 ± 8.7 to 267.8 ± 4.3 beats/min, respectively. Bopindolol and its two metabolites (intravenous injection) had no significant effect on the blood pressure. However, they increased the heart rate dose- dependently (Fig. 9). The increase in heart rate was most remarkable with 18-502. These chronotropic effects were attenuated significantly by pretreatment of propranolol (1 mg/kg, i.v.). Time to the maximal response as well as duration of the action were longer with bopindolol than with its two metabolites (Fig. 10).

Antihypertensive effect of bopindolol in spontaneously hypertensive rat (SHR)
Single dose administration SHRs of ten weeks of age were used in this study. Three days before experiment, a cannula had been implanted into the aorta by operation. The blood pressure was measured directly and the heart rate was measured by the use of tachometer under conscious unrestrained state (Ishihara *et al.*, 1989). Pindolol (15 mg/kg, *p.o.*) decreased the blood pressure markedly; the decrease was significant even after 24 hr (Fig. 11). Bopindolol also reduced the blood pressure. However, the fall of blood pressure was smaller and the duration of action was shorter than those of pindolol. Figure 12 shows the effects of bopindolol and pindolol on the heart rate. Although the heart rate rose throughout the first 8 hr after administration of pindolol (15 mg/kg), the changes were not significant. By contrast, bopindolol (5 and 15 mg/kg) reduced the heart rate. The decrease was dose-dependent at 8 hr. The heart rate of bopindolol groups was significantly lower than that of pindolol groups during 8 hr after administration.

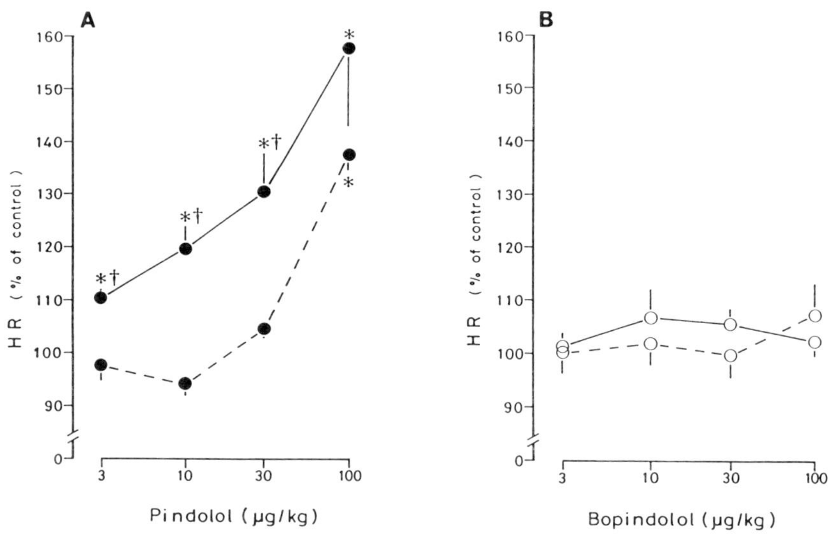

Figure 6. The effects of pindolol and bopindolol on the heart rate of chronically instrumented conscious beagle dogs. Values are obtained after 20 min of infusion and expressed as per cent of the control (pre-infusion values). Left panel: Effect of pindolol on the heart rate in the absence (solid line) or presence (dotted line) of propranolol (3 mg/kg). Right panel: Effect of bopindolol on the heart rate in the absence (solid line) or presence (dotted line) of propranolol (3 mg/kg).
* Significantly (p <0.05) different from the control values. † Significantly (p <0.05) different from the non-treated group (solid line). [From Ishibashi *et al.* (1989) with permission.]

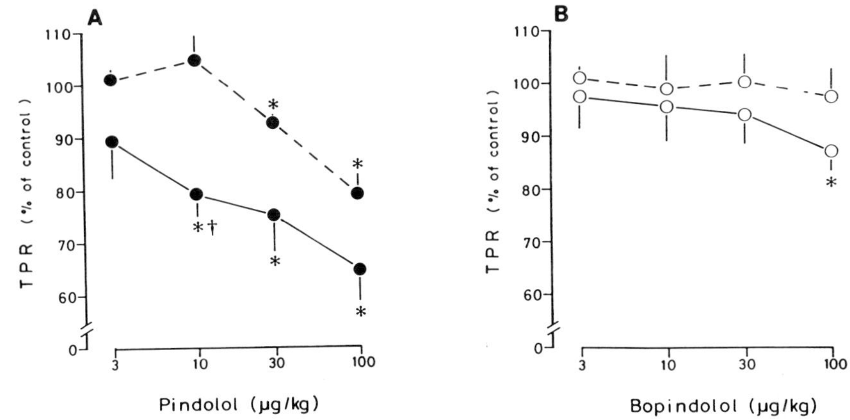

Figure 7. The effects of pindolol and bopindolol on the total peripheral resistance in chronically instrumented conscious dogs. Values are obtained after 20 min of infusion and expressed as per cent of the control (pre-infusion values). Left panel: Effect of pindolol on the total peripheral resistance in the absence (solid line) or presence (dotted line) of propranolol (3 mg/kg). Right panel: Effects of bopindolol on the total peripheral resistance in the absence (solid line) or presence (dotted line) of propranolol (3 mg/kg). * Significantly ($p < 0.05$) different from the control values. † Significantly ($p < 0.05$) different from the non-treated group (solid line). (From Ishibashi *et al.* (1989) with permission.)

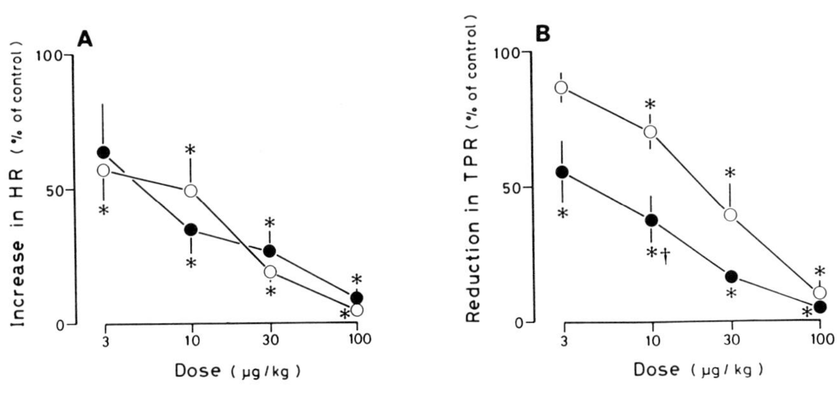

Figure 8. Inhibitory effects of pindolol and bopindolol on the increase in heart rate and the reduction of total peripheral resistance produced by isoproterenol. Left panel: Inhibitory effect of pindolol (●) and bopindolol (○) on the isoproterenol (0.1 μg/kg)-induced tachycardia. Right panel: Inhibitory effect of pindolol (●) and bopindolol (○) on the reduction of total peripheral resistance produced by isoproterenol (0.1 μg/kg). *: Significantly ($p < 0.05$) different from the control (100%). † Significantly ($p < 0.05$) different from the bopindolol group. [From Ishibashi *et al.* (1989) with permission.]

Repeated administration Starting from the 5th week of age on, bopindolol and other reference β-blockers were administered orally by gavage twice daily (10 a.m. and 4 p.m.). Systolic blood pressure was measured in unanesthetized state by a tail cuff plethysmographic method and heart rate was counted with a digital tachometer triggered by tail pressure pulses (Ishihara *et al.*, 1989). Figure 13 illustrates the effects of chronic oral administration of β-blockers on the development of hypertension in SHR. The blood pressure of the control group increased markedly during the course of the experiment. Bopindolol (5 and 15 mg/kg) and propranolol (15 mg/kg) attenuated the increase in blood pressure and a significant difference against the control group was observed after the 5th week of administration. There were no significant differences between propranolol and bopindolol groups in the degree of attenuation of increases in blood pressure. Two weeks after withdrawal, blood pressure in the propranolol and bopindolol groups had returned to the level of the control group.

In the control group, the heart rate increased markedly around the 2nd week of experiment, and remained high thereafter (Fig. 14). Bopindolol (5 and 15 mg/kg) and propranolol (15 mg/kg) attenuated the heart rate significantly from the 3rd week of experiment on. Pindolol (15 mg/kg) also attenuated the heart rate. However, a significant difference between the pindolol and the control groups was observed only after the 6th week. Two weeks after withdrawal, heart rate had returned to the level of the control group, except for the pindolol group. All agents had no significant effect

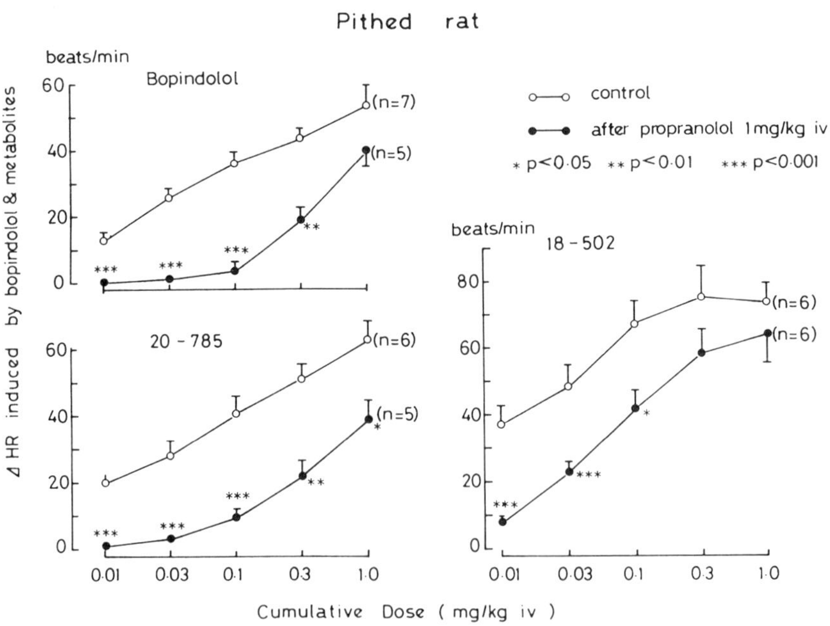

Figure 9. Antagonistic effects of propranolol (1 mg/kg, *i.v.*) on the positive chronotropic action induced by pindolol and its two metabolites in the pithed rat. Vertical bars indicate the mean±S.E. *p <0.05, **p <0.01, *vs.* control. [From Chin *et al.* (1989) with permission.]

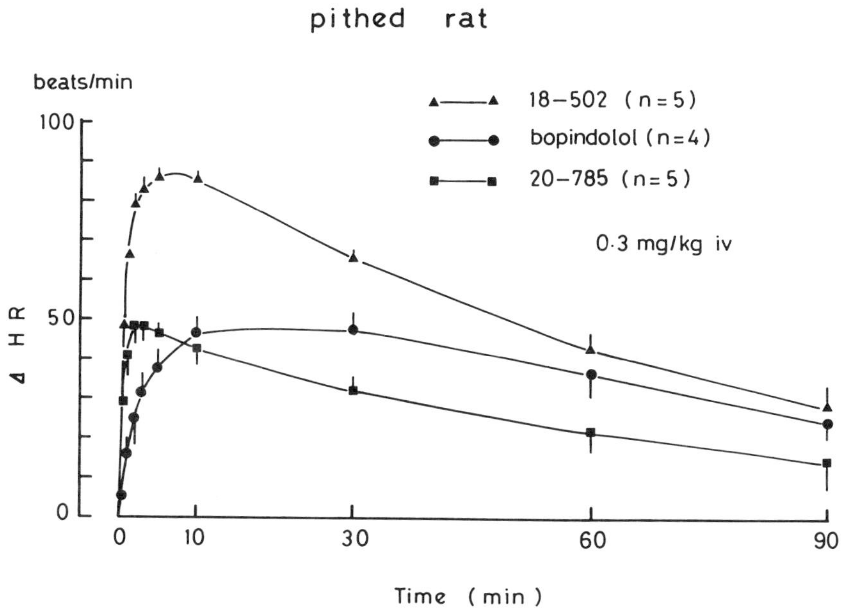

Figure 10. The time course of positive chronotropic action in pithed rat induced by a single dose of 0.3 mg/kg. *i.v.* of bopindolol, 18-502 and 20-785. [From Chin *et al.* (1989) with permission.]

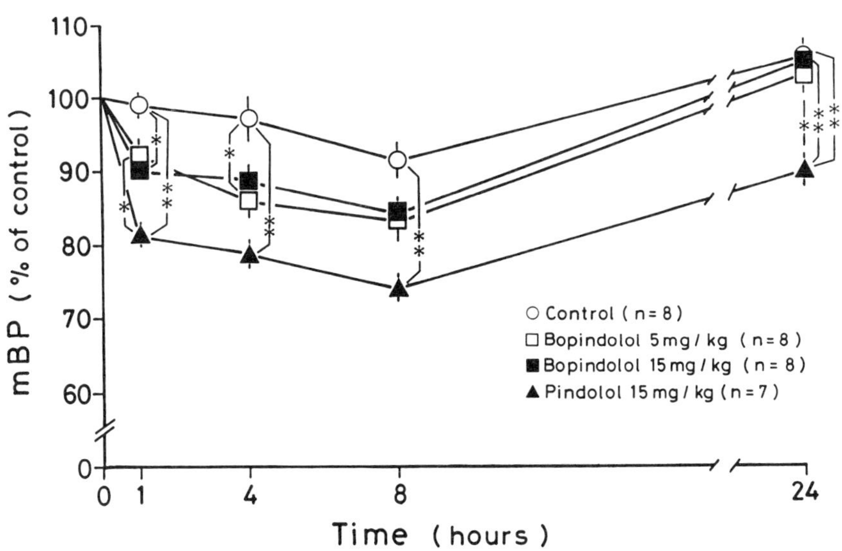

Figure 11. Acute effects of bopindolol and pindolol on the mean blood pressure in SHR. [From Ishihara *et al.* (1989) with permission.]

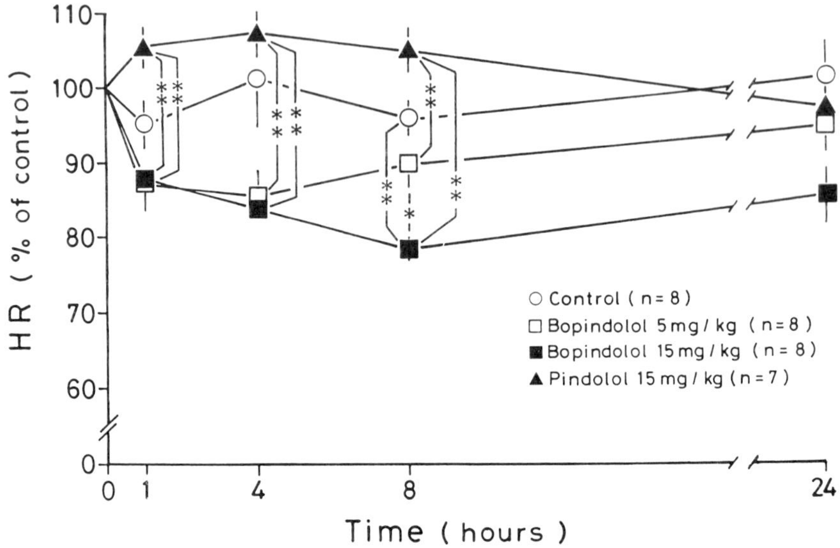

Figure 12. Acute effects of bopindolol and pindolol on the heart rate in SHR. [From Ishihara *et al.* (1989) with permission.]

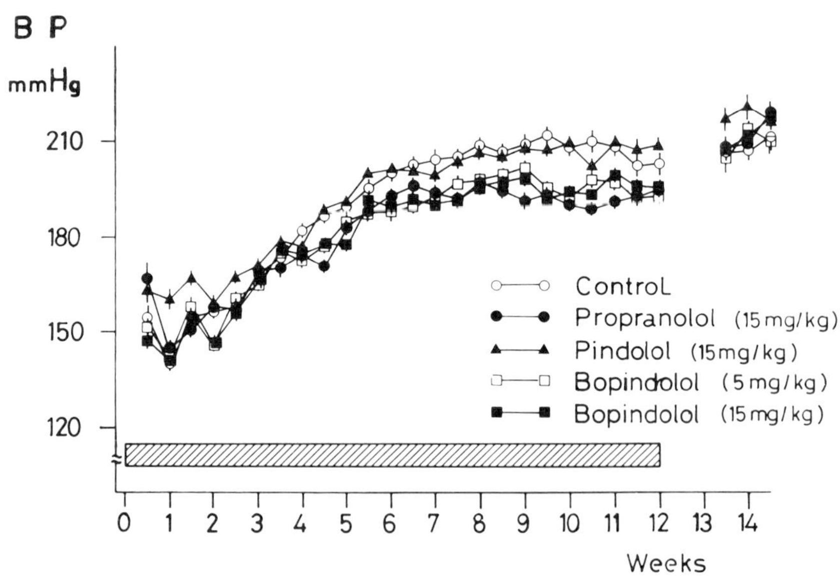

Figure 13. Antihypertensive effects of chronic oral administration of β-blockers in SHR. Ordinate: Systolic blood pressure (BP) in mmHg. Abscissa: Weeks after onset of administration of β-blockers, which was initiated at 5 weeks of age and continued during the period depicted as hatched bars. Each point represents the mean±S.E.M. (n =13-15). [From Ishihara *et al.* (1989) with permission.]

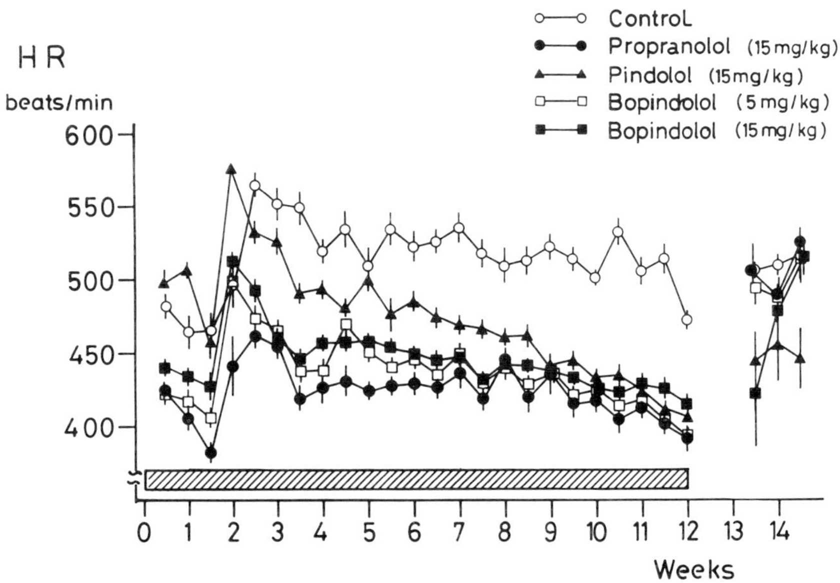

Figure 14. Attenuation of the increase in heart rate (HR) in SHR by β-blockers. Ordinate: Heart rate (HR) in beats/min. Abscissa: Weeks after onset of administration of β-blockers, which was initiated at 5 weeks of age and continued during the period depicted as hatched bars. Each point represents the mean±S.E.M. (*n* =13-15). [From Ishihara *et al.* (1989) with permission.]

on the ratio of heart weight to body weight and that of kidney weight to body weight.

After 12 weeks of treatment and following the withdrawal period (2 weeks), the concentrations of the serum lipid fractions and creatine phosphokinase (CPK) activities were examined. Significant increases in HDL cholesterol concentrations and decreases in LDL+VLDL/HDL cholesterol ratios were observed 12 weeks after bopindolol (5 and 15 mg/kg) and pindolol (15 mg/kg) (Fig. 15). Two weeks after withdrawal, differences in serum lipid fractions were no longer observed (Fig. 16). However, the LDL+VLDL/HDL cholesterol ratio of the bopindolol (5 mg/ kg) group was significantly higher as compared with that of the control group. There were no changes in CPK activity after 12 weeks of treatment and 2 weeks after withdrawal, but the activity was significantly higher in the propranolol group than in the pindolol and the bopindolol (15 mg/kg) groups 2 weeks after withdrawal.

DISCUSSION

β-blocking activity of bopindolol and its metabolites.

The first report characterizing bopindolol (Berthold *et al.*, 1981) demonstrated that this substance is 17 times more potent than propranolol as a β-adrenoceptor blocker in

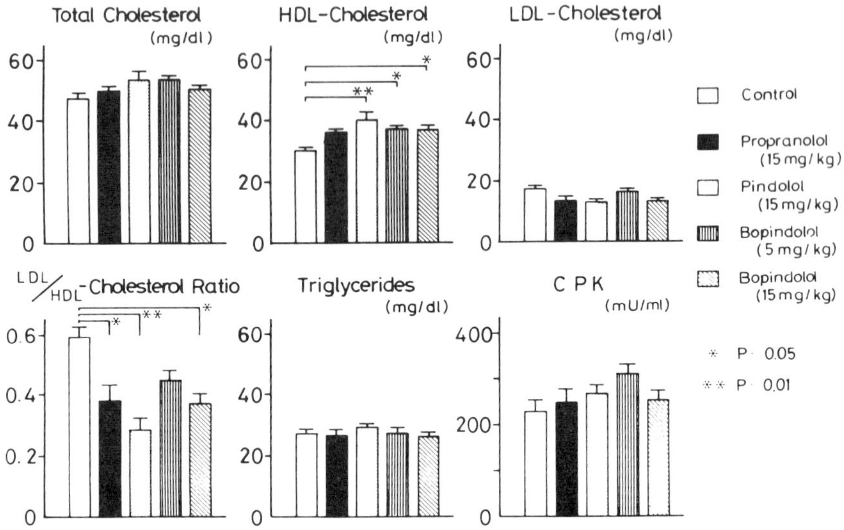

Figure 15. Effect of chronic administration (12 weeks) of β-blockers on the concentrations of serum lipids, lipoprotein fractions and serum creatine phosphokinase (CPK) activity. Each bar represents the mean±S.E.M. (n = 8-10). [From Ishihara *et al.* (1989) with permission.]

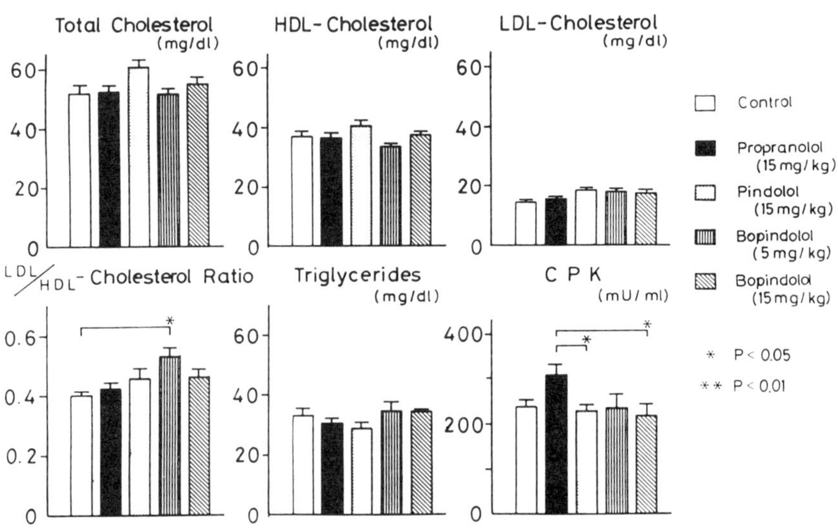

Figure 16. Effect of β-blockers (2 weeks after withdrawal) on the concentrations of serum lipids, lipoprotein fractions and CPK activity. Each bar represents the mean±S.E.M. (n = 5-6). [From Ishihara *et al.* (1989) with permission.]

Table 1. PA_2 values of bopindolol and its two metabolites (n = 5-7)

	Atria		Trachea
	Right	Left	
Bopindolol	8.81±0.12	8.45±0.10	9.56±0.07
18-502	9.94±0.07	9.95±0.20	9.88±0.04
20-785	8.03±0.05	8.05±0.05	7.38±0.16
Propranolol	8.36±0.06	8.41±0.08	8.41±0.05
Pindolol	8.79±0.12	9.04±0.11	8.93±0.06

From Chin *et al.* (1989) with permission.

isolated guinea-pig atria (agonist both chronotropic and inotropic effects of adrenaline). However, we found the relative potency of bopindolol to propranolol to be 1.1 and 2.8 in the right and the left atria and to be 14.1 in the trachea. Thus, bopindolol is nearly equipotent to propranolol as a β_1-antagonist and has β_2-antagonist activity one order more potent than that of propranolol. Caution has to be paid to the fact that these results were obtained in the isolated organs where transformation of bopindolol to active metabolites was considered to be minimal. In clinical trials, it is reported that no measurable amounts of bopindolol are present in plasma after oral intake. Only hydrolyzed metabolite, 18-502, was found (Platzer *et al.*, 1984; Aellig *et al.*, 1986) under this condition. Therefore, it is necessary to examine the β-blocking activities of active metabolites, 18-502 and 20-785, *in vitro*. Though 20-785 is less potent than propranolol and rather β_1-selective, main metabolite 18-502 is found to be non-selective β-antagonist with an extremely high potency. The potency ratios to propranolol (38.0 in right atria, 34.7 in left atria and 29.5 in trachea) are the highest we have ever experienced. However, these data may not be taken to directly represent the activity in clinical situations because there is a possibility that the metabolic pathway may not be the same in different species. Indeed, though Berthold *et al.* (1981) reported a high potency (17 fold of propranolol) in isolated guinea-pig atria, the potency ratio to propranolol obtained in the anesthetized dog was 4 in the same paper, a value not so much different from that we obtained in conscious beagle dogs. Furthermore, no definite selectivity was found in either of dog experiments. These results indicate that, in dogs, production of 18-502 is not large, or, if 18-502 is the first product even in dogs, most of this product is metabolized rapidly into other less potent metabolites (e.g. 20-785) or inactive metabolites. The pharmacological activity observed is the sum of activities of these fractions. Further study is required.

Membrane stabilizing activity

Chin *et al.* (1989) clearly showed the membrane stabilizing activity of bopindolol and 18-502 (potency order is: bopindolol>18-502). The metabolite with the lowest potency, 20-785, has no effect on Vmax. These results together with those on β-blocking activity indicate that the membrane stabilizing activity does not correlate to

Table 2. Effects of bopindolol and its metabolites on the parameters of action potential in guinea-pig papillary muscles.

Drugs	APA (mV)		APD_0 (msec)		APD_{90} (msec)		RMP (mV)		Vmax
	control	after drug	control	after drug	control	after drug	control	after drug	(% of decrease after drug)
Propranolol									
10^{-5}M	123±2.7	119±1.5**	133±5.8	113±6.2**	174±4.7	157±7.5*	83±1.8	83±1.3	19.6±3.15**
Bopindolol									
$3x10^{-6}$M	126±1.7	119±2.2*	129±4.5	120±6.0*	167±3.8	164±3.3*	86±1.5	84±1.4	25.3±1.78**
10^{-5}M	118±1.6	113±1.6**	124±4.3	109±3.7**	171±3.0	165±3.5**	80±0.7	79±0.4	44.6±2.35
18-502									
$3x10^{-6}$M	129±0.4	129±0.4	126±4.5	115±7.7	169±4.3	154±8.5	80±0.5	80±0.3	12.6±2.87**
10^{-5}M	127±0.7	125±0.8	128±3.7	115±2.6**	172±3.3	158±2.0**	80±0.4	80±0.5	29.1±1.30**
20-785									
10^{-5}M	129±0.5	129±0.5	131±3.7	132±4.5	173±3.9	171±5.3	80±0.5	79±1.2	0.8±0.83
$3x10^{-5}$M	129±0.5	129±1.1	129±5.1	126±6.9	170±5.2	165±6.7	78±0.5	79±0.4	0.3±0.28
Pindolol									
10^{-5}M	127±0.8	125±1.3	114±4.5	104±4.8**	158±4.1	148±4.5**	82±0.5	82±0.2	10.5±1.58**
$3x10^{-5}$M	126±0.8	123±0.6**	121±4.7	102±4.4**	166±4.1	148±4.5**	82±0.9	83±0.5	35.7±1.35**

Values are means ± SEM (n = 5-7). APA, action potential amplitude; APD_0, action potential duration at zero membrane potential; APD_{90}, action potential duration at 90% repolarization; RMP, resting membrane potential; Vmax, the maximal rate of upstroke of action potential. * $p < 0.05$, ** $p < 0.01$: significantly different from control values. Modified from Chin et al. (1989) with permission.

β-blocking activity. From the viewpoint of metabolic degradation, it may be said that the membrane stabilizing activity of bopindolol was decreased as the metabolic breakdown progressed. Anyway, this ancillary effect would contribute partly to the reduction in heart rate and there is possibility that bopindolol may have some beneficial effects on the ischemic derangement of myocardium.

Partial agonist activity

In the present study only the isoproterenol-induced increase in heart rate was inhibited by lower doses of bopindolol; isoproterenol-induced decrease in peripheral vascular resistance was not affected, indicating the β_1-selective nature of β-blocking action of bopindolol. Thus, we must conclude that somewhat β_1-selective metabolite of bopindolol, 20-785, is mainly responsible for the β-blocking action of bopindolol observed in conscious dogs.

In agreement with the report by Brodde *et al.* (1986), bopindolol produced a significant decrease in total peripheral vascular resistance in conscious dogs, though only at the highest dose tested, indicating the presence of a partial agonist activity. However, increases in heart rate were not observed with this compound, while a prototype β-blocker with partial agonist activity, pindolol, produced dose-related increases in heart rate as well as dose-related decreases in total peripheral vascular resistance. Thus, unlike the β-blocking action, the partial agonist activity of bopindolol is β_2-selective. In the light of the β_1-selective nature of the β-blocking action of bopindolol, the selective manifestation of partial agonist activity at β_2-receptor deserves special comment. Presumably, the intrinsic activity of 20-785 at β_1-receptor is low, while the affinity to this receptor is high (thus, 20-785 is a rather pure antagonist at this receptor in this animal species). In the pithed rat in which only the increase in heart rate (an effect mainly mediated by β_1-receptor) can be used for evaluation of the partial agonist activity because of the extremely low vascular tone, bopindolol and its two metabolites produced dose-related increases in heart rate. Pretreatment with propranolol resulted in an inhibition of the increases in this parameter. However, this is not necessarily contradictory to the findings in conscious dogs, as it is known that there is no formation of 20-785 in this species. In this species non-selective 18-502 was mainly responsible for the observed partial agonist activity.

Antihypertensive effect

The present study clearly demonstrates that single and repetitive administrations of bopindolol lower the blood pressure and the heart rate of spontaneous hypertensive rats. In single administration study, however, the duration of action was shorter and the potency was smaller than those of pindolol. Because acute reduction of blood pressure was considered to be due, in most part, to vasodilation (as a consequence of partial agonist activity at β_2-adrenoceptor), weaker hypotensive effect in single administration study is compatible with weaker partial agonist activity of bopindolol in conscious beagle dog experiment (Ishibashi *et al.*, 1989 or previous section).

There has been no published data on acute hemodynamic effect of bopindolol in hypertensive experimental model before our report (Ishihara *et al.*, 1989). Though a long duration of action was reported in conscious dogs with regard to the β-blocking activity (Berthold *et al.*, 1981), it dose not necessarily mean that the duration of blood pressure lowering effect should also be long. This is because of the fact that many kinds of metabolites are responsible for the effects of bopindolol and the β-blocking activities of these metabolites do not correlate with partial agonist activities (see previous section).

The chronic antihypertensive effects of β-blockers without partial agonist activity are usually associated with a reduction of cardiac output, mainly due to a reduced heart rate, while the effects of β-blockers with a strong partial agonist activity are associated with a reduction of total peripheral resistance (van Baak *et al.*, 1985). In the present study, the increase in heart rate observed with the maturation of the SHR was attenuated by bopindolol, indicating that bopindolol behaved in this circumstances as a β-blocker endowed with weak partial agonist activity. These chronic effects of bopindolol are in agreement with those reported by Yamashita *et al.* (1989).

Effect of bopindolol on serum lipoproteins and pathological changes in organs

One of the main concerns in the long term β-blocker administration is the possibility of producing an adverse effect on the blood lipid profile (Helgeland *et al.*, 1987). The possible relevance of partial agonist activity in the prevention of β-blocker-induced hypertriglyceridemia and of a decrease in HDL cholesterol has been reviewed by van Brummelen (1983). In our study, bopindolol as well as pindolol induced a significant increase in HDL cholesterol concentrations, although no changes were found in the concentration of triglycerides. The increase in the concentration of HDL cholesterol tended to be greater with pindolol than with bopindolol. This increase was in proportion to the partial agonist activity of these two compounds. Decreases in serum triglyceride and phospholipid by bopindolol were reported from another laboratory using stroke-prone spontaneously hypertensive rats (SHRSP) (Yamashita *et al.*, 1989). These results indicate the beneficial effects of bopindolol on the lipid metabolism.

Because muscular aches and cramps with an increase in serum creatine phosphokinase (CPK) activity related to the partial agonist activity were reported during treatment with pindolol (Frishman, 1988; Shima *et al.*, 1986), we examined CPK in our experiment. However, CPK activity showed only a tendency to rise with pindolol and bopindolol, which is in agreement with the low incidence of this side effect with bopindolol (Schiess *et al.*, 1984)

Chronic administration of bopindolol was also reported to prevent functional and morphological changes in various organs that developed during the progress of hypertension. Improved renal function and attenuation of proliferative arteritis in the kidney were observed in SHRSP (Yamashita *et al.*, 1989). In addition, not only the increase in organ weight to body weight ratios (brain, heart, liver and adrenal gland), but also pathological findings of respective organs were attenuated by treatment of bopindolol (Yamashita *et al.*, 1989).

Clinical correlations
In human trials, bopindolol itself was not detected after oral intake suggesting that the pro-drug, bopindolol, is rapidly hydrolysed to its active metabolite *in vivo* (Oddie *et al.*, 1983; Humbert *et al.*, 1987). Indeed, hydrolysed active metabolite, 18-502, appears in plasma with a half life of 0.3 h (Aellig *et al.*, 1986). As the peak plasma concentration of 18-502 correlates well to the degree of attenuation of exercise-induced tachycardia (Aellig *et al.*, 1986), it may be reasonable to regard that 18-502, a potent non-selective β-blocker with membrane stabilizing effect and partial agonist activity (Chin *et al.*, 1989), is largely responsible for the expressed effects of bopindolol. In fact, the potency of bopindolol in man (Aellig, 1985) is very close to that obtained by 18-502 in *in vitro* experiments (Table 1 and Berthold *et al.*, 1981). The potency was lower in beagle dogs (Fig. 8 and Berthold *et al.*, 1981). However, it may be due to the species difference in metabolic pathway.

Bioavailability of bopindolol is reported to be 60-70% (Aellig,1982;Aellig *et al.*, 1986) and half life for α and β phase are 3-4 h and 7-8 h, respectively (Aellig, 1982; Aellig *et al.*, 1986). For more detailed pharmacokinetics see Aellig *et al.*(1986). We have no data on the duration of β-blocking activity in dogs comparable to those in humans.

Contribution of the partial agonist activity which was proved in animal experiments may be conceivable also in the human. Some degree of the partial agonist activity to β_1-adrenoceptor was suggested to be responsible for prevention of a marked fall in heart rate (van Brummelen *et al.*, 1982; Hulthén *et al.*, 1983; Colnago *et al.*, 1987). Weak partial agonist activity at β_2-adrenoceptor is also suggested to have contributed to prevention of an increase in airway resistance (Dorow and Schiess, 1986) and in a reduction of systemic vascular resistance (Fitscha *et al.*, 1988b). Although the partial agonist activity is not easy to observe under normal sympathetic tone (van den Meiracker *et al.*, 1987), definite partial agonist activity of bopindolol at β_1-adrenoceptor was shown by Nyberg (1985) in comparison with propranolol under treatment of atropine. He concluded that bopindolol approximately replaced the resting sympathetic tone through its partial agonist activity.

The antihypertensive efficacy of bopindolol has been examined in hypertensive patients (mainly in stage WHO I-II) (van Brummelen *et al.*, 1982 and 1984; Hulthén *et al.*, 1983; Kirch and Axthelm, 1983; Andersson *et al.*, 1985; Westerman *et al.*, 1986; Colnago *et al.*, 1987; van den Meiracker *et al.*, 1987; Fitscha *et al.*, 1988a, b). Due to a higher potency and a longer duration than those observed in SHR it was possible to reduce the effective dose to 1-4 mg once a day, or less. A similar antihypertensive effect as that observed in SHR was obtained by 8 mg of bopindolol once a week (Andersson *et al.*, 1985). The mechanism of action in humans was regarded to be a reduction of systemic vascular resistance in acute phase and a reduction of cardiac output in chronic phase as we indicated in animal experiments. Though partial agonist activity of bopindolol possibly participates in the acute reduction of the systemic vascular resistance (Fitscha *et al.*, 1988b), to what extent the partial agonist activity contributes is obscure (van der Meiracker *et al.*, 1987; Fitscha *et al.*, 1988b).

After withdrawal of bopindolol, no rebound hypertension was observed (Andersson *et al.*, 1985; van den Meiracker *et al.*, 1987). These observations are consistent with those observed in our SHR study. A decrease in β-adrenoceptor density during treatment (De Blasi *et al.*, 1988) and after withdrawal of bopindolol has indeed been described (Brodde *et al.*, 1986). However, contribution of the partial agonist activity of bopindolol to this phenomena is not defined.

A decrease in total cholesterol (Colnago *et al.*, 1987; van Brummelen *et al.*, 1982) and prevention of decrease in HDL cholesterol (van Brummelen *et al.*, 1984) were observed in man with bopindolol. A similar beneficial effect of bopindolol on the serum lipoprotein profile was observed in the present animal experiments. This ancillary effect will also be helpful for a long-term treatment with bopindolol.

CONCLUSION

Bopindolol is a potent, long acting and non-selective β-adrenoceptor blocking agent with membrane stabilizing effect and moderate partial agonist activity. These characteristics appear with varying degree depending on the animals. From the results of our experiments and from the pharmacokinetic examinations in humans, it may be concluded that 18-502 is mainly responsible for these activities. However, the presence of many other active metabolites with different degrees of activities result in diverse effects in different animals. Anyway, its high potency, long duration of action and ancillary effects seem to promise a good compliance and benefits to patients for bopindolol.

REFERENCES

Aellig, W. H. (1982). Clinical pharmacological experiments with bopindolol (LT31-200): a long acting β-adrenoceptor blocking drug with partial agonist activity. *Br. J. Clin. Pharmacol.* **13,** 267P-268P.

Aellig, W.H. (1985). Pharmacological experiments in healthy volunteers with bopindolol, a long-acting β-adrenoceptor blocking drug with partial agonist activity. *Br. J. Clin. Parmacol.* **19,** 775-781.

Aellig, W. H., Nüesch, E., Engel, G., Grevel, J., Niederberger, W. and Rosenthaler, J. (1986). Relationship between plasma concentrations and cardiac β-adrenoceptor blockade - a study with oral and intravenous bopindolol. *Br. J. Clin. Pharmacol.* **21,** 45-51.

Andersson, O. K., Hedner, T., Nyberg, G. (1985). Acute and long-term antihypertensive β-adrenergic blocker with ISA. *Eur. J. Clin. Pharmacol.* **29,** 281-285.

Berthold, R., Waite, R. and Weber, H. (1981). Pharmacological studies with bopindolol: a new long acting β-adrenoceptor antagonist. *Br. J. Pharmacol.* **74,** 829P.

Brodde, O. E., Schemuth, R., Brinkmann, M., Wang, X. L., Daul, A. and Borchard, U. (1986). β-adrenoceptor antagonists (non-selective as well as β_1-selective) with partial agonistic activity decrease β2-adrenoceptor density in human lymphocytes: Evidence for a β_2-agonistic component of the partial agonistic activity. *Naunyn-Schmiedeberg's Arch. Pharmacol.* **333,** 130-138.

Chin, W. P., Sun, H. T., Kawada, T. and Imai, S. (1989). β-blocking actions and the partial agonist activity of bopindolol, a new β-adrenoceptor antagonist, and its two metabolites. *Folia Pharmacol. Japon.* **94,** 27-33.

Colnago, R., Meregalli, M., Turri, D., Nazzari, M., Fiorella, G., Guffanti E. E. and Mazzola C.

(1987). Double-blind comparison of once-daily bopindolol and atenolol for mild-to-moderate hypertension. *Int. J. Clin. Pharmacol. Ther. Toxicol.* **25,** 347-352.

Dayer, P., Leemman, T., Narmy, A. and Rosenthaler, J. (1985). Interindividual variation of beta-adrenoceptor blocking drugs, plasma concentration and effect: Influence of genetic status on behavior of atenolol, bopindolol and metoprolol. *Eur. J. Clin. Pharmocol.* **28,** 149-153.

Dorow, P. and Schiess, W. (1986). Bopindolol, pindolol and atenolol in patients with chronic obstructive lung disease. *Klin. Wschr.* **64,** 366-369.

De Blasi, A., Fratelli, M. and Marsco, O. (1988). Certain β-blockers can decrease β-adrenargic receptor number: I. Acute reduction in receptor number by tertatolol and bopindolol *Circ. Res.* **63,** 273-278.

Fitscha, P., Meisner, W. and Tiso, B. (1988a). Comparison of the acute haemodynamic effects of bopindolol and propranolol at rest and during supine exercise. *Eur. J. Clin. Pharmacol.* **34,** 359-362.

Fitscha, P., Meisner, W., Brandstetter, G. and Tiso, B. (1988b). Haemodynamic effects of short-term treatment with bopindolol in essential hypertension. *Eur. J. Clin. Pharmacol.* **34,** 411-413.

Frishman, W. H. (1988). β-adrenoceptor receptor blockers. Adverse effects and drug interactions. *Hypertension* **11** (Suppl. II), II21-II29.

Gillespie, J. S. and Muir, T. C. (1967). A method of stimulating the complete sympathetic outflow from the spinal cord to blood vessels in the pithed rat. *Br. J. Pharmacol.* **30,** 78-87.

Helgeland, A., Hjermann, I., Leren, P., Enger, S. and Holme, I. (1987). High-density lipoprotein cholesterol and antihypertensive drug: the Oslo study. *Br. Med. J.* **II,** 403.

Hulthén, U. L., Van Brummelen, P., Amann, D. W. and Bühler, F. R. (1983). Antihypertensive effecacy of the new long-acting β-blocker bopindolol as related to age. *J.Cardiovasc. Pharmacol.* **5,** 426-429.

Humbert, H., Denouel, J. and Keller, H. P. (1987). Column liquid chromatographic determination of hydrolysed bopindolol, in the picogram per millilitre range in plasma, using cartridge extraction and dual electrochemical detection. *J. Chromatography* **422,** 205-215.

Ishibashi, T., Mitomi, A., Tatebe, S. and Imai, S. (1989). Cardiohemodynamic effects of bopindolol in normotensive conscious dogs: A comparative study with pindolol. *Arznim. Forsch.* **39,** 454-457.

Ishihara, T., Chin, W. P., Yoshida, Y., Sun, H. T., Mitomi, A., Ishibashi, T., Tamura, K. and Imai, S. (1989). Effects of chronic oral administration of a new β-blocker, bopindolol, on serum lipoprotein concentrations and blood pressure of spontaneously hypertensive rats. *Arch. Int. Pharmacodyn.* **302,** 145-157.

Kirch, W. and Axthelm, T. (1983). Bopindolol, a new long-acting beta-adrenoceptor antagonist - A randomized comparison against propranolol in hypertensive patients. *Int. J. Clin. Pharmacol. Res.* **3,**89-93.

Nyberg, G. (1985). Intrinsic sympathomimetic activity of bopindolol, a non-selective β-adrenoceptor antagonist. *Br. J. Clin. Pharmacol.* **20,** 242P-243P.

Oddie, C. J., Jackman, G. P. and Bobik, A. (1983). Analysis of bopindolol and its active matabolite 18-502 in human plasma by high-performance liquid chromatography. *J. Chromatography.* **273,** 469-474.

Platzer, R., Galeazzi, R.L., Niederberger, W. and Rosenthaler, J. (1984). Simultaneous modeing of bopindolol kinetics and dynamics. *Clin. Pharmacol. Ther.* **36,** 5-13.

Schiess, W., Welzel, D. and Gugler, R. (1984). Double-blind comparison of once-daily bopindolol, pindolol and atenolol in essential hypertension. *Eur. J. Clin. Pharmacol.* **27,** 529-534.

Shima, K., Koyama, T., Yamashita, K., Namba, K., Hosaka, T. and Maezawa, H. (1986). Increased creatine phosphokinase (CPK) level and muscle symptom during β-blocker therapy. *J. Jpn. Soc. Intern. Med.* **75,** 1400-1404.

van Baak, M. A., Struyker-Boudier, H.A.J. and Smith, J.F.M. (1985). Antihypertensive mechanisms of beta-adrenoceptor blockade: A review. *Clin. Exp. Hypert.* **A7,** 1-72.

van Brummelen, P., Bühler, F.R., Amann, F.W. and Bolli, P. (1982). Effects of a new long-acting beta-blocker Bopindolol (LT 31-200) on blood pressure, plasma catecholamines, renin and cholesterol in patients with arterial hypertension. *Eur. J. Clin. Pharmacol.* **22,** 491-493.

van Brummelen, P. (1983). The relevance of intrinsic sympathomimetic activity for β-blocker-induced changes in plasma lipids. *J. Cardiovasc. Pharmacol.* **5** (Suppl. 1), S51-S55.

van Brummelen, P., Bolli, P., Koolen, M.I., Staehelin, H.B. and Bühler, F.R. (1984). Plasma lipid fractions during long-term monotherapy with the ISA-containing β-adrenoceptor blocker bopindolol in hypertensive patients. *Br. J. Clin. Pharmacol.* **17,** 86-88.

van den Meiracker, A. H., Man in't Veld, A. J., Ritsenma van Eck, H. J., Boomsma, F., Derkx, F.H.M., Mulder, P. and Schalekamp, M.A.D.H. (1987). The clinical pharmacology of bopindolol, a new long-acting beta-adrenoceptor antagonist, in hypertension. *Clin. Pharmacol. Ther.* **42,** 411-419.

Westerman, R. F., Christensen, C., Westheim, A. and Donker, A.J.M. (1986). Bopindolol versus metoprolol in hypertension. *Eur. J. Clin. Pharmacol.* **31,** 379-380.

Yamashita, K., Shigematsu, K., Maeda, T., Ishida, Y., Niwa, M. and Ozaki, M. (1989). Antihypertensive effect of bopindolol on stroke-prone spontaneously hypertensive rats. (SHRSP). *Folia Pharmacol. Japon.* **93,** 7-15.

Progress in Hypertension, Vol. 2, pp. 93-105.
Saito *et al.* (Eds)

NEWLY DEVELOPED BETA-BLOCKERS: BASIC AND CLINICAL INVESTIGATIONS

KAZUAKI SHIMAMOTO and OSAMU IIMURA

The Second Department of Internal Medicine, Sapporo Medical College, S-1 W-16, Chuo-ku, Sapporo 060, Japan

INTRODUCTION

The beta-adrenocepter blocking agents have been the most important development in hypertension and cardiovascular therapeutics in the past 20 years. Today, they assume a paramount role in the treatment of hypertension and a variety of other cardiovascular diseases.

Beta-adrenocepter blocking agent, propranolol, was first introduced by Prichard and Gillam in 1964 to decrease the blood pressure of hypertensive patients. Propranolol,the prototype, was the first beta-blocking drug to be approved and widely used throughout the world. Thereafter, many beta-blockers are now available, including several newly developed drugs. In fact, they have been evaluated as the drugs of first choice for hypertension treatment from the traditional stepped care program of the WHO reports of 1978 (Report of a WHO expert committee, 1978) to the most recent stepped care program proposed by the Joint National Committee on Detection, Prevention and Treatment of High Blood Pressure in United States of America (The 1988 report of the JNC).

Recently, individualized stepped care therapy has been recommended for hypertension treatment (The 1988 report of the JNC), as it also provides for more emphasis on the control of other risk factors for cardiovascular diseases including hyperlipidemia, glucose intolerance, hyperuricemia and so on.

It has been pointed out that beta-adrenoceptor blockades have some adverse effects on cardiac output, peripheral resistance, lipid metabolism and glucose metabolism (Table 1) (Prichard and Gillam, 1964; Leren, 1987; Pollare *et al.,* 1989). However, these adverse effects are very different for each beta-blocker.

Recently, several new beta-blockers, which have some special additional action on vasodilation and lipid and glucose metabolism, have been developed to eliminate these adverse effects.

Now they can be used properly depending on the background of each individual patient.

This chapter reviews the newly developed beta-blockers with vasodilating action.

Table 1. Disadvantages of beta-blockade therapy

1. They may adversely affect a range of common disease states:
   Asthma and obstructive lung disease
   Cardiac failure and conduction defect
   Peripheral vascular disease
   Insulin-dependent diabetes
2. Withdrawal may cause rebound effects in angina pectoris.
3. Beta-blockers may not prevent progression of atheroma or coronary heart disease:
   Effects on triglycerides and HDL-cholesterol
   Interaction with cigarette smoking
4. Greater impairment in the quality of life by nonselective beta-blockers compared with newer drugs such as converting enzyme inhibitors.
5. The hemodynamic effects of beta-blockade may not reserve the primary changes of essential hypertentsion.

CLASSIFICATION OF BETA-BLOCKERS

At the present time, over 20 beta-blockers are available on the world market (Table 2). It is well known that beta-blocking agents have various pharmacological properties in addition to their beta-1 adrenoceptor antagonizing effect. As for the classification of beta-blockers, the proposals by Fitzgerald (1969) and Prichard (1978) are very well known and widely used. Their classifications are based mainly on the intrinsic sympathomimetic activity (ISA), membrane stabilizing activity (MSA), the degree of selectivity for beta-1 receptor and alpha receptor blocking action. In the classification of Prichard, beta-adrenoceptor blockers are divided into 3 main groups, non-selective beta-adrenoceptor blocker (group 1), beta-1 selective beta-blocker (group 2), and beta-adrenoceptor blocker with post-synaptic alpha-1 blocking action (group 3). Then, each group is further divided into four subgroups, subgroup 1-possessing both intrinsic sympathomimetic activity and membrane stabilizing activity, subgroup 2-possessing membrane stabilizing activity without intrinsic sympathomimetic activity, subgroup 3-possessing intrinsic sympathomimetic activity without membrane stabilizing activity, and subgroup 4-with neither intrinsic sympathomimetic activity nor membrane stabilizing activity. Recently, it has been reported that bunitrolol has post-synaptic alpha-1 blocking action (Hohki and Kimihara, 1982). However, it should be classified as group 1-subgroup 1, because its alpha-1 blocking action is very weak. Labetalol, which was previously classified into group 3-subgroup 2, should be placed in group 3-subgroup 1 because of its intrinsic sympathomimetic activity (Katano *et al.*, 1978).

Regarding the new beta-blockers, tilislol is classified as group 1-subgroup 4, celiprolol, dilevalol and bopindolol in group 2-subgroup 3, carvedilol as group 3-subgroup 2 and nipradilol, amosulalol and urapidil as group 3-subgroup 4.

VASODILATING EFFECTS OF BETA-BLOCKERS: MECHANISM AND CHARACTERIZATION

During the therapeutic course of beta-adrenoceptor blocker administration, there is an initial increase in peripheral vascular resistance which is followed by, in what may be an adaptive process, a return of peripheral resistance to pretreatment levels (usually over a period of weeks) (Tarazi and Dustan, 1972). The blocking action of beta-mediated vasodilation in skeletal muscle vasculature may play a role in the initial increase in peripheral resistance as well as in the potential for beta-blockers to cause exacerbation of peripheral vascular syndromes which are characterized by arteriolar vasospasms (probably from unopposed alpha-agonistic activity). Indeed the aggravation of peripheral vascular insufficiency, intermittent claudication, and Raynaud's phenomenon have been reported as rather common adverse effects after non-specific beta-blocker therapy, although they are less likely to follow beta-1 blockers.

Because of these adverse effects, new beta-blockers with vasodilating action have been developed. For the mechanisms of vasodilating effects by beta-blockers, three

Table 2. Pharmacodynamic properties of beta-blocking agents

Drug	MSA	ISA	Beta-1 Selectivity	Alpha Blocking Action
propranolol	++	-	-	-
bufetolol	+	-	-	-
bupranolol	+	-	-	-
carvedilol	+	-	-	+ -
nipradilol	+ -	-	-	+ -
timolol	-	-	-	-
tilisolol	-	-	-	-
nadolol	-	-	-	-
pindolol	+ -	+	-	-
oxprenolol	+	+	-	-
alprenolol	+	+	-	-
carteolol	+ -	+	-	-
bunitrolol	+	+	-	+ -
penbutolol	+	+	-	-
bopindolol	+	+	-	-
acebutolol	+	+	+	-
celiprolol	-	+	+	-
metoprolol	-	-	+	-
atenolol	-	-	+	-
dilevalol	-	+ -	-	+ -
bevantolol	-	-	++	+
arotinolol	-	-	-	+
labetalol	+ -	+ -	-	++
amosulalol	-	-	-	++
urapidil*				+++

*Urapidil should be classified to alpha-adrenoceptor blockade because of its weak beta-blocking action.

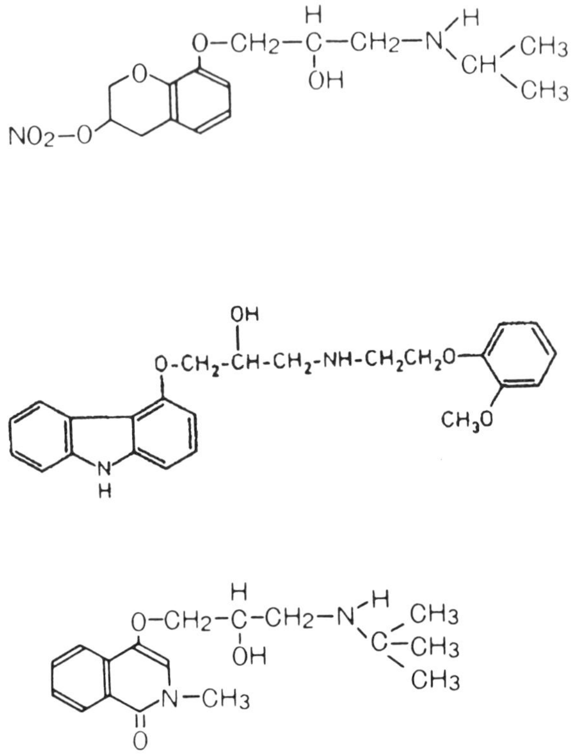

nipradilol

carvedilol

tilisolol

N-696

Figure 1. Chemical Structure of nipradilol, narvedilol and tilisolol.

possibilities are considered,1) a direct vasodilating action in addition to beta-blocking action, 2) alpha-1 blocking action in addition to beta-blocking action, and 3) intrinsic sympathomimetic activity to the vascular beta-2 receptor.

Beta-blockades possessing direct vasodilating action

Nipradilol

Nipradilol (Fig. 1) is a newly synthesized chemical agent designed to possess both beta-adrenoceptor blocking and vasodilating actions. The agent has both beta-1 and -2 blocking action. The aminopropanol side chain and nitro-ester group of this molecule are expected to exert beta-adrenoceptor blocking and vasodilating actions, respectively. Vasodilation induced by nipradilol was not accompanied by any obvious changes in the electrical properties of the smooth muscle membrane, indicating that the vasodilator action of nipradilol resembles that of nitroglycerin (Kou and Suzuki, 1982, Winbury *et al.*, 1969). In addition to nitroglycerin-like vasodilator action, nipradilol also has alpha-adrenoceptor blocking activity (Kou and Suzuki, 1982). It has been reported that in pithed rats nipradilol reduces blood pressure mainly through direct vasodilator action rather than by blocking the alpha-adrenoreceptor (Nakagawa *et al.*, 1985). Nipradilol decreases venous return in anesthetized dogs mainly by increasing venous capacitance through a direct venodilator action, which is similar to that of nitroglycerin. Precipitation of heart failure is one of the most serious adverse effects of beta-adrenoceptor antagonists. In this respect, the favorable properties of nipradilol, especially its effects on the left ventricular preload, appear to be beneficial in patients with low cardiac reserve. It has been reported in human essential hypertension (Walsh *et al.*, 1969) and in hypertensive animals (Simom, 1976), that venous compliance is decreased and venous contractility is enhanced. This data shows that venous compliance is already impaired in the prehypertensive stage, and nipradilol can partly contribute to the improvement of venous compliance. In the treatment of essential hypertension, nipradilol reduces the heart rate and cardiac output, but does not increase peripheral resistance. This is a very different effect from that of propranolol (Tsukiyama *et al.*, 1985).

Thus, it has been suggested that nipradilol decreases not only the afterload, but also venous dilation, and improves the hemodynamics of hypertensive patients.

Carvedilol

Carvedilol (Fig. 1) is a potent antihypertensive agent in animals and humans (Ruffolo *et al.*, 1990, Sponer *et al.*, 1987, Sundberg *et al.*, 1987). Studies on its mechanism of action indicate that the compound is a beta-adrenoceptor antagonist and a vasodilator, and both of these activities contribute to the antihypertensive response produced by the compound. The reduction in blood pressure produced by carvedilol is associated with a decrease in peripheral vascular resistance, indicative of arterial vasodilation (Sundberg *et al.*, 1987), and the lack of reflex tachycardia is a consequence of the beta-adrenoceptor blocking properties of the compound. The mechanism of action responsible for the vasodilator activity of carvedilol has not been completely established, in as much as this compound has a variety of pharmacologic actions over a wide range of concentrations. It has been proposed that the vasodilation produced by carvedilol results from nonspecific relaxation of vascular smooth muscles, selective alpha-1 adrenoceptor blockade, calcium channel blockade, and/or some combination of these actions.

In the treatment of essential hypertension, the agent reduces heart rate and cardiac output in addition to the blood pressure fall, but the suppression of left ventricular contractility is not clear as indicated by the changes of left ventricular ejection fraction and max systolic-dV/dt, and the slight reduction of peripheral resistance.

Tilisolol (N-696)

Tilisolol (Fig. 1) is a newly synthesized, non-selective beta-adrenoceptor blockade (Sasaki *et al.*, 1987). Experimental studies indicate that this agent is a unique compound which exerts a direct vasodilatory effect in addition to its beta-blocking action. It has been shown that, in pithed and adrenalectomized rats, tilisolol attenuates the pressor response to norepinephrine, phenylephrine, angiotensin II and the electrical stimulation of preganglionic sympathetic fibers (Nakagawa *et al.*, 1984). However, these effects cannot be accounted for by the alpha-blocking effect, since the alpha-blocking effect of tilisolol is very weak.

In the treatment of essential hypertension, this drug decreases heart rate and cardiac output, but does not decrease the left ventricular ejection fraction nor increase peripheral resistance.

Beta adrenoceptor blocking agents possessing alpha blocking action

Alpha and beta adrenoceptor blocking agents have vasodilating action through the alpha-1 adrenergic blocking properties. Labetalol (Richards and Prichard, 1979) which is a nonselective beta-adrenoceptor blocker possessing alpha-1 blocking action, and arotinolol (Hara, 1978) which has milder alpha-1 blocking action than labetalol, have been used in hypertension treatment. Now, new alpha- and beta-adrenoceptor blocking agents with a different ratio of alpha blocking activity/ beta blocking activity have been developed.

Amosulalol

Amosulalol (Fig. 2) (Takenaka *et al.*, 1982) is a nonselective beta-adrenoceptor blocker with alpha-1 blocking action. The ratio of beta-blocking activity/ alpha blocking activity is almost 1. In the treatment of essential hypertension, a decreased peripheral resistance and increased cardiac output were observed by amosulalol administration, but heart rate showed no change. Thus, as to the hypotensive mechanism of this drug, the reduction of afterload by alpha-1 adrenoceptor blocking action may be suggested as contributing the most.

Urapidil

The main hypotensive mechanism of urapidil (Fig.2) is considered to be an alpha-1 adrenergic blocking action and central alpha-2 adrenoceptor stimulating action (Van Zwieten *et al.*, 1985). Its beta-blocking activity is very weak. As to the

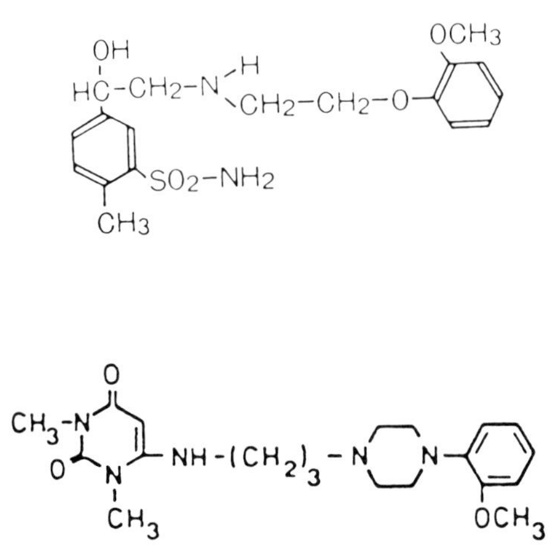

amosulalol

urapidil

Figure 2. Chemical structure of amosulalol and urapidil.

hemodynamic effects of urapidil in essential hypertension, peripheral resistance is decreased with the drugs administration similar to that of alpha-1 blocking agents. However, left ventricular contractility indicated by heart rate, cardiac output and left ventricular ejection fraction, showed no significant changes, and it is different from the typical alpha-1 blocking agent, prazosin, because the alpha-1 adrenoceptor blocker increases cardiac output and left ventricular ejection fraction. This difference may be explained by its weak beta-blocking effects.

Beta blockers with intrinsic sympathomimetic activity

At present, beta-receptors are divided into beta-1 and -2 subtypes (Nahorski, 1979). The postsynaptic beta-1 adrenoceptors are involved mainly in the regulation of cardiac function. Their stimulation causes an increase in cardiac frequency and contractility. Excitation of beta-2 adrenoceptors by the appropriate agonists induces bronchodilatation, vasodilation, and hyperglycemia. Pindolol is unique among the approved beta-blockers in that it possesses intrinsic sympathomimetic activity (ISA). Intrinsic sympathomimetic activity of beta-blockade induces vasodilation through the stimulation of vascular beta-2 receptor activity. It appears to have little effect on resting cardiac output while decreasing total peripheral resistance. The chemical nature of pindolol allows it to combine with and activate beta-receptors, thus giving it partial agonist activity. Recently, celiprolol, dilevalol and bopindolol, new beta-blockers with beta-2 selective intrinsic sympathomimetic activity, have been developed.

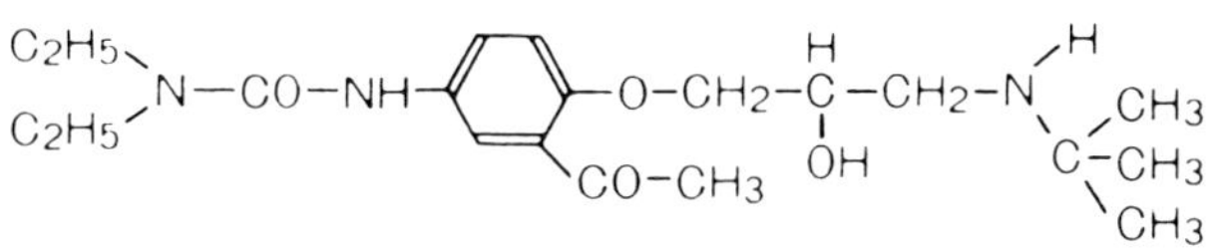

celiprolol

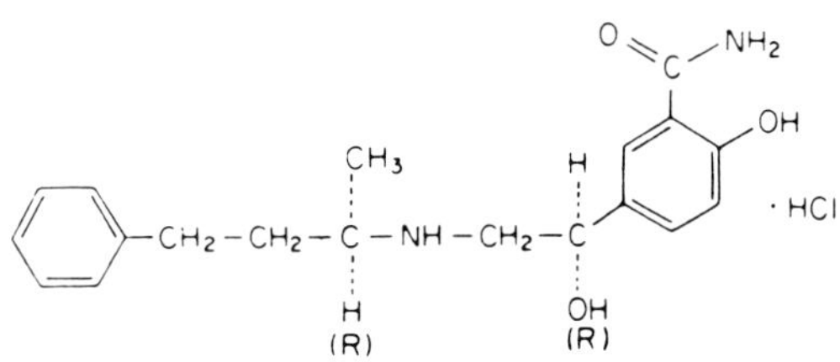

dilevalol

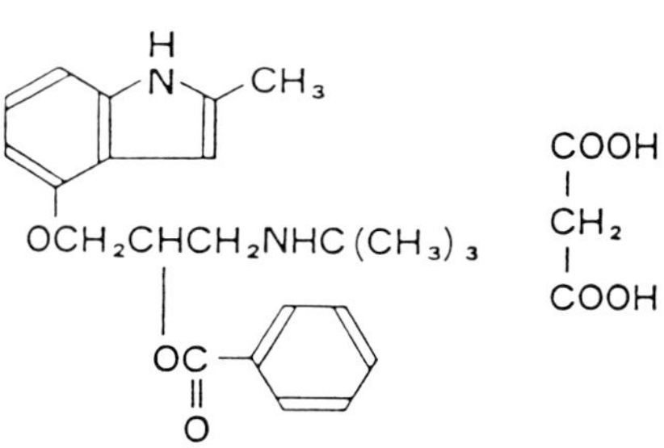

bopindolol

Figure 3. Chemical structure of celiprolol, dilevalol and bopindolol.

Celiprolol

Celiprolol (Fig.3) combines highly beta-1 selective beta-adrenoceptor blockade with beta-2 selective partial agonism and alpha-2 receptor blocking action on the post synaptic area (Capone and Mayol, 1986). The vasodilating action of this agent depends on its beta-2 agonistic action. Thus, the vasodilating action of celiprolol means that it acts by a different, more physiological mechanism composed to conventional beta-blockers while retaining the cardioselective potential of beta-1 antagonists. The vasodilation may also be largely responsible for the low incidence of typical beta-blockade side effects, such as fatigue and cold extremities. Celiprolol reduces blood pressure and peripheral resistance. The reduction of heart rate and cardiac output is very small, and left ventricular contractility indicated by left ventricular ejection fraction does not decrease with its administration.

Dilevalol

Dilevalol (Fig. 3) (Baum *et al.*, 1981) is one of the optical isomer of labetalol (RR-isomer of labetalol) which is an alpha- and beta-adrenoceptor blocking agent with beta-2 selective intrinsic sympathomimetic activity. However, its alpha blocking action is very gentle and its vasodilating action is speculated to be caused by its intrinsic sympathomimetic activity on the vascular beta-2 receptor. It reduces blood pressure and peripheral resistance, and the degree of the reduction of heart rate and cardiac output is not so remarkable.

Bopindolol

Bopindolol (Fig. 3) is a new long-acting non-selective beta adrenergic blockade (Aellig, 1985). Animal pharmacological studies have indicated a mild membrane stabilizing effect and some intrinsic sympathomimetic activity (Berthold, 1981), but the selectivity of its intrinsic sympathomimetic activity is not yet clarified. By its administration, heart rate and cardiac output are reduced and peripheral resistance increased. With long-term administration, the degree of both the reduction of heart rate and cardiac output and of increased peripheral resistance are attenuated. The duration of its hypotensive activity is very long, and it can be used once a week.

CLINICAL APPLICATION AND SIGNIFICANCE OF NEW BETA-ADRENOCEPTOR BLOCKADES

Depressor effect

It is now well recognized that beta-adrenoceptor blockades are effective in reducing the blood pressure of many patients with systemic hypertension. At the present time there is no consensus of opinion as to the mechanisms whereby these drugs lower blood pressure. It is probable that some or all of the mechanisms proposed in the following play a part: negative chronotropic and inotropic effects, central nervous system effects, suppression of plasma renin activity, reduction in venous return, resetting of baroreceptor levels and so on.

Vascular beta-2 adrenoceptor blocking action induces passive vasoconstriction and vasoconstriction by vascular alpha-1 receptor stimulation. Thus, beta-2 blocking action by nonselective beta adrenoceptor blocking agents may attenuate the depressor effect. Therefore, the effects of the new beta-blockades possessing a vasodilating effect may augment the depressor effect of beta-blockers through the reduction of peripheral vascular resistance.

In considering the depressor effect of nipradilol in double-blind clinical trials, the antihypertensive effect of nipradilol given at 6-18 mg/day b.i.d. was superior to that of propranolol (60-120 mg/day t.i.d.) and equivalent to that of pindolol (15-30 mg/day t.i.d.) (Uchida *et al.*, 1987). Of the beta-blockers with alpha receptor antagonistic action, there are no differences between amosulalol and labetalol (Ikeda *et al.*, 1985), and between dilevalol and labetalol (Kaneko *et al.*, 1988). As the beta-1 selective beta-blockades, celiprolol, possessing intrinsic sympathomimetic and direct vasodilating actions, is a highly effective and safe drug for the treatment of mild to moderate essential hypertension compared with propranolol (Abe *et al.*,1989). In double-blind clinical studies reduction in blood pressure by bopindolol was comparable to that obtained with metoprolol, atenolol and propranolol (Westerman *et al.*, 1987). Thus, it is concluded that these new beta-blockers with vasodilating action have higher or the same effect on blood pressure lowering compared to the conventional drugs used presently.

Application for patients with angina pectoris

In patients with hypertension, coronary heart disease is one of the very significant complications. Beta-blockades are used widely in patients with effort angina. New beta-blockers with direct vasodilating action or alpha-1 blocking action is reported to be very effective in these angina patients. Recently, it has been reported that coronary vasospasms relates not only to the variant angina, but also in part to effort angina, and that beta-adrenoceptor blockers may induce coronary vasospasms by vascular beta-2 adrenoceptor antagonism and/or alpha-adrenoceptor stimulation (Yasue *et al.*,1979).

Nipradilol is effective in 70% of the angina patients and is more effective than propranolol (Kato *et al.*, 1986). Nipradilol seems to augment the antianginal effect through the reduction of preload and coronary artery dilation by its direct nitroglycerin-like vasodilating action. Therefore, the agent is effective even in the case of coronary spastic angina. In double-blind clinical trials of celiprolol, it has been shown that celiprolol is a useful antianginal drug for which a dosage of 200 to 400mg once a day has usefulness equal to or superior to that of atenolol at 50 to 100 mg once a day (Yui *et al.*, 1990).

Thus, it has suggested that the vasodilating action of beta-adrenoceptor blockades may have prepared many beneficial effects on angina patients, and these new beta-blockers seem to be very effective in those hypertension patients with angina pectoris.

In general, it is considered that the intrinsic sympathomimetic activity of beta-adrenoceptor blockers does not relate to the antianginal effect of beta-blockers. However, it was reported that atenolol is more effective in patients with angina pectoris than pindolol, and that beta-blockers with a strong effect on heart rate reduction are more effective. Thus, the beta-blockers with strong intrinsic sympathomimetic activity have not been used in patients with angina pectoris, although beta-adrenoceptor blockers with beta-2 selective intrinsic sympathomimetic activity are useful for angina patients with bradycardia.

Lipid Metabolism

Elevated blood pressure is accompanied by an increased risk of cerebrovascular disease, a risk that is reduced when the blood pressure is lowered. From studies on primary prevention, it has not, however, been possible to conclude that pharmacological treatment of blood pressure by beta-adrenoceptor blockades in a corresponding manner reduces the risk of coronary heart disease (Houston, 1988). Beta-blockers generally have been found to reduce HDL-cholesterol by 5-115% and to increase triglycerides 10-30% and therefore may have long-term adverse effects on cardiovascular disease in spite of the benefit from blood pressure reduction. This has led to an increased interest in concurrent metabolic disturbances and their significance in coronary heart disease in hypertensive patients. However, the worsening effects of beta-blockers on lipid metabolism is different for each beta-blockers, and depends on the degree of intrinsic sympathomimetic activity, beta-1 selectivity and alpha-1 blocking action.

In nipradilol administration for 16 weeks, total cholesterol, HDL-cholesterol and triglyceride levels does not induce any significant changes (Kajiyama *et al.*, 1985). These results are very different from propranolol, a comparative drug in the double blind study, which induces an increase of triglyceride and decrease of HDL-cholesterol. This is explained by the alpha-1 blocking effect of nipradilol. Carvedilol, which has alpha-1 blocking action, also reveals no significant change of plasma triglyceride, total cholesterol, nor HDL-cholesterol (Seguchi *et al.*, 1990). Dilevalol does not induce a significant change of total cholesterol nor increase the HDL-cholesterol (Materson *et al.*, 1989). Amosulalol decreases triglyceride, but does not show a change of triglyceride nor HDL-cholesterol (Ohta *et al.*, 1986). Celiprolol improves lipid status by significantly decreasing triglyceride and increasing HDL-cholesterol levels (Fogari *et al.*, 1989; Hitzenberger, 1985).

Thus, these new beta-blockers have no adverse effects on lipid metabolism, and at times show a slightly favorable effect on lipid metabolism. These effects of new beta-blockers on lipid metabolism can be explained by the intrinsic sympathomimetic activity and/or alpha-1 blocking action of these new beta-blockers.

CONCLUSION

Numerous beta-adrenoceptor blocking agents have been developed and used as antihypertensive or antianginal drugs. However, most of the clinically used beta-adrenoceptor antagonists possess side effects, such as transient elevation of peripheral vascular resistance, disturbance of lipid or glucose metabolism and increase of left ventricular endo-diastolic pressure. In order to eliminate or lessen these adverse effects, beta-adrenoceptor antagonist with intrinsic sympathomimetic action, cardioselectivity, alpha-adrenoceptor antagonist action or direct vasodilating action have been developed. These new beta-blockers, possessing vasodilating action, may be more effective in hypertension and coronary heart disease treatment, because of their superior effects.

REFERENCES

Abe, H., Kaneyo, Y., Omae, T., Kokubu, T., Iimura, O., Yoshinaga, K., Ishii, M., Inagaki, Y., Mizuno, Y., Arakawa, K., Yoneda, S. and Ebihara, A. (1988). Clinical evaluation of celiprolol hydrochlotide (NBP-582) on essential hypertension. Double blind group comparative study vs propranolol hydrochloride. *Rynsho-Iyaku* **5,** 2321-2365 (in Japanese).

Aellig, W.H. (1985). Pharmacological experiments in healthy volunteers with bopindolol, a long-acting beta-adrenoceptor blocking drug with partial agonist activity. *Br. J. Clin. Pharmacol.* **19,** 775-781.

Baum , T., Watkins, R. W., Sybertz, E. J., Vemulapalli, S., Pula, K. K., Eynon, E., Nelson, S., Vliet, G. V., Glennon, J. and Moran, R. M. (1981). Antihypertensive and hemodynamic actions of SCH 19927, the R,R-isomer of labetalol. *J. Pharmacol. Exp. Ther.* **218,** 444-452.

Berthold, R., Waite, R. and Weber, H. (1981). Pharmacological studies with bopindolol: A new long acting beta-adrenoceptor antagonist. *Br. J. Pharmacol.* **74,** 829P.

Capone, P. and Mayol, R. (1986). A placebo-controlled double-blind multi-center study of celiprolol in the treatment of mild and moderate hypertension. *J. Cardiovasc. Pharmacol.* **8** (suppl. 4), s19-s21.

Fizgerald, J. D. (1969). Perspectives in adrenergic beta-receptor blockade. *Clin. Pharmacol. Ther.* **10,** 292-306.

Fogari, R., Zoppi, A., Pasotti, C., Poletti, L., Tettamanti, F., Malamani, G. and Corradi, L. (1989). Plasma lipids during chronic antihypertensive therapy with different beta-blockers. *J. Cardiovasuc. Pharmacol.* **14** (Suppl. 7), s28-s32.

Hara ,Y. (1978). Synthesis and beta-adrenergic blocking action of a new thiazolyl-thiopanolamine derivative. *J. Pharmceut. Sci.* **67,** 1334-1335.

Hitzenberger, G. (1985). The effects of the beta-adrenoceptor blocking agent, celiprolol, on blood lipids. *Br. J. Clin. Proc.* **39** (suppl. 40), 85-87.

Hohki, N. and Kimihira, H. (1982). Postsynaptic alpha adrenoceptor blocking action of beta blockade, bunitrolol. *Igaku To Yakugaku* **8,** 1781-1788 (in Japanese).

Houston, M.C. (1988): The effects of antihypertensive drugs on glucose intolerance in hypertensive nondiabetics and diabetics. *Am. Heart. J.* **115,** 640-656.

Ikeda, M., Masuayma, Y., Yoshinaga, K., Inagaki, Y., Isii, M., Kajiwara, T., Kuramoto, K., Saruta, T., Ogihara, T., Kubo, S., Nishio, I. and Kuramoti, M. (1985). Clinical evaluation of amosulalol in essential hypertension. *Igaku no Ayumi* **139,** 53-81 (in Japanese).

Kajiyama, G., Horiuti, I., Kawamoto, T., Ohki, M., Hino, F., Fujiyama, M., Yuasa, A., Tsumaru, K., Kubota, S., Maruhashi, A., Yamada, K., Fukuhara, A., Takizawa, A., Nakagawa, K., Mizuno, T., Oyamada, K., Nakao, S. and Tanaka, T. (1985). The effects of long-term adminstration of K-351 (Nipradilol) on serum lipids and blood pressure in patients with essential hypertention. *Ther. Res.* **3,** 847-862 (in Japanese) (Engl. Abst.).

Kaneko, Y., Isii, M., Iimura, O., Yoshinaga, K., Inagaki, Y., Kajiwara, T., Yamada, K., Omae, T., Abe, H., Takeda, T., Masuyama, Y., Kokubu, T., Arakawa, K., Fujishima, M. and Ebihara, A. (1988). Double-blind study of dilevalol versus labetalol in monotherapy and in combination therapy with thiazide diuretics in patients with essential hypertention. *Saishin-Igaku* **43,** 607-647 (in Japanese).

Katano, Y., Takeda, K. and Nakagawa, Y. (1978). Alpha- and beta-adrenoceptor blocking action and the effects on coronary circulation, cardiac function, myocardial energy metabolism by labetalol. *Jpn. J. Pharmacol.* **74,** 819-832 (in Japanese) (Engl Abst).

Kato, K., Kishida, H., Odashima, H., Osada, H., Inoue, K., Sugai, J., Kobayashi, T., Ikeda, N., Kinoshita, M., Hiramori, K. and Yoshida, H. (1986). Clinical effect and safety on long term treatment of nipradilol (K-351) in angina pectoris. *Ther. Res.* **5,** 415-426.

Kou, K. and Suzuki, H. (1969). Effects of 3,4-dihydro-8-(2-hydroxy-3-isopropyl-aminopropoxy)-3-nitroxy-2H-1-benzopyran (K-351) on smooth muscle cells and neuromuscular transmission in the canine mesenteric artery. *Br. J. Pharmacol.* **77,** 679-689.

Leren, P. (1987). Comparison of effects on lipid metabolism of antihypertensive drugs with alpha and beta adrenergic antagonist profile. *Am. J. Med.* **82** (suppl.1A), 31-36.

Materson, B. J., Vlachakis, N. D., Glasser, S. P., Lucas, C., Ramanathan, K. B., Ahmad, S., Morledge, J. H., Saunders, E., Lutz, L. J., Schnaper. H. W., Maxwell. M. and Poland. M. P (1989). Influence of beta-2 agonism and beta-1 and beta-2 antagonism on adverse effects and plasma lipoproteins: Results of a multicenter comparison of dilevalol and metoprolol. *Am. J. Cardiol.* **63,** 58I-63I.

Nahorski. S.R. (1979). Identification and significance of beta-adrenoceptor subtypes. *Am. Heart J.* **97,** 663-670.

Nakagawa, Y., Sugai, T., Chin, W.R., Shibuya, T., Hashimoto, K. and Imai, S. (1984). Pharmacological profile of a new beta-adrenoceptor blocker, 4,(3(tert butylamino)-2-hydroxy-propoxy)-N metylisocarbostyril hydrochloride (N-696). *Arzneim-Forsch/Drug Res.* **34,** 194-199.

Nakagawa, Y., Nakahara, H., Chin, W.P. and Imai, S. (1985). Alpha blockades and vasodilation induced by nipradilol, arotinolol and labetalol in pithed rats. *Jpn. J. Pharmacol.* **39,** 481-485.

Ohta, K., Hikuma, K., Asahina, S., Nagasawa, T. and Hiroto, M. (1986). The effects of alpha- and beta-blockade, amosulalol on hemodynamics, renal function, lipid metabolism and renin-angiotensin-aldosterone system in essential hypertensives. *Kiso To Rynsho* **20**:6028-6034 (in Japanese)

Pollare, T., Lithell, H., Selinus, I. and Berne, C. (1989). Sensitivity to insulin during treatment with atenolol and metoprolol: a randomized, double-blind study of effects on carbohydrate and lipoprein metabolism in hypertensive patients. *Br. Med. J.* **298,** 1152-1157.

Prichard, B.N.C. and Gillam, P.M.S. (1964). The use of propranolol in the treatment of hypertension. *Br. Med. J.* **2,** 725-727.

Prichard, B.N.C. (1978). Beta-adrenergic receptor blockade in hypertension, past, present and future. *Br. J. Clin. Pharmacol.* **5,** 379-399.

Report of a WHO expert committee. Arterial hypertension. In*WHO technical report series 35.* WHO, Geneva, 1978.

Richards, D.A. and Prichard, B.N.C. (1979). Clinical pharmacology of labetalol. *Br. J. Clin. Pharmacol.* **8** (suppl. 2), 89s-93s.

Ruffolo, Jr. R.R., Gellai, M., Hieble, J. P., Willette, R. N. and Nichols, A. J. (1990). The pharmacology of carvedilol. *Eur. J. Clin. Pharmacol.* **38,** s82-s88.

Sasaki, T., Sekino, M., Munakata, K., Kishida, K., Hayakawa, H., Satoh, K. and Yonezawa, K. (1987). Clinical evaluation of new beta-blocker, N-696 (first phase clinical trial). *Jpn. J. Clin. Pharmacol. Ther.* **18,** 497 (in Japanese).

Seguchi, H., Nakamura, H., Aosaki, N., Mikami, Y. and Takahashi, S. (1990). Effects of carvedilol on serum lipids in hypertensive and normotensive subjects. *Eur. J. Clin. Pharmacol.* **38,** s139-s142.

Simom, G. (1976). Altered venous function in hypertensive rats. *Cir. Res.* **38,** 412-418.

Sponer, G., Strein, K., Muller-Beckman, B. and Bartsh, W. (1987). Studies on the mode of vasodilating action of carvedilol. *J. Cardiovasc. Pharmacol.* **10** (Suppl. 11), 42-48.

Sundberg, S., Tiihonen, K. and Gordin, A. (1987). Vasodilatory effects of carvedilol and pindolol. *J. Cardiovasc. Pharmacol.* **10** (Suppl. 11), 76-80.

Takenaka. T.(1982). Antihypertensive and adrenoceptor blocking properties of new sulfonamide substituted phenylethylamines. *Clin. Exp. Hypertens.* **A4,** 125-137.

Tarazi, R.C. and Dustan, H.P. (1972). Beta-adrenergic blockade in hypertension. Practical and theoretical implications of long-term hemodynamic variations. *Am. J. Cardiol.* **29,** 633.

The 1988 report of the Joint National Committee on Detection (1988). Evaluation and Treatment of High blood Pressure. *Arch. Intern. Med.* **148,** 1023-1038.

Tsukiyama, H., Otsuka, K. and Higuma, K. (1985). Hemodynamic effects of short term treatment with nipradilol in essential hypertension. *Ther. Res.* **3,** 1121-1129 (in Japanese) (Engl. Abst.).

Uchida, Y., Nakamura, M., Tsuruta, T., and Yoshimura, M. (1987). Nipradilol. In *New Cardiovascular Drugs* (Eds). Scriabine, A., Raven Press, New York, pp. 95-115.

Van Zwieten, P. A., de Jonge. A. and Wilffer. B. (1985). Cardiovascular effects and interaction with adrenoceptors of urapidil. *Arch. Int. Pharmacodyn.* **276,** 180-201.

Walsh, M. E., Hyman, C. and Moronde, R. F. (1969). Venous distensibility in essential hypertension. *Cardiovasc. Res.* **3,** 338-349.

Westerman, R. F., Christensen, C., Westheim, A. and Donker, A.J.M. (1987) Bopindolol *vs* metoprelol in hypertension: Comparison of clinical efficacy and tolerance in a double-blind, parallel study. In: *A new beta-blocker, Royal Society of Medicine Services International Congress and Symposium series No. 106.* (Eds). van Zwieten, P.A., Royal Society of Medicine Services Limited, pp. 75-82.

Winbury, M. M., Howe, B. B. and Hefner, M. A. (1969). Effect of nitrates and other coronary dilators on large and small coronary vessels. *J. Pharmacol. Exp. Ther.* **168,** 79-94.

Yasue, H., Omote, S., Takizawa, A., Nakano, M., Miwa, K. and Tanaka, S. (1979). Circadian variation of exercise capacity in patients with Prinzmetal's variant angina: Role of exercise induced coronary arterial spasm. *Circulation* **59,** 938.

Yui, Y., Kawai, T., Sugimoto, K., Yasuda, T., Kanazawa, T., Hosoda, S., Fukuzaki, H., Kusukawa, R., Toshima, H., Kawakubo, K. and Nakajima, M. (1990). Clinical evaluation of celiprolol on angina pectoris-Multicenter double-blind comparative study vs atenolol. *Rynsho-Iyaku* **6,** 337-367 (in Japanese).

Progress in Hypertension, Vol. 2, pp. 107-129.
Saito *et al.* (Eds)

CLINICAL ASPECTS OF ORAL ANGIOTENSIN CONVERTING ENZYME INHIBITORS

MICHAEL BURSZTYN* and HARALAMBOS GAVRAS**

*Department of Medicine, Hypertension Unit, Hadassah University Hospital on Mount Scopus, Jerusalem, Israel; and **Section of Hypertension and Atherosclerosis, Boston University School of Medicine, Boston, Massachusetts 02118, U.S.A.

INTRODUCTION

Since Tigerstedt and Bergman showed in the previous century that kidney extracts may elevate blood pressure (Tigerstedt and Bergman, 1898) and after Goldblat's classic experiments (Goldblat *et al.*, 1934) with renal ischemia, activation of the renin angiotensin system was, and still is, one of the most frequently used experimental methods to produce hypertension. The interaction of this system with the sympathetic nervous system (Braun-Menendez *et al.*, 1940), aldosterone (Ames *et al.*, 1964) and vasopressin and the much more recently demonstrated effects of angiotensin II on vascular smooth muscle cell hypertrophy (Geisterfer *et al.*, 1988) and induction of growth factors and oncogenes (Naftilan *et al.*, 1989) indicate that its role is pivotal in the development and maintenance of hypertension. Hence, it has become the focus of much interest for pharmacologic intervention in the treatment of hypertension.

The renin-angiotensin dependency of blood pressure is considerably augmented by sodium depletion both in experimental (Gavras *et al.*, 1975) and human hypertension (Gavras *et al.*, 1976). Our understanding of the role of this system in hypertension and other pathologic conditions was greatly enhanced since the discovery of inhibitors of the angiotensin converting enzyme, which generates angiotensin II, the active effector of the system from its inactive precursor, angiotensin I.

It was from the venom of a Brazilian snake, bothrops jararaca, that Ferreira identified a number of bradykinin potentiating factors (Ferreira, 1965) which were subsequently shown to be inhibitors of the angiotensin converting enzyme (Yang *et al.*, 1970) and used to demonstrate overactivity of the renin angiotensin system (Krieger *et al.*, 1971). Soon teprotide was identified as one of the active components and synthesized (Ondetti *et al.*, 1971). It served to demonstrate the role of the renin angiotensin system in normal cardiovascular homeostasis (Sancho *et al.*, 1976) in essential hypertension (Gavras *et al.*, 1974; Case *et al.*, 1977), its effects on glomerular filtration rate (Hollenberg *et al.*, 1979), its contribution in hypertensive emergencies (Tifft *et al.*, 1979), in congestive heart failure (Gavras *et al.*, 1978b) and in the renal crisis of scleroderma (Lopez-Ovejero *et al.*, 1979). Subsequently an orally active angiotensin converting enzyme inhibitor, captopril, was designed (Ondetti *et al.*, 1977), and was shown to be extremely effective in the treatment of hypertension (Gavras *et al.*, 1978a). Later, other so called "second generation" oral converting enzyme inhibitors, such as enalapril (Gavras *et al.*, 1981) and lisinopril became available. Currently more than a dozen newer converting enzyme inhibitors are in various stages of clinical development.

Different Angiotensin Converting Enzyme Inhibitors
It is beyond the scope of this chapter to provide detailed information on each of the converting enzyme inhibitors. In fact, there are probably no major differences in pharmacodynamics between the various active drugs. Their main difference is the presence of either a sulphydryl group (captopril, alacepril, zofenopril and others), a carboxylic group (enalapril, lisinopril, ramipril, benazepril, cilazapril, perindopril, quinapril and others) or a phosphinylic group (fosenopril and others) for binding to the active zinc-containing moiety of the converting enzyme (Kostis, 1989). They also differ in their potency and kinetics, i.e. absorption in active form (captopril, lisinopril) or as a prodrug that need to be activated in vivo (enalapril, ramipril, cilazapril, benazepril and others) or their duration of action (longer half life of enalapril, benazepril, lisinopril, ramipril and others), permitting once daily dosing. But in practice these differences are minimized by adjusting the doses and dosing intervals.

PHARMACODYNAMICS

Mechanism of Action
Angiotensin converting enzyme (ACE) inhibitors interfere with the renin angiotensin system by reducing formation of angiotensin II and, subsequently, aldosterone levels. However, it soon became apparent that blood pressure response is unrelated to baseline plasma renin activity, because ACE inhibition is effective in many patients with low renin (Gavras *et al.*, 1978a; Veterans Administration Cooperative Study Group on Antihypertensive Agents, 1984). Moreover, it was shown that ACE inhibition is more effective than competitive angiotensin II blockade and that actually ACE inhibitors have antihypertensive actions even in patients without circulating renin, such as anephric subjects (Man in't Veld *et al.*, 1980; Wenting *et al.*, 1987), or patients submitted to prior renin inhibition (Bursztyn *et al.*, 1990).

These observations led to suggestions of three possible additional mechanisms of action: (1) Effects on bradykinin, which is degraded by the kininase II, shown to be identical to ACE (Yang *et al.*, 1970). Indeed, aprotinin, which blocks bradykinin production, attenuated the antihypertensive effects of captopril (Mimran *et al.*, 1980). A partial reversal of the blood pressure lowering induced by ACE inhibition was obtained also with antibradykinin antibodies (Carretero *et al.*, 1981) and bradykinin antagonists (Benetos *et al.*, 1986). (2) Increase in vasodilatory prostaglandins (Swartz *et al.*, 1980; Moore *et al.*, 1981), which might also be a reflection of bradykinin's effect (McGiff *et al.*, 1972). 3) Identification of various components of the renin-angiotensin system in the vessel wall suggests that some effects occur independently of the circulating pool (Ganten *et al.*, 1970; Thurston *et al.*, 1979; Kowaloff *et al.*, 1980; Dzau, 1989). Indeed, angiotensin II was shown to be generated in perfused peripheral arteries (Mizuno *et al.*, 1988) and suppressed by ACE inhibition (Hilgers *et al.*, 1989).

Hemodynamics

Experiments with the angiotensin II receptor antagonist saralasin demonstrated that blockade of angiotensin II induced vasodilation preferentially in vital organs such as the kidney, heart and brain, whose vasculature was far more sensitive to angiotensin II (Liang *et al.*, 1978a; b). Redistribution of regional blood flow in favor of vital organs was reproduced with angiotensin converting enzyme inhibition (Gavras *et al.*, 1978c; Liang and Gavras, 1978) and was also confirmed in congestive heart failure (Watkins *et al.*, 1976). This indicated that, even in the face of falling blood pressure, vital organ perfusion could be not only maintained but even augmented. Moreover, because of angiotensin II elimination, the secondary hyperaldosteronism, so characteristic of diuretics and non-specific vasodilators, was abolished.

The ability to preferentially affect "angiotensin II sensitive organs" allowed reversal of experimentally induced cerebral vasospasm without deleterious systemic vasodilation (Gavras *et al.*, 1981; Andrews *et al.*, 1982). Lowering of blood pressure through converting enzyme inhibition allows resetting of cerebral autoregulation in normotensive and hypertensive man (Waldmar *et al.*, 1989). This principle is probably even better demonstrated in patients with congestive heart failure and relative hypotension whose cerebral circulation is improved after angiotensin converting enzyme inhibition (Paulson *et al.*, 1986). Similarly, coronary blood flow in normal man is improved (Faxon *et al.*, 1982) as it is in patients with dilated cardiomyopathy (Foult *et al.*, 1988) and renal blood flow is improved with maintenance of glomerular filtration rate despite the fall in systemic blood pressure (Dunn *et al.*, 1984).

One of the most attractive aspects of angiotensin converting enzyme inhibition is its demonstrated ability to improve vascular compliance (Creager *et al.*, 1984; Asmar *et al.*, 1988). Since this determines the pulse pressure which, in turn, is an independent determinant of left ventricular hypertrophy in the population (Darne *et al.*, 1989) it should not be surprising that left ventricular hypertrophy can be reduced with these agents (Dunn *et al.*, 1984; Nakishima *et al.*, 1989; Asmar *et al.*, 1988; Eichenstaedt *et al.*, 1989). The effect on the compliance of the large arteries and on left ventricular hypertrophy could well explain the differential effects of various antihypertensive medications in this respect (Safar, 1989; Sen and Tarazi, 1983).

A recently clinically proven benefit of angiotensin converting enzyme inhibition is the prevention of the global remodelling of the heart after anterior myocardial infarction, where end diastolic volume index deterioration and the reduction of the ejection fraction have been arrested or even improved over one year of treatment (Sharpe *et al.*, 1988; Pfeffer *et al.*, 1988).

CLINICAL EXPERIENCE IN THE TREATMENT OF HYPERTENSION

Initially, when angiotensin converting enzyme inhibitors were introduced into clinical practice, they were aimed at patients with refractory, renovascular and complicated (scleroderma, renal and heart failure patients) hypertension. However, it soon became apparent that treatment of hypertension of all stages is not only feasible (Williams, 1988) but also very well tolerated (Croog *et al.*, 1986; Gavras and Gavras, 1988) and, indeed ACE inhibitors are now recommended as first step therapy together with diuretics, beta blockers and calcium antagonists (JNC, 1988). As monotherapy, they can achieve a good blood pressure response in up to 70 per cent of unselected patients (Sassano *et al.*, 1987; Bolzano *et al.*, 1987; Morlin *et al.*, 1987; Pool *et al.*, 1987; Lessem *et al.*, 1985; Weinberger, 1983; Corea *et al.*, 1984; Lees *et al.*, 1989; Gavras, 1988; Vidt, 1984; Veteran Administration Cooperative Study Group on Antihypertensive Agents, 1984; Kogler, 1989).

Part of the variability in blood pressure response may be related to different patient selection but, probably most of all, different sodium balances, which is a critical determinant in the contribution of the renin angiotensin system to cardiovascular homeostasis (Gavras *et al.*, 1975; 1976; Sancho *et al.*, 1976; Shoback *et al.*, 1983). In any event, angiotensin converting enzyme inhibitors as monotherapy were found to compare favorably in terms of effectiveness with thiazide diuretics (Sassano *et al.*, 1987; Pool *et al.*, 1987; Weinberger, 1983; Vidt, 1984) and beta blockers (Bolzano *et al.*, 1987; Veteran Administration Cooperative Study Group on Antihypertensive Agents, 1984; Helgland *et al.*, 1986; van Shaick *et al.*, 1986; Enalapril in Hypertension Study Group, 1984; Zacharia *et al.*, 1987; Belz *et al.*, 1989; Herrick *et al.*, 1989). These studies confirm that, despite the different pharmacokinetics and potency, the pharmacodynamics of different angiotensin converting enzyme inhibitors are remarkably similar when the abovementioned differences are minimized by adjusting the doses and dosing intervals. Comparative studies of the newer agents showed similar antihypertensive effects when captopril was compared to enalapril (Vlasses *et al.*, 1986; Chrysant *et al.*, 1985; Lewis *et al.*, 1985), spirapril to captopril (Weber *et al.*, 1986), perindopril and ramipril to captopril (Lees *et al.*, 1989; Witte and Walter, 1987) or ramipril to enalapril (Zabludowski *et al.*, 1988).

In combination therapy, thiazides demonstrate excellent synergism with angiotensin converting enzyme inhibitors, as about only five per cent of mild to moderate hypertensive patients are unresponsive to the combination (Sassano *et al.*, 1987; Pool *et al.*, 1987; Vidt, 1984; Veteran Administration Cooperative Study Group on Antihypertensive Agents, 1984; Captopril Research Group of Japan, 1985; Atkinson *et al.*, 1980). When using such a combination, quite low doses of hydrochlorothiazide are sufficient (Dahlof *et al.*, 1988). Moreover, some of the undesired metabolic side-effects of diuretics (e.g. hypokalemia, hyperglycemia, hypercholesterolemia and hyperuricemia) have been blunted (Weinberger, 1983), which makes the combination even more attractive.

Combination with calcium channel blockers has also been gratifying (Guazzi *et al.*, 1984; White *et al.*, 1984; Morgan *et al.*, 1988; Eggersten *et al.*, 1989; Singer *et al.*, 1987; McGregor *et al.*, 1985; Brouwer *et al.*, 1985). Indeed, captopril may attenuate some of the tachycardia and ankle edema produced by nifedipine (Guazzi *et al.*, 1984). More controversial is the combination with beta-blockers, where there have been quite disparate results: in some cases beta blockers and angiotensin converting enzyme inhibitors had synergistic action (Belz *et al.*, 1989; Muisen *et al.*, 1987), while in others there was no effect (McGregor *et al.*, 1982; Bursztyn *et al.*, 1990) or very marginal effect (McGregor *et al.*, 1985; Staessen *et al.*, 1985). In any case, it seems advisable to add a diuretic rather than a beta blockers to converting enzyme inhibitors (McGregor *et al.*, 1982; 1985; Captopril Study Group of Japan, 1985). Indeed, since the dose response of angiotensin converting enzyme inhibitors reaches a plateau (Davies *et al.*, 1984), it makes more sense again to add a diuretic (Sassano *et al.*, 1989; Salvetti and Arzilli, 1989) than push the dose upwards unnecessarily (Guyene *et al.*, 1989). Often the early response may predict long term response (Waeber *et al.*, 1981) (Donnelly *et al.*, 1990).

This last report recently suggested that the relationship of angiotensin converting enzyme inhibition and blood pressure response in the individual patients is best expressed with the use of kinetic-dynamic modelling. This might explain some previously unexplained observations, such as the discrepancy in duration of ACE suppression versus duration of antihypertensive effect. (Waeber *et al.*, 1980; Brunner *et al.*, 1988). It also reconfirmed that neither age nor baseline plasma renin activity predict blood pressure response (Donnelly *et al.*, 1990; Bidiville *et al.*, 1988).

EFFECTS ON SPECIFIC PATIENT GROUPS

The Elderly

Contrary to earlier claims that angiotensin converting enzyme inhibitors are not effective in the elderly, low-renin population, it has now been demonstrated that angiotensin converting enzyme inhibitors are quite effective and safe in the elderly (Jenkins *et al.*, 1985; Woo *et al.*, 1987; Forrette *et al.*, 1986; Mulinari *et al.*, 1987; Reid, 1988; Corea *et al.*, 1984; Cox *et al.*, 1989) and comparable with diuretics, the "traditional" antihypertensive agent for the elderly. In fact, the general tolerability and well-being conferred by converting enzyme inhibitors (Croog *et al.*, 1986), the paucity of side-effects, the absence of propensity towards orthostatic hypotension, favorable effects on cerebrovascular, coronary and renal blood flow, together with absence of contra-indication in coexisting conditions common in the elderly such as gout, hyperglycemia, chronic lung disease, etc., render ACE inhibition an excellent first choice treatment for this particularly complicated and ever increasing segment of the hypertensive population (Bursztyn *et al.*, 1989).

Diabetes Mellitus

In this common condition, beta-blockers and diuretics are relatively contraindicated in insulin dependent and non-insulin dependent diabetes mellitus, respectively, because of possible masking of hypoglycemic symptoms in the former, and worsening of hyperglycemia in the latter condition. Actually, it has been shown recently that glucose disposition, i.e. insulin sensitivity, is better with captopril than either beta-blockers (Pollare *et al.*, 1989a) or hydrochlorothiazide (Pollare *et al.*, 1989b), and that the improved glucose disposition in diabetics could be related to the effects on bradykinin (Dietze *et al.*, 1987; Jauch *et al.*, 1987; Rett *et al.*, 1988). This is of great importance since so many hypertensives are glucose intolerant (Modan *et al.*, 1985) and hyper-insulimenia and insulin resistance are often associated with essential hypertension (Modan *et al.*, 1985; Ferannini *et al.*, 1987; Pollare *et al.*, 1990).

Predisposition to hypertension is a major determinant in the development of diabetic nephropathy (Krolewski *et al.*, 1988) as well as one of its earliest manifestations (together with microalbuminuria) (Parving *et al.*, 1981). It has been suggested that an angiotensin II dependent increase in intraglomerular pressure (Hostetter *et al.*, 1982) is the culprit in this condition. Indeed, massive proteinuria can be reduced (Taguma *et al.*, 1985; Valvo *et al.*, 1988) with converting enzyme inhibitors in both hypertensive and normotensive patients with either type I or type II diabetes and renal insufficiency (Bjorck *et al.*, 1986; Parving *et al.*, 1988; 1989; Hommel *et al.*, 1986; Marre *et al.*, 1988; Valvo *et al.*, 1988; Romanelli *et al.*, 1989; Stronello *et al.*, 1989). We also have observed a reduction of proteinuria with improvement in creatinine clearance and phosphate clearance in a homogeneous group of 15 elderly overweight hypertensives with type II diabetes mellitus and chronic renal failure after cilazapril treatment (Bursztyn *et al.*, unpublished data).

Chronic Renal Failure

Ample experimental and clinical evidence has demonstrated the benefits of angiotensin converting enzyme inhibition in chronic renal failure (Keane *et al.*, 1989). Indeed, while several years ago it was questioned if angiotensin converting enzyme inhibition is safe in patients with chronic renal failure (Borniade, 1983), these fears were not borne out (Captopril Collaborative Study Group, 1982). More recently, an authoritative panel of nephrologists actually recommended angiotensin converting enzyme inhibition as a means of arresting the progressive nature of chronic renal failure (Keane *et al.*, 1989). These recommendations are based on numerous studies showing absence of deterioration of renal function, despite a marked blood pressure reduction and, in some selected groups, even an improvement in renal function was seen (Parving *et al.*, 1989; Bjork *et al.*, 1986; Hommel *et al.*, 1986; Reams and Bauer, 1986; Bauer and Reams, 1987; Reisch *et al.*, 1987; Abrahams *et al.*, 1988; Heeg *et al.*, 1987). This was shown in chronic renal failure of diverse etiologies ranging from diabetic nephropathy to systemic lupus erythematosus (Herlitz *et al.*, 1984) but the majority of patients in most of the abovementioned studies had renal failure of unspecified origin, which is also most often the case in clinical practice.

The beneficial effects of angiotensin converting enzyme inhibitors could be non-specific in part, namely, related to improved hypertension control analogous to that obtained with minoxidil (Pettinger *et al.*, 1989) or to specific effects. The specific effects of angiotensin converting enzyme inhibitors are related to the perceived role of increased intraglomerular pressure in the progression of renal disease (Hostetter *et al.*, 1982). It is angiotensin II, through its preferential effect on the efferent glomerular arterioles, which maintains the intraglomerular pressure. Abolishing this effect thus decreases the intraglomerular pressure, increases renal blood flow and, consequently, maintains glomerular filtration while the filtration fraction decreases with possible reduction in long term hyperfiltration, the perceived culprit in the progression of renal disease (Hostetter *et al.*, 1982; Keane *et al.*, 1989). Converting enzyme inhibition had been useful also in hemodialysis patients not controlled by dialysis itself. An additional benefit in dialysis patients could be diminished thirst and interdialysis weight gain reduction (Oldenberg *et al.*, 1988), possibly because of the dependency of blood pressure on renin and norepinephrine under these circumstances (Textor *et al.*, 1981).

In rare cases, captopril was reported to be the cause of renal insufficiency due to interstitial nephritis or membranous nephropathy (Hoorntje *et al.*, 1980; Textor *et al.*, 1983). These cases have been attributed to idiosyncracy to this particular ACE inhibitor (i.e. captopril), rather than a "class effect" of ACE inhibition in general. This distinction is of practical importance as class effects are expected to recur with other agents of the same class, whereas idiosyncratic reactions are associated with the molecular structure of a particular drug, but are unrelated to its pharmacodynamics.

Renovascular Hypertension

Functional renal artery stenosis is characterized by hyperreninemia, hyperangiotensinemia and hyperalderonism (Vaughan, 1985) and thus is particularly amenable to angiotensin converting enzyme inhibition. As a matter of fact, in some patients with renovascular hypertension the responsiveness of blood pressure to this treatment is such that very small doses are sufficient and, indeed, episodes of abrupt hypotension to the first dose have been described (Hodssman *et al.*, 1983), especially when patients are con-comitantly salt-depleted.

The sensitivity of hypertension to converting enzyme inhibition in patients with renal artery stenosis has been used for the prediction of surgical outcome (Atkinson *et al.*, 1982; Staessen *et al.*, 1988) and for enhancement of lateralization of renin output during renal vein sampling for the assessment of the functional importance of the renal artery stenosis (Re *et al.*, 1978; Thibonnier *et al.*, 1984). However, as mentioned earlier in patients with renal artery stenosis in a solitary kidney or bilateral renal artery stenoses, angiotensin converting enzyme inhibition may result in rapid deterioration of renal function (Hricik *et al.*, 1983; Curtis *et al.*, 1983; Ying *et al.*, 1984). Nevertheless, these conditions are rather rare and renal failure is readily reversible upon discontinuation of inhibitors and saline infusions (Hricik, 1985). Actually, in the absence of sodium depletion, even patients with bilateral renal artery stenoses or a stenosis to a solitary kidney may not experience renal failure (Bursztyn *et al.*, 1984), especially when azotemia is not present (Hollenberg, 1983).

The long term consequences of angiotensin converting enzyme inhibition in renovascular hypertension have not been determined yet. There is some evidence that there may be unilateral reduction in glomerular filtration rate, compensated by the improved hemodynamics of the other kidney and the reduction of systemic hypertension (Pederson *et al.*, 1989). There are some reports of renal artery occlusion in renovascular hypertensive patients treated with captopril (Hoefnagels and Thien, 1986), but this could relate to the natural history of the disease. Indeed, it was reported that 39 per cent of subjects with severe renal artery stenosis developed total occlusion within a mean follow-up period of 13 months (Schreiber *et al.*, 1984).

Congestive Heart Failure

The first demonstration of benefit by angiotensin inhibition in a patient with congestive heart failure was obtained with the use of the angiotensin II competitive antagonist, saralasin. Cardiac output, left ventricular stroke volume index and coronary blood flow were shown to increase, while pulmonary wedge pressure, left ventricular end-diastolic pressure, heart rate and pulmonary and systemic vascular resistance decreased (Gavras *et al.*, 1977). These observations were soon confirmed with angiotensin converting enzyme inhibitors (Gavras *et al.*, 1978b). Such studies were repeated and amplified by others and all confirmed the improved systemic and regional hemodynamics and enhanced exercise tolerance, as well as correction of metabolic abnormalities (such as hyponatremia and diuretic-induced hypokalemia and in some cases reduction of ventricular arrhythmias (Webster *et al.*, 1985). Similar results have been observed with all of the newer converting enzyme inhibitors tested to date (Cleland *et al.*, 1985; Insel *et al.*, 1989; Captopril-Digoxin Multicenter Research Group, 1988; Giles *et al.*, 1989; Drexler *et al.*, 1989).

Most importantly, it was shown for the first time that mortality in severe heart failure may be significantly reduced (The CONSENSUS Trial Study Group, 1987). Since it is not uncommon to have hypertension combined with heart failure (especially in the elderly), such patients could derive much benefit from receiving one medication for both conditions. This might also decrease the amount of drugs prescribed to some of these subjects as polypharmacy is one of the main sources of adverse events in such patients (Montamat *et al.*, 1989).

Coronary Artery Disease

Coronary insufficiency is the inability of stenosed coronary arteries to supply the myocardium with increased oxygen upon demand. Unlike many arteriolar dilators which, while reducing systemic vascular resistance, increase coronary tone and/or heart rate, angiotensin converting enzyme inhibitors reduce arterial and venous tone and, thus, combine increase of coronary blood flow with absence of compensatory tachycardia (Creager *et al.*, 1984; Faxon *et al.*, 1982). Indeed, in patients with stable angina pectoris captopril was shown to decrease blood pressure without affecting heart rate or transmyocardial norepinephrine balance, while decreasing myocardial oxygen consumption and severity of angina over time (Daly *et al.*, 1985). In a different group of patients these authors found that captopril significantly prolonged exercise time to a similar double product. In another study captopril was found to improve double product and decrease ischemic S-T changes in comparison to alpha-methyldopa, despite a significant and similar reduction of blood pressure in hypertensives with angiographically proven coronary artery disease and stable angina pectoris (Strozzi *et al.*, 1985).

Moreover, the capacity to regress left ventricular hypertrophy (Dunn *et al.*, 1984; Nakashima *et al.*, 1984; Asmar *et al.*, 1988), a major determinant of mortality from coronary artery disease in the Framingham heart study, should be of great benefit to the hypertensive patient. Whether angiotensin converting enzyme inhibitors may prevent left ventricular hypertrophy or death from coronary artery disease remains to be shown in prospective trials. However, once a myocardial infarction was sustained, angiotensin converting enzyme inhibition was shown to prevent or retard the inexorable progression of left ventricular dilation to cardiomyopathy (Sharpe *et al.*, 1988; Pfeffer *et al.*, 1988).

Cerebrovascular Insufficiency

As presented in the section on hemodynamics, the cerebrovascular circulation is sensitive to angiotensin converting enzyme inhibition (Gavras *et al.*, 1978; Gavras *et al.*, 1981; Andrews *et al.*, 1982). These experimental findings were confirmed in patients with essential hypertension, where angiotensin converting enzyme inhibition resulted in increased diameter and a fall in resistance of the common carotid circulation (Bouthier *et al.*, 1986). The benefits of this resetting in the autoregulation of cerebral blood flow have now been demonstrated in normotensives, hypertensives (Waldemar *et al.*, 1989) and in patients with congestive heart failure (Paulson *et al.*, 1986).

Chronic Obstructive Pulmonary Disease
Chronic obstructive pulmonary disease is common, especially in the older population, and thus can frequently coexist with hypertension. This sometimes poses management problems since sympatholytic antihypertensive agents may precipitate bronchospasm.

In our extensive experience, just removing these agents could help. Indeed, chronic angiotensin converting enzyme inhibition is well tolerated by the majority of patients with chronic obstructive pulmonary diseases (Bertoli *et al.*, 1986; Oliver *et al.*, 1989). Moreover, hypoxia-induced pulmonary vasoconstriction may be blunted (Borschetti *et al.*, 1985). However, one of the side effects of angiotensin converting enzyme inhibitors is a dry non productive cough (Chalmers *et al.*, 1987; Strozzi *et al.*, 1989). It has been suggested that this cough could be due to bronchial hyperreactivity (Kaufman *et al.*, 1989; Selena, 1986). More recently it was shown that it is related to sensitization of vagal C fibers, probably through bradykinin accumulation (McEwan *et al.*, 1989). Thus, at times ACE inhibition may exacerbate chronic cough from pre-existing stimuli and rarely it can aggravate bronchial asthma.

Peripheral Vascular Disease
The coexistence of chronic peripheral vascular disease with hypertension is not uncommon as atherosclerosis accompanies both. Other antihypertensive agents, such as beta-blockers can aggravate symptoms of chronic vascular insufficiency or predispose vulnerable individuals to peripheral vasospasm and Raynaud's phenomenon. Angiotensin converting enzyme inhibition tends to improve this symptomatology (Catalano and Libertti, 1985; Truberstein *et al.*, 1984). Although musculoskeletal vascular beds are less sensitive to the effects of angiotensin converting enzyme inhibition (Gavras *et al.*, 1978; Creager *et al.*, 1984), they still benefit from the improved systemic hemodynamics (Drexler *et al.*, 1989), decreased sympathetic tone and, last but not least, decreased sensitivity to the vasoconstrictor effects of norepinephrine (Uhlinger *et al.*, 1989).

Athritis
Patients with arthritis are frequently treated with nonsteroidal antiinflammatory drugs. Though there have been some claims to be contrary (Gomes *et al.*, 1985), these agents were shown repeatedly to interfere with the antihypertensive actions of captopril (Moore *et al.*, 1981; Salvetti *et al.*, 1982), enalapril (Salvetti *et al.*, 1987) and lisinopril (Abdel-Haq *et al.*, 1989). Captopril may have some immunosuppressive activities (Johnson and Aurell, 1981), perhaps related to its sulphydryl group and structural similarity to penicillamine (which may also explain some of their common side-effects). Indeed, captopril was shown to benefit some patients with rheumatoid arthritis (Martin *et al.*, 1989). Such effects could have contributed to the efficacy in some patients with Takayushu's disease treated with

captopril (Grossman *et al.*, 1984).

USE OF ACE INHIBITORS IN DIAGNOSTIC PROCEDURES

Renal vein sampling for renin is needed at times to assess the functional significance of renal artery stenosis. It was first shown with teprotide that lateralization can be improved (Re *et al.*, 1978) and this was soon confirmed with captopril (Thibonnier *et al.*, 1982; 1984) and enalapril (Tunny *et al.*, 1988). Moreover, it was suggested in a retrospective series, that plasma renin activity increments after captopril could identify patients with renovascular hypertension (Muller *et al.*, 1986). A recently published prospective series identified plasma renin activity under converting enzyme inhibition as the most useful method of screening for renovascular hypertension (Svetsky *et al.*, 1989). However, the role of this test is still unclear in clinical practice (Ben-Ishay and Bursztyn, unpublished data; McCarthy and Weder, 1990).

Another new approach to the diagnosis of renovascular hypertension is the ^{99}Tc-diethylenetriaminepentaacetic acid (DTPA) scanning with separate analysis of each kidney before and after captopril (Hommel *et al.*, 1986; Dondi *et al.*, 1989). In this test, the stenotic kidney displays a reduction of glomerular filtration, presumably by the same mechanism angiotensin converting enzyme inhibition causes renal failure in bilateral renal artery stenoses or stenosis to a single functional kidney (Hricik *et al.*, 1983, Curtis *et al.*, 1983). This test could be used also to follow patients after successful (or unsuccessful) corrective procedure without exposing them again to angiography and, perhaps, renal vein sampling.

Since prompt creatinine rise after converting enzyme inhibition may reflect arterial stenosis in a single kidney it has been suggested as a screening test for renovascular hypertension in patients with post-transplant hypertension (Curtis *et al.*, 1985).

On the other extreme of renin-hypertension relationship, it was suggested that primary aldosteronism could be diagnosed from the inability to stimulate plasma renin activity and to reduce plasma aldosterone significantly by angiotensin converting enzyme inhibition (Thibonnier *et al.*, 1982; Luderer e*t al.*, 1982; Lyons *et al.*, 1983).

SIDE-EFFECTS

Side-effects of angiotensin converting enzyme inhibitors can be categorized as either related to their pharmacological action (effects on blood pressure, angiotensin II, bradykinin, etc.) or attributable to their specific chemical structure (such as the sulphydryl moiety of captopril). Class-related effects are predictable and expected to recur with different angiotensin converting enzyme inhibitors, while those related to the molecular structure of each compound can be avoided by switching to a different one.

Class Effects

Marked or severe hypotension is a rare occurrence in salt-replete hypertensives. Mostly it occurs with the first dose in high renin, accelerated hypertension, or in sodium depleted patients with intensive diuretic treatment (Hodsman *et al.*, 1983) and patients with congestive heart failure, especially those with hyponatremia (Cleland *et al.*, 1985). It was reported in 0.34 per cent of 13, 295 hypertensives in a post-marketing surveillance study of captopril (Chalmers *et al.*, 1987). Hyperkalemia is another potentially dangerous side-effect resulting from reduction of angiotensin II-mediated aldosterone secretion. Usually, it should be self-limited because of the angiotensin II-independent hyperkalemic stimulation of aldosterone secretion. However, in patients with advanced renal failure, in elderly diabetics, in patients taking nonsteroidal antiinflammatory drugs and in those taking potassium-sparing diuretics, a potentially dangerous hyperkalemia may develop (Zanella *et al.*, 1985; Ponce *et al.*, 1985; Speirs *et al.*, 1988).

The functional renal failure in patients with bilateral renal stenosis and with a stenosis to a solitary kidney (Hricik *et al.*, 1983; Curtis *et al.*, 1983; Ying *et al.*, 1984) was discussed above. It occurs within hours after angiotensin converting enzyme inhibition in high risk patients (with azotemia and extensive atherosclerotic vascular disease), who should be closely monitored.

Angioedema is a rare occurrence with an overall cumulative incidence is 0.1 per cent (out of 12, 000). It is probably related to effects of accumulated bradykinin and its higher risk period is in the first week. Although it is usually mild and self-limited upon discontinuation of the drug, there have been reports of four deaths from 38 documented cases with enalapril (Slater *et al.*, 1988).

Another peculiar side-effects is the dry persistent cough, mentioned earlier, also believed to be related to accumulation of bradykinin resulting in irritation of vagal C fibers (McEwan *et al.*, 1989; Morris *et al.*, 1989) or increased inflammatory state (Lindgren *et al.*, 1989). There seems to be a controversy whether or not this represents an asthmatic tendency (Boulet *et al.*, 1989). Its frequency was probably underestimated in the past, because its association with ACE inhibition went unrecognized for a long time. It may reach 7.39% with a higher incidence in women (10.5%) versus men (4.42%) (Strozzi *et al.*, 1989; Gibson, 1989) and may be more common in the elderly.

Toxic or Idiosyncractic Side-Effects

The most common side-effect in this category is rash at times with eosinophilia. All ACE inhibitors have been associated with such reactions, though they occurred more frequently with captopril (Williams, 1988). However, with the currently given smaller doses and the realization that side-effects are more common in patients with renal failure (Cooper *et al.*, 1984), these also became rather uncommon. 0.81per cent of 13, 295 patients stopped taking captopril because of rash (Chalmers *et al.*, 1987). At times enalapril replaced captopril with resolution of such side-effects and vice-versa (Gavras and Gavras, 1983; Smith *et al.*, 1983; Jackson *et al.*, 1988). Captopril has rarely been associated with more severe mucocutaneous febrile reactions and, occasionally with nephrotoxicity (Textor *et al.*, 1983; Smith *et al.*, 1989).

Agranulocytosis is another potentially fatal complication of captopril therapy (Gavras *et al.*, 1981; Weizman, 1983; Borniade, 1983). Again, this side-effect is extremely rare with the currently recommended lower doses and their adjustment in renal failure (Cooper *et al.*, 1983). A single controversial case of transient leucopenia was reported in association with enalapril. However, in a relatively large series most patients with serious captopril associated side-effects, such as neutropenia, rash, proteinuria and taste disturbance, did not experience recurrence of these side-effects with enalapril (Rucinska *et al.*, 1989; Jackson *et al.*, 1988).

Taste disturbance, namely, loss of taste or a metallic sensation, has been reported with captopril (Chalmers *et al.*, 1987). This side-effect is less common with the current lower doses and with the "second generation" ACE inhibitors.

A PERSONAL OVERVIEW

Angiotensin converting enzyme inhibitors are extremely well tolerated by the majority of patients, with a sensation of well being, improved social interactions and sexual function when compared to propranolol or alpha-methyl dopa (Croog *et al.*, 1986). Postmarketing surveillance studies with both captopril (Chalmers *et al.*, 1987) and enalapril (Speirs *et al.*, 1988; Inman *et al.*, 1988) (about 13, 000 patients each) have demonstrated an acceptable side-effect profile which is a major determinant of patient compliance with treatment of a life-long asymptomatic disease. One benefit of antihypertensive treatment that still remains to be shown is cardioprotection, which has eluded older blood pressure lowering regimens.

Very recently two laboratory reports suggested that angiotensin converting enzyme inhibition may prevent myointimal proliferation after vascular injury (Powell *et al.*, 1989) and exert antiatherogenic effects in hyperlipidemic rabbits (Chobanian *et al.*, 1990). Thus, for the first time there exists an antihypertensive agent which might favorably affect almost all components of cardiovascular risks and complications. Obviously, the role of angiotensin converting enzyme inhibitors in preventing atherosclerosis remains to be seen as is their efficacy in preventing mortality from coronary artery disease in hypertensive patients. However, we can now treat hypertension effectively without inducing or aggravating metabolic disturbances such as insulin resistance, glucose intolerance, lipid abnormalities, hypokalemia and hyperuricemia, as is the case with beta-blockers and diuretics. On the contrary, angiotensin converting enzyme inhibition may correct some of these metabolic abnormalities.

If left ventricular hypertrophy was already present, these agents may regress it. If the patient has angina, he or she could still benefit from improved double product and prolonged exercise capacity because of elevation in anginal threshold. If myocardial infarction has occurred, angiotensin converting enzyme inhibition could retard the progression of left ventricular dysfunction to congestive heart failure. If this grave condition is already present, still both symptoms and survival may improve. This versatility of effects, in addition to the proposed ability to retard progression of renal disease, as well as paucity of side-effects, makes angiotensin converting enzyme inhibitors able to not only treat hypertension but possibly prevent and also treat most of its complications and confer the desirable and hitherto elusive cardioprotection (Gavras, 1988; Gavras, 1990).

REFERENCES

Abdel-Haq, B., Lucarini, A. R., Magagna, A., Arrighi, P., Favilla, S. and Salvetti, A. (1989). Indomethacin reduces the hypotensive effect of lisinopril but does not modify its renal hemodynamic effects. *The International Symposium on ACE-inhibition*, London, February 14-17, 1989.

Abrahams, P. A., Opsahl, J. A., Halstenson, L. E. and Keane, W. F. (1988). Efficacy and renal effects of enalapril therapy for hypertensive patients with chronic renal insufficiency. *Arch. Int. Med.* **148,** 2358-2362.

Ames, R. P., Borkoneski, A. J., Sicinski, A. M. and Laragh, J. H. (1964). Prolonged infusions of angiotensin II and norepinephrine on blood pressure, electrolyte balance, aldosterone and cortisol in normal man and in cirrhosis with ascites. *J. Clin. Invest.* **44,** 1171-1186.

Andren, L., Karlberg, B. E., Svensson, A., Ohman, P., Nilsson, O. R.and Hansson, L. (1985). Long term effects of captopril and atenolol in essential hypertension. *Acta. Med. Scand.* **217,** 155-160.

Andrews, P., Papadakis, N. and Gavras, H. (1982). Reversal of acute cerebral vasospasm by angiotensin converting enzyme inhibition. *Stroke* **13,** 480-483.

Asmar, R. G., Pannier, B., Santoni, J. P., Laurent, S., London, G. M., Levi, B. I. and Safar, M. E. (1988). Reversion of cardiac hypertrophy and reduced arterial compliance after converting enzyme inhibition in essential hypertension. *Circulation* **78,** 941-950.

Atkinson, A. B., Brown, J. J., Cumming, A. M., Fraser, R., Lever, A. F., Leckie, B. J., Morton, J. and Robertson, J.I.S. (1982). Captopril in renovascular hypertension: Long term use in predicting surgical outcome. *Br. Med. J.* **284,** 689-693.

Atkinson, A. B., Brown, J. J., Lever, A. F. and Robertson, J.I.S. (1980). Combined treatment of severe intractable hypertension with captopril and diuretic. *Lancet* **2,** 105-110.

Belz, G. G., Breithaup, K., Erb, K., Kleinbloesem, C. and Wolf, G. K. (1989). Influence of the angiotensin converting enzyme cilazapril, the beta-blocker propranolol and their combination on hemodynamics of hypertension. *J. Hypertens.* **7,** 817-824.

Benetos, A., Gavras, H., Stewart, J. M., Vavrek, R. J., Hatinoglou, S. and Gavras, I. (1986). Vasodepressor role of endogenous bradykinin assessed by a bradykinin antagonist. *Hypertension* **8,** 971-974.

Borniade, V. (1983). Severe side effects of captopril in advanced chronic renal insufficiency. *Proc. Eur. Dial Transplant Assoc.* **20,** 530-537.

Borschetti, E., Tantucci, C., Cocchieri, M., Fornari, G., Grassi, V. and Sorbini, C. A. (1985). Acute effects of captopril in hypoxic pulmonary hypertension. Comparison with transient oxygen administration. *Respiration* **48,** 296-302.

Bertoli, L., Fusco, M., Micallef, E. and Busnardo, I. (1986). Effects of captopril on hemodynamics and blood gases in chronic obstructive lung disease with pulmonary hypertension. *Respiration* **48,** 296-302.

Bidiville, J., Nussberger, J., Waeber, G., Porchet, M., Waeber, B .and Brunner, H. R. (1988). Individual responses to converting enzyme inhibitors and calcium antagonists. *Hypertension* **II,** 166-173.

Bjorck, S., Nyberg, G., Mulec, H., Granerus, G., Herlitz, H. and Aurell, M. (1986). Beneficial effects of angiotensin converting enzyme inhibition on renal function in patients with diabetic nephropathy. *Br. Med. J.* **293,** 471-474.

Bolzano, K., Arriga, J., Bernal, R., Bernandes, H., Calderon, J. L., Debruya, J., Dienstle, F., Drayer, J., Goodfriend, T. L. and Gross, W. (1987). The antihypertensive effect of lisinopril compared to atenolol in patients with mild to moderate hypertension. *J. Cardiovas. Pharmacol.* **9** (Suppl 3), s43-s47.

Boulet, L. P., Milot, J., Lampron, N. and Lacourciere, Y. (1989). Pulmonary function and airway responsiveness during long term therapy with captopril. *JAMA* **261,** 413-416.

Bouthier, J. D., Safar, M. E., Benetos, A., Simon, A. C., Levenson, J. A. and Hughes, C. M. (1986). Haemodynamic effects of vasodilating drugs on the common carotid and brachial circulations of patients with essential hypertension. *Br. J. Clin. Pharmacol.* **21,** 137-142.

Braun-Menendez, E., Fasciolo, J. and Leloire, L. F. (1940). Farmacologia de la hipertensione. *Rev. Soc. Argent. Biol.* **16,** 398-426.

Brouwer, R.M.L., Bolli, P., Erne, P., Conen, D., Kiowski, W. and Buhler, F. R. (1985). Antihypertensive treatment using calcium antagonists in combination with captopril rather than diuretics. *J. Cardiovas. Pharmacol.* **7** (Suppl), s88-s91.

Brunner, H. R., Waeber, B. and Nussberger, J. (1988). What we would like to know about the antihypertensive mechanisms of angiotensin converting enzyme inhibition. *J. Hypertens.* **6** (Suppl 3), s1-s5.

Bucknall, C. E., Meilly, J. B., Carter, R., Stevenson, R. D. and Semple, P. F. (1988). Bronchial hyperactivity in patients who cough after receiving converting enzyme inhibitors. *Br. Med. J.* **296,** 86-88.

Bursztyn, M., Knecht, A., Rosenthal, T., Grossman, E., Boichis and H., Rubinstein, Z. (1984). Bilateral renal artery stenosis. *Arch. Int. Med.* **144,** 2282.

Bursztyn, M., Gavras, I. and Gavras, H. (1989). Hypertension in the aging patient: Implications for the selection of drug therapy. *J. Am. Ger. Soc.* **37,** 814-818.

Bursztyn, M., Gavras, I., Gourley, L., Whaling, J. J., Gomez, H. V. and Gavras, H. (1990). Effect of combination therapy with atenolol and the novel angiotensin converting enzyme inhibitor, benazepril. (Submitted).

Bursztyn, M., Gavras, I., Tifft, C.P., Boger, R., Luther, R. and Gavras, H. (1990). Effects of a novel renin inhibitor on blood pressure in patients with essential hypertension. *J. Cardiovas. Pharmacol.* **15,** 493-500.

Captopril Collaborative Study Group. (1982). Does captopril cause renal damage in hypertensive patients? *Lancet* **i,** 988-990.

Captopril-digoxin Multicenter Research Group. (1988). Comparative effects of therapy with captopril and digoxin in patients with mild to moderate heart failure. *JAMA* **259,** 539-544.

Captopril Research Group of Japan. (1985). Clinical effects of low dose captopril plus a thiazide diuretic on mild to moderate essential hypertension: Comparison with propranolol. *J. Cardiovas. Pharmacol.* **7** (Suppl 1), s77-s81.

Carretero, O.A., Miyazaki, S. and Scicli, A.G. (1981). Role of kinins in the acute antihypertensive effect of the converting enzyme inhibitor, captopril. *Hypertension* **3,** 18-22.

Case, D. B., Wallance, J. M., Keim, H. J., Sealey, J. and Lalagh, J. H. (1977). Possible role of renin in hypertension as suggested by renin sodium profiling and inhibition of converting enzyme. *N. Engl. J. Med.* **296,** 641-646.

Catalino, M. and Liberti, A. (1985). Captopril for the treatment of hypertension with peripheral vascular disease. *Angiology* **36,** 293-296.

Chalmers, D., Domberg, S. L. and Lawson, D. H. (1987). Post-marketing surveillance of captopril (for hypertension): A preliminary report. *Br. J. Clin. Pharmacol.* **24,** 343-349.

Chobanian, A. V., Haudenschild, C. C., Nickerson, C .and Drago, R. (1990). Aniatherogenic effect of captopril in the Watanabe Heritable hyperlipidemic rabbit. *Hypertension* **15,** 327-331.

Chrysant, S.G., Bal, I.S., Johnson, B. and McPherson, M. (1985). A comparative study of captopril and enalapril in patients with severe hypertension. *J. Clin. Pharmacol.* **25,** 149-151.

Cleland, J. G., Dargie, H. J., Ball, S. G., Gillen, G., Hodsman, G. P., Morton, J. J., East, B. W., Robertson, I., Ford, I. and Robertson, J.I.S. (1985). Effects of enalapril in heart failure: A double blind study of effects on exercise performance, renal function, hormones and metabolic state. *Br. Heart J.* **54,** 305-312.

Cleland, J.G.F., Dargie, H. J., McAlpine, H., Ball, S. G., Morton, J. J., Robertson, J.I.S. and Ford, I. (1985). Severe hypotension after first dose of enalapril in heart failure. *Br. Med. J.* **291,** 1309-1311.

Corea, L., Bentivoglio, M., Verdecchia, P. and Provvidenza, M. (1984). Converting enzyme inhibition *vs* diuretic therapy as first therapeutic approach to the elderly hypertensive patient. *Curr. Ther. Res.* **36,** 347-351.

Cox, J. P., Duggen, J., O'Boyle, C. A., Mee, F., Walsh, J. B., Coakley, D., O'Brien, E. and O'Malley, K.(1989). A double-blind evaluation of captopril in elderly hypertensives. *J. Hypertens.* **7,** 299-303.

Creager, M. A., Faxon, D. P., Rockwell, S. M., Melby, J. C., Gavras, H. and Coffman, J. D. (1984). Effects of renin angiotensin system on limb circulation in normal subjects. *Am. J. Physiol.* **246,** H239-H244.

Croog, S. H., Levine, S., Testa, M. A., Broun, B., Bulpitt, C. J., Jenkins, C. D., Klerman, G. L. and Williams, G. H. (1986). The effects of antihypertensive therapy on the quality of life. *N. Engl. J. Med.* **314,** 1657-1664.

Curtis, J. J. (1986). Hypertension and kidney transplantation. *Am. J. Kidney Dis.* **7,** 181-196.

Curtis, J. J., Luke, R. G., Welchel, J. D., Diethelm, A. G., Jones, P. and Dustan, H. P. (1983). Inhibition of angiotensin converting enzyme in renal transplant recipients with hypertension. *N. Engl. J. Med.* **308,** 377-381.

Dalhof, B., Hansson, L., Acosta, J.H., Bolzano, K., Fairhurst, G., Ferreira, C., Kaarsalo, E., Silva, M.C. and Simone, A. (1988). Controlled trial of enalapril and hydrochlorothiazide in 200 hypertensive patients. *Am. J. Hypertens.* **1,** 38-41.

Daly, P., Mettauer, B., Roulleau, J. L., Cousineau, D. and Burgess, J. H. (1985). Lack of reflex increase in myocardial sympathetic tone after captopril: Potential antianginal effect. *Circulation* **71,** 317-325.

Darne, B., Girard, X., Safar, M.E., Cambien, F. and Giuze, L. (1989). Pulsatile versus steady component of blood pressure: A cross-sectional analysis and a prospective analysis on cardiovascular mortality. *Hypertension* **13,** 392-400.

Davies, R. O., Gomez, H. J., Irvin, J. D. and Walker, J. F. (1984). An overview of the clinical pharmacology of enalapril. *Br. J. Clin. Pharmacol.* **18,** 215s-229s.

Dietze, G. J., Rett, K., Jauch, K. W., Wicklmayr, M., Fink, E., Hartl, W., Gaunther, B., Fritz, H. and Mehnert, H. (1987). Captopril bei hypertoniken mit Diabetes Mellitus typ II. *Herz* **12** (Suppl 1), 16-21.

Dondi, M., Franchi, R., Levorato, M., Zucalla, A., Geggi, R., Nirelli, M., Stella, A., Marchetta, F., Losinno, F. and Monetti, N. (1989). Evaluation of hypertensive patients by means of captopril enhanced renal scintigraphy with technetium -99m DPTA. *J.Mucl. Med.* **30,** 615-621.

Donnelly, R., Meredith, P. A., Elliot, H. L. and Reid, J. L. (1990). Kinetic-dynamic relations and individual responses to enalapril. *Hypertension* **15,** 301-309.

Drexler, H., Bernhardt, V., Meinertz, T., Wollschlager, H., Lehman, M. and Just, H. (1989). Contrasting peripheral short-term and long-term effects of converting enzyme inhibition in patients with congestive heart failure. A double blind placebo controlled trial. *Circulation* **79,** 491-502.

Dunn, F. G., Oigman, W., Ventura, H. O., Messerli, F. H., Kobrin, I. and Frohlich, E. D. (1984). Enalapril improves systemic and renal hemodynamics and allows regression of left ventricular mass in essential hypertension. *Am. J. Cardiol.* **53,** 105-108.

Dzau, V. J. (1989). Multiple pathways of angiotensin production in the blood vessel wall: Evidence, possibilities and hypotheses. *J. Hypertens.* **7,** 933-936.

Eggersten, R., Svensson, A., Dalhof, B. and Hansson, L. (1989). Additive effect of isradipine in combination with captopril in hypertensive patients. *Am. J. Med.* **86,** 124-126.

Eichenstaedt, H., Danne, O., Langer, M., Cordes, M. and Schubert, M. (1989). Regression of left ventricular hypertrophy under ramipril treatment investigated by nuclear magnetic resonance imaging. *J. Cardiovas. Pharmacol.* **13,** (Suppl 3), s75-s80.

Enalapril in Hypertension Study Group (UK). (1986). Enalapril in essential hypertension: A comparative study with propranolol. *Br. J. Clin. Pharmacol.* **18,** 51-56.

Faxon, D. P., Creager, M. A., Halperin, J. L., Sussman, H. A., Gavras, H. and Ryan, T. J. (1982). The effect of angiotensin converting enzyme inhibition on coronary blood flow and hemodynamics in patients without coronary artery disease. *Int. J. Cardiol.* **2,** 251-262.

Ferannini, E., Buzzigoli, G., Bonnadona, R., Giorico, M. A., Oleginni, M., Graziadei, L., Pedrinelli, R., Brandi, L. .and Bevilacqua, S. (1987). Insulin resistance in essential hypertension. *N. Engl. J. Med.* **317,** 350-357.

Ferreira, S. (1965). A bradykinin-potentiating factor (BPF) present in the venom of Bothrops Jararaca. *Br. J. Pharmacol.* **24,** 163-169.

Forrette, F., Handfield-Jones, R., Henry-Amar, M., Fouchard, M., Bouchacourt, P., Hervey, M.P., Henry, J.F., Mesguisch-Billaud, E. and Alexandre, J.M. (1985). Treatment of arterial hypertension in the aged subject with converting enzyme inhibitor: Enalapril. *Presse. Med.* **14,** 2237-2241.

Foult, J. M., Tavoloro, O., Antary, I. .and Nitenberg, A. (1988). Direct myocardial and coronary effects of enalapril in patients with dilated cardiomyopathy: Assessment by a bilateral intracoronary infusion technique. *Circulation* **77,** 337-344.

Ganten, D., Hayduk, K., Brecht, H.M., Boucher, R. and Genest, J. (1970). Evidence of renin release or production in splanchnic territory. *Nature* **226**, 551-552.

Gavras, H. (1986). A multicenter trial of enalapril in the treatment of essential hypertension. *Clin. her.* **31,** 24-30.

Gavras, H. (1988). The place of angiotensin-converting enzyme inhibition in the treatment of cardiovascular diseases. *N. Engl. J. Med.* **319,** 1541-1543.

Gavras, H. (1990). Angiotensin inhibition and its impact on cardiovascular disease. *Circulation* **81,** 381-388.

Gavras, H., Andrews, P. and Papadakis, N. (1981). Reversal of experimental delayed cerebral vasopressin by angiotensin converting enzyme inhibition. *J. Neurosurg.* **55,** 884-888.

Gavras, H., Brunner, H.R., Lalagh, J.H., Sealey, J.E., Gavras, I. and Vukovich, R.A. (1974). An angiotensin converting enzyme inhibitor to identify and treat vasoconstrictor and volume factor in hypertensive patients. *N. Engl. J. Med.* **291,** 817-821.

Gavras, H., Brunner, H. R., Thurston, H. and Lalagh, J. H. (1975). Reciprocation of renin dependency with sodium volume dependency in renal hypertension. *Science* **188**, 1316-1317.

Gavras, H., Brunner, H. R., Turini, G. A., Kershaw, G. R., Tifft, C. P., Cuttelod, S., Gavras, I.,Vukovich, R. A. and McKinstry, D. N. (1978a). Antihypertensive effect of oral angiotensin converting enzyme inhibitor SQ 14225 in man. *N. Engl. J. Med.* **298,** 991-995.

Gavras, H., Faxon, D. P., Berkoben, J., Brunner, H. R. and Ryan, T. J. (1978b). Angiotensin converting enzyme inhibition in patients with congestive heart failure. *Circulation* **58,** 770-775.

Gavras, H., Flessas, A., Ryan, T. J., Brunner, H. R., Faxon, D. P. and Gavras, I. (1977). Angiotensin II inhibition: Treatment of congestive cardiac failure in high renin hypertension. *JAMA* **283,** 880-882.

Gavras, H. and Gavras, I. (1988). Angiotensin converting enzyme inhibitors: Properties and side-effects. *Hypertension* **11,** II37-II41.

Gavras, H., Liang, C.S. and Brunner, H.R. (1978c). Redistribution of blood flow after inhibition of the angiotensin converting enzyme. *Circ. Res.* **43** (Suppl 1), 59-63.

Gavras, H., Ribeiro, A. B., Gavras, I. and Brunner, H. R. (1976). Reciprocal relation between renin dependency and sodium dependency in essential hypertension. *N. Engl. J. Med.* **295,** 1278-1283.

Geisterfer, A.A.T., Peach, M. J.,and Owens, G. (1988). Angiotensin II induces hypertrophy not hyperplasia of cultured rat aortic smooth muscle cells. *Circ. Res.* **62,** 749-756.

Gibson, G. R. (1989). Enalapril induced cough. *Arch. Int. Med.* **149,** 2701-2703.

Giles, T. D., Katz, R., Sullivan, J. M., Wolfson, P., Haugland, M., Kirlin, P., Powers, E., Rich, E., Hackshaw, B. and Chiaramida, A. (1989). Short and long acting angiotensin converting enzyme inhibitors: A randomized trial of lisinopril versus captopril in the treatment of congestive heart failure. The Multicenter Lisinopril Captopril Congestive Heart Failure Study Group. *J. Am. Coll.Cardiol.* **13,** 1240-1247.

Goldblat, H., Lynch, J., Hanzal, R., Ranar, F. and Summerville, W. (1934). Studies on experimental hypertension. I. The production of persistent elevation of systolic pressure by means of renal ischemia. *J. Exp. Med.* **59,** 347-379.

Gomez, H. J., Cirillo, V. J.,and Irvin, J. D. (1985). Enalapril: A review of human pharmacology. *Drugs* **30** (Suppl 1), 13-24.

Grossman, E., Morag, B., Nussinovitch, N., Boichis, H., Knecht, A. and Rosenthal, T. (1984). Clinical use of captopril in Takayasu's disease. *Arch. Int. Med.* **144,** 95-96.

Guazzi, M. D., De, Cesare, N., Galli, C., Salvioni, A., Tramontana, C., Tamborini, G. and Bartorelli, A. (1984). Calcium-channel blockade with nifedipine and angiotensin converting enzyme inhibition with captopril in the therapy of patients with severe primary hypertension.

Circulation **70,** 279-284.

Gyene, T. T., Bellet, M., Sassano, P., Serrurier, D., Corvol, P. and Menard, J. (1989). Cross-over design for the dose determination of an angiotensin converting enzyme inhibitor in hypertension. *J. Hypertens.* **7,** 1005-1012.

Heeg, J. E., de, Jong, P. E., Van, Der, Hem, G. K. and de Zeeuw, D. (1987). Reduction of proteinuria by angiotensin converting enzyme inhibitor. *Kidney Int.* **32,** 78-83.

Helgland, A., Helglund, C. H., Strommen, R. and Tretly, S. (1986). Enalapril, atenolol and hydrochlorothiazide in mild to moderate hypertension. *Lancet* **1,** 872-875.

Herlitz, H., Edeno, C., Mulec, H., Westberg, G. and Aurell, M. (1984). Captopril treatment of hypertension and renal failure in systemic lupus erythematous. *Nephron* **38,** 253-256.

Herrick, A. L., Waller, P. C., Berkin, K. E., Pringle, S. D., Callender, J. S., Robertson, M. P., Findlay, J. G., Murray, G. D., Reid, J. C. and Lorimer, A. R. (1989). Comparison of enalapril and atenolol in mild to moderate hypertension. *Am. J. Med.* **86,** 421-426.

Hilgers, K.F., Kuczera, M., Wilhelm, M.J., Wiecek, A., Ritz, E., Ganten, D. and Mann, J.F.E. (1989). Angiotensin formation in the isolated rat hindlimb. *J. Hypertens.* **7,** 789-798.

Hodssman, G. P., Isles, C. G., Murray, G. D., Usherwood, T. P., Webb, D. J. and Robertson, J.I.S. (1983). Factors related to first dose hypotensive effect of captopril: Prediction and treatment. *Br. Med. J.* **286,** 832-834.

Hoefnagels, W.H.C. and Thien, T. (1986). Renal artery occlusion in patients with renovascular hypertension treated with captopril. *Br. Med. J.* **292,** 24-25.

Hollenberg, N. K. (1983). Medical therapy of renovascular hypertension: Efficacy and safety of captopril in 269 patients. *Cardiovas. Rev. Rep.* **4,** 852-876.

Hollenberg, N. K., Swartz, S. L., Passan, D. R. and Williams, G. H. (1979). Increased glomerular filtration rate after converting enzyme inhibition in essential hypertension. *N. Engl. J. Med.* **301,** 9-12.

Hommel, E., Parving, H. H., Mathiesen, E., Edsberg, B., Damkjaer-Nielsen, M. and Giese, J. (1986). Effect of captopril on kidney function in insulin dependent diabetic patients with nephropathy. *Br. Med. J.* **293,** 467-470.

Hoontje, S. J., Donker, A. J., Prins, E. J. and Weening, J. J. (1980). Membranous glomerulopathy in a patient on captopril. *Acta. Med. Scand.* **208,** 154-156.

Hostetter, R. H., Renneke, H. G. and Brenner, B. M. (1982). The case for intrarenal hypertension in the initiation and progression of diabetic and other glomerulopathies. *Am. J. Med.* **72,** 375-380.

Hricik, D. E., Browning, P. J., Kopelman, R., Groono, W. E., Madias, N. E. and Dzau, V. J. (1983). Captopril induced functional renal insufficiency in patients with bilateral renal artery stenoses or renal artery stenosis in a solitary kidney. *N. Engl. J. Med.* **308,** 373-376.

Inman, W. H., Rawson, N. S., Wilton, L. V. and Speirs, C. J. (1988). Postmarketing surveillance of enalapril. I. Results of prescription-event monitoring. *Br. Med. J.* **297,** 826-829.

Insel, J., Mirvis, D. M., Boland, M. J., Cinquegrani, M. P., Shanes, J., Rubin, S. A. and Whalen, J. J. (1989). A multicenter study of the safety and efficacy of benazepril hydrochloride, a long acting angiotensin converting enzyme inhibitor in patients with chronic congestive heart failure. *Clin. Pharmacol. Therap.* **45,** 312-320.

Jackson, B., Maher, D., Matthews, P. G., McGrath, B. P. and Johnston, C. I. (1988). Lack of cross sensitivity between captopril and enalapril. *Aust. N.Z. Med. J.* **18,** 21-27.

Jauch, K. W., Hartl, W., Gunther, B., Wicklmayr, M., Rett, K. and Dietze, G. (1987). Captopril enhances insulin responsiveness of forearm muscle tissue in non-insulin dependent diabetes mellitus. *Eur. J.Clin. Invest.* **17,** 448-454.

Jenkins, A. C., Knill, J. R. and Dreslinski, G. R. (1985). Captopril in the treatment of the elderly hypertensive patient. *Arch. Int. Med.* **145,** 2029-2031.

Johnsen, S. A. and Aurell, M. (1981). Immunosuppressive action of captopril blocked by prostaglandin synthetase inhibitor. *Lancet* **1,** 1005.

Joint National Committee. (1988). The 1988 report of the Joint National Committee for the detection, evaluation and treatment of high blood pressure. *Arch. Int. Med.* **148,** 1023-1038.

Kaufman, J., Casanova, J.E., Riendl, P. and Schlueter, D.P. (1988). Bronchial hypersensitivity and

cough due to angiotensin converting enzyme inhibitors. *Chest.* **95,** 544-548.
Keane, W. F., Anderson, S., Aurell, M., de Zeeuw, D., Narins, R. G. and Povar, G. (1989). Angiotensin converting enzyme inhibitors and progressive renal insufficiency. Current experience and future directions. *Ann. Int. Med.* **111,** 503-516.
Kogler, P. (1989). Cilazapril: A New non-thiol-containing angiotensin converting enzyme inhibitor: Worldwide clinical experience in hypertension. *Am. J. Med.* **87** (6B), 50s-55s.
Kostis, J.B. (1989). Angiotensin-converting-enzyme inhibitors: Emerging differences and new compounds. *Am. J. Hypertens.* **2,** 57-64.
Kowaloff, H., Gavras, H. and Brecher, P. (1980). Renin like enzymatic activity in the cerebral microvessels of the rat. *Am. J. Physiol.* **238,** H384-H388.
Krieger, E. M., Salgado, H. C., Assan, C. J., Green, K. J. and Ferreira, S. H. (1971). Potential screening test for detection of overactivity of the renin angiotensin system. *Lancet* **1,** 269-272.
Krolewski, A., Canessa, M., Warram, J. H., Laffel, L. M., Christlieb, A. R., Knowler, W. C. and Rand, L. I. (1988). Predisposition to hypertension and susceptibility to renal disease in insulin dependent diabetes mellitus. *N. Engl. J. Med.* **318,** 140-145.
Lees, K. R., Reid, J. L., Scott, M. G., Hosie, J., Herpin, D., Santoni, J. P. (1989). Captopril versus perindopril: A double blind study in essential hypertension. *J. Hum. Hypertens.* **3,** 17-22.
Lessem, J. N., Rucinska, E., Vickers, F. F., Espie, J. and Sromovsky, J. A. (1985). Long term enalapril-A new converting enzyme inhibitor in the treatment of mild to moderate hypertension: Results of a worldwide multiclinic study: Comparing two ways of analyzing data. *Clin. Exp. Hypertens.* **7** (A), 1515-1539.
Lewis, R. A., Baker, K. M., Ayers, C. R., Weaver, B. A. and Lehman, M. R. (1985). Captopril versus enalapril maleate: A comparison of antihypertensive and hormonal effects. *J. Cardiovas. Pharmacol.* **7** (Suppl 1), s12-s15.
Liang, C. S. and Gavras, H. (1978). Renin angiotensin system inhibition in conscious dogs during acute hypoxemia: Effects on systemic hemodynamics, regional blood flow and tissue metabolism. *J. Clin. Invest.* **62,** 961-970.
Liang, C. S., Gavras, H. and Hood, W. B. (1978a). Renin angiotensin system inhibition in conscious sodium depleted dogs: Effects on systemic and coronary hemodynamics. *J. Clin. Invest.* **61,** 874-883.
Liang, C. S., Gavras, H. and Brunner, H. R. (1978b). Redistribution of regional flow after administration of saralasin in salt depleted dogs. *Clin. Sci. Mol. Med.* **55,** 243s-246s.
Lindgren, B. R., Rosenqvist, U., Ekstrom, T., Gronneberg, R., Karlberg, B. E. and Andersson, R. G. (1989). Increased bronchial reactivity and potentiated skin responses in hypertensive subjects suffering from cough during ACE-inhibitor therapy. *Chest.* **95,** 1225-1230.
Lopez-Ovejero, J. A., Soal, C. D., D'Angelo, W.A., Chligh, H. S., Stenzel, K. H. and Lalagh, J. H. (1979). Reversal of vascular and renal crisis of scleroderma by oral angiotensin converting enzyme blockade. *N. Engl. J. Med.* **300,** 1417-1419.
Luderer, J. R., Demers, L. M., Harrison, T. S. and Hayes, A. H. (1982). Converting enzyme inhibition with captopril in patients with primary hyperaldosteronism. *Clin. Pharmacol. Ther.* **31,** 305-311.
Lyons, D. F., Kem, D. C., Brown, R. D., Hanson, C. S. and Corollo, M. L. (1983). Single dose captopril as a diagnostic test for primary aldosteronism. *J. Clin. Endocrinol. Metab.* **57,** 892-896.
Man, in't, Veld, A. J., Schmidt, I.M., Derkx, F.H.M., de, Bryen, J.H.B. and Schalekamp, M.A.D.H. (1980). Effect of an angiotensin converting enzyme inhibitor (captopril) on blood pressure of anephric subjects. *Br. Med. J.* **280,** 288-290.
Marre, M., Chatellier, G., Leblanc, H., Guyene, T. T., Menard, J. and Passa, P. (1988). Prevention of diabetic nephropathy with enalapril in normotensive diabetes with persistent microalbuminuria. *Br. Med. J.* **297,** 1092-1095.
Martin, M. F., Surrall, K. E., McKenna, F., Dixon, J. S., Bird, H. A. and Wright, V. (1984). Captopril: A new treatment for rheumatoid arthritis? *Lancet* **1,** 1325-1328.
McCarthy, J. E., Weder, A. B. (1990). The captopril test and renovascular hypertension: A

cautionary tale. *Arch. Int. Med.* **150,** 493-495.
McEwan, J. R., Choudry, N., Street, R. and Fuller, R. W. (1989). Change in cough reflex after treatment with enalapril and ramipril. *Br. Med. J.* **299,** 13-16.
McGiff, J. C., Terragno, N., Malik, K. U. and Lonigro, A.J. (1972). Release of a prostaglandin like substance from canine kidney by bradykinin. *Circ. Res.* **31,** 36-43.
McGregor, G. A., Markandu, N. D., Banks, R. A., Bayliss, J., Roulston, J. E. and Jones, J. C. (1982). Captopril in essential hypertension: Contrasting effects of adding hydrochlorothiazide or propranolol. *Br. Med. J.* **284,** 693-696.
McGregor, G. A., Markandu, N. D., Smith, S. J. and Sagnella, G. A. (1985). Captopril: Contrasting effects of adding hydrochlorothiazide, propranolol or nifedipine. *J. Cardiovas. Pharmacol.*, **7** (Suppl 1), s82-s87.
Mimran, A., Targhetta, R. and Laroche, B. (1980). The antihypertensive effect of captopril: Evidence for the effect of kinins. *Hypertension* **2,** 732-737.
Mizuno, K., Nakamura, M., Higashimori, K. and Inagami, T. (1988). Local generation and release of angiotensin II in peripheral vascular tissue. *Hypertension* **11,** 223-229.
Modan, M., Halkin, H., Almog, S., Lusky, A., Eshkol, A., Shefi, M., Shitrit, A. and Fuchs, Z. (1985). Hyperinsulinemia: A link between hypertension obesity and glucose intolerance. *J. Clin. Invest.* **75,** 809-817.
Montamat, S. C., Cusack, B. J. and Vestal, R. E. (1989). Current concepts - geriatrics: Management of drug therapy in the elderly. *N. Engl. Med.* **321,** 303-308.
Moore, T. J., Gantz, F. R., Hollenberg, N. K., Koletsky, R. J., Leboff, M. S., Swaertz, S. L., Levine, L., Podolsky, S., Dluhy, R. G. and Williams, G. H. (1981). Contribution of prostaglandins to the antihypertensive action of captopril in essential hypertension. *Hypertension* **3,** 168-173.
Morgan, T., Anderson, S. and Hopper, J. (1988). Enalapril and nifedipine in essential hypertension: Synergism of the hypertensive effect in combination. *Clin. Exp. Hypertens.* (A) **10,** 779-789.
Morlin, C., Baglivo, H. and Boeijinga, K. (1987). Comparative trial of lisinopril and nifedipine in mild to severe essential hypertension. *J. Cardiovas. Pharmacol.* **9** (Suppl 3), s49-s52.
Morris, A.H., Brown, M.J. and Hingenbottom, T. (1989). Cough associated with angiotensin converting enzyme inhibition. *J. Cardiovas. Pharmacol.* **13** (Suppl 3), s59-s62.
Muiesan, G., Alicandri, C. and Agabiti-Rosei, E. (1987). A multicenter trial of low dose captopril administered twice daily in patients with essential hypertension unresponsive to beta blockers diuretic treatment. *J. Clin. Hypertens.* **3,** 144-152.
Mulinari, R., Gavras, I. and Gavras, H. (1987). Efficacy and tolerability of enalapril monotherapy in mild to moderate hypertension in older patients compared to younger patients. *Clin. Ther.* **9,** 678-682.
Muller, F. B., Sealey, J. E., Case, D. B., Atras, S. A., Pickering, T. G., Pecker, M. S., Preibisz, J. J. and Laragh, J. H. (1986). The captopril test for identifying renovascular disease in hypertensive patients. *Am. J. Med.* **80,** 633-644.
Naftilan, A. J., Pratt, R. E. and Dzau, V. J. (1989). Induction of platelet derived growth factor A chain and c-myc gene expressions by angiotensin II in cultured vascular smooth muscle cells. *J. Clin. Invest.* **83,** 1419-1424.
Nakashima, Y., Fouad, F. M. and Tarazi, R. C. (1984). Regression of left ventricular hypertrophy from systemic hypertension by enalapril. *Am. J. Cardiol.*, **53,** 1044-1049.
Oldenburg, B., McDonald, G.J. and Shelley,S. (1988). Controlled trial of enalapril in patients with chronic fluid overload undergoing dialysis. *Br. Med. J.* **296,** 1089-1091.
Oliver, R. M., Peacock, A. J., Fleming, J. S. and Waller, D. G. (1989). Renal and pulmonary effects of angiotensin converting enzyme inhibition in chronic hypoxic lung disease. *Thorax.* **44,** 513-515.
Ondetti, M. A., Rubin, B. and Cushman, D. W. (1977). Design of specific inhibitors of angiotensin converting enzyme: New class of orally active antihypertensive agents. *Science* **196,** 441-444.
Ondetti, M. A., Williams, N. J., Sabo, E. F., Pulsec, J., Weaver, E. R. and Koig, O. (1971). Angiotensin-converting enzyme inhibitors from the venom of bothrops jararaca: Isolation, elucidation of structure and synthesis. *Biochemistry* **19,** 4033-4041.

Parving, H. H., Smidt, U. M., Frusberg, B., Bennerie-Nielsen, V. and Andersen, A. R. (1981). A prospective study of glomerular filtration rate and arterial blood pressure in insulin-dependent diabetics with diabetic nephropathy. *Diabetologia* **20,** 457-461.

Parving, H. H., Hommel, E. and Smidt, U. M. (1988). Protection of kidney function and decrease in albuminuria by captopril in insulin dependent diabetics with nephropathy. *Br. Med. J.* **297,** 1086-1091.

Parving, H. H., Hommel, E., Damkjaer-Nielsen, M. and Giese, J. (1989). Effect of captopril on blood pressure and kidney function in normotensive insulin dependent diabetics with nephropathy. *Br. Med. J.* **299,** 533-536.

Paulson, O. B., Jarden, J. O., Vorstrup, S., Holm, S. and Godtfredsen, J. (1986). Effects of captopril on cerebral circulation in chronic heart failure. *Eur. J. Clin. Invest.* **16,** 124-132.

Pedersen, E. B., Sorensen, S. S., Amdisen, A., Danielsen, H., Eiskjaer, H., Hansen, H. H., Jensen, F. T., Jespersen, B., Madsen, B. and Nielsen, H. K. (1989). Abnormal glomerular and tubular function during angiotensin converting enzyme inhibition in renovascular hypertension evaluated by lithium clearance. *Eur. J. Clin. Invest.* **19,** 135-141.

Pettinger, W. A., Lee, H. C., Reisch, J. and Mitchell, H. C. (1989). Long term improvement in renal function after short term strict blood pressure control in hypertension nephrosclerosis. *Hypertension* **13,** 766-772.

Pfeffer, M. A., Lamas, G. A., Vaughan, D. E., Parisi, A. F. and Braunwald, E. (1988). Effect of captopril on progressive ventricular dilatation after anterior myocardial infarction. *N. Engl. J. Med.* **319,** 80-86.

Pollare, T., Lithell, H. and Berne, C. (1989). A comparison of the effects of hydrochlorothiazide and captopril on glucose and lipid metabolism in patients with hypertension. *N. Engl. J. Med.* **321,** 868-873.

Pollare, T., Lithell, H., Selinus, I. and Berne, C. (1989). Sensitivity to insulin during treatment with atenolol and metoprolol: A randomized double blind study on carbohydrate and lipoprotein metabolism in hypertensive patients. *Br. Med. J.* **298,** 1152-1157.

Pollare, T., Lithell, H. and Berne, C. (1990). Insulin resistance is a characteristic feature of primary hypertension independent of obesity. *Metabolism* **39**, 167-174.

Ponce, S. P., Jennings, A. E., Madias, N. E. and Harrington, J. T. (1985). Drug induced hyperkalemia. *Medicine* **64,** 357-370.

Pool, J. L., Gennari, J., Goldstein, R., Kochar, M. S., Levin, A. J., Maxwell, M. H., McCherney, J. A., Melita, J., Nash, D. T. and Nelson, E. B. (1987). Controlled multicenter trial of the antihypertensive effects of lisinopril, hydrochlorothiazide and lisinopril plus hydrochlorothiazide in the treatment of 394 patients with mild to moderate essential hypertension. *J. Cardiovas. Pharmacol.* **9** (Suppl 3), s36-s42.

Powell, J.S., Clozel, J.P., Muller, R.K.M., Kuhn, H., Hefti, F., Hosang, M. and Baumgartner, H.R. (1989). Inhibitors of angiotensin converting enzyme prevent myointimal proliferation after vascular injury. *Science* **245,** 186-188.

Re, R., Novelline, R., Escourou, M.T., Athanasoulis, C., Burton, J. and Haber, E. (1978). Inhibition of angiotensin converting enzyme for diagnosis of renal artery stenosis. *N. Engl. J. Med.* **298,** 582-586.

Reams, G.P. and Bauer, J.H. (1986). Effect of enalapril in subjects with hypertension associated with moderate to severe renal dysfunction. *Arch. Int. Med.* **146,** 2145-2148.

Reisch, C., Mann, J. and Ritz, E. (1987). Conversion enzyme inhibitors in the antihypertensive treatment of patients with renal failure. *Dtsch. Med. Wochenschrift.* **33,** 1259-1262.

Rett, K., Lotz, N., Wicklmayr, M., Jauch, K. W., Hartl, W. and Dietze, G. J. (1988). Verbesserte insulinwirkung druch ACE-hemmung beim typ-II diabetiker. *Dtch. Med. Wochenschr.* **113,** 243-249.

Ried, J. L. (1988). Angiotensin converting enzyme inhibition in the elderly. *Br. Med. J.* **295,** 943-944.

Romanelli, G., Giustina, A., Cimino, A., Valentin, U., Agabiti-Rosei, E., Muisan, G. and Giustina, G. (1989). Short term effect of captopril on microalbuminuria induced by exercise in normotensive diabetics. *Br. J. Med.*, **298,** 284-288.

Rucinska, E. J., Small, R., Mulcahy, W. S., Snyder, D. L., Rodel, P. V., Rush, J. E., Smith, R. D., Walker, J. F. and Irving, J. D. (1989). Tolerability of long term enalapril maleate in patients resistant to other therapies and intolerant to captopril. *Med. Toxicol. Adverse. Drug. Exp.* **4,** 144-152.

Safar, M. E. (1989). Pulse pressure in essential hypertension: Clinical and therapeutic implications. *J. Hypertens.* **7,** 769-776.

Salvetti, A., Abdel-Haq, B., Magagna, A. and Pedrinelli, R. (1987). Indomethacin reduces the antihypertensive action of enalapril. *Clin. Exp. Hypertens.* (A) **9,** 559-567.

Salvetti, A. and Arzilli, F. (1989). Chronic dose response curve of enalapril in essential hypertension. An Italian multicenter study. *Am. J. Hypertens.* **2,** 352-354.

Salvetti, A., Pedrinelli, R., Magagna, A. and Ugenti, P. (1982). Differential effects of selective and non-selective prostaglandin synthesis inhibition on the pharmacological response to captopril in patients with essential hypertension. *Clin. Sci.* **63,** 261s-263s.

Sancho, J., Re, R., Burton, J., Barger, A. L. and Haber, E. (1976). The role of the renin-angiotensin-aldosterone system in cardiovascular homeostasis in normal human subjects. *Circulation* **53,** 400-405.

Sassano, P., Chatellier, G., Billaud, E., Alhenc-Gelas, F., Corvol, P.,and Menard, J. (1987). Treatment of mild to moderate hypertension with or without the converting enzyme inhibitor enalapril. *Am. J. Med.* **83,** 227-235.

Sassano, P., Chatellier, G., Billaud, E., Corvol, P. and Menard, J. (1989). Comparison of increase in the enalapril dose and addition of hydrochlorothiazide as second step treatment of hypertensive subjects not controlled with enalapril alone. *J. Cardiovas. Pharmacol.* **13,** 314-319.

Schreiber, M. J., Pohl, M. A. and Novick, A. L. (1984). The natural history of atherosclerotic and fibrous renal artery disease. *Urol. Clin. North. Am.* **11,** 383-392.

Selena, B. J. (1986). Chronic cough and the use of captopril: Unmasking asthma. *Arch. Int. Med.* **146,** 202-203.

Sen, S. and Tarazi, R. C. (1983). Regression of myocardial hypertrophy and influence of adrenergic system. *Am. J. Physiol.* **244,** H97-H101.

Sharpe, N., Murphy, J., Smith, H.and Hannan, S. (1988). Treatment of patients with symptomless left ventricular dysfunction after myocardial infarction. *Lancet* **6** (1), 255-259.

Shoback, D. M., Williams, G. H., Swartz, S. L., Davies, R. O. and Hollenberg, N. K. (1983). Time course and effect of sodium intake on vascular and hormonal response to enalapril (MK-421) in normal subjects. *J. Cardiovas. Pharmacol.* **5,** 1010-1018.

Singer, D. R., Markandu, N. D., Shore, A. C. and McGregor, G. A. (1987). Captopril and nifedipine in combination for moderate to severe essential hypertension. *Hypertension* **9,** 629-633.

Slater, E. E., Merrill, D. D., Roylance, P. J., Cooper, W. D., Inman, W.H.W. and Ewan, P. W. (1988). Clinical profile of angioedema associated with angiotensin converting enzyme inhibition. *JAMA* **260,** 967-970.

Smith,W. R., Neill, J., Cushman, W. C. and Butkus, D. E. (1989). Captopril associated acute interstitial nephritis. *Am.J. Nephrol.* **9,** 230-235.

Speirs, C. J., Dollery, C. T., Inman, W. H., Rawson, N. S. and Wilton, L. V. (1988). Postmarketing surveillance of enalapril II: Investigation of the potential role of enalapril in death with renal failure. *Br. Med. J.* **297,** 830-832.

Staessen, J., Fagard, R., Lijnen, P., Versdurren, L. J. and Amery, A. (1983).Double blind comparison between propranolol and bendroflumethiazide in captopril treated resistant hypertensive patients. *Am. Heart. J.* **106,** 321-328.

Staessen, J., Wilms, G., Baert, A., Fagard, R., Lijnen, P., Suy, R. and Amery, A. (1988). Blood pressure during long term converting-enzyme inhibition predicts the curability of renovascular hypertension. *Am. J. Hyperten.* **1,** 208-214.

Stronello, M., Valvo, E. V. and Scopellato, L. (1989). Angiotensin converting enzyme inhibition in normotensive type II diabetics with persistent mild proteinuria. *J. Hypertens.* **7** (Suppl 6), s314-s315.

Strozzi, C., Cocco, G., Portaluppi, F., Paola, A., Alfiero, R., Rizzo, A. and Tasini, T. R. (1985). Ergometric evaluation of the effects of captopril in hypertensive patients with stable angina. *J.*

Hypertens. **3** (Suppl 2), s147-s148.
Svetkey, L. P., Himmelstein, S. I., Dunnick, N. R., Wilkinson, R. H., Bollinger, R. R., McCann, R. L., Beytas, E. M. and Ktotman, P. E. (1989). Prospective analysis of strategies for diagnosis of renovascular hypertension. *Hypertension* **14,** 258-260.
Swartz, S. L., Williams, G. H., Hollenberg, N. K., Levine, L., Dluhy, R. G. and Moore, T. J. (1980). Captopril-induced changes in prostaglandin production: Relationship to vascular responses in normal man. *J. Clin. Invest.* **65,** 1257-1264.
Taguma, Y., Kitamoto, Y., Futaki, G., Ueda, H., Monma, H., Ishizaki, M., Takahashi, H., Sekino, H. and Sasaki, Y. (1985). Effect of captopril on heavy proteinuria in azotemic diabetics. *N. Engl. J. Med.* **313,** 1617-1620.
Textor, S. C., Gavras, H., Tifft, C. P., Bernard, D. B., Idelson, B. and Brunner, H. R. (1981). Norepinephrine and renin activity in chronic renal failure: Evidence for interacting roles in hemodialysis hypertension. *Hypertension* **3,** 294-300.
Textor, S. C., Gephardt, G. N., Bravo, E., Tarazi, R. C., Fouad, F. M., Tubbs, R. and McMahon, J. T. (1983). Membranous glomerulopathy associated with captopril therapy. *Am. J. Med.*, **74,** 705-712.
The CONSENSUS Trial Study Group (1987). Effect of enalapril on mortality in severe congestive heart failure: Results of the Cooperative North Scandinavian Enalapril Survival Study (CONSENSUS). *N. Engl. J. Med.* **316,** 1429-1435.
Thibonnier, M., Joseph, A., Sassano, P., Guyenne, T.T., Corvol, P., Raynaud, A., Serot, M. and Gaux, J.C. (1984). Improved diagnosis of unilateral renal artery stenosis after captopril administration. *JAMA.* **251,** 56-60.
Thibonnier, M., Sassano, P., Joseph, A., Plouin, P. F., Corvol, P. and Menard, J. (1982). Diagnostic value of a single dose of captopril in renin and aldosterone dependent surgically curable hypertension. *Cardiovas. Rev. Rep.* **3,** 1659-1668.
Thurston, H., Swales, J. D., Bing, R. F., Hurst, B. C. and Marks, E. S. (1979). Vascular renin-like activity and blood pressure maintenance in the rat: Studies of the effect of changes in sodium balance, hypertension and nephrectomy. *Hypertension* **1,** 643-649.
Tifft, C. P., Gavras, H., Kershaw, G. R., Gavras, I., Brunner, H. R., Liang, C. S. and Chobanian, A. V. (1979). Converting enzyme inhibition in hypertensive emergency. *Ann. Int. Med.* **90,** 43-47.
Tigerstedt, R. and Bergman, P. G. (1898). Niere und Kreislauf. *Scand. Arch. Physiol.* **8,** 223-271.
Truberstein, G., Wigger, E., Truberstein, R., Ludwig, M., Wilgalis, M., Sumpe, O.K. (1984). Treatment of Raynaud's phenomenon with captopril. *Dtch. Med. Wochenschr.* **109,** 857-860.
Tunny, T. J., Klemm, S. A., Hamlet, S. M. and Gordon, R. D. (1988). Diagnosis of unilateral renovascular hypertension: Comparative effect of intravenous enalapril and oral captopril. *J. Urol.* **140,** 713-715.
Uhlinger, D.E., Ferrier, C.P., Mathien, R., Rentec, K., Gnaldinger, M.P., Saxenhofer, H., Shaw, S. and Weidmann, P. (1989). Antihypertensive contribution of sodium depletion and sympathetic axis during chronic angiotensin II converting enzyme inhibition. *Clin. Hyperten.* **7,** 901-908.
Valvo, E., Bedogna, V., Casagrande, P., Antiga, L., Zamboni, N., Bommartini, F., Oldrizzi, L., Rugin, C. and Maschio, G. (1988). Captopril in patients with type II diabetes and renal insufficiency: Systemic and renal hemodynamic alterations. *Am. J. Med.* **85,** 344-348.
van Schaick, B. A., Geyskes, G. G., Ketner, N., Boer, P. and Dorhout-Mees, E. J. (1986). Comparison of enalapril and propranolol in essential hypertension. *Eur. J. Clin. Pharmacol.* **29,** 511-516.
Vaughan, E. D. (1985). Renovascular hypertension. *Kidney. Int.* **27,** 811-827.
Veteran Administration Cooperative Study Group on Antihypertensive Agents (1984). Low dose captopril for the treatment of mild to moderate hypertension. I. Results of a 14 week trial. *Arch. Int. Med.* **144,** 1947-1953.
Vidt, D. G. (1984). A controlled multiclinic study to compare the antihypertensive effects of MK-421, hydrochlorothiazide and MK-421 combined with hydrochlorothiazide in patients with mild to moderate essential hypertension. *J. Hypertens.* **2** (Suppl I), s81-s88.
Vlasses, P. H., Connor, D. P., Rotmensch, H. H., Frunicillo, R. J., Dangeisen, J. R., Shepley, K.

J. and Ferguson, R. K. (1986). A double blind comparison of captopril and enalapril in mild to moderate hypertension. *J. Am. Coll. Cardiol.* **7,** 651-660.

Waeber, B., Brunner, H. R., Brunner, D. R., Curtet, A. L., Turini, G. and Gavras, H. (1980). Discrepancy between antihypertensive effect and angiotensin converting enzyme inhibition by captopril. *Hypertension* **2,** 236-242.

Waeber, B., Gavras, I., Brunner, H. R. and Gavras, H. (1981). Safety and efficacy of chronic therapy with captopril in hypertensive patients: an update. *J. Clin. Pharmacol.* **21,** 508-516.

Waldemar, G., Schmidt, J. F., Andersen, A. R., Vorstrup, S., Ibsen, H. and Paulson, O. B. (1989). Angiotensin converting enzyme inhibition and cerebral blood flow autoregulation in normotensive and hypertensive man. *J. Hypertens.* **7,** 229-235.

Watkins, L., Burton, J. A., Haber, E., Cant, J. R., Smith, F. W. and Barger, A. C. (1976). The renin-angiotensin aldosterone system in congestive heart failure in conscious dogs. *J. Clin. Invest.* **57,** 1606-1617.

Weber, M. A., Ferraresi, R. W. and Poorvin, D. (1980). Evaluation of a new converting enzyme inhibitor spirapril in mild to moderate essential hypertension. *J. Hypertens.*, **4** (Suppl 6), s171-s173.

Webster, M.W.I., Fitzpatrick, M. A., Nicholls, M. G., Ikram, H. and Wells, J. E. (1985). Effect of enalapril on arrhythmias in congestive heart failure. *Am. J. Cardiol.* **56,** 566-571.

Weinberger, M.H. (1983). Influence of angiotensin converting enzyme inhibitor on diuretic induced metabolic effects in hypertension. *Hypertension* **5** (Suppl 3), 132-138.

Wenting, G. J., Blankstyn, P. J., Poldermans, D., van Geelen, J., Derkx, F. H., Man in't Veld, A. J. and Schalekamp, M.A.D. (1987). Blood pressure response in nephrectomized subjects and subjects with essential hypertension to ramipril: Indirect evidence that inhibition of tissue angiotensin converting enzyme is important. *Am. J. Cardiol.* **59,** 92D-97D.

White, W. B., Viadero, J. J., Lane, T. J. and Podelsa, S. (1986). Effects of combination therapy with captopril and niedipine in severe or resistant hypertension. *Clin. Pharmacol. Therap.* **39,** 43-48.

Williams, G. H. (1988). Converting enzyme inhibitors in the treatment of hypertension. *N. Engl. J. Med.* **319,** 1517-1525.

Witte, P. U. and Walter, U. (1987). Comparative double blind study of ramipril and captopril in mild to moderate essential hypertension. *Am. J. Cardiol.* **59,** 115D-120D.

Woo, J., Woo, K. S., Kin, T. and Vallence-Owen, J. (1987). A single blinded, randomized crossover study of angiotensin converting enzyme inhibitor and triamterene and hydrochlorothiazide in the treatment of mild to moderate hypertension in the elderly. *Arch. Int. Med.* **147,** 1386-1389

Yang, H.Y.T., Erdos, E. G. and Levin, Y. (1970). A dipeptidyl carboxypeptidase that converts angiotensin I and inactivates bradykinin. *Biochem. Biophys. Acta.* **214,** 374-376.

Ying, C. Y., Tifft, C. P., Gavras, H. and Chobanian, A. V. (1984). Renal revascularization in the azotemic hypertensive patient resistant to therapy. *N. Engl. J. Med.* **311,** 1070-1075.

Zabludowski, J., Rosenfeld, J., Akbary, M. A. and Schinzel, S. (1988). A multicenter randomized double-blind study comparing the antihypertensive efficacy and safety of ramipril with enalapril in mild to moderate hypertension. *Eur. Heart. J.* **9** (Suppl 1), 76-89.

Zacharia, P. K., Bonnet, G., Chrysant, S. G., De Backer, G., Goldstein, R., Herrera, J., Lindner, A., Materson, B. J., Maxwell, M. H. and McMahon, F. G. (1987). Evaluation of antihypertensive efficacy of lisinopril compared to metoprolol in moderate to severe hypertension. *J. Cardiovas. Pharmacol.* **9** (Suppl 3), s53-s58.

Zenella, M. T., Mattei, E., Draibe, S. A., Kater, C. E. and Ajzen, H. (1985). Inadequate aldosterone response to hyperkalemia during angiotensin converting enzyme inhibition in chronic renal failure. *Clin. Pharmacol. Therap.* **38,** 613-617.

Progress in Hypertension, Vol. 2, pp. 131-144.
Saito *et al.* (Eds)

MOLECULAR PHARMACOLOGY OF INTRA-CELLULAR Ca^{2+} ANTAGONISTS

HIROYOSHI HIDAKA, TOMOHIKO ISHIKAWA and TAKASHI CHIJIWA*

Department of Pharmacology, Nagoya University School of Medicine, Tsurumai-cho 65, Showa-ku, Nagoya 466, Japan; and * Pharmacology, 3rd. Tokushima Institute of New Drug Research, Ohtuka Pharmaceutical Co. Ltd., Kagasuno, Kawauchi-cho, Tokushima 771-01, Japan.

INTRODUCTION

Calcium ion (Ca^{2+}) plays a critical role in various cell functions as a second messenger (Ebashi, 1985; Abdel-Latif, 1986). Thus, it is necessary to clarify the molecular mechanism of intracellular Ca^{2+} regulation in the cell. Inevitably, the developments in the therapeutic exploitation of Ca^{2+} modulation will be paralleled by developments in our basic understanding of the molecular events underlying Ca^{2+} regulatory events. Since there are many drugs acting on cellular Ca^{2+} function as well as Ca^{2+} movements (Hidaka and Tanaka, 1985; Janis *et al.*, 1987; Tanaka and Hidaka, 1989), they may be useful to elucidate the regulation mechanisms of Ca^{2+}-related intracellular messenger systems.

In this chapter, we review molecular pharmacology of "intracellular Ca^{2+} antagonists" which modulate cellular Ca^{2+} function, especially in vascular and non-vascular smooth muscle contraction. In addition, we also introduced a clinically applicable, novel intracellular Ca^{2+} antagonist, HA1077, which has been proved to be an effective anti-cerebral vasospasm agent (Takayasu *et al.*, 1986).

REGULATION OF Ca^{2+} IN THE CELL

It is well recognized the importance of Ca^{2+} for the transmission of information inside living cells (Rasmussen and Barret, 1984; Gill, 1985). In the unstimulated cell, the cytosolic Ca^{2+} concentration is very low (approximately 0.1μM), and following a stimulus it rises to 1 to 10 μM. The source of Ca^{2+} could be either extracellular or intracellular. Stimulation of cell-surface receptors initiates hydrolysis of a membrane-bound inositol lipid, which produces second messenger, inositol 1,4,5-trisphosphate (IP_3) which releases Ca^{2+} from internal stores (Berridge and Irvine, 1989). The mobilized Ca^{2+} binds to the intracellular Ca^{2+} receptors (Ca^{2+} binding proteins), calmodulin, troponin C, calcyclin, parvalbunins, calbindins, S-100 proteins, and

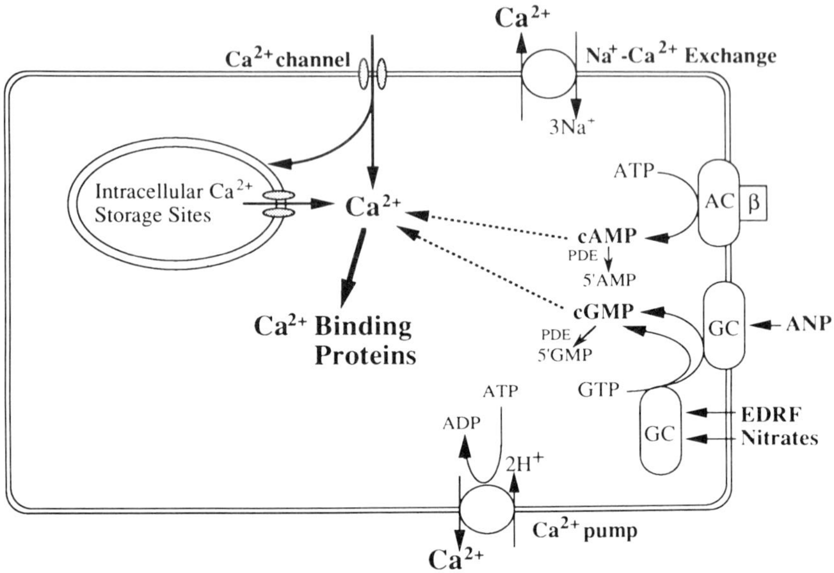

Figure 1. Schematic representation of the cellular Ca^{2+} mobilization in the vascular smooth muscle. Cyclic nucleotides are also known to affect the Ca^{2+} mobilization.

calcineurin, etc. (Moore and Dedman, 1982; Heizmann and Hunziker, 1990). These proteins exhibit a general structural principle, the EF-hand model, which bind Ca^{2+} selectively and with high affinity.

These pathways form the cornerstone of an ubiquitous transduction mechanism now known to regulate a large array of cellular processes including metabolism, secretion, contraction, neural activity and cell proliferation.

The mobilized Ca^{2+} is then extruded and/or sequestrated, so the intracellular concentration of Ca^{2+} returns to resting levels (Fig. 1).

DEFINITION AND CLASSIFICATION OF INTRACELLUAR Ca^{2+} ANTAGONISTS

About 20 years have past since the term "calcium antagonists" was introduced by Fleckenstein *et al.* (1969). Ca^{2+} antagonists (= Ca^{2+} channel blocker or Ca^{2+} entry blocker) are defined as agents that act on Ca^{2+} channels operated by membrane depolarization or other stimuli and inhibit Ca^{2+} influx. The pharmacological basis of the therapeutic effects of Ca^{2+} antagonists is their interference with Ca^{2+} transport over excitable membranes in different tissues. They are currently employed successfully in a number of cardiovascular disorders including hypertension, variant and exertional angina and certain types of cardiac arrythmias (Godfraind *et al.*, 1986).

Recently, a WHO committee of experts have classified calcium channel blockers into six groups taking into consideration their effects in a number of tests: Group I includes verapamil-like drugs (i.e. phenylalkylamine derivatives such as verapamil or gallopamil), Group II is made up of nifedipine-like drugs (i.e. dihydropyridine derivatives such as nifedipine, nicardipine, nimodipine or nitrendipine), Group III is constituted by diltiazem (a benzothiazepine derivative), Group IV is formed by flunarizine-like drugs (i.e. diphenylpiperazine derivatives such as flunarizine or cinnarizine), Group V includes prenylamine-like agents (e.g. prenylamine or fendiline), and finally Group VI is made up of other calcium channel blockers (e.g. caroverine or perhexiline) (Vanhoutte and Paoletti, 1987).

A Ca^{2+} antagonist, a compound that does not necessarily (and in fact is unlikely to) compete with Ca^{2+} for a binding site, may now be defined as a drug that alters the cellular function of Ca^{2+} by inhibiting its entry and/or its release and/or by interfering with one of its intracellular actions.

The term Ca^{2+} overload inhibitors has also been utilized in some classifications, but has not gained wide currency.

On the other hand, it is also possible to affect the cell function via the inhibition of intracellular Ca^{2+}-dependent events, without preventing Ca^{2+} influx, in the exited cells.

Such agents can be classified to be intracellular Ca^{2+} antagonists which, for example, antagonize the effects of Ca^{2+} in smooth muscle intracellularly by directly

Table 1. Classification of Intracellular Ca^{2+} Antagonists

Ca^{2+} release Inhibitors	Calmodulin Antagonists	Calmodulin-dependent Enzyme Inhibitors	
		Direct	Indirect
TMB-8	Naphthalenesulfonamides	MLCK Inhibitors	Calmodulin
Ryanodine	W-7	ML-9	Antagonists
Dantrolene	W-13	ML-7	
Procaine	No. 233	Thyroxine	
MDIs	Phenothiazine	KT5926	
KH	Chlorpromazine	Kaempferol	
KT-362	Trifluoperazine	Amiloride	
	Diphenylpropylamines	K252a	
HA-1004	Prenylamine	CaM KII Inhibitor	
HA-1077	Phendiline	KM-62	
	Benzylisoquinoline	Ca^{2+} PDE Inhibitor	
	Calmidazolium(R24571)	Vinpocetine	
	Compound 48/80	(TCV-3B)	
	Ro22-4839		
	HT-74		
	GMCHA		

modulating Ca^{2+}-regulated actin-myosin interactions. Especially calmodulin antagonist has been first introduced as a intracellular Ca^{2+} antagonist (Hidaka *et al.*, 1978; see later for details of myosin phosphorylation theory in smooth muscle contraction).

Taking into account the action of the agents at the molecular level, intracellular Ca^{2+} antagonists can be divided into two large categories. The first can be thought of as those agents which may interfere with the release of Ca^{2+} from intracellular storage sites (such as mitochondria or sarcoplasmic reticulum); the second category comprises those agents which may interfere with the effectors of Ca^{2+}. This latter category can include agents which directly modify interactions between Ca^{2+} and the intracellular Ca^{2+} binding proteins (such as calmodulin or troponin C), agents which modify the interactions between the Ca^{2+}-Ca^{2+} binding protein complex and target system (usually an enzyme), or agents which may directly modify the target system (such as direct protein kinase inhibitors) (summarized in Table 1).

In this article, agents which have been reported to modify intracellular Ca^{2+} release will be reviewed. This will be followed by sections describing agents which inhibit the actions of Ca^{2+} through competition with calmodulin, agents which may directly modify some Ca^{2+}/calmodulin messenger systems.

Intracellular Ca^{2+} release inhibitor

This class of drugs differs markedly from the traditional Ca^{2+} antagonists in their mechanism of action. Specifically, it appears that release of intracellular Ca^{2+} which occurs in response to agonists, such as norepinephrine or caffein is inhibited by these drugs. A large body of information can be found in the excellent review (Janis *et al.*, 1987).

One of the first agents to be described as a "intracellular Ca^{2+} antagonist " was 8-(N,N-dimethylamino)octyl-3,4,5,-trimethoxybenzoate hydrochloride (TMB-8). An early report showed that TMB-8 had inhibitory effects in smooth muscle by interfering with the availability of Ca^{2+} for muscle contraction by blocking the Ca^{2+} release from intracellular bound stores, and inhibited caffeine-induced Ca^{2+} release from skeletal muscle sarcoplasmic reticulum (Malagodi and Chiou, 1974; Chiou and Malagodi, 1975). However, it is also known that the agents inhibited the K^+-depolarized contractions in smooth muscle preparations (Malagodi and Chiou, 1974; Weichman *et al.*, 1983).

Ryanodine (insecticide alkaloid) has been demonstrated to alter specifically Ca^{2+} movements across sarcoplasmic reticulum membranes in cardiac and skeletal muscles (Sutko *et al.* 1985). They also reported that ryanodine caused a diminished release of Ca^{2+} from the intracellular store in vascular smooth muscle (Ito *et al.*, 1986). More recently, Iino *et al.* (1988) reported that treatment of skinned fiber bundles of smooth muscle with ryanodine depletes Ca^{2+} store (caffeine-sensitive but not IP_3-sensitive) due to open lock of the Ca^{2+}-induced Ca^{2+} release channels with the drug.

Another group of agents which have been suggested to function as intracellular Ca^{2+} antagonists are the methylenedioxyindenes (MDI compounds). The MDI can relax vascular and non-vascular smooth muscles, inhibit cardiac contractile activity, and exhibit some antiarrhythmic activity (Rahwan, 1985; Rahwan *et al.* 1979).

Itoh *et al.* (1981; 1982) have demonstrated that procaine, a local anesthetic, can inhibit the caffeine-induced contraction in skinned muscles of vascular tissues. These studies have implied that procaine probably inhibited the activation of the Ca^{2+}-induced Ca^{2+}-release mechanism from the Ca^{2+} storage sites.

Recently, Shibata *et al.* (1987) reported that 5-[3[[2-(3,4-dimethoxy phenyl)-ethyl]amino]-1-oxopropyl]-2,3,4,5-tetrahydro-1,5-benzothiazepine fumarate (KT-362), an anti-arrhythmic agent with vasodilatory effects, depressed contractile response to α-adrenoceptor agonists without affecting KCl-induced contractions. It is suggested that the primarily inhibitory action of KT-362 on vascular smooth muscle is due to the interference with the release of intracellular Ca^{2+} from norepinephrine- and caffeine-sensitive Ca^{2+} stores.

N-(2-guanidinoethyl)-5-isoquinolinesulfonamide (HA1004) was found to be representative of a novel class of pharmacological agents that have been characterized as Ca^{2+} antagonists acting at an intracellular site rather than at cell surface sites (Asano and Hidaka, 1984). HA1004 produced a dose-dependent relaxation of vascular strips contracted by various agonists such as KCl, phenylephrine, histamine, 5HT, angiotensin II and $PGF_{2\alpha}$, with similar potent. Neither propranolol, atropine, theophylline nor indomethacin interfered significantly with the relaxant action of HA1004. Although HA1004 had no effect on the Ca^{2+}-induced contraction in skinned smooth muscle cells, the drug can relax A23187-induced contraction in the normal bathing solution, and phenylephrine-induced contraction in the Ca^{2+} -free solution (Ishikawa *et al.*, 1985). HA1004 will be a potent useful tool to elucidate the molecular mechanism of intracellular Ca^{2+} utilization.

Calmodulin antagonists

It is generally accepted that calmodulin acts as an intracellular Ca^{2+} receptor and regulates many biological functions in cells (Dedman *et al.*, 1978; Klee *et al.*, 1980; Weiss and Wallace, 1980; Cheung, 1980, 1982a,b; Means and Dedman, 1980; Means *et al.*, 1982).

There are many agents which have been shown to inhibit calmodulin-dependent enzymes through binding of the Ca^{2+}-calmodulin complex.

At the present time, considerable evidence supports the theory that Ca^{2+} antagonists (Ca^{2+} channel blockers) inhibit plasmalemmal Ca^{2+} channel that are activated by depolarization and receptor occupation. The definite difference in molecular mechanisms of Ca^{2+} antagonists and calmodulin antagonists can be demonstrated clearly by the different effects of these agents on Ca^{2+}-induced contraction in potassium-depolarized vascular strips (Kanamori *et al.*, 1981).

Phenothiazines and a series of naphthalenesulfonamides were the first compounds demonstrated to antagonize the activity of calmodulin. These agents inhibit contractions in vascular strips elicited by various contractile agonists and by depolarization.

We have firstly reported that calmodulin antagonist such as N-(6-aminohexyl)-5-chrolo-1-naphthalenesulfonamide (W-7) which we synthesized, relaxed vascular strips and inhibited aortic actomyosin superprecipitation thereby predicting the involvement of calmodulin in vascular smooth muscle contraction (Hidaka *et al.* 1978). This hypothesis was given support in two different types of experiments, reported by Dabrowska *et al.* (1978) and Hidaka *et al.* (1979).

The naphthalenesulfonamide series of calmodulin antagonists, typified by W-7, have been extensively characterized primarily by us. Structure-activity relationships for W-7 derivatives (naphthalenesulfonamides) as calmodulin antagonists, vasorelaxants and platelet aggregation inhibitors have also been determined. Halogenation of the naphthalene ring is important for potency, and the length of the alkyl chain separating the naphthalene ring from the terminal nitrogen group also greatly affects potency. A good correlation between potency in relaxation of vascular strips and affinity to calmodulin was obtained (Hidaka *et al.* 1981).

The Ca^{2+} antagonist-induced relaxation was completely reversed by the addition of an excess amount of Ca^{2+} ion and Ca^{2+} antagonists shifted the dose-response curve of Ca^{2+} ion to the right in a competitive manner. Calmodulin antagonists such as W-7, however, induced relaxation that was only partially reversed by 10 mM $CaCl_2$, and W-7 suppressed the maximum contractile response, suggesting that relaxation by W-7 is noncompetitive for Ca^{2+}-induced contraction in potassium-depolarized strips. W-7 inhibits several Ca^{2+}/calmodulin dependent enzymes, including Ca^{2+}-dependent cyclic nucleotide phosphodiesterase, myosin light chain kinase, Ca^{2+}/calmodulin kinase II and Ca^{2+}/Mg^{2+}-ATPase activity, by Ca^{2+}-dependent binding to calmodulin with two equilibrium constants (K) of 11 μM and 200 μM (Hidaka *et al.*, 1978a; 1979a,b; 1980).

The concept of "calmodulin antagonism" was firstly introduced by us (Hidaka *et al.*, 1978). And we proposed that calmodulin antagonists must meet definite criteria; (1) selectively inhibit the calmodulin-dependent enzymes activated by Ca^{2+}-calmodulin, (2) bind with calmodulin in a Ca^{2+}-dependent manner. (3) the inhibition is overcome by excess amount of calmodulin but not by Ca^{2+}, (4) significant cell membrane permeability.

Some Ca^{2+} channel blockers were shown to possess calmodulin antagonistic action. Diphenylalkylamines such as bepridil, cinnarizine, flunarizine, and the dihydropyridines, felodipine and nicardipine, bind to calmodulin and inhibit enzyme activation by calmodulin (Bostrom *et al.*, 1981; Itoh *et al.*, 1984; Johnson *et al.*, 1983; Lugnier *et al.*, 1984; Silver *et al.*, 1985).

The binding of the dihydropyridine, felodipine, to calmodulin was allosterically modulated by D600 and diltiazem (Johnson *et al.*, 1983; Johnson, 1984), in a manner analogous to dihydropyridine binding to Ca^{2+} channels (De Pover *et al.*, 1982; Ferry and Glossmann, 1983). Furthermore, Luchowski *et al.* (1984) showed that the affinity of calmodulin antagonists for nitrendipine binding sites correlates well with their binding to calmodulin.

Bepridil inhibited the Ca^{2+}-induced contraction of saponin-skinned mesenteric arteries, and the addition of excess calmodulin partly reversed the relaxant effects of this agent (Itoh *et al.*, 1984). Felodipine, a dihydropyridine Ca^{2+} antagonist, interacted with calmodulin (Bostrom *et al.*, 1981). Felodipine, prenylamine, fendiline, D600, verapamil and diltiazem inhibited calmodulin-induced activation of myosin light chain kinase, however, the concentrations of diltiazem, verapamil and D600 needed for the binding to calmodulin and inhibition of calmodulin-induced activation of myosin light chain kinase are higher that those of their pharmacological action (Johnson *et al.*, 1983). However, verapamil and diltiazem, as well as most dihydropyridine derivatives, have a very low affinity for calmodulin and do not inhibit contraction in skinned smooth muscle fibers (Itoh *et al.*, 1984; Kreye *et al.*, 1983; Metzger *et al.*, 1982; Spedding, 1983), and the effectiveness of many dihydropyridines as inhibition of purified phosphodiesterase is generally about 3 orders of magnitude less than their effectiveness at displacing [^3H]-nitrendipine binding (Van Inwegen *et al.*, 1984). Furthermore, Ca^{2+} entry blockers do not inhibit contraction mediated by the release of intracellular Ca^{2+} stores even though some of them, such as bepridil and cinnarizine (Godfraind and Morei, 1977; Mras and Sperelakis, 1982; Pang and Sperelakis, 1983), appear to be concentrated in cells to an extent severalfold (>10) that of the normal extracellular concentration.

Unlike Ca^{2+} channel blockers, calmodulin antagonists do not generally exhibit stereoselectivity for binding or inhibition (Levin and Weiss, 1979; Norman *et al.*, 1979).

Direct calmodulin-dependent enzyme inhibitors

As mentioned above, calmodulin is now recognized as a cellular Ca^{2+}-binding protein and regulates several enzymes activity. Ca^{2+} in turn binds to calmodulin and the Ca^{2+}/calmodulin complex activates myosin light chain kinase (MLCK), calmodulin-dependent protein kinase II (CaM KII) and Ca^{2+}-dependent cyclic nucleotide phosphodiesterase (Ca^{2+}-PDE) (Fig. 2). That is, MLCK rapidly phosphorylates

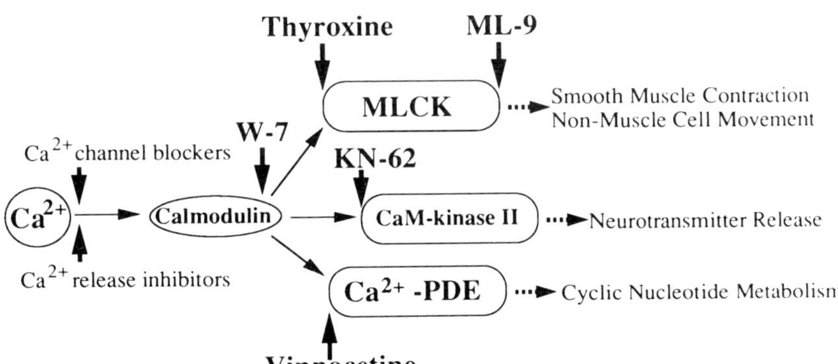

Figure 2. Intracellular Ca^{2+}/calmodulin messenger system and specific inhibitors of calmodulin-dependent enzymes. MLCK, Myosin Light Chain Kinase; CaM-kinase II, calmodulin-dependent protein kinase II; Ca^{2+}-PDE, Ca^{2+}/calmodulin-dependent cyclic nucleotide phosphodiesterase. Thyroxine, KN-62 and vinpocetine act on the regulatory site of the enzymes, while ML-9 act on the catalytic site of the protein kinase (see text for details).

smooth muscle light chain which results in the initiation of smooth muscle contraction (Kamm and Stull, 1985).

Since calmodulin exerted pleiotropic effects on various cellular function by activating multiple enzymes, calmodulin antagonists were not adequate to elucidate the specific role for the Ca^{2+}/calmodulin-dependent enzymes, *in vivo.*

Ca^{2+}-PDE inhibitors

Calmodulin was found to be a Ca^{2+}-dependent stimulator of cyclic nucleotide phosphodiesterase, but it was difficult to evaluate the physiological significance of the enzyme by using calmodulin antagonists. Therefore selective inhibitors of Ca^{2+}-PDE should be potential tools for characterizing the properties and physiological functions of this enzyme. Vinpocetine, (TCV-3B , 14-ethoxycarbonyl-(3α, 16α ethyl)-14,15-eburnamenine, inhibited Ca^{2+}-PDE not by interacting with calmodulin but rather by interacting directly with the enzyme (Hagiwara *et al.* 1984). The concentration of vinpocetine which produced 50% inhibition of the Ca^{2+}-PDE activity was approximately 21 μM, both in the presence and absence of Ca^{2+}/calmodulin complex. This compound was only a weak inhibitor of cGMP-PDE and cAMP-PDE, with IC_{50} values above 500 μM.

The psychotropic agents haloperidol and clozapine, which have been shown to inhibit Ca^{2+}-PDE activity with IC_{50} values 40 μM and 20 μM respectively (Norman *et al.*, 1979), did not significantly affect phosphorylation at 100 μM (Silver *et al.*, 1986).

MLCK inhibitors

Secondly, to elucidate the physiological significance of protein phosphorylation, we continued efforts to develop selective inhibitors for each protein kinase (Hidaka *et al.* 1984). These agents, isoquinolinesulfonamides, were structurally related to W-7, but differ by having an isoquinoline ring instead of a naphthalene ring, and were no longer calmodulin nor phospholipid interacting agents but rather compounds which directly inhibited kinase activity by competing for ATP binding (Hidaka *et al.* 1984).

As described above, the Ca^{2+}-dependent phosphorylation of 20 kDa light chain of myosin is necessary for the smooth muscle contraction. Ca^{2+}/calmodulin complex activate myosin light chain kinase (MLCK), which catalyzes the phosphorylation of a serine (and threonine in certain conditions) residue on myosin light chain. When this light chain is phosphorylated, increases in the activity of actomyosin ATPase and in the rate of actin-myosin interactions occurs (Ikebe and Hartshorne, 1985; 1986).

Because myosin light chain phosphorylation is important in the regulation of contractile activity in smooth muscle, physiologic or pharmacologic interventions might conceivably influence regulation of extent of the myosin light chain phosphorylation. These interventions might indirectly influence phosphorylation by altering the amount of Ca^{2+} (Ca^{2+} channel blockers and Ca^{2+} release inhibitors) or calmodulin available for activation of the MLCK (calmodulin antagonists) or might also act directly by influencing the enzymatic properties of MLCK.

Calmodulin antagonists such as W-7 or phenothiazine derivatives inhibit MLCK activity through their interaction with Ca^{2+}/calmodulin complex and suppress vascular contractions (Hidaka *et al.*, 1978; 1980). A derivative of W-7, named ML-9, 1-(5-chloronaphthalene-1-sulfonyl)-1H-hexahydro-1,4-diazepine, exhibited the selective inhibition toward smooth muscle MLCK (Ki=3.8 μM) (Saitoh *et al.* 1987). The mode of interaction of ML-9 with MLCK was investigated by double reciprocal plots, and it revealed that ML-9 was a competitive inhibitor with ATP and noncompetitive inhibitor with the phosphate acceptor of the catalytic reaction. ML-9 had no or little effect on any other enzymes tested *in vitro* over similar concentration range to inhibit MLCK. It is suggested that ML-9 interacts with MLCK at binding site or near the site of ATP (Saitoh *et al*, 1987).

Then, we investigated the effects of ML-9 on superprecipitation of actomyosin, isometric tension development, and phosphorylation of myosin light chain in vascular smooth muscle. Preincubation of ML-9 significantly inhibited both the maximal rate and extent of KCl-induced contraction and the phosphorylation of myosin light chain, in a dose-dependent manner (Ishikawa *et al.* 1988). Furthermore, ML-9 not only nonspecifically antagonized the contraction induced by various contractile agonists, but also inhibited the Ca^{2+}-independent contraction provoked by application of trypsin-treated MLCK in chemically skinned vascular smooth muscle cells (Ishikawa *et al.* 1988). ML-9 was then used as a pertinent pharmacological tool for studying physiological roles of MLCK.

We also found that thyroid hormones inhibit MLCK activity and have direct effects on vascular smooth muscle. Kinetic analysis indicated that thyroxine inhibited MLCK competitively with respect to calmodulin and the Ki value was 2.5 μM (Hagiwara *et al.* 1989). Contrary to the potent inhibition by thyroxine of MLCK, enzyme activities of other protein kinases were affected only weakly by thyroxine. Thyroxine was thought to bind at or near calmodulin-binding domain of MLCK and inhibit calmodulin-induced activation of the enzyme. Pretreatment of thyroxine inhibited the contractile response concomitant with the inhibition of the myosin light chain phosphorylation without significant suppression of Ca^{2+} influx from extracellular fluid (Ishikawa *et al.*, 1989).

These results clearly demonstrate the involvement of Ca^{2+} /calmodulin-dependent myosin light chain phosphorylation in the regulation of vascular contractile activity. We also presented some data that W-7, ML-9, and thyroxine (binds to calmodulin, the catalytic domain of MLCK, and the regulatory domain of MLCK, respectively) are selective MLCK inhibitors, which are novel classes of vascular relaxants and these modulators can be classified into three types on the basis of sites of drug action.

Moreover, some other compounds have been found to inhibit MLCK. Amiloride (Higashihara, 1989), Kaempferol (Rogers and Williams, 1989) and KT5926 (Nakanishi *et al.*, 1990) inhibited the protein kinase activity with Ki values of 0.95, 0.5 and 0.018 μM, respectively. The mode of inhibition of these compounds was competitive with respect to ATP.

CaM KII inhibitor

In 1980, we first reported the existence of Ca^{2+}/calmodulin-dependent protein kinase phosphorylated myelin basic protein in rabbit brain (Endo and Hidaka, 1980).

Yamauchi and Fujisawa (1981) reported Ca^{2+}/calmodulin-dependent protein kinase II (CaM KII) prepared from rat brain cytosol. CaM KII was found to have relatively broad substrate specificities in the nervous system and to be a large multimetic enzyme composed of two related subunits with α (50 kDa) and β (60 kDa). Because of the complexity of activation mechanism and broad specificities of substrates, the physiological role of this kinase was not fully elucidated.

Recently, we reported a newly developed synthetic inhibitor, specific for CaM KII (Tokumitsu *et al.* 1990). 1-[N,O-bis(5-isoquinolinesulfonyl)-N-methyl-L-tyrosyl]-4-phenylpiperazine (KN-62) inhibited CaM KII activity with Ki value of 0.9 μM, but no significant effect up to 100 μM on activities of MLCK, protein kinase C, cAMP-dependent protein kinase. KN-62 also inhibited the Ca^{2+}/calmodulin-dependent autophosphorylation of CaM KII dose dependently in the presence or absence of exogenous substrate. Kinetic analysis indicated that this inhibitory effect of KN-62 was competitive with respect to calmodulin. However, KN-62 did not inhibit the activity of autophosphorylated CaM KII.

KN-62 affects the interaction between calmodulin and CaM KII following inhibition of this kinase activity by direct binding to the calmodulin binding site of the enzyme. In the recent work, we have examined the physiological function of CaM KII, using KN-62. CaM KII in PC12D cells was detected by immunoprecipitation, and the phosphorylation of the enzyme was increased by A23187 which increases intracellular Ca^{2+}. KN-62 blocked the phosphorylation of immunoprecipitated CaM KII (53 kDa) dose dependently in PC12D cells. The function of CaM KII is thought to be involved in various Ca^{2+} signal transduction in various tissues and cells. Present study shows that the specific inhibitor against CaM KII, KN-62, is a useful tool for elucidation of the function of the enzyme in relation to cellular responses.

CLINICALLY APPLICABLE INTRACELLULAR Ca^{2+} ANTAGONIST

During a search of the various sulfonamide derivatives, HA-1077, a intracellular Ca^{2+} antagonist, was created by modifying the structure of W-7. HA-1077 is a novel and potent Ca^{2+} antagonist vasodilator which is considered to act by employing different mechanisms from the usual Ca^{2+} channel blockers, such as nifedipine, verapamil or diltiazem (Asano *et al.* 1987). We found that intrathecal administration of isoquinoline sulfonamides reversed completely the cerebral vasospasm in the two-hemorrhage canine model of SAH (Takayasu *et al.*, 1986; Shibuya *et al.*, 1988). Among these derivatives, HA-1077 had the most potent vasodilator action in dogs. HA-1077 given i.v. at relative high doses produced cerebral vasodilation in the canine model of SAH (Takayasu *et al.*, 1986; Shibuya *et al.*, 1988). This is the first drug which angiographically dilated the spastic basilar artery on SAH on day 7 via i.v. administration of doses which caused the systemic blood pressure to fall only slightly in a two-hemorrhage canine model.

CONCLUSION

Pharmacologically, several classes of agents with differing chemical structures have been identified as "intracellular Ca^{2+} antagonists". We discussed that some vasorelaxants, which directly inhibit Ca^{2+}/calmodulin-dependent processes and indirectly affect the intracellular Ca^{2+} messenger system in vascular smooth muscle, may be classified as "intracellular Ca^{2+} antagonists" and new classes of vascular relaxants. These agents clearly demonstrated the involvement of Ca^{2+}/calmodulin-dependent myosin light chain phosphorylation in the regulation of vascular contractile activity. Future research on the Ca^{2+} signalling in various cells may lead to new insights and potential therapeutic applications of intracellular Ca^{2+} antagonists.

REFERENCES

Abdel-Latif, A. A. (1986). Calcium-mobilizing receptors, polyphosphoinositides, and the generation of second messengers. *Pharmacol. Rev.* **38,** 227-272.

Asano, T. and Hidaka, H. (1984). Vasodilatory action of HA1004 [N-(2-guanidinoethyl)-5-isoquinolinesulfonamide], a novel calcium antagonist with no effect on cardiac function. *J. Pharmacol. Exp. Ther.* **231,** 141-145.

Asano, T., Ikegaki, I., Satoh, S., Suzuki, Y., Shibuya, M., Takayasu, M. and Hidaka, H. (1987). Mechanism of action of a novel antivasospasm drug, HA1077. *J. Pharmacol. Exp. Ther.* **241,** 1033-1040.

Berridge, M. J. and Irvine, R. F. (1989). Inositol phosphates and cell signalling. *Nature (London)* **341,** 197-205.

Boström, S. L., Ljung, B., Mardh, S., Forsen, S. and Thulin, E. (1981). Interaction of the antihypertensive drug felodipine with calmodulin. *Nature (London)* **292,** 777-778.

Cheung, W. Y. (1980). Calmodulin plays a pivotal role in cellular regulation. *Science* **207,** 19-27.

Cheung, W. Y. (1982a). Calmodulin. *Sci. Am.* **246,** 62-70.

Cheung, W. Y. (1982b). Calmodulin: an overview. *Fed. Proc.* **41,** 2253-2257.

Chiou, C. Y. and Malagodi, M. H. (1975). Studies on the mechanism of action of a new Ca^{2+} antagonist, 8-(N,N-diethylamino)octyl 3,4,5-trimethoxybenzoate hydrochloride in smooth and skeletal muscles. *Br. J. Pharmacol.* **53,** 279-285.

Dabrowska, R., Rherry, J.M.F., Aromatorio, D. K. and Hartshorne, D. J. (1978). Modulator protein as a component of the myosin light chain kinase from chicken gizzard. *Biochemistry* **17,** 253-258.

De Pover, A., Matlib, M. A., Lee, S. W., Dube, G. P., Grupp, I. L., Grupp, G. and Schwartz, A. (1982). Specific binding of [^{3}H]nitrendipine to membranes from coronary arteries and heart in relation to pharmacological effects. Paradoxical stimulation by diltiazem. *Biochem. Biophys. Res. Commun.* **108,** 110-117.

Dedman, J. R., Welsh, M. J. and Means, A. R. (1978). Ca^{2+}-dependent regulator. Production and characterization of a monospecific antibody. *J.Biol. Chem.* **253,** 7517-7521.

Ebashi, S. (1985). In: *Cell Motility: Mechanism and Regulation* (Eds) Ishikawa, H., Hatano, S. and Sato, H.), Univ. of Tokyo Press, Tokyo, pp.3-12.

Endo, T. and Hidaka, H. (1980). Ca^{2+}-calmodulin dependent phosphorylation of myelin isolated from rabbit brain. Biochem. Biophys. *Biochem. Biophys. Res. Commun.* **97,** 553-558.

Ferry, D. R. and Glossmann, H. (1983). *Br. J. Pharmacol.* **78-80** (suppl.), 81p.

Fleckenstein, A., Tritthart, H., Fleckenstein, B., Herbst, A. and Grün, G. (1969). A new group of competitive Ca antagonists with highly potent inhibitory effects on excitation-contraction coupling in mammalian myocardium. *Pflugers Arch.* **307,** R25.

Gill, D. L. (1985). Receptors coupled to calcium mobilization. *Adv. Cyclic Nucleotide Protein Phosphorylation Res.* **19,** 307-324.

Godfraind, T. and Morei, N. (1977). An upper limit to the number of calcium channels in smooth muscle as estimated with cinnarizine. *Eur. J. Pharmacol.* **41,** 245-246.

Godfraind, T., Miller, R. and Wibo, M. (1986). Calcium antagonism and calcium entry blockade. *Pharmacol. Rev.* **38**, 321-416.

Hagiwara, M., Endo, T. and Hidaka, H. (1984). Effects of vinpocetine on cyclic nucleotide metabolism in vascular smooth muscle. *Biochem. Pharmacol.* **33,** 453-457.

Hagiwara, M., Mamiya, S. and Hidaka, H. (1989). Selective binding of L-thyroxine by myosin light chain kinase. *J. Biol. Chem.* **264,** 40-44.

Heizmann, C.W. and Hunziker, W. (1990). In: *Intracellular Calcium Regulation*, Alan R. Liss, Inc., pp.211-248.

Hidaka, H., Asano, M. and Tanaka, T. (1981). Activity-structure relationship of calmodulin antagonists. Naphthalenesulfonamide derivatives. *Mol. Pharmacol.* **20,** 571-578.

Hidaka, H., Asano, M., Iwadare, S., Matsumoto, I., Totsuka, T. and Aoki, N. (1978). A novel vascular relaxing agent, N-(6-aminohexyl)-5-chloro-1-naphthalensulfonamide which affects vascular smooth muscle actomyosin. *J. Pharmacol. Exp. Ther.* **207,** 8-15.

Hidaka, H., Inagaki, M., Kawamoto, S. and Sasaki, Y. (1984). Isoquinolinesulfonamides, novel and potent inhibitors of cyclic nucleotide dependent protein kinase and protein kinase C. *Biochemistry* **23,** 5036-5041.

Hidaka, H., Naka, M. and Yamaki, T. (1979b). Effect of novel specific myosin light chain kinase inhibitors on Ca^{2+}-activated Mg^{2+}-ATPase of chicken gizzard actomyosin. *Biochem. Biophys. Res. Commun.* **90,** 694-699.

Hidaka, H., Sasaki, Y., Tanaka, T., Endo, T., Ohno, S., Fujii, Y. and Nagata, T. (1981). N-(6-aminohexyl)-5-chloro-1-naphthalenesulfonamide, a calmodulin antagonist, inhibits cell proliferation. *Proc. Natl. Acad. Sci. USA.* **78,** 4354-4357.

Hidaka, H. and Tanaka, T. (1985). In: *Calmodulin Antagonists And Cellular Physiology* (Eds) Hidaka and Hartshorne), Academic Press, New York, pp.13-23.

Hidaka, H., Yamaki, T., Asano, M. and Totsuka, T. (1978). Involvement of calcium in cyclic nucleotide metabolism in human vascular smooth muscle. *Blood Vessels* **15,** 55-64.

Hidaka, H., Yamaki, T., Naka, M., Tanaka, T., Hayashi, H. and Kobayashi, R. (1980). Calcium-regulated modulator protein interacting agents inhibit smooth muscle calcium-stimulated protein kinase and ATPase. *Mol. Pharmacol.* **17,** 66-72.

Hidaka, H., Yamaki, T., Totsuka, T. and Asano, M. (1979a). Selective inhibitors of Ca^{2+}-binding modulator of phosphodiesterase produce vascular relaxation and inhibit actin-myosin interaction. *Mol. Pharmacol.* **15,** 49-59.

Higashihara, M. (1989). Inhibition of myosin light chain kinase by amiloride. *Biochem. Biophys. Res. Commun.* **162,** 1253-1259.

Iino, M., Kobayashi, T. and Endo, M. (1988). Use of ryanodine for functional removal of the calcium store in smooth muscle cells of the guinea-pig. *Biochem. Biophys. Res. Commun.* **152,** 417-422.

Ikebe, M. and Hartshorne, D.J. (1985). Effects of Ca^{2+} on the conformation and enzymatic activity of smooth muscle myosin. *J. Biol. Chem.* **260,** 13146-13153.

Ikebe, M. and Hartshorne, D.J. (1986). Reverse reaction of smooth muscle myosin light chain kinase. Formation of ATP from phosphorylated light chain plus ADP. *J. Biol. Chem.* **261,** 8249-8253.

Ishikawa, T., Chijiwa, T., Hagiwara, M., Mamiya, S. and Hidaka, H. (1989). Thyroid hormones directly interact with vascular smooth muscle strips. *Mol. Pharmacol.* **35,** 760-765.

Ishikawa, T., Chijiwa, T., Hagiwara, M., Mamiya, S., Saitoh, M. and Hidaka, H. (1988). ML-9 inhibits the vascular contraction via the inhibition of myosin light chain phosphorylation. *Mol. Pharmacol.* **33,** 598-603.

Ishikawa, T., Inagaki, M., Watanabe, M. and Hidaka, H. (1985). Relaxation of vascular smooth muscle by HA-1004, an inhibitor of cyclic nucleotide-dependent protein kinase. *J. Pharmacol.*

Exp. Ther. **235,** 495-499.

Ito, K., Takakura, S., Sato, K. and Sutko, J. L. (1986). Ryanodine inhibits the release of calcium from intracellular stores in guinea pig aortic smooth muscle. *Circ. Res.* **58,** 730-734.

Itoh, H., Ishikawa, T. and Hidaka, H. (1984). Effects on calmodulin of bepridil, an antianginal agent. *J. Pharmacol. Exp. Ther.* **230,** 737-741.

Itoh,T., Kajiwara, M., Kitamura, K. and Kuriyama, H. (1982). Roles of stored calcium on the mechanical response evoked in smooth muscle cells of the porcine coronary artery. *J. Physiol.* **322,** 107-125.

Itoh, T., Kanmura, Y., Kuriyama, H. and Suzuki, H. (1984). Nisoldipine-induced relaxation in intact and skinned smooth muscles of rabbit coronary arteries. *Br. J. Pharmacol.* **83,** 243-258.

Itoh, T., Kuriyama, H. and Suzuki, H. (1981). Excitation--contraction coupling in smooth muscle cells of the guinea-pig mesenteric artery. *J. Physiol.* **321,** 513-535.

Janis, R. A., Silver, P. J. and Triggle, D. J. (1987). *Adv. in Drug Res.* **16,** 309-591.

Johnson, J. D. (1984). *Biophys. J.* **45,** 134-136.

Johnson, J. D. and Fugman D. A. (1983). Calcium and calmodulin antagonists bindings to calmodulin and relaxation of coronary segments. *J. Pharmacol. Exp. Ther.* **226,** 330-334.

Johnson, J. D. and Willenauer, L. A. (1983). A fluorescent calmodulin that reports the binding of hydrophobic inhibitory ligands. *Biochem. J.* **211,** 473-479.

Johnson, J. D., Wittenuer, L. A. and Nathan, R. D. (1983). Calmodulin, Ca^{2+} antagonists and Ca^{2+}-transporters in nerve and muscle. *J. Neural Transm.* (Suppl. 18), **7,** 97-111.

Kamm, K. E. and Stull, J. T. (1985). The function of myosin and myosin light chain kinase phosphorylation in smooth muscle. *Annu. Rev. Pharmacol. Toxicol.* **25,** 593-620.

Kanamori, M., Naka, M., Asano, M. and Hidaka, H. (1981). Effects of N-(6-aminohexyl)-5-chloro-1-naphthalenesulfonamide and other calmodulin antagonists (calmodulin interacting agents) on calcium-induced contraction of rabbit aortic strips. *J. Pharmacol. Exp. Ther.* **217,** 494-499.

Klee, C.B., Crouch, T.H. and Richman, P.G. (1980). Calmodulin. *Annu. Rev. Biochem.* **49,** 489-515.

Kreye, V.A.W., Ruegg, J.C. and Hofmann, F. (1983). Effect of calcium-antagonist and calmodulin-antagonist drugs on calmodulin-dependent contractions of chemically skinned vascular smooth muscle from rabbit renal arteries. *Naunyn-Schmiedeberg's Arch. Pharmacol.* **323,** 85-89.

Levin, R. M. and Weiss, B. (1979). Selective binding of antipsychotics and other psychoactive agents to the calcium-dependent activator of cyclic nucleotide phosphodiesterase. *J. Pharmacol. Exp. Ther.* **208,** 454-459.

Luchowski, E. M., Yousif, F., Triggle, D. J., Maurer, S. C., Sarmiento, J. G. and Jannis, R. A. (1984). Effects of metal cations and calmodulin antagonists on [^{3}H] nitrendipine binding in smooth and cardiac muscle. *J. Pharmacol. Exp. Ther.* **230,** 607-613.

Lugnier, C., Follennius, A., Gerard, D. and Stochet, J.C. (1984). Bepridil and flunarizine as calmodulin inhibitors. *Eur. J. Pharmacol.* **98,** 157-158.

Malagodi, M. H. and Chiou, C. Y. (1974). Pharmacological evaluation of a new Ca^{2+} antagonist 8-(N,N-diethylamino)-octyl-3,4,5-trimethoxybenzoate hydrochloride (TMB-8): studies in smooth muscles. *Eur. J. Pharmacol.* **27,** 25-33.

Means, A. R. and Dedman, J. R. (1980). Calmodulin--an intracellular calcium receptor. *Nature* **285,** 73-77.

Means, A. R. , Lagace, L., Guerriero, V. Jr. and Chafouleas, J. G. (1982). Calmodulin as a mediator of hormone action and cell regulation. *J. Cell. Biochem.* **20,** 317-330.

Metzger, H., Stern, H. O., Pfitzer, G. and Ruegg, J. C. (1982). Calcium antagonists affect calmodulin-dependent contractility of a skinned smooth muscle. *Arzneim-Forsch.* **32,** 1425-1427.

Moore, P. B. and Dedman, J. R. (1982). Calcium binding proteins and cellular regulation. *Life Sci.* **31,** 2937-2946.

Mras, S. and Sperelakis, N. (1982). Comparison of [^{3}H]bepridil and [^{3}H]verapamil uptake into rabbit aortic rings. *J. Cardiovasc. Pharmacol.* **4,** 777-783.

Nakanishi, S., Yamada, K., Iwahashi, K., Kuroda, K. and Kase, H. (1990). KT5926, a potent and selective inhibitor of myosin light chain kinase. *Mol. Pharmacol.* **37,** 482-488.

Norman, J. A., Drummond, A. H. and Moser, P. (1979). Inhibition of calcium-dependent regulator-stimulated phosphodiesterase activity by neuroleptic drugs is unrelated to their clinical efficacy. *Mol. Pharmacol.* **16,** 1089-1094.

Pang, D.C. and Sperelakis, N. (1983). Nifedipine, diltiazem, bepridil and verapamil uptakes into cardiac and smooth muscles. *Eur. J. Pharmacol.* **87,** 199-207.

Rahwan, R. G. (1985). The methylenedioxyindene calcium antagonists. *Life Sci.* **37**, 687-692.

Rahwan, R. G., Piascik, M. F. and Witiak, D. T. (1979). The role of calcium antagonism in the therapeutic action of drugs. *Can. J. Physiol. Pharmacol.* **57,** 443-460.

Rasmussen, H. and Barret, P. R. (1984). Calcium messenger system: an integrated view. *Physiol. Rev.* **64,** 938-984.

Rogers, J.C. and Williams, Jr. D.L. (1989). Kaempferol inhibits myosin light chain kinase. *Biochem. Biophys. Res. Commun.* **164,** 419-425.

Saitoh, M., Ishikawa, T., Matsushima, S., Naka, M. and Hidaka, H. (1987). Selective inhibition of catalytic activity of smooth muscle myosin light chain kinase. *J. Biol. Chem.* **262,** 7796-7801.

Sasaki, Y. and Hidaka, H. (1982). Calmodulin and cell proliferation. *Biochem. Biophys. Res. Commun.* **104,** 451-456.

Shibata, S., Wakabayashi, S., Satake, N., Hester, R.K., Ueda, S. and Tomiyama, A. (1987). Mode of vasorelaxing action of 5-[3-[[2-(3,4-dimethoxyphenyl)-ethyl]amino]-1-oxopropyl]-2,3,4,5-tetrahydro-1,5-benzothiazepine fumarate (KT-362), a new intracellular calcium antagonist. *J. Pharmacol. Exp. Ther.* **240,** 16-22.

Shibuya, M., Suzuki, Y., Takayasu, M., Asano, T., Harada, T., Ikegaki, I., Satoh, S. and Hidaka, H. (1988). The effects of an intracellular calcium antagonist HA 1077 on delayed cerebral vasospasm in dogs. *Acta Neurochir.* **90,** 53-59.

Silver, P. J., Dachiw, J., Ambrose, J. M. and Pinto, P. G. (1985). Effects of the calcium antagonists perhexiline and cinnarizine on vasuclar and cardiac contractile protein function. *J. Pharmacol. Exp. Ther.* **234,** 629-635.

Silver, P. J., Pinto, P. B. and Dachiw, J. (1986). Modulation of vascular and cardiac contractile protein regulatory mechanisms by calmodulin inhibitors and related compounds. *Biochem. Pharmacol.* **35,** 2545-2551.

Spedding, M. (1983). Direct inhibitory effects of some calcium-antagonists and trifluoperazine on the contractile proteins in smooth muscle. *Br. J. Pharmacol.* **79,** 225-231.

Sutko, J. L., Ito, K. and Kenyon, J.L. (1985). Ryanodine: a modifier of sarcoplasmic reticulum calcium release in striated muscle. *Fed. Proc.* **44,** 2984-2988.

Takayasu, M., Suzuki, Y., Shibuya, M., Asano, T., Kanamori, M., Okada, T., Kageyama, N. and Hidaka, H. (1986). The effects of HA compound calcium antagonists on delayed cerebral vasospasm in dogs. *J. Neurosurg.* **65,** 80-85.

Tanaka, T. and Hidaka, H. (1989). In: *Calcium Protein Signaling* (Eds) Hidaka, H., Carafoli, E., Means, A.R. and Tanaka, T.), Adv. Exp. Med. Biol., Plenum Press, New York, pp165-171.

Tokumitsu, H., Chijiwa, T., Hagiwara, M., Mizutani, A., Terasawa, M. and Hidaka, H. (1990). KN-62, 1-[N,O-bis(5-isoquinolinesulfonyl)-N-methyl-L-tyrosyl]-4-phenylpiperazine, a specific inhibitor of Ca^{2+}/calmodulin-dependent protein kinase II. *J. Biol. Chem.* **265,** 4315-4320.

Van Inwegen, R. G., Weinryb, I., Jones, H. and Khandwala, A. (1984). Comparative structure-activity relationships for dihydropyridines as inhibitors of [^{3}H]-nitrendipine bindings vs cyclic nucleotide phosphodiesterases. *Res. Commun. Chem. Pathol. Pharmacol.* **45**, 191-206.

Vanhoutte, P. M. and Paoletti, R. (1987). The WHO classification of calcium antagonists. *Trends Pharmacol. Sci.* **8,** 4-5.

Weichman, B. M., Muccitelli, R. M., Tucker, S. S. and Wasserman, M. A. (1983). Effect of calcium angatonists on leukotriene D4-induced contractions of the guinea-pig trachea and lung parenchyma. *J. Pharmacol. Exp. Ther.* **225,** 310-315.

Weiss, B. and Wallace, T. L. (1980). In: *Calcium and Cell Function* (Ed) Cheung, W.Y., Academic Press, New York, Vol. 1, 329-379.

Yamauchi, T. and Fujisawa, H. (1981). A calmodulin-dependent protein kinase that is involved in the activation of tryptophan 5-monooxygenase in specifically distributed in brain tissues. *FEBS Lett.* **129,** 117-119.

Progress in Hypertension, Vol. 2, pp. 145-155.
Saito *et al.* (Eds)

CLINICAL ASPECTS OF CALCIUM ANTAGONISTS

AKIKAZU NOMURA and HISAKAZU YASUDA

Department of Cardiovascular Medicine, Hokkaido University School of Medicine, Sapporo 060, Japan

INTRODUCTION

Treatment of hypertension is widely accepted today. One reason for this is that effective antihypertensive drugs with fewer adverse side effects have been developed. There are many drugs with differing mechanisms used in the reduction of elevated blood pressure. It is said that almost all hypertension can be controlled today. Diuretics, beta blockers, calcium antagonists, and angiotensin converting enzyme inhibitors are recommended as initial pharmacological therapy (Joint National Committee, 1988). Calcium antagonists were developed many years ago. However, their use as antihypertensives is relatively recent. In Japan, verapamil was officially approved in 1965 as a coronary artery vasodilator, for which it has successfully been used clinically. Diltiazem and nifedipine were introduced into clinical practice in 1974 and 1976, respectively. However, their antihypertensive effects were not widely accepted until the 80s, though some papers reported the antihypertensive effectiveness of nifedipine (Murakami *et al.*, 1972) and of diltiazem (Tanaka, 1974). Today calcium antagonists are the No.1 selling drug among antihypertensives in Japan.

PATHOPHYSIOLOGY OF HYPERTENSION

In the early stage of essential hypertension or at younger age, cardiac output increases and total peripheral vascular resistance remains normal; plasma catecholamines and renin-aldosterone are normal or slightly high. In later stages or at older ages, cardiac output is normal to low and peripheral vascular resistance increases; plasma catecholamines increase and renin-aldosterone is low to normal (Messerli, 1981). Vasoconstriction is characteristic once essential hypertension is established and is due to postjunctional alpha adrenoceptor activation (Amann *et al.*,1981) and/or enhanced calcium influx (Hulthen *et al.*,1982). This activation results in elevation of the intracellular free calcium concentration in vascular smooth muscle cells. This increase in calcium concentration is also observed in the platelets of patients with essential hypertension and correlates with the elevation in blood pressure (Erne *et al.*,1984).

The kidney is also involved in essential hypertension. Renal blood flow and glomerular filtration begin to decrease after 25 years of age. The reduction of renal blood flow is larger than that of glomerular filtration, resulting in an increase in the filtration fraction. This increase results in a rise in hydrostatic pressure in the proximal tubule. This change accelerates sodium reabsorption in the proximal tubule (Nomura *et al.*,1980). However, fluid volume retension is not seen (Messerli, 1981), which would be the result of hydrostatic pressure diuresis.

CALCIUM ANTAGONIST SUBCLASS

Many calcium antagonists are currently under development and investigation. These drugs are chemically divided into three main classes (Table 1). This chemical classification also relates to the binding site of the calcium antagonist. There are at least three binding sites in the voltage-dependent L-channel and they influence each other allosterically (Opie *et al.*,1987).

Dihydropyridines
This subclass is represented by nifedipine, which is the most commonly used calcium antagonist for the treatment of hypertension. Because of a short elimination half-life (Miyazaki *et al.*,1984), three times daily dosing is usually prescribed. In order to obtain improved compliance, sustained release granules have been developed (Kohri *et al.*,1987) (Fig.1). Changing the formulations have resulted in less complaints of headaches. This is probably due to the reduction in peak plasma concentration. Recently, drugs with different pharmacokinetic profiles have been introduced into clinical practice to overcome their disadvantages.

Phenylalkylamines
This subclass is represented by verapamil, which is also used for the treatment of supraventricular tachycardia (Sakurai *et al.*,1983). Three or four times daily dosing is usually recomended. Prolongation of the ECG PQ and QTc intervals should be watched for when verapamil is used as an antihypertensive.

Benzothiazepines
This subclass is represented by diltiazem, which has features of both the dihydropyridines and phenylalkylamines. Three times daily dosing is usually prescribed. 8-chloro-diltiazem (TA-3090, clentiazem) is under investigation as a cerebral artery vasodilator (Bevan *et al.*, 1989).

Table 1. List of calcium antagonists.

Dihydropyridine	nifedipine, felodipine, isradipine, nicardipine, nilvadipine nimodipine, nisoldipine, nitrendipine
Phenylalkylamine	verapamil, galopamil, tiapamil
Benzothiazepine	diltiazem, clentiazem

ANTIHYPERTENSIVE MECHANISMS

During the stage of established essential hypertension or older age, cardiac output is normal to low and peripheral vascular resistance is increased (Messerli, 1981). This vasoconstriction is a characteristic pathophysiological change and is the result of postjunctional alpha adrenoceptor activation and/or enhanced intracellular calcium influx. These activations elevate the intracellular free calcium concentration in vascular smooth muscle cells. Because of this elevated vascular resistance, it is a logical and accepted approach to dilate the constricted vascular bed through manipulation of the intracellular calcium concentration with calcium antagonists. There are other types of arterial vasodilating agents which act directly on vascular smooth muscle. These agents often cause tachycardia and expansion of the extracellular volume, which diminishes their antihypertensive effects (Koch-Weser, 1974). Typically these are then used in combination with sympatholytic drugs and/or diuretics. However, calcium antagonists cause less of this phenomena than do vasodilators and are used as monotherapy as well as in combination therapy.

The antihypertensive mechanism of nifedipine may be different in the acute versus chronic phase. After first dose of nifedipine, blood pressure decreased and heart rate increased with the initial dose of nifedepine. The cardiac index is increased and

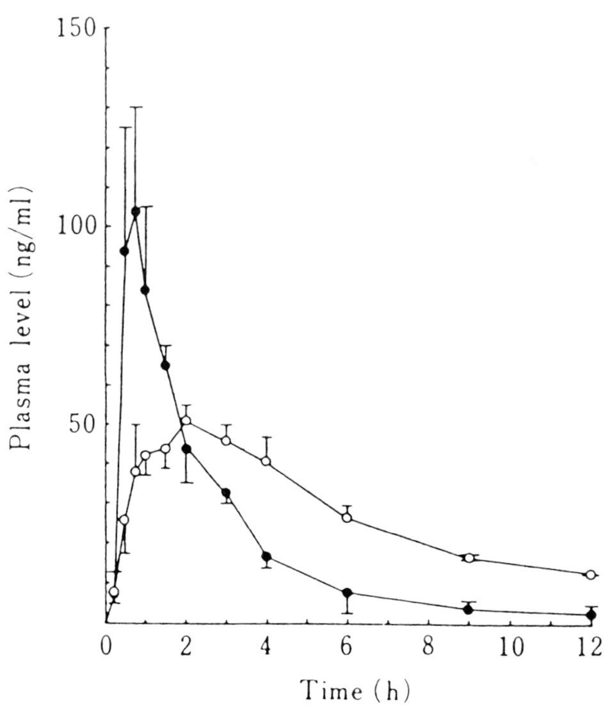

Figure 1. Mean plasma nifedipine profiles after oral administration of two formulations: (●), soft gelatin capsule containing 10mg of nifedipine; (○), sustained release granules containing 20 mg of nifedipine. Each point represents a mean value ± S.E.M.

systemic vascular resistance decreases. The reduction in vascular resistance is greater than the augmentation in cardiac index; resulting in a reduction in blood pressure. After 6 weeks of nifedipine therapy, the blood pressure is decreased much more than after the initial dose. The heart rate returns to its rate prior to treatment. The cardiac index remains elevated but at a level less than following the initial dose. Systemic vascular resistance remains decreased at the level seen after the initial dose (Kiowski *et al.*, 1989). These hemodynamic changes explain why the patients complaint of palpitations and throbbing headaches after initiation of calcium antagonists antihypertensive therapy tend to decrease or disappear with chronic treatment.

The hypotensive effects of calcium antagonists are widely accepted in hypertensive patients. However, these effects are not seen in normotensive volunteers. In a pharmacokinetic study of normotensive volunteers (Fig.1), ingestion of nifedipine 10 mg caused no change in blood pressure (less than 10/4 mmHg) or heart rate (less than 5 bpm). This differs from other antihypertensive drugs including alpha blockers, beta-blockers, angiotensin converting enzyme inhibitors, clonidine, alpha-methyldopa and guanethidine.

IN COMBINATION WITH OTHER DRUGS

Calcium antagonists have been used in combination with other antihypertensive agents. They are effective in combination with diuretics (Hallin *et al.,* 1983), betablockers, alpha-methyldopa (Guazzi *et al.,* 1980), alpha-1-blocker (Pasanisi *et al.,* 1984), and angiotensin converting enzyme inhibitors (Guazzi *et al.,* 1984). The combination of dihydropyridines (nifedipine) with betablockers is most often used.

Nifedipine causes reflex-induced tachycardia in some patients. Normalization of heart rate is advisable for such patients because of typical life long therapy. Additionally, dihydropyridine decreases systemic vascular resistance and increases cardiac output. This increase often blunts the reduction in blood pressure resulting from decrease in systemic vascular resistance. Therefore, reduction in this increased cardiac output by betablockers intensifies the antihypertensive effect. This combination is not frequently used with phenylalkylamines (verapamil) or benzothiazepines (diltiazem) since they induce less reflex tachycardia and the combination has a negative effect on atrioventricular conduction. The combination of phenylamines or benzothiazepines with betablockers may result in atrioventricular conduction delay since then all affect atrioventricular conduction negatively. Three drug combination therapy has been used in patients with severe or accelerated hypertension. Calcium antagonists, especially nifedipine, have been used as a vasodilator in such patients (Evans *et al.,*1984).

HYPERTENSIVE EMERGENCY

The usefulness of nifedipine has been established for rapid reduction of severe hypertension. This drug easily absorbed and effective through either sublingual (Beer *et al.,* 1981), buccal (Haft and Litterer, 1984), or oral routes. There is controversy as to which route most rapidly reduces blood pressure. If one capsule of nifedipine (10 mg) is not effective in reducing the blood pressure to a desired level in 30 to 60 minutes, an additional capsule may be given. However, this drug works rapidly and effectively and therefore the blood pressure should be measured every 15 minutes for the first hour after administration. Nifedipine is accepted today as an initial choice in hypertensive emergency therapy.

PATIENTS

Elderly
Peripheral vascular resistance in elderly hypertensive patients increases and cardiac output decreases. Renal and cerebral blood flow is reduced. Responsiveness to beta-adrenoceptor decrease while that of the alpha-adrenoceptor does not change. The activity of the renin-angiotensin system is attenuated in the elderly (North *et al.*, 1977).

Calcium antagonists achieve their desired effect without involving the status of these hormonal states and correct the hemodynamics, that is improvement of vascular resistance and cardiac output. These drugs also increase cerebral, cardiac and renal blood flow. Therefore, calcium antagonists seem to be an ideal drug for the elderly.

Left ventricular hypertrophy
Epidemiological studies have reported left ventricular hypertrophy as an important risk factor (Kannel *et al.*, 1969). Left ventricular hypertrophy results from not only hemodynamic factors, e.g. hypertension, but also from nonhemodynamic factors, e.g. hormones (Frohlich, 1983). It can be preventable in hypertensive patients if control of hypertension begins early and is adequate. Established treatment must therefore consider reduction of hypertrophy as well as control of blood pressure. However, all antihypertensive does not prevent or reduce hypertrophy, even if blood pressure is adequately controlled (Tarazi and Frohlich, 1987). Nifedipine is one drug known to reduce increased left ventricular mass.

There is some controversy as to whether reversal of cardiac hypertrophy results in improvement in the increased risk or recovery of cardiac function (Tarazi, 1987). Some animal studies have failed to show improvement in cardiac contractility with reduction in cardiac hypertrophy. There is not enough data in humans to be conclusive. However, as of now, vigorous treatment aimed at reduction is recommended.

Concomitant disease
Most hypertensive patients are either middle-aged adults or elderly. They often have other diseases, which influence prognosis. Attention to coexisting disease is important when developing therapy.

Coronary artery disease
Calcium antagonists were originally used as coronary vasodilating agents and are widely accepted for the treatment of angina pectoris. There are many papers identifying the beneficial effects in patients with hypertension and coronary artery disease through reduction of afterload and improvement of myocardial oxygen supply.

Antihypertensive treatment has reduced cerebral vascular accidents. However, it is still unknown as to whether it decreases the incidence of ischemic coronary heart disease. Animal studies have shown that calcium antagonists inhibit atherosclerosis, prevent ischemia, inhibit platelet aggregation and limit ischemic damage (Hall and Ball, 1989). The initial three factors mentioned would be expected to prevent angina pectoris and myocardial infarction. The last one is thought to be beneficial if myocardial infarction occurs. A survey of all randomized trials involving calcium antagonists in patients with myocardial infarction and angina pectoris revealed 19, 000 patients in 28 trials (Held *et al.*, 1989). Death occurred in 873 of 8, 870 patients on calcium antagonists compared to 825 deaths in 8, 889 control patients. There were no differences in incidence or size of myocardial infarction or rate of reinfarction. On the other hand, the diltiazem secondary prevention study showed that patients without pulmonary congestion after myocardial infarction had a reduced incidence of cardiac events if they were treated with diltiazem (Multicenter diltiazem postinfarction trial research group,1988). It remained unknown as to whether myocardial infarction can be prevented in patients with uncomplicated hypertension treated with calcium antagonists.

Congestive heart failure

In congestive heart failure, the cardiac output is reduced with reduced left ventricular contractility, and elevated afterload and preload. Calcium antagonists improve preload and afterload with impairment of the reduced contractility (Cohn, 1986). The net effect of these drugs is favorable for the hypertensive heart, since reduction of afterload is mandatory in such patients.

Cardiac arrhythmias

Verapamil and diltiazem affect the sinus and atrioventricular node while nifedipine does not. Verapamil and diltiazem are used for the treatment of supraventricular tachycardia due to AV nodal re-entry (Sakurai *et al.*, 1983) and control of ventricular response through the AV node during atrial tachycardia. These agents do not have any effect on accessory pathway conduction and ventricular tachycardia cannot be controlled by these drugs.

Raynaud's phenomenon

Functional or mechanical vascular obstruction causes ischemic symptoms such as parasthesia, numbness and pain. Raynaud's phenomenon is characterized by reduction of blood flow. This decrease may result from decreased perfusion pressure, increased blood viscosity, or narrowing of the arterial lumen, which is the most common cause (Smith and Rodeheffer, 1985). Calcium antagonists relax arterial smooth muscle and stop arterial vasospasm. They decrease peripheral vascular resistance and increase peripheral blood flow. These drugs are effective in patients with idiopathic disease, but not with secondary Raynaud's phenomenon (Tietze *et al.*, 1987).

Nocturnal leg cramps

Some patients compleint nocturnal leg cramps are probably due to increased blood viscosity and decreased muscle blood flow. Verapamil was administered to patients with angina pectoris and it was noted that some had relief of their leg cramps (Baltodano *et al.*, 1988). These agents are expected to be effective for such patients with this condition.

Renal disease
Chronic renal disease is progressive and eventually results in loss of function. Various factors alter the clinical course to renal failure. Restriction of dietary protein, control of systemic blood pressure, normalization of elevated glomerular capillary pressure, correction of plasma cholesterol, and anticoagulant therapy slow the progression of disease (Klahr *et al.*, 1988). The calcium antagonist nisoldipine or a combination of diuretics, beta blocker and vasodilator were administered to patients with chronic renal failure of various etiology. The rate of progression improved in 14 of 17 patients in the nisoldipine group and in 6 of 17 in the combination group (Eliahou *et al.*, 1988). There is not enough data to conclude the preservation of renal function with calcium antagonists.

Diabetes mellitus
There are many patients with hypertension and diabetes mellitus. Secretion of insulin is coupled with calcium at the islet cell. Therefore calcium antagonists might influence the secretion of insulin. Verapamil, as well as nifedipine, reduced insulin secretion (Lewis, 1983) and nifedipine was found to change glucose tolerance (Charles *et al.*, 1981). However, diabetes mellitus patients may be treated with a calcium antagonist at a dose sufficient to control the blood pressure without deleterious effects on glucose tolerance or lipids. Further clinical observation is needed to develop a conclusion on this. This argument might not be applicable to patients treated with insulin. Patients with diabetes mellitus should be treated carefully with calcium antagonists.

Pulmonary disease
Asthma and chronic obstructive pulmonary disease reduce ventilatory function due to bronchospasm or mechanical change. Calcium antagonists do not adversely affect ventilation (Anavekar *et al.*, 1982; Tinkelman, 1985). Calcium antagonist are the drug of choice for these patients.

Migraines
The mechanism of migraine is not fully understand although a vascular abnormal is suspected; some unidentified substances trigger the cascade of events resulting in platelet aggregation and release of vasoactive substances such as thromboxane A2, serotonin, prostaglandins and kinins. These substances increase calcium influx into vascular smooth muscle and cause vasoconstriction and reduction of blood flow. Following this reactive vasodilatation occurs and increases blood flow and the headache occurs (Tietze *et al.*, 1987). Calcium antagonists are expected to be effective due to their ability to block calcium entry into the cells. Though the frequency of attacks is decreased by these drugs, the severity and duration of headaches often changed. It is thought that tolerance may occur and switching to another calcium antagonist has been recommended. A large scale study is needed for conclusive results.

Central nervous system disease
Drugs used to treat mania-depression and psychoses often influence the cardiovascular system. Reserpine, betablockers and alpha-2-stimulants may influence the psychological state to some degree. Calcium antagonists are less likely to cause this effect and are preferable drugs for these patients.

Surgery
Hypertension after operation is an important problem, especially in elderly patients, who frequently have both cardiac and noncardiac surgery. Bleeding from fresh anastomoses, cardiovascular accidents and myocardial ischemia may be serious complications. Correctable factors such as pain and respiratory inadequacy should be treated. Nitroglycerin and nitroprusside have been used for such patients. Nicardipine, a dihydropridine derivative, is stable in solution and effective when given intravenously (Kaplan, 1990).

Post-transplantation
The mechanism of hypertension after organ transplantation is not fully understood. Drugs used for prevention of rejection also influence blood pressure. Calcium antagonists are expected to be effective (Curtis *et al.*, 1988).

ADVERSE EFFECTS

Many adverse reactions can be categorized into six groups: vasodilatation, negative inotropic effects, conduction disturbances, gastrointestinal effects, metabolic effects, and drug interactions (Russell, 1988). Common side effects are listed in Table 2 (Lewis, 1983). Nifedipine often causes symptoms related to vasodilatation. The frequency of side effects is higher with this drug than with verapamil or diltiazem.

LIPIDS

No changes occur in plasma triglycerides, and there is no change or a decrease in total and LDL cholesterol. HDL cholesterol does not change or may increase (Schoen *et al.*, 1988).

Table.2 Common side effects of calcium antagonists

	Nifedipine	Verapamil	Diltiazem
Frequent (> 5%)		headaches flushing	constipation
Occasional (< 5%)	dizziness nausea heartburn palpitation edema	dizziness hypotension headaches edema nausea	nausea edema headaches

DRUG INTERACTION

Drug interactions with digoxin, quinidine, phenytoin, prazosin, beta blockers, lithium and theophylline have been reported (Halperin and Cubeddu, 1986).

QUALITY OF LIFE

Patients with hypertension should have a normal daily life with as little sense of illness as possible. Antihypertensive drugs often interfere with a patient's sense of well being. Calcium antagonists cause less problems than do others (Materson and Preston, 1989).

CONCLUSION

Calcium antagonists are potent vasodilators and effective for mild to severe hypertension. They are also useful in patients with concomitant disease. They cause less interference in patients daily life. Calcium antagonists are important drugs for the treatment of hypertension.

REFERENCES

Amann, F. W., Bolli, P., Kiowski, W. and Buhler, F. R. (1981). Enhanced alpha-adrenoceptor-mediated vasoconstriction in essential hypertension. *Hypertension* **3,** (suppl I), I-119-I-123.

Anavekar, S. N., Barter, C., Adam, W. R. and Doyle, A. E. (1982). A double-blind comparison of verapamil and labetalol in patients with coexisting chronic obstructive airways disease. *J. Cardiovasc. Pharmacol.* **4** (suppl 3), S374-S377.

Baltodano, N., Gallo, B. V. and Weider, D. J. (1988). Verapamil vs Quinine in recumbent nocturnal leg cramps in the elderly. *Arch. Intern. Med.* **148,** 1969-1970.

Beer, N., Gallegos, I., Cohen, A., Klein, N., Sonnenblick, E. and Frishman, W. (1981). Efficacy of sublingual nifedipine in the acute treatment of systemic hypertension. *Chest* **79,** 571-574.

Bevan, J. A., Kaminow, L., Laher, I. and Thompson, L. P. (1989). Pharmacology of TA-3090 (8-chloro diltiazem) related to its cerebrovascular protective properties. *Circulation* **80** (suppl IV), IV-178-IV-183.

Charles, S., Ketelslegers, J., Buysschaert, M. and Lambert, A. (1981). Hyperglycaemic effect of nifedipine. *Br. J. Med.* **283,** 19-20.

Cohn, J. N. (1986). Calcium antagonists and left ventricular function: effects of nifedipine in congestive heart failure. *Am. J. Cardiol.* **58,** 27D-39D.

Curtis, J. J., Luke, R. G., Jone, P. and Diethelm, A. G. (1988). Hypertension in cyclosporine-treated renal transplant recipients is dependent. *Am. J. Med.* **85,** 134-138.

Eliahou, H. E., Cohen, D., Hellberg, B., Ben-David, A., Herzog, D., Schechter P., Kapuler S. and Kogan N. (1988). Effect of the calcium channel blocker nisoldipine on the progression of chronic renal failure in man. *Am. J. Nephrol.* **8,** 285-290.

Erne, P., Bolli, P., Burgisser, E. and Buhler, F. R. (1984). Correlation of platelet calcium with blood pressure. *N. Engl. J. Med.* **310,** 1084-1088.

Evans, M. G., Olanoff, L. S., Hurwitz, G., Cowart, T. D. Conradi, E. C. (1984). Use of nifedipine as an adjunct to current antihypertensive therapy. *Arch. Intern. Med.* **144,** 985-987.

Frohlich, E. D. (1983). Hemodynamics and other determinants in development of left ventricular hypertrophy. *Federation proc.* **42,** 2709-2715.

Guazzi, M. D., Fiorentini, C., Olivari, M. T., Bartorelli, A., Necchi, G. and Polese, A. (1980).

Short- and long-term efficacy of a calcium-antagonistic agent (nifedipine) combined with methyldopa in the treatment of severe hypertension. *Circulation* **61,** 913-919.

Guazzi, M. D., Decesare, N., Galli, C., Salvioni, A., Tramontana, C., Tamborini, G. and Bartorelli, A. (1984). Calcium-channel blockade with nifedipine and angiotensin converting-enzyme inhibition with captopril in the therapy of patients with severe primary hypertension. *Circulation* **70,** 279-284.

Haft, J. I. and Litterer, W. E. (1984). Chewing nifedipine to rapidly treat hypertension. *Arch. Intern. Med.* **144,** 2357-2359.

Hall, A. S. and Ball, S. G. (1989). Calcium-channel antagonists, angiotensin-converting enzyme inhibitors, and serotonin antagonists. *Curr. Opin. Cardiol.* **4,** 647-654.

Hallin, L., Andren, L. and Hansson, L (1983). Control trial of nifedipine and bendroflumethiazide in hypertension. *J. Cardiovascu. Pharmacolo.* **5,** 1083-1085.

Halperin, A. K. and Cubeddu, L. X. (1986). The roll of calcium channel blocker in the treatment of hypertension. *Am. Heart. J.* **111,** 363-382.

Held, P. H., Yusuf, S. and Furberg, C. D. (1989). Calcium channel blockers in acute myocardial infarction and unstable angina : an overview. *Br. Med. J.* **299,** 1187-1192.

Hulthen, U. L., Bolli, P., Amann, F. W. and Buhler, F. R. (1982). Enhanced vasodilation in essential hypertension by calcium channel blockade with verapamil. *Hypertension* **4** (suppl II), II-26-II-31.

Kannel, W. B., Gordon, T. and Offutt, D. (1969). Left ventricular hypertrophy by electrocardiogram. *Ann. Intern. Med.* **71,** 89-105.

Keplan, J.A. (1990). Clinical consideration for the use of intravenous nicardipine in the treatment of postoperative hypertension. *Am. Heart. J.* **119,** 443-446.

Kiowski, W., Bolli, P., Erne, P., Muller, F. B., Hulthen, U. L. and Buhler, F. R. (1989). *Circulation* **80** (suppl IV), IV-136-IV-144.

Klahr, S., Schreiner, G. and Ichikawa, I. (1988). The progression of renal disease. *N. Engl. J. Med.* **318,** 1657-1666.

Kohri, N., Miyazaki, K., Arita, T., Shimono, H., Nomura, A. and Yasuda, H. (1987). Release characteristic of nifedipine sustained-release granules in vitro and in healthy subjects. *Chem. Pharm. Bull.* **35,** 2504-2509.

Koch-Weser, J. (1974). Vasodilator drugs in the treatment of hypertension. *Arch. Intern. Med.* 133, 1017-1027.

Lewis J.G. (1983). Adverse reaction to calcium antagonists. *Drug* **25,** 196-222.

Materson, B.J. and Preston, R.A. (1989). Newer principles of patient profiling for antihypertensive therapy. *Circulation* **80** (suppl IV), IV-128-IV-135.

Multicenter diltiazem postinfarction trial research group (1988). The effect of diltiazem on mortality and reinfarction after myocardial infarction. *N. Engl. J. Med.* **319,** 385-392.

Murakami, M., Murakami, E., Takekoshi, Y., Tsuchiya, M., Kin, T., Onoe, T., Takeuchi, N., Funatsu, T., Haro, S., Ishise, S., Mifune, J. and Maeda, S. (1972). Antihypertensive effect of 4(-2'-nitrophenyl)-2,6-dimethyl-1,4-dihydropyridine-3,5-dicarbonic acid dimethylester (nifedipine, Bay-a1040), a new coronary dilator. *Jpn. Heart J.* **13,** 128-135.

Messerli, F. H. (1981). Individualization of antihypertensive therapy: an approach based on hemodynamics and age. *J. Clin. Pharmacol.* **21,** 517-528.

Miyazaki, K., Kohri, N., Arita, T., Shimono, Y., Katoh, K., Nomura, A. and Yasuda, H. (1984). High-performance liquid chromatographic determination of nifedipine in plasma. *J. Chromatogr.* **310**, 219-222.

1988 joint national committee (1988). The 1988 report of the joint national committee on detection, and treatment of high blood pressure. *Arch. Intern. Med.* **148,** 1023-1038.

Nomura, A., Kato, K., Minami, M., Yasuda, H., Akimoto, T., Miyazaki K. and Arita T. (1981). Renal effects of hydralazine in patients with cardiac disease. *Jpn. J. Cli. Pharmaco. Ther.* **11,** 349-350 (in Japanese).

North, R. H., Lassman, N., Fernandez-Cruz, A. and Mulrow, P. (1977). Age and the renin-aldosterone system. *Arch. Intern. Med.* **137,** 1414-1417.

Opie, L. H., Buhler, F. R., Fleckenstein, A., Hansson, L., Harrison, D. C., Poole-Wilson, P. A.,

Schwartz, A. and Vanhoutte, P. M. (1987). International society and federation of cardiology: working group on classification of calcium antagonists for cardiovascular disease. *Am. J. Cardiol.* **60,** 630-632.

Pasanisi, F., Elliott, H. L., Meredith, P.A., McSharry, D.R. and Reid, J.L. (1984). Combined alpha adrenoceptor antagonism and calcium channel blockade in normal subjects. *Cli. Pharmaco. Ther.* **36,** 716-723.

Russell, R.P. (1988). Side effects of calcium channel blockers. *Hypertension* **11** (suppl II), II-42-II-44.

Sakurai, M., Yasuda, H., Kato, N., Nomura, A., Fujita, M., Nishino, T., Fujita, K., Koike, Y. and Saito, H. (1983). Acute and chronic effects of verapamil in patients with paroxysmal supraventricular tachycardia. *Am. Heart J.* **105,** 619-628.

Schoen, R. E., Frishman, W. H. and Shamoon, H. (1988). Hormonal and metabolic effects of calcium channel antagonist in man. *Am. J. Med.* **84,** 482-504.

Smith, C. R. and Rodeheffer, R. J. (1985). Treatment of Raynaud's phenomenon with calcium blocker. *Am. J. Med.* **78** (suppl 2B), 39-42.

Tarazi, R. C. and Frohlich, E. D. (1987). Is reversal of cardiac hypertrophy a desirable goal of antihypertensive therapy? *Circulation* **75** (suppl I), I-113-I-117.

Tanaka, H. (1974). Effect of diltiazem (Herbesser) to the patients with hypertension and angina pectoris. *Shinyaku to rinsho.* **23**, 2149-2154 (in Japanese).

Tietze, K. J., Schwartz, M. L. and Vlasses, P. H. (1987). Calcium antagonists in cerebral/peripheral vascular disorders. *Drugs* **32,** 531-538.

Tinkelman, D. G. (1985). Calcium channel blocking agents in prophylaxis of asthma. *Am. J. Med.* **78** (suppl 2B), 35-38.

Progress in Hypertension, Vol. 2, pp. 157-188.
Saito *et al.* (Eds)

PHARMACOLOGY OF CENTRAL DOPAMINERGIC MODULATION IN HYPERTENSION

MAARTEN VAN DEN BUUSE and WYBREN DE JONG

Merrell Dow Research Institute, Strasbourg Research Center, 16, rue d'Ankara, B.P. 447 R/9, 67009 Strasbourg, France.

INTRODUCTION

There is a growing interest in the use of dopaminergic drugs to influence cardiovascular regulation. It is now well described that in the periphery, dopamine D_2 receptors are located presynaptically on sympathetic nerve endings and in autonomic ganglia, whereas D_1 receptors are found on vascular smooth muscle cells and in the kidney. Dopamine D_2 agonists induce a decrease in blood pressure by inhibition of the release of noradrenaline from sympathetic nerve endings, whereas D_1 agonists induce a decrease in blood pressure by direct vasodilatation (for details and references, see Goldberg, 1972; Hilditch and Drew, 1985; Lokhandwala *et al.,* 1987). Compared to the role of peripheral dopaminergic mechanisms in cardiovascular regulation, less in known of the involvement of central dopamine systems in this respect.

The anatomy of the central dopamine system is extensively described (Björklund and Lindvall, 1984; Moore and Bloom, 1978; Ungerstedt, 1971). As part of the nomenclature concerning monoamines and indolamines in the brain, cellgroups synthesizing dopamine are coded as A8 to A14. The major groups are the A9 and A10 in the substantia nigra and ventral tegmental area, respectively, and the A12 in the nucleus arcuatus. From the substantia nigra, the nigrostriatal system innervates the striatum. The mesolimbic/mesocortical system has its perikarya in the A10 group and innervates limbic (e.g. nucleus accumbens and septum) and cortical structures. From the A12, the tuberoinfundibular and tuberohypophyseal systems innervate the median eminence and the pituitary. From other cell-groups in the hypothalamus various hypothalamic nuclei receive dopaminergic innervation. Spinal cord dopamine originates from hypothalamic dopamine cell-groups. Dopamine D_1 and D_2 receptors are unequally distributed throughout the brain, with highest concentrations in the striatum (including the caudate and accumbens nuclei), moderate to low concentrations in hypothalamic, limbic and cortical regions, and low to undetectable levels in medullary and cerebellar regions. Dopamine D_2 receptors are present postsynaptically to dopamine neurons, but also presynaptically on dopamine neuronal endings and

Table 1. Dopaminergic drugs: receptors and mode of action.

Name	DA-receptor involved	Action
Amphetamine	-	DA-release
Apomorphine	D_1 and D_2	agonist
Bromocriptine	D_2	agonist
Butaclamol	D_1 and D_2	antagonist
Domperidone	D_2	antagonist[1]
L-DOPA	-	DA-release
Dopamine	D_1 and D_2	agonist
GBR 12909	-	uptake inhibitor
Haloperidol	Preferential D_2	antagonist
6-Hydroxydopamine	-	neurotoxin
Metoclopramide	Preferential D_2	antagonist
Pergolide	D_2	agonist
Quinpirole/LY171555	D_2	agonist
SCH 23390	D_1	antagonist
SKF 38393	D_1	agonist
Sulpiride	D_2	antagonist
Spiroperidol/spiperone	D_2	antagonist

(1) Does not cross the blood-brain-barrier.

dopamine cell bodies. D_1 receptors are present predominantly postsynaptic to dopaminergic neurons.

Brain dopamine is involved in a wide range of physiological processes, including behavioural and endocrine regulation (see e.g. Bannon and Roth, 1983; Mehta and Tolis, 1979; Ungerstedt, 1971; Waddington, 1989). Central dopamine systems may also influence autonomic regulation, and there is increasing evidence to suggest that brain dopamine may play a direct role in the regulation of cardiovascular function and the development of hypertension (see Van den Buuse *et al.*, 1986b; Versteeg and Van den Buuse, 1989). In the present paper, we give an overview regarding the role of brain dopamine in cardiovascular regulation. The first section focuses on the effects of central dopamine or dopaminergic drugs on cardiovascular regulation. A number of dopaminergic compounds used in such studies have been described in Table 1 and are shown in Fig. 1. The second section deals with the role of central dopamine systems in cardiovascular regulation and the development of hypertension. Finally, we have included a section on cardiovascular studies in Parkinson's disease.

CENTRAL EFFECTS OF DOPAMINERGIC DRUGS ON CARDIOVASCULAR REGULATION

Dopamine

A number of authors have described the effect of a central injection of dopamine on

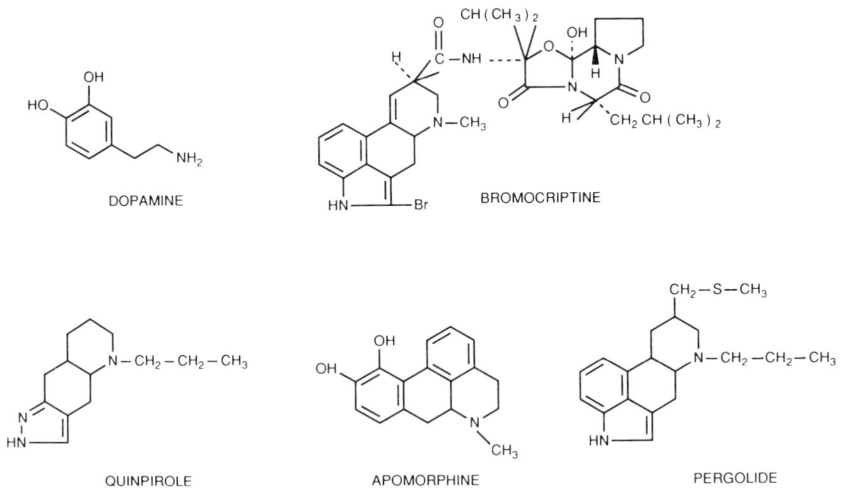

Figure 1. Chemical structure of dopamine and four dopamine agonists. Quinpirole and bromocriptine are selective D_2 agonists, whereas apomorphine and pergolide, and especially dopamine, have D_1 activity as well. For more details and full chemical names, see references in section 1.

blood pressure (Van den Buuse *et al.*, 1986b). Already in 1960, McCubbin, Kaneko and Page described a decrease in blood pressure, heart rate and carotid sinus pressor reflex in anesthetized dogs upon intracerebroventricular (*i.c.v.*) injection of dopamine. In conscious dogs, *i.c.v.* dopamine caused an increase in blood pressure (Lang and Woodman, 1979). In conscious cats, *i.c.v.* dopamine caused an increase in blood pressure and heart rate, which in some animals was followed by a decrease. Interestingly, the depressor response was more predominant after intracisternal administration of the catecholamine (Day and Roach, 1976).

In rats, a decrease in blood pressure and heart rate after *i.c.v.* administration of dopamine has been described (Baum and Shropshire, 1973; DePalma and Ackerman, 1983; Kondo *et al.*, 1981; Pazos *et al.*, 1982). This effect appears to be more pronounced in spontaneously hypertensive rats (SHR) (Hutchinson *et al.*, 1983; Kawabe *et al.*, 1983; Mok and Sim, 1987). In all these studies relatively large doses, in the order of 50-500 μg, of dopamine were needed, probably because the administered dopamine is normally rapidly metabolized in the brain. In fact, McCubbin *et al.* (1960) and Pazos *et al.* (1982) could only find effects of central dopamine injection after inhibition of monoamine oxidase, which prevents dopamine degradation.

Also local injection of dopamine in the brain of anaesthetized animals has been shown to affect cardiovascular regulation. Micro-injection of 0.3 to 20 nmol of dopamine in the area of the nucleus tractus solitarii (NTS) of normotensive rats induced a dose-dependent decrease in blood pressure and heart rate (Zandberg *et al.*,

Table 2. Bromocriptine and blood pressure: selected references.

Species	Reference	Proposed site of action
Rat	Hamilton, 1981	peripheral
	Hutchinson *et al.*, 1981	central
	McMurtry *et al.*, 1979	central
	Nagahama *et al.*, 1984, 1985	central
	Roquebert *et al.*, 1990	peripheral
	Sowers, 1981	central
	Struyker-Boudier *et al.*, 1984	peripheral
	Tadepalli and Novak, 1983	peripheral
	Van den Buuse *et al.*, 1989b	peripheral
Dog	Chamontin *et al.*, 1984	central
	Lokhandwala *et al.*, 1979	peripheral
	Montastruc and Montastruc, 1981	peripheral
Man	Esposti *et al.*, 1985	?
	Greenacre *et al.*, 1976	?
	Johns *et al.*, 1984	peripheral
	Kolloch *et al.*, 1980, 1981	central
	Montastruc *et al.*, 1985b	central
	Nilsson and Hökfelt, 1978	?
	Quinn *et al.*, 1981	central
	Sowers *et al.*, 1982c	central
	Stumpe *et al.*, 1977	central

induced a dose-dependent decrease in blood pressure and heart rate (Zandberg *et al.*, 1979). On the other hand, intra-NTS injection of 5-100 nmol dopamine was found to increase blood pressure and heart rate, while micro-iontophoretically administered dopamine inhibited NTS cellular activity (Granata and Woodruff, 1979; 1982). Micro-injection of dopamine (20 μg/kg) in the nucleus arcuatus of the hypothalamus decreased blood pressure, but not heart rate, in SHR. No effect was evident in normotensive rats (Mok *et al.*, 1990). Superfusion with dopamine of the posterior hypothalamus of the anesthetized cat enhanced pressor responses resulting from electrical stimulation of this region (Philippu, 1988). This effect could be antagonized with locally administered haloperidol. Micro-injection of 20 μg of dopamine into the caudate nucleus of conscious, paralyzed cats failed to change blood pressure (Pazo and Medina, 1983).

Bromocriptine

Bromocriptine is an ergot alkaloid (see Fig. 1), which after peripheral administration may act as an agonist at central dopamine D_2 receptors (Fuxe *et al.*, 1978; Calne, 1978; Mehta and Tolis, 1979). While there is general agreement that bromocriptine may cause a decrease in blood pressure, there is disagreement regarding its mode of action (for references, see Table 2 and Lokhandwala *et al.*, 1979; Mehta and Tolis, 1979; Montastruc and Montastruc, 1981). Clark and colleagues (1978) suggested that at least three mechanisms are involved in the cardiovascular effects of bromocriptine: "relaxation of vascular smooth muscle, inhibition of transmitter release from sympathetic nerve endings, and central inhibition of sympathetic nerve activity". Later studies have detailed that bromocriptine may inhibit noradrenaline release from sympathetic nerve endings through an action at presynaptic or ganglionic dopamine receptors (see Lokhandwala *et al.*, 1979). In addition, some authors have suggested that bromocriptine is an α-adrenoreceptor antagonist (Gibson and Samini, 1978).

According to the dose of bromocriptine used, species, and experimental preparation (e.g. anesthetized or non-anesthetized), some of the above-mentioned mechanisms may predominate. Nevertheless, using virtually identical protocols, authors have come to opposite conclusions regarding the site of action of bromocriptine. For instance, Nagahama *et al.* (1984) as well as Van den Buuse and Lambrechts (1989b) studied the effect of an intraperitoneal dose of 0.5 mg/kg bromocriptine on blood pressure of SHR. In both studies a decrease in blood pressure was found, being slightly greater in the latter study. Nagahama and colleagues did not observe an effect of pretreatment with 0.5-5 mg/kg of the specific dopamine D_2 antagonist domperidone on the action of bromocriptine. Domperidone does not cross the blood-brain-barrier and as such is an important tool to distinguish between central and peripheral effects (Laduron and Leysen, 1979). Thus, Nagahama and colleagues suggested that bromocriptine was acting centrally. However, in the study of Van den Buuse and Lambrechts (1989b) 0.5 mg/kg domperidone completely prevented the decrease in blood pressure, suggesting a predominantly peripheral site of action.

Few studies have actually employed central injection of bromocriptine. We observed only a small decrease in blood pressure in SHR after intracerebroventricular (*i.c.v.*) injection of 50 μg (Van den Buuse and Lambrechts, 1989b). Superfusion of the cat posterior hypothalamus with bromocriptine enhanced the pressor response to electrical stimulation of this region (Philippu, 1988). This effect was also found with dopamine and apomorphine.

Despite problems in interpreting its effects, bromocriptine has been shown to attenuate the development of hypertension in SHR (Richer *et al.*, 1984; Tan and Hutchinson, 1987) and of DOCA-salt hypertension (Yoshimura *et al.*, 1987).

Quinpirole

Quinpirole (also known as LY 171555, the levo-enantiomer of LY 141865) is a selective dopamine D_2 agonist (Ruffolo and Shaar, 1983; Tsuruta *et al.*, 1981) with a structure which resembles the ergoline nucleus of bromocriptine (see Fig. 1). Because of its selectivity, quinpirole is a useful tool for the study of dopamine D_2 function, including cardiovascular studies (see Lokhandwala *et al.*, 1987). Upon peripheral administration, quinpirole induces dual effects. Thus, a decrease in blood pressure may be caused by peripheral mechanisms similar to that apparent with bromocriptine (see above). Such an effect has been described in anesthetized and conscious normotensive (Cavero, 1983; Nagahama *et al.*, 1986b) and hypertensive rats (Hahn *et al.*, 1983; Igarashi *et al.*, 1987).

Under certain circumstances, however, a centrally mediated rise in blood pressure can be observed as well. Intravenous bolus injections of quinpirole cause an immediate increase in blood pressure in conscious normotensive rats (Nagahama *et al.*, 1986a; Van den Buuse, unpublished results, see also Fig. 2) and humans (McNay *et al.*, 1987). In SHR, such an increase in blood pressure was observed only after blockade of the peripherally mediated decrease in blood pressure with domperidone (Kurz *et al.*, 1986).

An increase in sympathetic outflow and vasopressin release are involved in the pressor response to quinpirole (Nagahama *et al.*, 1986a; 1987; Damase-Michel *et al.*, 1990). From further studies it was suggested that enhanced posterior hypothalamic noradrenaline release, measured by a reduction in noradrenaline content, could also play a role in the pressor response (Chen *et al.*, 1987a). However, no cardiovascular effects were found after injection of 5-10 μg/kg quinpirole in different regions of the hypothalamus of conscious, normotensive rats (Yang *et al.*, 1990). In contrast, injection of 5 μg/kg quinpirole in the posterior region of the NTS of unanesthetized, decerebrate rats induced a pressor effect (Yang *et al.*, 1990). Injections of quinpirole in the anterior NTS, the area postrema, C1/A1 region, raphe nuclei or locus coeruleus were without effect on blood pressure. Therefore, the site of action of quinpirole to exert its pressor effect may be the posterior NTS, a finding which is in line with the pressor response after intra-NTS injections of high doses of dopamine (see above). Thus, a dopaminergic pressor system might exist in the NTS.

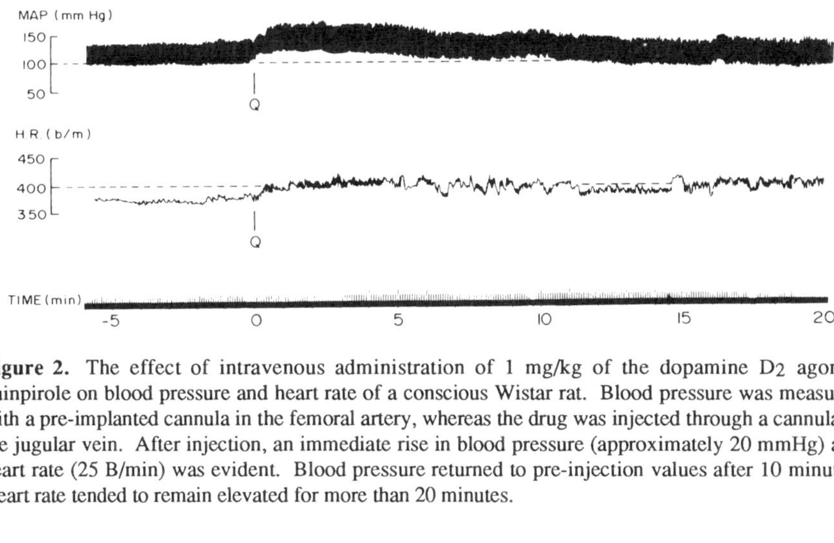

Figure 2. The effect of intravenous administration of 1 mg/kg of the dopamine D_2 agonist quinpirole on blood pressure and heart rate of a conscious Wistar rat. Blood pressure was measured with a pre-implanted cannula in the femoral artery, whereas the drug was injected through a cannula in the jugular vein. After injection, an immediate rise in blood pressure (approximately 20 mmHg) and heart rate (25 B/min) was evident. Blood pressure returned to pre-injection values after 10 minutes. Heart rate tended to remain elevated for more than 20 minutes.

Interestingly, a rapid desensitization develops to the pressor effect of quinpirole. Intravenous infusion of quinpirole in rats causes only a short-lasting pressor effect which will disappear or reverse to a depressor response despite continued administration of the compound (Igarashi *et al.*, 1987). We have observed that this desensitization also occurs with multiple bolus injections (Van den Buuse, unpublished results). This desensitization phenomenon may be one reason why the pressor action of quinpirole has not been generally found.

Apomorphine and other compounds

Intravenous administration of the dopamine receptor agonist apomorphine (Fig. 1) induced a short-lasting decrease in blood pressure in anesthetized cats and rats (Barnett and Fiore, 1971; Finch and Haeusler, 1973). This effect was also observed in sympathectomized rats, but was blocked by vagotomy, atropine administration or midcollicular transection. In line with these findings, sympathetic activity was found not to be altered by apomorphine, whereas vagal nerve activity was markedly increased (Finch and Haeusler, 1973). Dutta and colleagues (1975) confirmed these findings and, in addition, showed that micro-injection of apomorphine in the dorsal motor nucleus of the vagus and the dorsomedial nucleus of the hypothalamus had similar effects. Petitjean *et al.* (1984) showed that intrathecal injections of apomorphine induced a decrease in blood pressure. Casto *et al.* (1988) reported that intracisternal injection of apomorphine induced greater falls in pressure than intravenous injections, whereas *i.c.v.* injections were without effect. However, Bogaert *et al.* (1978) observed that intravenous injection of apomorphine decreased blood pressure, but that intravertebral or intracisternal injection actually increased blood pressure. Also other authors emphasized peripheral effects of apomorphine on blood pressure, possibly through facilitation of vagal output (Montastruc *et al.*, 1985a; 1989). In conclusion, apomorphine mimicks other dopamine receptor agonists in that it appears to act at multiple sites in and outside the central nervous system to induce primarily a decrease in blood pressure.

As an ergoline derivative, the dopamine D_2 agonist pergolide resembles bromocriptine and quinpirole in its structure (Fig. 1). As other dopamine receptor agonists, pergolide decreases blood pressure by both peripheral and central mechanisms (Barrett and Lokhandwala, 1983; Jadhav *et al.*, 1983; Cavero *et al.*, 1984; Lokhandwala *et al.*, 1984; Yen *et al.*, 1979). In anesthetized, normotensive dogs intracisternal administration of pergolide resulted in a decrease in blood pressure and heart rate and in illiac vasodilatation (Barrett and Lokhandwala, 1983). Moreover, in neurogenic hypertensive dogs the antihypertensive, bradycardic and vasodilatory effect of intravenously administered pergolide could be attenuated by intracisternal pretreatment with sulpiride (Barrett and Lokhandwala, 1983). Other authors suggested that pergolide inhibited sympathetic outflow by a predominantly peripheral site of action (Cavero *et al.*, 1984). The marked vagally mediated decrease in heart rate could have a major central component, though (Cavero *et al.*, 1984).

During the development of new inhibitors of dopamine-β-hydroxylase, it became apparent that these drugs may exert central antihypertensive effects through dopamine receptors (for references, see Ohlstein *et al.*, 1988). For instance, oral administration of 50 mg/kg SK&F 102698 decreased blood pressure markedly in SHR and DOCA-salt hypertensive rats, but not in renal hypertensive rats (Ohlstein *et al.*, 1988). By inhibiting noradrenaline formation the compound increased central and peripheral dopamine levels and the dopamine:noradrenaline ratio (Berkowitz *et al.*, 1988). Administration of 150 μg/kg SK&F 102698 into the fourth ventricle, but not lateral ventricle, induced a marked decrease in blood pressure and heart rate of conscious SHR (Ohlstein *et al.*, 1988).

CENTRAL DOPAMINE SYSTEMS AND HYPERTENSION

Blood pressure and central dopamine systems.
A reciprocal relationship appears to exist between brain dopamine and blood pressure. Changes in blood pressure influence the activity of central dopamine systems, whereas activation of central dopaminergic neurons causes changes in blood pressure.

Dopamine release from the anterior hypothalamus/preoptic region was markedly enhanced by a shortlasting increase in blood pressure caused by injection of blood (Elghozi *et al.*, 1981). In a series of experiments, Philippu and colleagues measured dopamine levels in push-pull perfusates of the posterior or anterior hypothalamus of anesthetized cats or conscious rabbits (for references, see Philippu, 1988). Dopamine release in the posterior hypothalamus was increased after a fall in blood pressure, induced either by nitroprusside, bleeding, chlorisondamine or spinal transection. No effect or a decrease in dopamine release was observed in the anterior hypothalamus. An increase in blood pressure by noradrenaline, tramazoline or splanchic nerve stimulation induced no effect or a decrease in the posterior hypothalamus, whereas dopamine release in the anterior hypothalamus was increased (see Philippu, 1988). Dopamine release in the NTS was decreased following a fall in blood pressure and increased by a rise in pressure (Kobilansky *et al.*, 1988). These responses could be interpreted as compensatory changes to the blood pressure variations. Thus, dopamine could play a hypertensive role in the posterior hypothalamus, but a hypotensive role in the anterior hypothalamus. Also, dopamine could play a hypotensive role in the NTS, an interpretation which could be in line with the observed fall in pressure after intra-NTS injections of dopamine (Zandberg *et al.*, 1979, see section 1.a).

In contrast to the above-mentioned findings in the hypothalamus, release of dopamine in the striatum was not influenced by changes in blood pressure (Keller *et al.*, 1983). Thus, it appears that striatal dopamine does not directly affect blood pressure, which could corroborate the finding that intra-striatal injections of dopamine did not alter blood pressure (Pazo and Medina, 1983, see section 1.a).

Of particular interest also are a number of studies by Alexander and colleagues on the sino-aortic-denervated (SAD) rat. In this model baroreceptor input to the brain is interrupted by sectioning of afferent nerves from the carotid sinus and aortic arch. Tyrosine hydroxylase activity and dopamine concentration were markedly decreased in the striatum of SAD-rats (Alexander *et al.*, 1984). Conversely, lesions of the striatum produced increased lability of blood pressure (Alexander *et al.*, 1986). Recently, using the in vivo microdialysis technique, these authors found a decreased release of dopamine, DOPAC and HVA in the striatum of conscious SAD-rats (Alexander *et al.*, 1988). The authors suggested that the striatum might play a role in the integration of behaviour and movement and cardiovascular reflex regulation. These experimental findings could be of importance in interpreting the cardiovascular changes observed in Parkinson's disease (see section 3).

Stimulation of central dopamine systems may change blood pressure. Stimulation of the anterior hypothalamus of anesthetized, normotensive rats induced a decrease in blood pressure which could be inhibited by intravenous pretreatment with the dopamine antagonist haloperidol (Earnhardt *et al.*, 1982). This suggests that activation of anterior hypothalamic dopamine neurons may cause a fall in blood pressure. In the anesthetized cat, electrical stimulation of the ventral midbrain area including the substantia nigra and ventral tegmental area caused increases in blood pressure which could be blocked by haloperidol, or by dopamine depletion with α-methyl-paratyrosine or 6-OHDA (Spring and Winkelmüller, 1975). Also electrical stimulation of the rat ventral tegmental area (Van den Buuse *et al.*, 1988) or rabbit ventral tegmentum (Tan *et al.*, 1983) induced pressor responses. Interestingly, hemisection in the posterior hypothalamus did not inhibit the pressor responses to electrical stimulation of the ventral midbrain (Spring and Winkelmüller, 1983). This suggested that ascending striatal or mesolimbic projections of the ventral midbrain may not be involved in the pressor effects elicited from this region, but that descending dopaminergic pathways to medullary and/or spinal cord regions could be involved. However, forebrain dopaminergic projections may be involved in the regulation of cardiovascular reflexes. Electrical stimulation of the substantia nigra facilitated reflex bradycardia to intravenous adrenaline injections (Lin *et al.*, 1982). Injection of dopamine antagonists, either systemically or locally in the striatum (Lin *et al.*, 1982), or kainic acid induced lesions in the striatum inhibited these responses (Wu *et al.*, 1984). Thus, substantia nigra projections to the striatum could play a role in baroreflex regulation rather than exerting a direct role in blood pressure regulation. In line with this conclusion are the findings of Chen *et al.* (1980) indicating that in rats, *i.c.v.* injection of dopamine agonists enhances and dopamine antagonists reduces reflex bradycardia. In patients with essential hypertension, the dopamine antagonist metoclopramide decreased baroreflex sensitivity (Alföldi and Farsang, 1987).

Transient occlusion of blood vessels to the brain induces a very marked increase in the release of dopamine in the striatum. This has been found with *in vivo* microdialysis (Globus *et al.*, 1988; Kawano *et al.*, 1988) as well as with *in vivo* voltammetry (Ogura *et al.*, 1989). A unilateral lesion of the substantia nigra prevented ischemic neuronal cell damage in the ipsilateral striatum (see Globus *et al.*, 1988). Together with the results of earlier studies, these findings suggest that central dopamine plays a role in the neuropathological effects of central ischemia (for references, see the above mentioned studies).

Levels and metabolites, receptors, release

A large number of studies has been performed on central monoamines in experimental hypertension (for reviews, see Howes, 1984b; Philippu, 1988; Versteeg *et al.*, 1984). Multiple differences appear to exist in the levels of noradrenaline, adrenaline, serotonin, and also dopamine, and their metabolites, their receptors and dynamic release characteristics, between normotensive controls and various genetical or experimental hypertensive models. As discussed before (Howes, 1984b; Philippu, 1988; Versteeg and Van den Buuse, 1989) the large variety of experimental protocols, techniques and models make it difficult to draw an overall picture of the involvement of central monoamines, including dopamine, in hypertension. Nevertheless, we would like to review a number of observations which together suggest a role of central dopamine systems in the development of hypertension.

Most studies were performed in SHR, although in some cases other models of experimental hypertension provide interesting comparative information. Table 3 summarizes the findings in SHR. Dopaminergic transmission appears to be increased

Table 3. Comparison of central dopaminergic neurochemistry in SHR *vs* WKY.
(1) Tyrosine hydroxylase activity

brain region	SHR *vs* WKY	reference
striatum	unchanged	Ikeda *et al.*, 1984
caudate nucleus	increased	Meyer and Carr, 1983
AV3V[1]	increased	
substantia nigra	unchanged	Nagatsu *et al.*, 1977
caudate nucleus	unchanged	
locus coeruleus	unchanged	
hypothalamus	unchanged	
striatum	increased	Nagaoka and Lovenberg, 1977
hypothalamus	increased	
A1 region	unchanged	Renaud *et al.*, 1979
A2 region	unchanged	
locus coeruleus	unchanged	
brainstem	increased	Lovenberg *et al.*, 1973
brainstem	unchanged	Yamabe *et al.*, 1973

[1]AV3V = Anteroventral part of the third ventricle.

(2) Dopamine concentrations

brain region	SHR vs. WKY	reference
striatum	unchanged	Fuller *et al.*, 1983
tub. olfactorium	unchanged	Le Fur *et al.*, 1981
striatum	unchanged	
frontal cortex	unchanged	Williams McKeon & Hendley, 1988
striatum	unchanged	
hypothalamus	unchanged	
cortex	unchanged	Howes *et al.*, 1983, 1984a
hypothalamus	unchanged	
hippocampus	unchanged	
striatum	unchanged	
midbrain	increased	
lower brainstem	increased	
spinal cord	unchanged	
telencephalon	increased	Yamabe *et al.*, 1973

continued

brainstem	unchanged	
hypothalamus	decreased	Fuxe *et al.*, 1979
hypothalamus	unchanged	Morris *et al.*, 1981a,b
neurohypophysis	decreased	
adenohypophysis	unchanged	Sim and Hsu, 1990
frontal cortex	increased	Van den Buuse *et al.*, 1984b
hypothalamus	unchanged	
hippocampus	unchanged	
medulla	unchanged	
frontal cortex	increased	Versteeg *et al.*, 1976
NIST[2]	increased	
NTS	increased	
forebrain nuclei	unchanged	
hypothalamic nuclei	unchanged	
medullary nuclei	unchanged	
A1-area	unchanged	Fujino, 1984
A2-area	unchanged	
NTS	unchanged	
locus coeruleus	unchanged	
Median eminence	unchanged	
Paraventricular nucleus	unchanged	
NTS	unchanged	Koulu *et al.*, 1986
A1-C1 area	unchanged	
locus coeruleus	increased	

[2]NIST = Nucleus interstitialis striae terminalis.

(3) Dopamine metabolite concentrations

metabolite	brain region	SHR *vs* WKY	reference
DOPAC[3]	striatum	increased	Fuller *et al.*, 1983
HVA[4]	striatum	unchanged	
DOPAC	cerebrospinal fluid	increased	Dawson *et al.*, 1988
HVA	cerebrospinal fluid	increased	
DOPAC	cortex	unchanged	Howes *et al.*, 1984a
	hypothalamus	unchanged	
	hippocampus	unchanged	
	striatum	increased	
	midbrain	increased	
	lower brainstem	unchanged	
	spinal cord	unchanged	

continued

DOPAC	frontal cortex striatum hypothalamus	unchanged increased increased	Williams Mckeon and Hendley, 1988
DOPAC	adenohypophysis	increased	Sim and Hsu, 1990

[3]DOPAC = 3,4-dihydroxyphenylacetic acid. [4]HVA = homovanillic acid

(4) Dopamine turnover

brain region	method	SHR *vs* WKY	reference
mesolimbic striatum	DOPAC, HVA[5]	unchanged unchanged	Fuller *et al.*, 1983
olfactory tub. striatum	α-MPT[6]	unchanged unchanged	Le Fur *et al.*, 1981
olfactory tub. striatum	DOPA acc.[7]	increased unchanged	
neurohypophysis	synthesis[8]	decreased	Morris *et al.*, 1981b
frontal cortex striatum	DA uptake	decreased decreased	Myers *et al.*, 1981
median eminence MBH[9] medial preoptic	DOPA acc.[7]	unchanged unchanged increased	Steger *et al.*, 1984
frontal cortex striatum hypothalamus	DOPAC/DA ratio	unchanged increased unchanged	Williams McKeon and Hendley, 1988
A1-area A2-area NTS locus coeruleus Median eminence Paraventricular nucleus	α-MPT[6]	unchanged unchanged unchanged unchanged unchanged unchanged	Fujino, 1984
Anterior hypothalamic area	unchanged		

[5]Decrease in DOPAC and HVA concentrations after increasing doses of Quinpirole. [6]Disappearance of dopamine after synthesis inhibition with α-methyl-para-tyrosine. [7]Accumulation of DOPA after DOPA-decarboxylase inhibition with NSD 1015. [8]Accumulation of labelled dopamine after administration of ^{3}H-tyrosine. [9]medial basal hypothalamus.

(5) Dopamine receptor binding

brain region	ligand	SHR vs. WKY	reference
striatum hypothalamus	spiroperidol	increased increased	Bhargava, 1984

striatum	spiroperidol	increased	Busch and Bhargava, 1983
striatum	apomorphine	unchanged	
hypothalamus		unchanged	
cortex		unchanged	
midbrain		unchanged	
pons/medulla		unchanged	
striatum	spiroperidol	increased	Chiu *et al.*, 1982
mesolimbic area	spiroperidol	increased	Chiu *et al.*, 1984
cortex		increased	
hippocampus		increased	
striatum	pergolide	unchanged	Fuller *et al.*, 1983
striatum	spiroperidol	unchanged	Hellstrand and Engel, 1980
limbic forebrain		unchanged	
tub. olfactorium	spiroperidol	increased	Le Fur *et al.*, 1981
striatum		increased	
frontal cortex		increased	
A9 region		unchanged	
A10 region		unchanged	
hypothalamus	spiperone	increased	Le Fur *et al.*, 1983
striatum	spiperone	increased	Lim *et al.*, 1989
hypothalamus		increased	
striatum	SCH-23390	increased	
hypothalamus		increased	
substantia nigra		increased	
striatum	spiperone	unchanged	Vaughn *et al.*, 1985
striatum	spiperone	unchanged	Wataneba *et al.*, 1990
striatum	SCH-23390	unchanged	

in many brain regions of this strain. Tyrosine hydroxylase activity, the concentrations of dopamine and its metabolites, dopamine turnover, and dopamine receptor binding were found to be higher in various brain regions, including striatum, of SHR when compared to WKY. Although a number of studies describe no difference between SHR and WKY in the respective parameters, actually hardly any studies suggest a decrease in dopaminergic activity (however, see also section 2.c).

With regard to tyrosine hydroxylase activity (see Table 3.1), Renaud *et al.* (1979) and Ikeda *et al.* (1984) found similar basal activity of tyrosine hydroxylase in SHR and WKY. However, Renaud and colleagues reported that repeated reserpine treatment increased enzyme activity in the A2 region of WKY but not of SHR. These authors interpret this finding as an inability of SHR to further increase tyrosine hydroxylase activity, possibly secondary to their high blood pressure. Ikeda *et al.* (1984) noted that swimming stress increased enzyme activity significantly in striatum of SHR, but not WKY. Thus, under certain circumstances, especially stressful conditions, dopamine synthesis and release is increased in SHR.

Williams McKeon and Hendley (1988) argued that the changes in DOPAC levels and DOPAC/DA ratios they found in the striatum of SHR are unrelated to the development of hypertension. The changes were not present in WK-HA and WK-HT, two substrains bred from SHR and WKY and selected for the presence of high activity/normal blood pressure and normal activity/high blood pressure, respectively. Thus, a genetical linkage between hypertension and changes in central dopaminergic parameters is not apparent. However, these two strains displayed particular changes in neurotransmitter levels themselves, which could indicate that during the process of breeding certain characteristics of SHR were lost and other characteristics were acquired. The use of F_2 backcross rats from SHR and WKY may circumvent this problem (for further discussion and references, see Versteeg and Van den Buuse, 1989).

In receptor binding studies, most authors have used the dopamine D_2 antagonist spiroperidol (spiperone) as radioactive ligand. However, this ligand shows significant binding to 5-HT_2 receptors also and only by using a proper definition of specific binding can the results be attributed to dopamine receptors. Wataneba *et al.* (1990) define specific binding with the specific dopamine D_2 receptor antagonist sulpiride and fail to find differences in [^{3}H]-spiroperidol binding between SHR and WKY. Most other, positive studies, used the non-selective dopamine receptor antagonist butaclamol to determine specific binding. Le Fur *et al.* (1981) argue that the binding of [^{3}H]-spiroperidol could be displaced with haloperidol, suggesting it was to dopamine receptors. Moreover, these authors did not find differences between SHR and WKY in [^{3}H]LSD binding, which could suggest that serotonin receptor density was similar in the two strains. Interestingly, by using the dopamine receptor agonist apomorphine, Busch and Bhargava (1983) failed to find differences in binding in any of the brain regions examined, whereas by using spiroperidol as a ligand they found that binding that increased in striatum of the SHR. By using the dopamine D_2 agonist pergolide, Fuller *et al.* (1983) also did not find differences in receptor density between SHR and WKY. Also in autoradiographic studies with ^{125}I-sulpiride as ligand no differences were found between SHR and WKY in D_2 dopamine receptor binding density in the brain (C.R. Jones, personal communication). Thus, choice of the ligand and experimental procedures importantly determines whether or not apparent differences in receptor binding characteristics between SHR and WKY are found.

Resting posterior hypothalamic dopamine release was higher in SHR than in WKY, as measured with a push-pull cannula system (Tuomisto *et al.*, 1983). However, the KCl-induced increase in dopamine release from this area was similar in the two strains. Ikeda *et al.* (1984) measured dopamine release in striatum of conscious SHR and WKY by *in vivo* voltammetry. A 5 minute swimming stress induced a much greater increase in the DOPAC signal in SHR than in WKY. This agrees well with findings of increased central DOPAC concentrations in SHR and possibly with decreased neuronal re-uptake (see Table 3.2). Dopamine release from striatal slices of SHR in vitro was lower than that from WKY (Linthorst *et al.*, 1990). Electrical stimulation induced a smaller increase in release from SHR than from WKY striatum, whereas inhibition of the electrically stimulated release by quinpirole was slightly, but significantly, greater in the SHR. These findings could be explained by an upregulation of presynaptic inhibitory elements, possibly as a result of the higher *in vivo* dopamine release.

Taken together, the available neurochemical data suggest that dopamine synthesis and turnover in a number of brain regions is increased in SHR when compared to WKY. Basal dopamine release is not always increased, however, but under certain circumstances, i.e. stress, dopamine release appears to be markedly enhanced in SHR when compared to WKY.

Also in other hypertension models changes in central dopamine parameters have been found (see Philippu, 1988; Versteeg *et al.*, 1984). Such findings indicate that hypertension by itself may influence central dopamine systems, and are in line with results of studies on the effects of acute changes in blood pressure on dopamine systems (see section 2.1). In a series of studies, Chen and colleagues have described the status of central dopamine function in DOCA-salt hypertension. Anterior pituitary dopamine content and in vitro hypothalamic dopamine release was reduced in 4 week DOCA-salt hypertensive rats (Chen *et al.*, 1985a,b). Administration of quinpirole reduced posterior hypothalamic noradrenaline content in controls, but not in DOCA-salt rats (Chen *et al.*, 1987a). In contrast, quinpirole reduced HVA outflow from striatum of DOCA-salt rats more than from controls. DOPAC changes and basal DOPAC and HVA release in the striatum were similar (Chen and Oparil, 1987b). From these and other findings Chen *et al.* (1987c) concluded that central dopamine receptor responsiveness is increased in DOCA-salt hypertension.

Endocrine and behavioural studies in hypertension

Rather than directly measuring the neurochemical parameters of central dopaminergic activity, it is also possible to obtain information on the role of central dopamine in hypertension through more indirect measures. A widely used parameter in this respect is the plasma concentration of prolactin, since it is generally established that the release of this hormone is inhibited by dopamine at the anterior pituitary level.

Basal plasma prolactin concentrations were shown to be increased in SHR (McMurtry *et al.*, 1979; Sowers *et al.*, 1980; 1981) and in patients with essential hypertension (Stumpe *et al.*, 1977; Kolloch *et al.*, 1981; Saruta *et al.*, 1983; Sowers *et al.*, 1982d). However, other authors were unable to find changes in basal plasma prolactin levels in experimental or human essential hypertension (Flack *et al.*, 1979; Horrobin *et al.*, 1975; Lengyel *et al.*, 1982; Musumeci *et al.*, 1989; Ogihara *et al.*, 1980; Os *et al.*, 1987; Steger *et al.*, 1984; Stern *et al.*, 1983; Tan and Hutchinson, 1987), whereas one report shows lower prolactin values (Holland and Gomez-Sanchez, 1977).

Administration of dopamine D_2 receptor antagonists induced an increase in plasma prolactin levels. Some authors showed this increase to be greater in hypertensive rats or patients (Os *et al.*, 1987; Sowers *et al.*, 1980; Stern *et al.*, 1983), but other authors reported no differences (Lengyel *et al.*, 1982; Ogihara *et al.*, 1980; Sowers *et al.*, 1982b). Interestingly, in normotensive rhesus monkeys the administration of domperidone also increased plasma prolactin levels, suggesting that this effect may be of peripheral origin (Sowers, 1982a).

Administration of dopamine receptor agonists decreased plasma prolactin levels to a greater extent in SHR than in WKY controls in some studies (Tan and Hutchinson, 1987), but other authors suggested that this effect was similar in SHR and WKY (Lengyel *et al.*, 1982; Ogihara *et al.*, 1980; Sowers *et al.*, 1980). Tan and Hutchinson (1988) suggested a genetical linkage between blood pressure and prolactin levels in the SHR. In an F_2 backcross generation of SHR and WKY they could classify rats into a low-, intermediate- and high-blood pressure group. Plasma prolactin levels were higher in hybrids from the high blood pressure group than in hybrids from the low blood pressure group. Values found in the intermediate blood pressure group were between the values found in the two other groups. Bromocriptine treatment decreased plasma prolactin in the high- and the intermediate blood pressure rats, but not in the low blood pressure rats (Tan and Hutchinson, 1988). Backcross studies can give useful information with regard to genetical linkage of different characteristics. However, in this case it is not clear whether prolactin levels affect blood pressure values or the reverse. Moreover, if there is indeed a linkage between the two factors, it is not clear through which (central?) mechanism this is mediated.

A number of methodological and experimental factors have been suggested to possibly explain the controversy between the results of the various studies mentioned above. Posture may be an important factor, such that prolactin levels were mildly elevated in essential hypertensives only when the patients were in the upright position, whereas supine values were normal (Meier *et al.*, 1980; Sowers *et al.*, 1982d). Renin status and salt homeostasis have been suggested to influence the prolactin levels and responsiveness (Saruta *et al.*, 1983; Stern *et al.*, 1983). In patients on a low-sodium diet, prolactin and renin responses to a dopamine antagonist were greater than in patients on high-sodium intake (Sowers *et al.*, 1982c). However, other authors found no correlation between salt intake, plasma renin and prolactin levels (Meier *et al.*, 1980; Lengyel *et al.*, 1982). Nevertheless, careful analysis of the human data is required, as changes in prolactin may occur only in particular subgroups of patients, different from others by their salt-sensitivity and activity of the renin-angiotensin-aldosterone system (see also Slater, 1983). Plasma prolactin levels are influenced by many other factors, including stress (Musumeci *et al.*, 1989; Sowers, 1981) and neuropeptides such as TRH (Kolloch *et al.*, 1981; Lengyel *et al.*, 1982).

Taken together, the above-mentioned reports make clear that plasma prolactin cannot serve as a good general index of central dopaminergic activity or of the role of central dopamine systems in the development of hypertension. The suggestion that "decreased brain dopaminergic activity" exists in hypertension can only count for the tuberoinfundibular dopamine system originating from the A12 group in the nucleus arcuatus and is unlikely to provide information about the other dopamine systems in the brain.

Surprisingly, few experimental hypertension studies deal with the direct effects of dopaminergic agents on prolactin release *in vitro*. In the study of Steger and colleagues (1984), basal plasma prolactin levels were similar in SHR and WKY. However, in vitro prolactin release was greater from hemipituitary glands of SHR than of WKY. Moreover, dopamine inhibited this release to a greater extent from SHR pituitaries, such that release was not different anymore between the strains. Although the authors suggest dopaminergic hypoactivity, these findings may indicate that in the *in vivo* situation dopaminergic inhibition of prolactin release in the SHR is actually greater than in WKY. Thus, removal of this inhibitory influence, as occurs in the in vitro situation, may explain the higher baseline release in vitro. In support of this conclusion is the observation that inhibition of *in vivo* dopamine synthesis caused a greater increase in plasma prolactin levels in SHR than in WKY, suggesting that normally an enhanced inhibition of prolactin secretion is present in SHR (Steger *et al.*, 1984). In this respect it is interesting to note that Sim and Hsu (1990) observed an increased turnover of dopamine in the anterior pituitary of SHR compared to WKY. Dopamine concentrations in the median eminence or anterior pituitary were normal, however (Fujino, 1984; Sim and Hsu, 1990). Further in vitro experiments and direct measurements of neurochemical parameters such as receptor density are necessary before definitive conclusions can be drawn with respect to the activity of the tuberoinfundibular or tuberohypophyseal dopamine system and its role in hypertension. Such data could provide better, albeit still indirect, proof for changes in tuberoinfundibular dopaminergic activity in hypertension. Changes of body temperature in response to the administration of dopaminergic agonists has been used as a parameter to obtain information concerning central dopaminergic mechanisms in hypertension. The decrease in body temperature to apomorphine or L-DOPA administration was greater in SHR than in WKY (Barlament *et al.*, 1984; Bhargava, 1983; 1984; Martin and Quock, 1984). These effects could indicate upregulation of postsynaptic dopamine receptors in the anterior hypothalamus of SHR, a finding which could be in line with the increased hypothalamic receptor density described by some authors in the SHR (see section 2.b, Table 5). However, the significance of changes in temperature regulation in hypertension needs further investigation.

Also the behavioural response of hypertensive subjects to dopaminergic drugs could give indirect information about the role of central dopaminergic mechanisms in hypertension. In the normotensive rat, dopamine receptor agonists at low to moderate doses produce an inhibition and a stimulation of locomotor activity, respectively. In some studies, baseline locomotor activity of SHR was increased and drugs such as apomorphine or quinpirole induced only an inhibition of activity in this strain (Fuller *et al.*, 1983; Hynes *et al.*, 1985). In WKY, the dopamine agonist bromocriptine induced an inhibition of activity at low doses and stimulation of activity at high doses. However, in SHR neither effect was observed (figure 3, Van den Buuse, unpublished observations). We recently reported similar levels of locomotor activity in SHR and WKY in an open-field test and a lack of responsiveness of SHR to the locomotor inhibition to apomorphine (Van den Buuse and De Jong, 1989a). In that same study, the dopamine antagonist haloperidol decreased locomotor activity in WKY, but only after higher doses in SHR. Furthermore, central injection of sulpiride inhibited activity of WKY, but not of SHR. In contrast, the increase of locomotor activity in response to administration of amphetamine or the dopamine uptake inhibitor GBR-12909 was similar in SHR and WKY. One explanation for these findings could be that endogenous dopamine release was enhanced in SHR in response to the mild stress of the open-field test (Van den Buuse and De Jong, 1989a).

McCarty *et al.* (1980) also studied the effect of amphetamine on open-field behaviour of SHR. These authors failed to observe a differential response to amphetamine on locomotor activity, but noted that at higher doses exploratory rearing

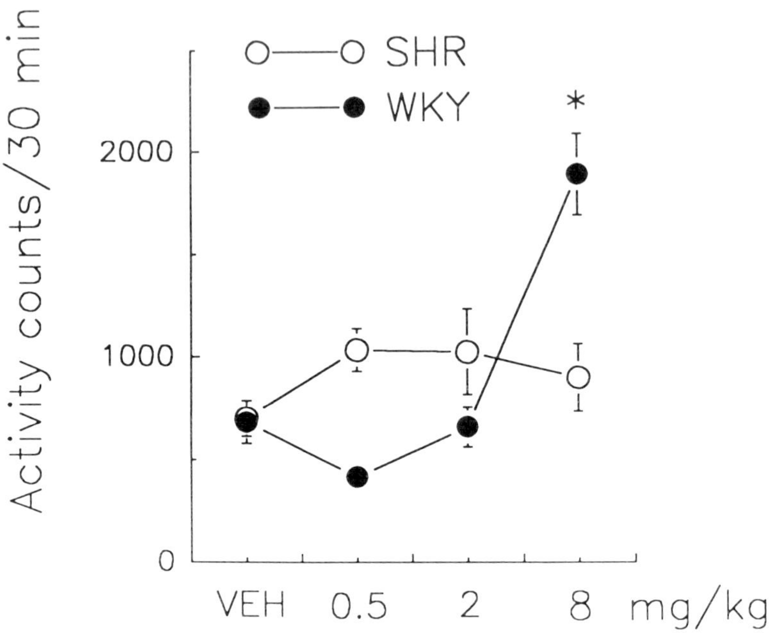

Figure 3. The effect of increasing doses of bromocriptine on locomotor activity of SHR and WKY. One hour after intraperitoneal injection the animals were placed in automated photocell cages and activity was measured for one hour. Data are mean ± SEM cumulative counts of 6 rats per group in the second half an hour of measurement. The first half an hour gave similar results although baseline activity of the SHR tended to be higher in this period. In the second half hour vehicle (50 % polyethyleneglycol in saline) injected SHR and WKY showed similar locomotor activity. Bromocriptine at 0.5 mg/kg induced a decrease and at 8 mg/kg an increase in activity of WKY. SHR showed neither effect.

activity increased in SHR whereas WKY responded with increased stereotyped headwaving. In contrast, apomorphine induced more rearing behavior in WKY than in SHR (McCarty *et al.*, 1980). Vaughn *et al.* (1985) reported no differences in apomorphine-induced stereotyped behaviour between SHR and WKY. Myers *et al.* (1982) found higher baseline locomotor activity, a smaller amphetamine-induced increase in activity, and under certain circumstances even an amphetamine-induced decrease in SHR when compared to WKY. We showed that exploratory rearing activity was markedly increased in SHR when compared to WKY also in an open-field test. Depletion of central dopamine by *i.c.v.* 6-hydroxydopamine or electrolytic lesions in the substantia nigra normalized this exaggerated behaviour (Van den Buuse *et al.*, 1986a; 1986d). In the open-field or after central injection of ACTH, SHR showed lower grooming activity than WKY (Van den Buuse *et al.*, 1987). In contrast, using animals acclimatized to their environment, Steel *et al.* (1988) found increased grooming activity in SHR. In this study, the dopamine D_1 agonist SKF 38393 increased grooming in WKY but not in SHR, suggesting a ceiling effect. Thus, the differences in behavioural responses of SHR and WKY may depend strongly on the experimental set-up.

Taken together, these studies clearly show differences in dopamine-mediated behavioural responses between SHR and WKY. More experiments are needed, however, to investigate the mechanism behind these changes and their relationship to the development of hypertension in these rats. For instance, local injection of dopaminergic drugs in particular brain areas could differentiate between the role of different brain dopamine systems in the behavioural differences between SHR and WKY. Preliminary results from Myers *et al.* (1982) suggested that injection of amphetamine into the nucleus accumbens inhibited the high levels of activity in SHR, whereas this treatment increased activity in WKY. Thus, the altered responsiveness of SHR to amphetamine could be due to changes in mesolimbic dopaminergic neurotransmission.

It must be noted that baseline behavioural responses of SHR may not necessarily reflect central changes involved in the development of high blood pressure (for discussion, see Versteeg and Van den Buuse, 1990; Williams McKeon and Hendley, 1988). However, the effects of dopaminergic drugs on behaviour of SHR when compared to WKY may provide useful information about central dopaminergic activity in these animals. This information needs to be combined with other findings, such as from neurochemical studies (see section 2.b) and lesion studies (see next section) and may then provide a basis of the role of central dopaminergic systems in the development of hypertension.

Dopaminergic lesions and hypertension

Direct evidence on the role of central dopamine systems in the development of hypertension can be obtained by the use of brain lesions in the pre-hypertensive state, i.e. in SHR of younger than 6 weeks of age or in normotensive rats before the induction of experimental (DOCA-salt, renal-clip) hypertension. Central treatment of perhypertensive SHR with 6-hydroxydopamine inhibited the development of hypertension (Haeusler *et al.*, 1972; Van den Buuse *et al.*, 1984a). This treatment caused widespread depletion of particularly noradrenaline, but also of dopamine and adrenaline throughout the central nervous system (Van den Buuse *et al.*, 1984a; 1984b). It was shown that the effect of central 6-hydroxydopamine on the development of hypertension was not due to depletion of spinal catecholamines (Loewy *et al.*, 1980), nor to depletion of forebrain noradrenaline (Erinoff *et al.*, 1975; Van den Buuse *et al.*, 1984a). Moreover, pretreatment of the animals with the noradrenaline uptake inhibitor desipramine blocked the central depletion of noradrenaline by 6-hydroxydopamine, but not the effect on the development of hypertension or on central dopamine (Van den Buuse *et al.*, 1984b). Conversely, pretreatment with the dopamine uptake inhibitor GBR-12909 attenuated the effect of 6-hydroxydopamine both on the development of hypertension as well as on central dopamine levels, but not on central noradrenaline levels (Van den Buuse *et al.*, 1986c). Thus, the inhibition of the development of hypertension by central treatment with 6-hydroxydopamine was caused by depletion of brain dopamine. In this respect it is interesting to note that there appear to be few descriptions of hypertension in patients with Parkinson's disease (Montastruc *et al.*, 1985). Blood pressure in these patients has been described to be lower than in controls (Brevetti *et al.*, 1990; see also section 3). Discrete electrolytic lesions in the substantia nigra of young SHR caused an attenuation of the development of hypertension as well as depletion of dopamine in the striatum, probably due to damage to the nigrostriatal dopamine system (Van den Buuse *et al.*, 1986c; 1986d). In SHR, lesions in the ventral tegmental area, origin of the mesolimbic/mesocortical dopamine system, did not markedly inhibit the development of hypertension, although depletion of dopamine in nucleus accumbens and frontal cortex was observed (Van den Buuse *et al.*, 1986c). These data suggest that the nigrostriatal dopamine system plays a role in the development of spontaneous hypertension in rats. It is unclear, however, by which mechanism the nigrostriatal system influences cardiovascular regulation. One possibility could be that it modulates cardiovascular reflexes (see section 2a, see also next section), such that in prehypertensive rats inadequate buffering of hypertensive responses to stress occurs. The behavioural and cardiovascular hyperreactivity to stress of SHR has been well-described (for references, see Van den Buuse *et al.*, 1986a; 1989a). Indeed, lesions of the substantia nigra not only inhibit the development of hypertension, but also attenuate behavioural hyperreactivity in SHR (Van den Buuse *et al.*, 1986d). As yet however, no measurements of cardiovascular regulatory mechanisms such as the baroreceptor reflex have been performed in SHR with central dopamine depletion. Moreover, further studies should also involve lesions in other dopamine systems, such as the tuberoinfundibular system (see section 2.c) and descending dopamine systems.

CARDIOVASCULAR REGULATION AND PARKINSON'S DISEASE

Ever since the first description by Parkinson in 1817, Parkinson's disease is known to involve autononomic disturbances such as sialorrhea, seborrhea, excessive sweating, constipation, bladder dysfunction and orthostatic hypotension (see Martignoni *et al.*, 1986). Relatively few systematic studies have been performed to analyse the mechanism underlying these disturbances, however. Here, we will focus on studies on blood pressure and cardiovascular reflex regulation.

In 1969, Barbeau and colleagues stated that "patients with Parkinson's disease (...) often have blood pressure lower than the mean for their age group. This relatively low blood pressure (...) is more evident in the standing position and contributes to a chronic state of fatigue and to occasional episodes of postural dizziness" (Barbeau *et al.*, 1969). These authors subsequently showed lower plasma renin activity in a group of patients with Parkinson's disease. Treatment with L-dopa further decreased plasma renin activity. The low levels of plasma renin activity were confirmed by Michelakis and Robertson (1970) although these authors did not find an effect of L-dopa treatment. Calne *et al.* (1970) and McDowell and Lee (1970) showed that L-dopa treatment in Parkinson patients may cause a lower resting blood pressure and greater fall in pressure upon getting up from the supine position. Unfortunately no proper controls were included in these studies.

Appenzeller and Goss (1971) reported a general deficit in compensatory vasoconstriction in response to the fall in cardiac output induced by Valsalva's maneuver, but no consistent orthostatic hypotension. Again no proper controls were included in this study. Reid and colleagues (1971) studied responses to head-up tilt and Valsalva's maneuver. Although reflex function diminished with age, no significant differences were found between patients with Parkinson's disease and age-matched controls. Also in a study of Gross *et al.* (1972) Valsalva responses were normal when patients with Parkinson's disease were compared with age-matched controls. However, the fall in blood pressure upon tilting was greater in Parkinson patients. In another report it was shown that L-dopa treatment significantly impaired the pressor response to Valsalva's maneuver and also induced a decrease in resting blood pressure (Reid *et al.*, 1972). Interestingly, in this study a peripheral decarboxylase inhibitor blocked the effects of L-dopa on reflex function, but not on resting blood pressure.

More recent studies usually involve age-matched controls, whereas treatment-free patients were used or treatment was withdrawn before the measurements. Thus, Orskov *et al.* (1987) found markedly decreased heart rate variability at rest, after deep breathing and after standing. Whereas resting blood pressures were similar, only long-term (>5 years) Parkinsonians displayed a fall in systolic blood pressure upon standing. Because pupillary responses were normal, these authors conclude that long-term Parkinson patients show a selective impairment of cholinergic control of the heart. This finding could be compatible with the earlier reports on attenuated cardiovascular reflex responses (see above). Moreover, the duration of the disease may be a factor which determines the occurrence of changes in cardiovascular regulation in Parkinson's disease. Nevertheless, studying long-term Parkinsonians, Sachs and colleagues (1985) found normal responses to Valsalva's maneuver, tilting and diving. However, these authors did observe a higher resting heart rate, attenuated respiratory sinus arrhythmia and attenuated changes in blood pressure and heart rate to isometric handgrip. Brevetti *et al.* (1990) used continuous intra-arterial recording of blood pressure in five patients with Parkinson's disease of varying duration and degree. The patients had blood pressure values lower than age-matched controls throughout the day, with the greatest difference occurring in the late afternoon and early evening. Heart rate tended to be higher in the patients with Parkinson's disease, but responses to Valsalva's maneuver, cold pressor test and mental arithmetic test were normal. The authors suggest that the inclusion of patients with mild autonomic impairment in some earlier studies might explain the discrepancy between their results and those of others.

The possible mechanism of the proposed changes in autonomic regulation in Parkinson's disease were studied by a number of authors. Histopathological studies (Rajput and Rozdilsky, 1976) showed Lewy bodies and cell loss in sympathetic ganglia of patients with Parkinson's disease, some of which had suffered from orthostatic hypotension. Turkka *et al.* (1985) observed lower plasma noradrenaline levels in Parkinson patients as compared to controls "of similar age". This difference was found after an overnight rest as well as immediately after standing up. Moreover, the rise in plasma noradrenaline induced by standing up was markedly diminished in Parkinson patients (5% versus 38% in controls). Orskov and colleagues (1987) also showed reduced plasma noradrenaline responses in long-term, but not in short-term Parkinson patients. Other authors were not able to find differences in catecholamine levels, but in those studies patients had a shorter duration of or less severe illness (Micieli *et al.*, 1987; Durrieu *et al.*, 1990).

Central α_2-adrenoreceptor sensitivity could be decreased in patients with Parkinson's disease. Villeneuve and colleagues (1985) measured platelet α_2-adrenoreceptor binding as a model of central receptors. They found lower specific binding in Parkinson patients as compared to age-matched controls. Treatment with L-dopa or bromocriptine normalized the lower binding. To further study α_2-adrenoreceptor function in Parkinson's disease, Berlan *et al.* (1989) studied the effect of clonidine. A much smaller decrease in systolic, but not diastolic blood pressure was observed in never-treated Parkinson patients. In a group of patients treated with L-dopa and/or bromocriptine the blood pressure response to clonidine was normal. However, the fall in heart rate induced by clonidine was absent in all Parkinson patients. Although the latter finding could not be explained, the authors conclude that in Parkinson's disease decreased central α_2-adrenoreceptor sensitivity exists.

Dopamine-β-hydroxylase activity in cerebrospinal fluid was found to be reduced in Parkinson patients, which is in line with reports showing decreased central noradrenaline concentrations in Parkinson's disease (Hurst *et al.*, 1985 and references therein). However, such decreased central noradrenergic function is difficult to reconcile with decreased α_2-adrenoreceptor number (see above).

Although Parkinson's disease may involve damage to a number of central neurotransmitter systems, the principal defect lies in the profound depletion of central dopamine caused by degeneration of the nigrostriatal dopamine system. As such, the study of autonomic (dys)regulation in Parkinson's disease may provide insight in the role of this system in cardiovascular control. However, in some of the earlier studies, age-matched controls were lacking. Moreover, because different forms of Parkinson's disease exist and in different grades of severity and of duration, it could be that autonomic defects exist in some subgroups of patients only (see Orskov *et al.*, 1987; Brevetti *et al.*, 1990). Taken together, the above-described studies suggest that central regulation of cardiovascular function, especially postural compensatory reflex mechanisms, is impaired in Parkinson's disease. Thus, central dopaminergic systems, especially the nigrostriatal pathway, may directly or indirectly influence cardiovascular reflexes and thereby cardiovascular homeostasis.

SUMMARY AND CONCLUSIONS

In this paper we presented an overview of the available information about the role of brain dopamine in blood pressure. It is clear that brain dopamine systems are involved in several aspects of cardiovascular regulation. The nigrostriatal system, originating in the substantia nigra, may play a role in the regulation of cardiovascular reflexes. This is evidenced by the work in SAD rats and the results of lesion and stimulation experiments in intact, normotensive animals (see section 2a), and the many (although sometimes contradictory) results of cardiovascular measurements in patients with Parkinson's disease (see section 3). It is suggested that through the effect on reflex regulation the nigrostriatal system may also be involved in the development of hypertension (see section 2b and 2d).

The mesocortical/mesolimbic dopamine system has been less studied.

Nevertheless, some data indicate a possible pressor role of mesolimbic pathways. Perhaps in response to certain stressors, this pressor mechanism is activated (see section 2a and 2c).

The tuberoinfundibular and tuberohypophyseal systems are obviously important for the regulation of endocrine function and thereby could affect cardiovascular regulation. However, by only measuring endocrine parameters in hypertension, insufficient information is obtained about the status of other dopamine systems in the brain, and no general statement can be made about the role of brain dopamine in the development of high blood pressure (see section 2c). Little if anything is known about the involvement of other dopamine systems in cardiovascular control.

Obviously much more work is needed to fill in all the details into this preliminary scheme. Nevertheless, centrally acting dopaminergic drugs (see section 1) may influence blood pressure and other cardiovascular parameters by interacting with each of the above-mentioned dopamine systems. It is important to further investigate these interactions to increase our understanding of the role of brain dopamine in cardiovascular control. This may explain or disclose certain side-effects of dopaminergic drugs used, e.g. in schizophrenia, Parkinson's disease or endocrine disorders, or it may possibly lead us to centrally acting cardiovascular drugs acting on dopamine receptors.

REFERENCES

Alexander, N., Hirata, Y. and Nagatsu, T. (1984). Reduced tyrosine hydroxylase activity in nigrostriatal system of sino-aortic denervated rats. *Brain Res.* **299,** 380-382.

Alexander, N., Ross-Cisneros, F., Kogosov, E., DeCuir, M., Haun, C. K. and Nagatsu, T. (1986). Nigrostriatal and cardiovascular systems interactions. In: *Brain and blood pressure control*, K. Nakamura (Ed.), Elsevier Sci. Publ., Excerpta Medica, Amsterdam, pp. 61-66.

Alexander, N., Nakahara, D., Ozaki, N., Kaneda, N., Sasaoka, T., Iwata, N. and Nagatsu, T. (1988). Striatal dopamine release and metabolism in sinoaortic-denervated rats by *in vivo* microdialysis. *Am. J. Physiol.* **254,** R396-R399.

Alföldi, A. and Farsang, C. (1987). Metoclopramide decreases baroreflex sensitivity in patients with essential hypertension. *Life Sci.* **40,** 1469-1472.

Aminoff, M.J. and Wilcox, C.S. (1971). Assessment of autonomic function in patients with a Parkinson Syndrome. *Brit. Med. J.* **4,** 80-84.

Appenzeller, O. and Goss, J.E. (1971). Autonomic deficits in Parkinson's syndrome. *Arch. Neurol.* **24,** 50-57.

Bannon, M. J. and Roth, R. H. (1983). Pharmacology of mesocortical dopamine neurons. *Pharmacol. Rev.* **35**, 53-68.

Barbeau, A., Gillo-Joffroy, L., Boucher, R., Nowaczynski, W. and Genest, J. (1969). Renin-aldosterone system in Parkinson's disease. *Science* **165,** 291-292.

Barlament, J., Quock, R.M. and Vaughn, L.K. (1984). Apomorphine-induced hypothermic effect in normotensive Wistar-Kyoto and spontaneously hypertensive rats of different ages. *Fed. Proc.* **43,** 583.

Barnett, A. and Fiore, J. W. (1971). Hypotensive effects of apomorphine in anaesthetised cats. *Eur. J. Pharmacol.* **14,** 206-208.

Barrett, R. J. and Lokhandwala, M. F. (1983). Central and peripheral dopaminergic mechanisms in the cardiovascular actions of pergolide in neurogenic hypertensive dogs. *Eur. J. Pharmacol.* **96,** 211-226.

Baum, T. and Shropshire, A. T. (1973). Reduction of sympathetic outflow by central administration of L-DOPA, dopamine and norepinephrine. *Neuropharmacology* **12,** 49-56.

Berkowitz, B. A., Arleth, A. J., Sung, C. P., Kruse, L. I., DeWolf, W. E. and Ohlstein, E. H. (1988). Effects of the novel dopamine β-hydroxylase inhibitor SK&F 102698 on catecholamines and blood pressure in spontaneously hypertensive rats. *J. Pharmacol. Exp. Ther.* **245,** 850-857.

Berlan, M., Rascol, O., Belin, J., Moatti, J. P., Rascol, A. and Montastruc, J. L. (1989). α-2 adrenergic sensitivity in Parkinson's disease. *Clin. Neuropharmacol.* **12,** 138-144.

Bhargava, H. N. (1983). Effects of cyclo(leu-gly) on the supersensitivity of dopamine receptors in SHR. *Life Sci.* **32,** 2131-2137.

Bhargava, H. N. (1984). Effect of cyclo(leucyl-glycine) on [^{3}H]spiroperidol binding in the corpus striatum and hypothalamus of spontaneously hypertensive rats. *Eur. J. Pharmacol.* **100**, 109-112.Björklund, A. and Lindvall, O. (1984). Dopamine-containing systems in the CNS. In: *Handbook of chemical neuroanatomy, Classical transmitters in the CNS, part I.* A. Björklund and T. Hökfelt (Eds). Elsevier, Amsterdam, pp. 55-122.

Brevetti, G., Bonaduce, D., Breglio, R., Perna, S., Simonelli, P., Marconi, R. and Campanella, G. (1990). Parkinson's disease and hypotension: 24-hour blood pressure recording in ambulant patients. *Clin. Cardiol.* **13,** 474-478.

Busch, T. C. and Bhargava, H. N. (1983). Binding of ^{3}H-ligands to neuroleptic/dopamine receptors in brain regions of spontaneously hypertensive rats. *Fed. Proc.* **42,** 893.

Calne, D. B., Brennan, J., Spiers, A.S.D. and Stern, G. M. (1970). Hypotension caused by L-DOPA. *Brit. Med. J.* **1,** 474-475.

Calne, D. B. (1978). Role of ergot derivatives in the treatment of Parkinsonism. *Fed. Proc.* **37,** 2207-2209.

Casto, R., Jödicke, A. and Stock, G. (1988). Central cardiovascular effects of dopamine receptor activation. *The FASEB journal* **2,** A500.

Cavero, I. (1983). Cardiovascular effects of LY141865, a new dopamine receptor agonist, in rats. *Brit. J. Pharmacol.* **s78,** 139P.

Cavero, I., Lefèvre-Borg, F., Lhoste, F., Sabatier, C., Richer, C. and Giudicelli, J. F. (1984). Pharmacological, hemodynamic and autonomic nervous system mechanisms responsible for the blood pressure and heart rate lowering effects of pergolide in rats. *J. Pharmacol. Exp. Ther.* **228,** 779-791.

Chamontin, B., Montastruc, J. L. and Rascol, A. (1984). How do we explain the antihypertensive effects of bromocriptine? *Arch. Mal. Coeur.* **77,** 1181-1185.

Chen, F. F. and Lin, M. T. (1980). Effects of dopamine, apomorphine, gamma-hydroxybutyrate, haloperidol and pimozide on reflex bradycardia in rats. *J. Pharmacol. Exp. Ther.* **214,** 427-432.

Chen, Y. F., Dawson, R., Gist, R. and Oparil, S. (1985a). Reduced tuberoinfundibular function in DOCA/NaCl hypertensive rats. *Endocr. Soc. 67th Ann. Meeting* **236,** abstr. 943.

Chen, Y. F., Nagahama, S., Winternitz, S. and Oparil, S. (1985b). Hyperresponsiveness of monoaminergic mechanisms in DOCA/NaCl hypertensive rats. *Am. J. Physiol.* **249,** H71-H79.

Chen, Y. F., Jin, H., Gist, R. and Oparil, S. (1987a). Blunted responsiveness of posterior hypothalamic norepinephrine to quinpirole in DOCA/NaCl hypertensive rats. *Brain Res. Bull.* **18,** 563-568.

Chen, Y. F. and Oparil, S. (1987b). Enhanced response to the inhibitory action of LY171555, a dopamine D_2 agonist, on *in vivo* striatal dopamine release in DOCA/NaCl hypertensive rats. *Brain Res.* **400,** 225-231.

Chen, Y. F., Jin, H., Gist, R. and Oparil, S. (1987c). Altered responsiveness of regional brain dopamine and DOPAC levels to systemic administration of quinpirole, a dopamine D_2 agonist, in DOCA/NaCl-hypertensive rats. *Brain Res.* **413,** 15-22.

Chiu, P., Rajakumar, G., Chiu, S., Kwan, C. Y. and Mishra, R. K. (1982). Enhanced [^{3}H]-spiroperodol binding in the striatum of spontaneously hypertensive rat (SHR). *Eur. J. Parmacol.* **83,** 243-244.

Chiu, P., Rajakumar, G., Chiu, S., Kwan, C. Y. and Mishra, R. K. (1984). Differential changes in central serotonin and dopamine receptors in spontaneous hypertensive rats. *Progr. Neuro-*

Psychopharmacol. Biol. Psychiatry **8,** 665-668.

Clark, B.J., Scholtysik, G. and Flückiger, E. (1978). Cardiovascular actions of bromocriptine. *Acta Endocr.* **s216,** 75-81.

Damase-Michel, C., Montastruc, J. L., Tran, M. A., Gharib, C., Geelen, G. and Montastruc, P. (1990). Involvement of vasopressin in the cardiovascular effects of quinpirole. *Eur. J. Pharmacol.* **184,** 179-183.

Dawson, R. and Meldrum, M. J. (1988). Increased 3,4-dihydroxyphenylacetic acid (DOPAC) levels in spontaneously hypertensive rats (SHR) during the development of hypertension. *Neurosci. Lett.* **86,** 346-350.

Day, M. D. and Roach, A. G. (1976). Cardiovascular effects of dopamine after central administration into conscious cats. *Brit. J. Pharmacol.* **58,** 505-515.

DePalma, P. D. and Ackerman, D. M. (1983). The effect of dopamine on mean arterial blood pressure in the spontaneously hypertensive rat. *J. Pharm. Pharmacol.* **35,** 536-537.

Durrieu, G., Senard, J. M., Rascol, O., Tran, M. A., Lataste, X., Rascol, A. and Montastruc, J. L. (1990). Blood pressure and plasma catecholamines in nevertreated Parkinsonian patients: Effect of a selective D1 agonist (CY 208-243). *Neurology* **40,** 707-709.

Dutta, S. N., Guha, D. and Pradhan, S. N. (1975). Cardiovascular effects of central microinjections of apomorphine in cats. *Arch. Int. Pharmacodyn. Ther.* **215,** 259-265.

Earnhardt, J. T., Elghozi, J. L., Le Quan-Bui, K. H. and Meyer, P. (1982). Implications for dopamine in central blood pressure regulation. *Eur. Heart. J.* **3** (suppl. C), 27-31.

Elghozi, J. L., Le Quan-Bui, K. H., Earnhardt, J. T., Meyer, P. and Devynck, M. A. (1981). *In vivo* dopamine release from the anterior hypothalamus of the rat. *Eur. J. Pharmacol.* **73,** 199-208.

Erinoff, L., Heller, A. and Oparil, S. (1975). Prevention of hypertension in the SH-rat: effect of differential depletion of central catecholamines. *Proc. Soc. Exp. Biol. Med.* **150,** 748-754.

Esposti, E. D., Sturani, A., Santoro, A., Zuccala, A., Chiarini, C. and Zucchelli, P. (1985). Effect of bromocriptine treatment of prolactin, noradrenaline and blood pressure in hypertensive haemodialysis patients. *Clin. Sci.* **69,** 51-56.

Finch, L. and Haeusler, G. (1973). The cardiovascular effects of apomorphine in the anesthetized rat. *Eur. J. Pharmacol.* **21,** 264-270.

Flack, J. D., Hamilton, T. C., McClelland, C. and Poyser, R. H. (1979). Plasma prolactin concentrations in hypertensive rats. *J. Pharm. Pharmacol.* **31,** 61.

Fujino, K. (1984). Brain catecholamines in spontaneously hypertensive and DOCA-salt hypertensive rats. *Acta Med. Okayama* **38,** 325-340.

Fuller, R. W., Hemrick-Luecke, S. K., Wong, D. T., Pearson, D., Threllkeld, P. G. and Hynes, M. D. III (1983). Altered behavioural response to a D_2 agonist, LY141865, in SHR exhibiting biochemical and endocrine responses similar to those in normotensive rats. *J. Pharmacol. Exp. Ther.* **227,** 354-359.

Fuxe, K., Ganten, D., Jonsson, G., Agnati, L.F., Andersson, K., Hökfelt, T., Bolme, P., Goldstein, M., Hallman, H., Unger, T. and Rascher, W. (1979). Catecholamine turnover changes in hypothalamus and dorsal midline area of the caudal medulla oblongata of spontaneously hypertensive rats. *Neurosci. Lett.* **15,** 283-288.

Fuxe, K., Fredholm, B. B., Ögren, S. O., Agnati, L. F., Hökfelt, T. and Gustafsson, J. Å. (1978). Pharmacological and biochemical evidence for the dopamine agonistic effect of bromocriptine. *Acta Endocr.* **s216,** 27-56.

Gibson, A. and Samini, M. (1978). Bromocriptine is a potent α-adrenoceptor antagonist in the perfused mesenteric blood vessels of the rat. *J. Pharm. Pharmacol.* **30,** 314-315.

Globus, M.Y.T., Busto, R., Dietrich, W. D., Martinez, E., Valdes, I. and Ginsberg, M. D. (1988). Effect of ischemia on the *in vivo* release of striatal dopamine, glutamate, and γ-aminobutyric acid studied by intracerebral microdialysis. *J. Neurochem.* **51,** 1455-1464.

Goldberg, L. I. (1972). Cardiovascular and renal actions of dopamine: potential clinical implications. *Pharmacol. Rev.* **24,** 1-29.

Granata, A. R. and Woodruff, G. N. (1979). The effects of dopamine on neurons in the NTS and on the central control of blood pressure in the rat. *J. Physiol.* **296,** 100P-101P.

Granata, A. R. and Woodruff, G. N. (1982). Dopaminergic mechanisms in the nucleus tractus solitarius and effects on blood pressure. *Brain Res. Bull.* **8,** 483-488.

Greenacre, J. K., Teychenne, P. F., Petrie, A., Calne, D. B., Leigh, P.N. and Reid, J. L. (1976). The cardiovascular effects of bromocriptine in Parkinsonism. *Brit. J. Clin. Pharmacol.* **3,** 571-574.

Gross, M., Bannister, R. and Godwin-Austen, R. (1972). Orthostatic hypotension in Parkinson's disease. *The Lancet* **January,** 174-176.

Haeusler, G., Gerold, M. and Thoenen, H. (1972). Central adrenergic neurons and the initiation and development of experimental hypertension. *Experientia* **28,** 1200-1203.

Hahn, R. A., McDonald, B. R. and Martin, M. A. (1983). Antihypertensive activity of LY141865, a selective presynaptic dopamine receptor agonist. *J. Pharmacol. Exp. Ther.* **224,** 206-214.

Hamilton, T. C. (1981). Involvement of the adrenal glands in the hypotensive response to bromocriptine in spontaneously hypertensive rats. *Brit. J. Pharmacol.* **72,** 419-425.

Hellstrand, K. and Engel, J. (1980). Locomotor activity and catecholamine receptor binding in adult normotensive and spontaneously hypertensive rats. *J. Neural Transm.* **48,** 57-63.

Hilditch, A. and Drew, G. M. (1985). Peripheral dopamine receptor subtypes a closer look. *TIPS* **6**, 396-400.

Holland, O. B. and Gomez-Sanchez, C. E. (1977). Prolactin in hypertension. *Lancet* **2,** 1033.

Horrobin, D. F., Nowaczynski, W., Messerli, F. H., Kuchel, O., Genest, J. and Guyda, H. (1975). Plasma prolactin: effect of posture in hypertensive patients. *New Engl. J. Med.* **292,** 318.

Howes, L. G., Rowe, P. R., Summers, R. J. and Louis, W. J. (1983). Dopamine and DOPAC distribution within the brain and spinal cord of SHR and WK-rats at various ages. *Clin. Exp. Pharmacol. Phys.* **10,** 642-643.

Howes, L. G., Rowe, P. R., Summers, R. J. and Louis, W. J. (1984a). Age-related changes of catecholamines and their metabolites in central nervous system regions of spontaneously hypertensive (SHR) and normotensive Wistar-Kyoto (WKY) rats. *Clin. Exp. Hypert.* **A6,** 2263-2277.

Howes, L. G. (1984b). Central catecholaminergic neurons and spontaneously hypertensive rats. *J. Auton. Pharmacol.* **4,** 207-217.

Hurst J. H., LeWitt, P. A., Burns, R. S., Foster, N. L. and Lovenberg, W. (1985). CSF dopamine-β-hydroxylase activity in Parkinson's disease. *Neurology* **35,** 565-568.

Hutchinson, J. S., DiNicolantonio, R., Lim, A., Clements, J. and Funder, J. W. (1981). Effect of bromocriptine on blood pressure and plasma β-endorphin in spontaneously hypertensive rats. *Clin. Sci.* **61,** 001s-003s.

Hutchinson, J. S., DiNicolantonio, R., Veroni, M. and Cleverdon, M. (1983). Evidence for a functional central dopaminergic insufficiency in the spontaneously hypertensive rat. *Clin. Exp. Pharmacol. Phys.* **10,** 311-314.

Hynes, M. D., Langer, D. H., Hymson, D. L., Pearson, D. V. and Fuller, R. W. (1985). Differential effects of selected dopaminergic agents on locomotor activity in normotensive and spontaneously hypertensive rats. *Pharmacol. Biochem. Behav.* **23,** 445-448.

Igarashi, Y., Chen, Y. F., Wyss, J. M., Lindheimer, M. D. and Oparil, S. (1987). Continuous intravenous infusion of LY171555, a potent selective D_2 receptor agonist, lowers blood pressure in the conscious rat. *Pharmacology* **35,** 194-202.

Ikeda, M., Hirata, Y., Fujita, K., Shinzato, M., Takahashi, H., Yagyu, S. and Nagatsu, T. (1984). Effects of stress on release of dopamine and serotonin in the striatum of spontaneously hypertensive rats: an *in vivo* voltammetric study. *Neurochem. Int.* **6,** 509-512.

Jadhav, A. L., Willett, R. N., Sapru, H. N. and Lokhandwala, M. F. (1983). Involvement of central dopamine receptors in the hypotensive action of pergolide. *Naunyn. Schmied. Arch. Pharmacol.* **324,** 281-286.

Johns, D. W., Ayers, C. R. and Carey, R. M. (1984). The dopamine agonist bromocriptine induces hypotension by venous and arteriolar dilation. *J. Cardiovasc. Pharmacol.* **6,** 582-587.

Kawabe, H., Kondo, K. and Saruta, T. (1983). Effect of the intracerebroventricular injection of dopamine on blood pressure in the spontaneously hypertensive rat. *Clin. Exp. Hypert.* **A5,** 1703-1716.

Kawano, T., Tsutsumi, K., Miyake, H. and Mori, K. (1988). Striatal dopamine in acute cerebral ischemia of stroke-resistant rats. *Stroke* **19,** 1540-1543.

Keller, R. W. Jr, Stricker, E. M. and Zigmond, M. J. (1983). Environmental stimuli but not homeostatic challenges produce apparant increases in dopaminergic activity in the striatum: an analysis by in vivo voltammetry. *Brain Res.* **279,** 159-170.

Kobilansky, C., Lanzinger, I. and Philippu, A. (1988). Release of endogenous catecholamines in the nucleus tractus solitarii during experimentally induced blood pressure changes. *Naunyn-Schmiedeberg's. Arch. Pharmacol.* **337,** 125-130.

Kolloch, R., Kobayashi, K. and DeQuattro, V. (1980). Dopaminergic control of sympathetic tone and blood pressure: evidence in primary hypertension. *Hypertension* **2,** 390-394.

Kolloch, R. E., Stumpe, K. O., Ismer, U., Kletzky, O. and DeQuattro, V. (1981). Central dopaminergic mechanisms in young patients with essential hypertension. *Clin. Sci.* **61,** 231s-234s.

Kondo, K., Suzuki, H., Handa, M., Hagahama, S. and Saruta, T. (1981). Effects of intracerebroventricular administration of dopamine and metoclopramide on blood pressure in rats. *Arch. Int. Pharmacodyn. Ther.* **250,** 273-278.

Koulu, M., Saavedra, J.M., Niwa, M. and Linnoila, M. (1986). Increased catecholamine metabolism in the locus coeruleus of young spontaneously hypertensive rats. *Brain Res.* **369,** 361-364.

Kurz, K., Main, B., Moore, R. and Smith, T. (1986). Differential effects of the selective dopamine (DA) agonist, quinpirole, on blood pressure in conscious normotensive and hypertensive rats. *Fed. Proc.* **45,** 1071.

Laduron, P. M. and Leysen, J. M. (1979). Domperidone, a specific in vitro dopamine antagonist, devoid of *in vivo* central dopaminergic activity. *Biochem. Pharmacol.* **28,** 2161-2165.

Lang, W. J. and Woodman, O. L. (1979). Cardiovascular responses produced by the injection of dopamine into the cerebral ventricles of the unanesthetized dog. *Brit. J. Pharmacol.* **66,** 235-240.

Le Fur, G., Guilloux, F., Kabouche, M., Mihani, N., Ferris, O. and Uzan, U. (1981). Central dopaminergic neurons during the development of genetic and DOCA-salt hypertension in the rat. *Brain Res.* **227,** 153-163.

Le Fur, G., Guilloux, F. and Uzan, U. (1983). Evidence for an increase in [^{3}H]spiperone binding in hypothalamic nuclei during the development of spontaneous hypertension in the rat. *Clin. Exp. Hypert.* **A5,** 1537-1542.

Lengyel, A. J., Vieira, J.G.H., Charra, A. R., Abucham-Filho, J. Z., Lima, A. B., Ribeiro, A. B. and Ramos, D. L. (1982). Dynamic evaluation of prolactin secretion in essential hypertension: evidence against hypothalamic-pituitary dopaminergic dysfunction. *J.Clin. Endocr. Metab.* **54,** 849-853.

Lim, D. K., Ito, Y., Hoskins, B., Rockhold, R. W. and Ho, I. K. (1989). Comparative studies of muscarinic and dopamine receptors in three strains of rats. *Eur. J. Pharmacol.* **165,** 279-287.

Lin, M. T., Tsay, B. L. and Chen, F. F. (1982). Activation of dopaminergic receptors within the caudate-putamen complex facilitates reflex bradycardia in the rat. *Jap. J. Physiol.* **32,** 431-442.

Linthorst, A.C.E., Van den Buuse, M., De Jong, W. and Versteeg, D.H.G. (1990). Electrically stimulated [^{3}H]dopamine and [^{14}C]acetylcholine release from nucleus caudatus slices: differences between spontaneously hypertensive rats and Wistar-Kyoto rats. *Brain Res.* **509,** 266-272.

Loewy, A. D., McKellar, S., Svensson, E. E. and Panneton, W. M. (1980). Onset of hypertension in spontaneously hypertensive rats despite depletion of spinal catecholamines. *Brain Res.* **185,** 449-454.

Lokhandwala, M. F., Bogaert, M. G. and Lefebvre, R. A. (1987). Proceedings of a symposium on dopaminergic systems in hypertension. *Clin. Exp. Hypert.* **A9,** 833-1119.

Lokhandwala, M. F., Kivlighn, S. D. and Jandhyala, B. . (1984). Characterization of the neurogenic vasodilation elicited by central dopamine receptor stimulation with pergolide. *J. Pharmacol. Exp. Ther.* **228,** 696-701.

Lokhandwala, M. F., Tadepalli, A. S. and Jandhyala, B. S. (1979). Cardiovascular actions of bromocriptine: evidence for a neurogenic mechanism. *J. Pharmacol. Exp. Ther.* **211,** 620-625.

Lovenberg, W., Yamabe, H., De Jong, W. and Hansen, C. T. (1973). Genetic variation of the catecholamine biosynthetic enzyme activities in various strains of rats including the spontaneously hypertensive rat. In *Frontiers in catecholamine research,* E. Udsin, S.H. Snyder (Eds). Pergamon Press, New York, pp. 891-895.

Martignoni, E., Micieli, G., Cavallini, A., Pacchetti, C., Magri, M. and Nappi, G. (1986). Autonomic disorders in idiopathic Parkinsonism. *J. Neural Transm.* **22,** 149-161.

Martin, J. R. and Quock, R. M. (1984). Pharmacological characterization of apomorphine-induced hypothermia in the spontaneously hypertensive rat. *Life Sci.* **35,** 929-936.

McCarty, R., Chiueh, C. C. and Kopin, I. J. (1980). Differential behavioral responses of spontaneously hypertensive (SHR) and normotensive (WKY) rats to d-amphetamine. *Pharmacol. Biochem. Behav.* **12,** 53-59.

McCubbin, J. W., Kaneko, Y. and Page, I. H. (1960). Ability of serotonin and norepinephrine to mimic the central effects of reserpine on vasomotor activity. *Circ. Res.* **8,** 849-858.

McDowell, F.H. and Lee, J.E. (1970). Levodopa, Parkinson's disease, and hypotension. *Ann. Int. Med.* **72,** 751-752.

McMurtry, J. P., Kazama, N. and Wexler, B. C. (1979). Effects of bromocriptine on hormone levels and blood pressure in the SHR. *Proc. Soc. Exp. Biol. Med.* **161,** 186-188.

McNay, J. L., Henry, D. P. and DeLong, A. F. (1987). LY171555, a selective dopamine (DA-2) agonist, increases the blood pressure of hypertensive patients. *Clin. Exp. Hypert.* **A9,** 1098.

Mehta, A. E. and Tolis, G. (1979). Pharmacology of bromocriptine in health and disease. *Drugs* **17,** 313-325.

Meier, A., Weidmann, P., Hennes, U. and Ziegler, W. H. (1980). Plasma prolactin in normal and hypertensive subjects: relationship with age, posture, blood pressure and catecholamines and renin. *J. Clin. Endocr. Metab.* **50,** 304-308.

Meyer, D. C. and Carr, L. A. (1983). Developmental patterns of blood pressure (BP) and regional tyrosine hydroxylase (TH) activity in hypertensive (SHR) and normotensive (WKY) rats. *Fed. Proc.* **42,** 895.

Michelakis, A. M. and Robertson, D. (1970). Plasma renin activity and levodopa in Parkinson's disease. *Am. Med. Ass.* **213,** 83-85.

Micieli, G., Martignoni, E., Cavallini, A., Sandrini, G. and Nappi, G. (1987). Postprandial and orthostatic hypotension in Parkinson's disease. *Neurology* **37,** 386-393.

Mok, J.S.L. and Sim, M. K. (1987). The actions of dopamine on the blood pressure and heart rate of conscious hypertensive rats: evidence for reduced dopaminergic activity in rats of the Japanese strains. *Clin. Exp. Hypert.* **A9,** 1615-1635.

Mok, J.S.L., Hutchinson, J. S. and Sim, M. K. (1990). Effects of dopamine on blood pressure and heart rate after microinjection into the hypothalamus of spontaneously hypertensive and normotensive wistar-Kyoto rats. *Clin. Exp. Hypert.* **A12,** 15-36.

Montastruc, J. L. and Montastruc, P. (1981). Antihypertensive action of bromocriptine in neurogenic hypertensive dogs. *Arch. Int. Pharamacodyn.* **252,** 210-218.

Montastruc, J. L., Guiol, C., Tran, M. A., Lhoste, F. and Montastruc, P. (1985a). Studies on the cardiovascular actions of apomorphine in dogs: central versus peripheral mechanisms and role of the adrenal medulla. *Arch. Int. Pharmacodyn.* **277,** 92-103.

Montastruc, J.L., Chamontin, B. and Rascal, A. (1985b). Parkinson's disease and hypertension: chronic bromocriptine treatment. *Neurology* **35,** 1644-1647.

Montastruc, P., Damase-Michel, C. and Montastruc, J. L. (1989). Apomorphine potentiates vagal bradycardia. *Eur. J. Pharmacol.* **166,** 511-514.

Moore, R. Y. and Bloom, F. E. (1978). Central catecholamine neurons systems: anatomy and physiology of the dopamine system. *Ann. Rev. Neurosci.* **1,** 129-169.

Morris, M. and Sundberg, D.K. (1981b). Neurohypophyseal dopamine biosynthesis in the spontaneously hypertensive rat. *Clin. Exp. Hypert.* **3,** 1165-1181.

Morris, M., Wren, J. A. and Sundberg, D. K. (1981a). Central neural peptides and catecholamines in spontaneous and DOCA/salt hypertension. *Peptides* **2,** 207-211.

Musumeci, V., Baroni, S., Cardillo, C., Zuppi, P. and Folli, G. (1989). Cardiovascular reactivity and plasma prolactin response to mental stress in normals and hypertensives. *Clin. Exp. Hypert.* **A11,** 277-293.

Myers, M. M., Whittemore, S. R. and Hendley, E. D. (1981). Changes in catecholamine neuronal uptake and receptor binding in the brains of spontaneously hypertensive rats (SHR). *Brain Res.* **220,** 325-338.
Myers, M. M., Musty, R. E. and Hendly, E. D. (1982). Attenuation of hyperactivity in the spontaneously hypertensive rat by amphetamine. *Behav. Neural Biol.* **34,** 42-54.
Nagahama, S., Chen, Y. F. and Oparil, S. (1984). Mechanism of action of the depressor effect of bromocriptine in the spontaneously hypertensive rat. *J. Pharmacol. Exp. Ther.* **228,** 370-375.
Nagahama, S., Chen, Y. F. and Oparil, S. (1985). Enhanced depressor effect of bromocriptine in the DOCA/NaCl hypertensive rat. *Am. J. Physiol.* **249,** H64-H70.
Nagahama, S., Chen, Y. F., Lindheimer, M. D. and Oparil, S. (1986a). Mechanism of the pressor action of LY171555, a specific dopamine D_2 receptor agonist, in the conscious rat. *J. Pharmacol. Exp. Ther.* **236,** 735-742.
Nagahama, S., Chen, Y. F., Lindheimer, M .D. and Oparil, S. (1986b). Mechanism of the depressor action of LY171555, a specific dopamine D_2 receptor agonist, in the anesthetized rat. *J. Pharmacol. Exp. Ther.* **239,** 426-432.
Nagahama, S., Soo Ann, H., Chen, Y. F., Lindheimer, M. D. and Oparil, S. (1987). Role of vasopressin in the cardiovascular effects of LY171555, a selective dopamine D_2 receptor agonist: studies in conscious Brattleboro and Long-Evans rats. *J. Pharmacol. Exp. Ther.* **242,** 143-151.
Nagaoka, A. and Lovenberg, W. (1977). Regional changes in the activities of aminergic biosynthetic enzymes in the brains of hypertensive rats. *Eur. J. Pharmacol.* **43,** 297-306.
Nagatsu, T., Kato, T., Namatasudo, Y., Ikuta, K., Sano, M., Nagatsu, I., Umezawa, H., Matsuzaki, M. and Takeuchi, T. (1977). Norepinephrine-synthesizing enzymes in brain, adrenals and peripheral sympathetic nerves of spontaneously hypertensive rats. *Jap. J. Pharmacol.* **27,** 531-535.
Nilsson, A. and Hökfelt, B. (1978). Effects of the dopamine agonist bromocriptine on blood pressure, catecholamines and renin activity in acromegalics at rest, following exercise and during insulin induced hypoglycemia. *Acta Endocr.* **s216,** 83-96.
Ogihara, T., Maruyama, A., Naka, T., Higaki, J., Kanda, T., Goto, M., Onishi, T. and Kumahara, Y. (1980). Is normoreninemic essential hypertension caused by disordered central dopaminergic regulation? *Clin. Exp. Hypert.* **2,** 89-97.
Ogura, K., Shibuya, M., Suzuki, Y., Kanamori, M. and Ikegaki, I. (1989). Changes in striatal dopamine metabolism measured by *in vivo* voltammetry during transient brain ischemia in rats. *Stroke* **20,** 783-787.
Ohlstein, E. H., Kruse, L. I., Ezekiel, M. and Willette, R. N. (1988). Further characterization of the cardiovascular effects of the dopamine-β-hydrozylase inhibitor SK&F 102698 in conscious hypertensive rats. *J. Pharmacol. Exp. Ther.* **247,** 186-195.
Orskov, L., Jakobsen, J., Dupont, E., de Fine Olivarius, B. and Christensen, N. J. (1987). Autonomic function in parkinsonian patients relates to duration of disease. *Neurology* **37,** 1173-1178.
Os, I., Kjedsen, S. E., Westheim, A., Aakeson, I., Norman, N., Enger, E., Hjermann, I. and Eide, I. (1987). Decreased central dopaminergic activity in essential hypertension. *J. Hypertension* **5,** 191-197.
Pazo, J. H. and Medina, J. H. (1983). Cholinergic mechanisms within the caudate nucleus mediate changes in blood pressure. *Neuropharmacology* **22,** 717-720.
Pazos, A., Mediaville, A. and Florez, J. (1982). Cardiovascular effects of intracerebroventricular injection of dopamine after selective MAO-inhibition in rats. *Neuropharmacology* **21,** 317-322.
Petitjean, P., Mouchet, P., Pellisier, G., Mainier, M., Feuerstein, C. and Demerge, P. (1984). Cardiovascular effects in the rat of intrathecal injections of apomorphine at the thoracic spinal cord level. *Eur. J. Pharmacol.* **105,** 355-359.
Philippu, A. (1988). Regulation of blood pressure by central neurotransmitters and neuropeptides. *Rev. Physiol. Biochem. Pharmacol.* **111,** 1-115.
Quinn, N., Illas, A., Lhermitte, F. and Agid, Y. (1981). Bromocriptine in Parkinson's disease: a study of cardiovascular effects. *J. Neurol. Neurosurg. Psych.* **44,** 426-429.
Rajput, A. H. and Rozdilsky, B. (1976). Dysautonomia in Parkinsonism: a clinicopathological

study. *J. Neurol. Neurosurg. Psych.* **39,** 1092-1100.
Reid, J. L., Calne, D. B., George, C. F., Pallis, C. and Vakil, S. D. (1971). Cardiovascular reflexes in Parkinsonism. *Clin. Sci.* **41,** 63-67.
Reid, J. L., Calne, D. B., George, C. F. and Vakil, S. D. (1972). The action of L(-)-DOPA on baroreflexes in Parkinsonism. *Clin. Sci.* **43,** 851-859.
Renaud, B., Joh, T. H. and Reis, D. J. (1979). An abnormal regulation of tyrosine hydroxylase restricted to one catecholamine nucleus in the brainstem of spontaneously hypertensive rats. *Brain Res.* **173,** 164-167.
Richer, C., Lhoste, F. and Giudicelli, J. F. (1984). Bromocriptine and genetic hypertension development in young spontaneously hypertensive rats. *J. Pharmacol.* **15,** 185-197.
Roquebert, J., Alaoui, K. Morán Benito, A. (1990). Cardiovascular effects of bromocriptine in rats: role of peripheral adrenergic and dopaminergic receptors. *J. Auton. Pharmacol.* **10,** 85-96.
Ruffalo, R .R.Jr. and Shaar, C. J. (1983). Relative potency of LY141865 at dopamine DA_2 and histamine H_2 receptors. *Eur. J. Pharmacol.* **92,** 295-306.
Sachs, C., Berglund, B. and Kaijser, L. (1985). Autonomic cardiovascular responses in Parkinsonism: effect of levodopa with dopa-decarboxylase inhibition. *Acta Neurol. Scand.* **71,** 37-42.
Saruta, T., Kawabe, H., Fujimaki, M., Nagahama, S., Saito, I. and Kondo, K. (1983). Prolactin, renin and catecholamines in essential hypertension. *Clin. Exp. Hypert.* **A5,** 531-541.
Sim, M. K. and Hsu, T. P. (1990). Levels of dopamine and 3,4,-dihydroxyphenylacetic acid in the adrenohypophysis of normo- and hypertensive rats. *Clin. Exp. Hypert.* **A12,** 343-353.
Slater, I. H. (1983). Editorial comment: prolactin, dopaminergic mechanisms and hypertension. *Clin. Exp. Hypert.* **A5,** 559-561.
Sowers, J. R., Sollars, E. G., Tuck, M. L. and Asp, N. D. (1980). Dopaminergic modulation of renin activity and aldosterone and prolactin secretion in the SHR. *Proc. Soc. Exp. Biol. Med.* **164,** 598-603.
Sowers, J. R. (1981). Effects of bromocriptine on responses to stress in spontaneously hypertensive rats. *Hypertension* **3,** 544-550.
Sowers, J.R. (1982a). Modulation of corticosteroid secretion by dopaminergic mechanisms in rhesus monkeys. *Am. J. Physiol.* **243,** E375-E379.
Sowers, J. R., Golub, M. S., Berger, M. E. and Whitfield, L. A. (1982b). Dopaminergic modulation of pressor and hormonal responses in essential hypertension. *Hypertension* **4,** 424-430.
Sowers, J. R., Golub, M. S., Eggena, P. and Catania, R. A. (1982c). Influence of sodium homeostasis on dopaminergic modulation of aldosterone, renin, and prolactin secretion in man. *J. Clin. Endocr. Metab.* **54,** 121-127.
Sowers, J. R., Nyby, M. and Jasberg, K. (1982d). Dopaminergic control of prolactin and blood pressure: altered control in essential hypertension. *Hypertension* **4,** 431-438.
Spring, A. and Winkelmüller, W. (1975). Ventral midbrain stimulation, blood pressure responses and their relation to the dopaminergic nigrostriatal pathway. *Pflügers Arch. (Eur. J. Physiol.)* **358,** 339-348.
Steel, L. H., Jackson, D. M. and Hambley, J. (1988). Behavioural effects of the dopamine D1 receptor agonist SKF 38393 in spontaneously hypertensive rats. In: *Abstracts 22nd Ann. Sci. Meeting of the Australian Soc. Clin. Exp. Pharmacol.*, Adelaide, 070.
Steger, R. W., Avila-Jimenez, R., Amador, A. and Johns, A. (1984). Altered hypothalamic monoamine metabolism and pituitary prolactin regulation in female SHR. *Life Sci.* **34,** 1691-1697.
Stern, N. Eshkol, A., Lunenfeld, B. and Rosenthal, T. (1983). Prolactin secretion in essential hypertension. *Clin. Exp. Hypert.* **A5,** 543-558.
Struyker-Boudier, H.A.J., Van Essen, H. and Smits, J.F.M. (1984). Hemodynamic effects of bromocriptine in the conscious spontaneously hypertensive rat. *J. Pharm. Pharmacol.* **36,** 123-125.
Stumpe, K. O., Kolloch, R., Higuchi, M., Krück, F. and Vetter, H. (1977). Hyperprolactinemia and antihypertensive effect of bromocriptine in essential hypertension. *Lancet* **2,** 211-214.

Tadepalli, A. S. and Novak, P. J. (1983). Cardiovascular effects of bromocriptine and lergotrile in renal and spontaneously hypertensive rats. *Arch. Int. Pharmacodyn.* **266,** 93-105.

Tan, B.K.H., Hutchinson, J. S. (1987). Plasma and pituitary prolactin and blood pressure in bromocriptine-treated spontaneously hypertensive and Wistar-Kyoto rats. *Clin. Exp. Pharmacol. Physiol.* **14,** 797-803.

Tan, B.K.H., Hutchinson, J. S. (1988). Blood pressure, plasma and pituitary prolactin levels and their responses to bromocriptine in the segregating F_2 generation of the SHR/WKY hybrid. *Clin. Exp. Hypert.* **A10,** 995-1007.

Tan, E., Goodchild, A. K. and Dampney, R.A.L. (1983). Intense vasoconstriction and bradycardia evoked by stimulation of neurons within the midbrain and ventral tegmentum of the rabbit. *Clin. Exp. Pharmacol. Physiol.* **10,** 305-309.

Tsuruta, K., Frey, E. A., Grewe, C. W., Cote, T. E., Eskay, R. L. and Kebabian, J. W. (1981). Evidence that LY141865 specifically stimulates the D-2 dopamine receptor. *Nature* **292,** 463-465.

Tuomisto, L., Yamatodani, A., Dietl, H., Waldmann, U. and Philippu, A. (1983). *In vivo* release of endogenous catecholamines, histamine and GABA in the hypothalamus of Wistar-Kyoto and spontaneously hypertensive rats. *Naunyn Schmied. Arch. Pharmacol.* 183-187.

Turkka, J. T., Juujärvi, K. K., Lapinlampi, T. O. and Myllylä, V. V. (1986). Serum noradrenaline response to standing up in patients with Parkinson's disease. *Eur. Neurol.* **25,** 355-361.

Ungerstedt, U. (1971). Stereotaxic mapping of monoamine pathways in rat brain. *Acta Physiol. Scand.* suppl. **367,** 1-48.

Van den Buuse, M., De Kloet, E.R., Versteeg, D.H.G. and De Jong, W. (1984a). Regional brain catecholamine levels and the development of hypertension in the spontaneously hypertensive rat: the effect of 6-hydroxydopamine. *Brain Res.* **301,** 221-229.

Van den Buuse, M., Versteeg, D.H.G. and De Jong, W. (1984b). Role of dopamine in the development of spontaneous hypertension. *Hypertension* **6,** 899-905.

Van den Buuse, M., De Boer, S., Veldhuis, H. D., Versteeg, D.H.G. and De Jong, W. (1986a). Central 6-OHDA affects both open-field exploratory behaviour and the development of hypertension in SHR. *Pharmacol. Biochem. Behav.* **24,** 15-21.

Van den Buuse, M., Versteeg, D.H.G. and De Jong, W. (1986b). Brain dopamine systems and hypertension. In: *Brain and blood pressure control*, K. Nakamura (Ed.), Elsevier Sci. Publ. Amsterdam, pp. 343-352.

Van den Buuse, M., Versteeg, D.H.G. and De Jong, W. (1986c). Brain dopamine depletion by lesions of the substantia nigra attenuates the development of hypertension in the spontaneously hypertensive rat. *Brain Res.* **368,** 69-78.

Van den Buuse, M., Veldhuis, H. D., Versteeg, D.H.G. and De Jong, W. (1986d). Substantia nigra lesions attenuate the development of hypertension and behavioural hyperreactivity in SHR. *Pharmacol. Biochem. Behav.* **25,** 317-324.

Van den Buuse, M. and De Jong, W. (1987). Grooming behaviour of spontaneously hypertensive rats. *Neurosci. Lett.* **77,** 71-76.

Van den Buuse, M., Lambrechts, C. and De Jong, W. (1988). Pressor effects of electrical stimulation of the rat ventral tegmental A10 dopamine system. In: *Pharmacology and functional regulation of dopaminergic neurons*, P.M. Beart, G.N. Woodruff, D.M. Jackson (Eds), MacMillan Press London, pp. 315-317.

Van den Buuse, M. and De Jong, W. (1989a). Differential effects of dopaminergic drugs on open-field behaviour of spontaneously hypertensive rats and normotensive Wistar-Kyoto rats. *J. Pharmacol. Exp. Ther.* **248,** 1189-1196.

Van den Buuse, M. and Lambrechts, A. C. (1989b). Bromocriptine-induced decrease in blood pressure in conscious spontaneously hypertensive rats: evidence for a peripheral site of action. *J. Pharm. Pharmacol.* **41,** 644-646.

Vaughn, L. K., Quock, R. M., Martin, J. R. and Zuengler, D. N. (1985). A lack of difference between normotensive Wistar-Kyoto rats (WKY) and spontaneously hypertensive rats (SHR) in behavioural response to apomorphine and striatal [^{3}H]spiperone binding. *Fed. Proc.* **44,** 2811.

Versteeg, D.H.G., Palkovits, M., Van den Gugten, J., Wijnen, H.J.L.M. and De Jong, W. (1976).

Catecholamine content of individual brain regions of spontaneously hypertensive rats (SH-rats). *Brain Res.* **122,** 429-434.

Versteeg, D.H.G., Petty, M. A., Bohus, B. and De Jong, W. (1984). The central nervous system and hypertension: the role of catecholamines and neuropeptides. In: *Handbook of hypertension, vol. 4: Experimental and genetic models of hypertension*, W. de Jong (Ed). Elsevier Sci. Publ., Amsterdam, pp. 398-430.

Versteeg, D.H.G. and Van den Buuse, M. (1989). Brain catecholamine systems and the development of spontaneous hypertension in rats: focus on dopamine. In: *Hypertension, Brain catecholamines and peptides*, F.P. Nijkamp, D. de Wied (Eds), Elsevier Sci. Publ., Amsterdam, pp. 1-10.

Villeneuve, A., Berlan, M., Lafontan, M., Caranobe., C., Boneu, B., Rascol, A. and Montastruc, J.L. (1985). Platelet $alpha_2$ adrenoreceptors in Parkinson's disease: decreased number in untreated patients and recovery after treatment. *Eur. J. Clin. Invest.* **15,** 403-407.

Waddington, J. L. (1989). Functional interactions between D-1 and D-2 dopamine receptor systems: their role in the regulation of psychomotor behaviour, putative mechanisms, and clinical relevance. *J. Psychoparmacol.* **3,** 54-63.

Wataneba, M., Tsuruta, S., Inoue, Y., Kinuya, M., Ogawa, K., Mamiya, G. and Tatsunuma, T. (1989). Dopamine D_1 and D_2 receptors in spontaneously hypertensive rat brain striatum. *Can. J. Physiol. Pharmacol.* **67,** 1596-1597.

Williams McKeon, T. and Hendley, E. D. (1988). Brain monoamines and metabolites in hypertensive and hyperactive rat strains. *Clin. Exp. Hypert.* **A10,** 971-994.

Wu, J. J., Shih, C. J. and Lin, M. T. (1984). Tachycardia, hypertension and decreased reflex bradycardia produced by striatal lesions induced by kainic acid. *Neuropharmacology* **23,** 1231-1235.

Yamabe, H., De Jong, W. and Lovenberg, W. (1973). Further studies on catecholamine synthesis in the spontaneously hypertensive rat: catecholamine synthesis in the central nervous system. *Eur. J. Pharmacol.* **22,** 91-98.

Yang, R. H., Igarashi, Y., Wyss, J. M. and Chen, Y. F. (1990). Dopamine D_2 receptors in the posterior region of the nucleus tractus solitarius mediate the central pressor action of quinpirole (LY171555). *Brain Res. Bull.* **24,** 97-103.

Yen, T. T., Stamm, N. B. and Clemens, J. A. (1979). Pergolide: a potent dopaminergic antihypertensive. *Life Sci.* **25,** 209-215.

Yoshimura, M., Kambara, S., Takahashi, H. and Ijichi, H. (1987). Attenuated development of hypertension by chronic administration of bromocriptine in DOCA-salt hypertensive rats. *Clin. Exp. Hypert.* **A9,** 1089.

Zandberg, P., De Jong, W. and De Wied, D. (1979). Effects of catecholamine-receptor stimulating agents on blood pressure after local application in the nucleus tractus solitarii of the medulla oblongata. *Eur. J. Pharmacol.* **55,** 43-56.

Progress in Hypertension, Vol. 2, pp. 189-205.
Saito *et al.* (Eds)

SEROTONIN-2 RECEPTOR ANTAGONIST: KETANSERIN
-FOCUS ON THE SYMPATHOINHIBITORY ACTION-

MACHIKO MATSUMOTO, MITSUHIRO YOSHIOKA, HIROKO TOGASHI, MASARU MINAMI* and HIDEYA SAITO

First Department of Pharmacology, Hokkaido University School of Medicine, Sapporo 060; and *Department of Pharmacology, Faculty of Pharmaceutical Sciences, Higashi-Nippon-Gakuen University, Ishikari-Tobetsu, Hokkaido 061-02, Japan.

INTRODUCTION

Ketanserin (3-[2-[4-(4-fluorobenzoyl)-1-piperidinyl]-ethyl]-2,4 (1H,3H)-quinazoline-dione is a recently developed antihypertensive agent, with the structure of both quinazoline and piperizine which acts as a serotonin (5-HT) antagonist (Fig. 1). The classical 5-HT antagonists such as methysergide, cyproheptadine and spiperone bind to both 5-HT_2 receptors and 5-HT_1 receptors, whereas ketanserin has an approximately 250-fold higher affinity for the 5-HT_2 binding site rather than the 5-HT_1 binding site. While ketanserin has low affinity for $alpha_1$-adrenoceptors, it has even less affinity for H_1-histaminergic, D_2-dopaminergic and $alpha_2$-adrenergic receptors. The most note-worthy property of this compound is a selective 5-HT_2 receptor antagonist devoid of partial agonistic actions (Leysen *et al.*, 1981). This suggests that 5-HT may play an important role in blood pressure (BP) regulation via 5-HT_2 receptors.

The involvement of 5-HT in the pathogenesis of hypertension has long been suspected because it is a potent vasoconstrictor that is comparable to norepinephrine (NE) or angiotensin II (A-II). However, the role of 5-HT in the BP regulation remains unclear for three reasons: (1) the complex properties of 5-HT in both peripheral and central nervous systems, *i.e.*, 5-HT not only causes constriction, but also dilation; (2) Few reports are available on the elevated plasma 5-HT levels in either hypertensive patients or hypertensive animals; (3) Lack of selective 5-HT antagonist cause an antihypertensive effect. Accordingly, the development of ketanserin has provided a useful tool for studying of BP regulation by 5-HT. Ketanserin has also contributed to both the classification of 5-HT receptor subtypes and the development of new selective agonists and antagonists. To date, ketanserin has been extensively tested as a new type of vasodilator for the clinical treatment of hypertension. However, its principal hypotensive mechanism is not completely understood. Questions remain because despite its being a vasodilator, it does not promote reflex tachycardia. Also, it possesses $alpha_1$-adrenoceptor antagonistic action.

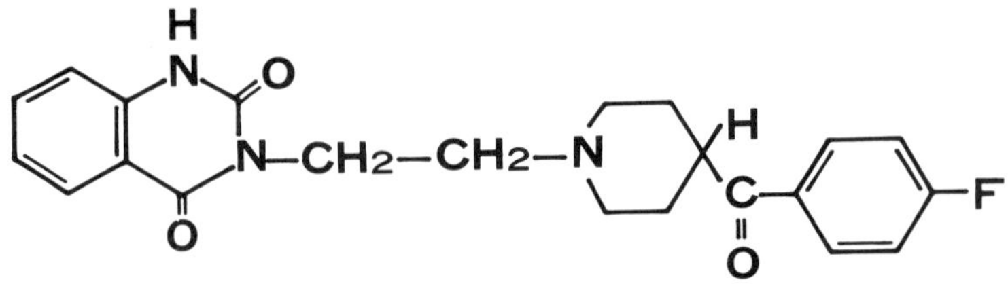

Figure 1. Chemical structure of ketanserin.

This chapter summarizes the hypotensive mechanism of ketanserin with a focus on central sympathetic nerve activity. Furthermore, based on the properties of ketanserin as a 5-HT_2 receptor antagonist, the role of 5-HT in the BP regulation is discussed.

PHARMACOLOGICAL PROFILE

Radioligand binding techniques and pharmacological studies have demonstrated that 5-HT_2 receptors exist widely throughout peripheral and central organs, *i.e.*, vascular smooth muscles, platelets, frontal cortex and spinal cord. It is recognized that 5-HT stimulates 5-HT_2 receptors to produce various responses. For instance, 5-HT-caused vaso-constriction in most arteries and veins is mediated by 5-HT_2 receptors. 5-HT has an amplifying effect on the responsiveness to other vasoactive agents such as NE, A-II and histamine, which are mediated via the 5-HT_2 receptors. Ketanserin inhibits 5-HT-induced vasoconstriction and amplifying effect dose-dependently *in vitro* (Van Nueten *et al.*, 1981). It has been proposed that 5-HT itself produce platelet aggregation that is mediated through the 5-HT_2 receptors. 5-HT also has an amplifying effect on 5-HT-induced platelet reactions to other aggregating substances such as NE, epinephrine (Ep) and adenosine diphosphate (ADP). Ketanserin prevents these reactions which are mediated via 5-HT_2 receptors. It has no effect, however, on the primary aggregation induced by NE, Ep and ADP (De Clerck *et al.*, 1982a). *In vivo* , intravenous (*i.v.*) administration of 5-HT produces triphasic effects on BP; an initial brief depressor response, followed by a pressor response and finally a longer-lasting depressor response. The initial depressor response, von Bezold-Jarish reflex, which is accompanied with bradycardia, result from the activation of 5-HT_3 receptors on afferent cardiac vagal nerves. The pressor response is mediated by the activation of 5-HT_2 receptors. The long-lasting depressor response seems either to be mediated by 5-HT_1 receptors or to result from the release of endothelium derivated relaxing factor (EDRF) via 5-HT_1 receptors in the vascular smooth muscles. Fig. 2 shows our experimental data elucidating the effect of ketanserin on exogenously administered 5-HT-induced BP response. Ketanserin prevented the pressor phase completely. This antagonistic action by ketanserin has been also reported in SHR (Pettersson *et al.*,

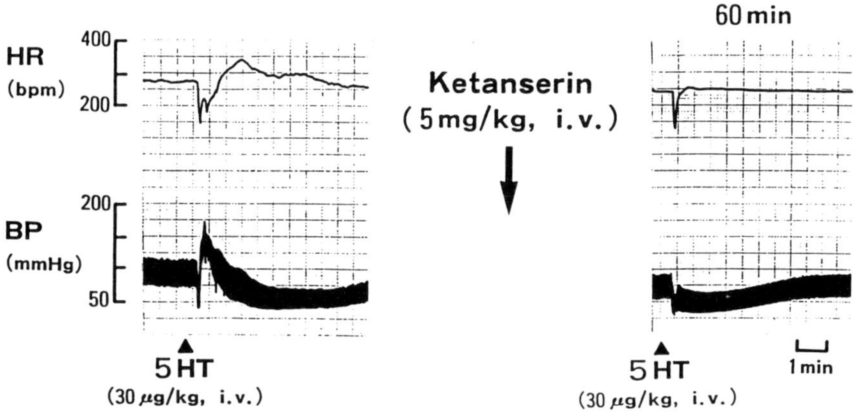

Figure 2. Effect of intravenous (*i.v.*) administration of ketanserin (5 mg/kg) on exogenously administered serotonin (5-HT) (30 μg/kg, *i.v.*) in normotensive rats.

1984) and pithed rats (Fozard *et al.*, 1987). In humans, intra-arterial infusion of ketanserin prevented vasoconstriction caused by 5-HT infused through the forearm (Blauw *et al.*, 1987). Likewise, a dose-dependent antagonistic action of ketanserin was observed when 5-HT was injected into a finger vein (Jageneau *et al.*, 1980). These pharmacological studies provide evidence that ketanserin is 5-HT_2 receptor antagonist.

Radioligand binding studies have indicated, on the other hand, that ketanserin has a low binding affinity for alpha_1-adrenoceptors. The Ki-value of ketanserin for alpha_1-adrenoceptors is approximately 25-30 times higher than that for 5-HT_2 receptors (Leysen *et al.*, 1981). *In vitro* and *in vivo* experiments show that ketanserin inhibits vasoconstriction induced by NE and selective alpha_1-adrenoceptor agonists such as methoxamine and phenylephrine (Kalkman *et al.*, 1982; Van Nueten, 1983). Moreover, ketanserin inhibits pressor responses induced by electric stimulation in adrenoectomized, pithed rats (Fozard, 1982). These results indicate that ketanserin has antagonistic action not only for 5-HT_2 receptors but also for alpha_1-adrenoceptors. The potency of the antagonistic action of ketanserin for 5-HT_2 receptors is more powerful than that for alpha_1-adrenoceptors. In addition, this antagonistic effect for alpha_1-adrenoceptors is less than that of selective alpha_1-adrenoceptor antagonist, prazosin (Zabludowski *et al.*, 1985).

PHARMACOKINETICS

After a single oral administration of ketanserin, a terminal half-life is 12-20 hours, which is slightly shorter than that during chronic administration. The absolute bioavailability of ketanserin is in the order of 50% (Trenk *et al.*, 1983). In elderly hypertensive patients over 65 years of age, its bioavailability is slightly higher compared with healthy volunteers (Kurowski, 1985). This may be partially related to that ketanserin decreases BP more effectively in older patients. Plasma levels of ketanserin can be determined by high-performance liquid chromatography (HPLC) with ultraviolet or fluorescence detection. At therapeutic doses with a dosing interval of 12 hours, plasma concentrations of ketanserin ranges between 15 and 140 ng/ml (Hedner *et al.*, 1983). There is no clear relationship between plasma concentrations and antihypertensive effect of ketanserin. Several animal studies indicate that ketanserin is rapidly absorbed from the gastrointestinal tract. The absolute bioavailability of oral administered ketanserin is more than 80% and its terminal plasma half-life is 2-5 hours in rats (Van Nueten *et al.*, 1987).

Ketanserin is mainly metabolized via three pathways, *i.e.*, ketone reduction, aromatic hydroxylation and oxidative N-dealkylation. Ketanserin-ol which is resulted from ketone reduction, the major metabolite of humans, is approximately 1000 time less biologically active than ketanserin (Frenken and Kaumann, 1984). In rats, most urinary metabolite is formed by the oxidative N-dealkylation of ketanserin. Whereas aromatic hydroxylation, resulting in the formation of 6-hydroxy-ketanserin, was the main metabolite pathway in biliary. None of the metabolites affects the pharmacological activity of ketanserin. It has been demonstrated that ketanserin is extensively bound to plasma protein, mainly to albumin, *i.e.*, 95.0 per cent in humans and 98.8 per cent in rats. Ketanserin does not displace any drugs from their plasma protein binding site at the chronic therapeutic doses. Pharmacokinetic studies, furthermore, indicate the interaction between ketanserin and propranolol, metoprolol and digoxin does not cause clinically significant changes (Ochs *et al.*, 1985; Williams *et al.*, 1986).

HYPOTENSIVE MECHANISM

Peripheral Vasodilating Actions

Clinical studies have been reported that ketanserin caused no significant changes in BP in normotensive volunteers. Acute and chronic administration of ketanserin, however, produce significant decreases in BP in patients with essential hypertension (Zabludowski *et al.*, 1984; Wenting *et al.*, 1982). Ketanserin also decreased BP in several experimental and genetic hypertensive animals, *i.e.*, spontaneously hypertensive rats (SHR) (Fozard, 1982), spontaneously hypertensive dogs (De Clerck *et al.*, 1982b) and DOCA hypertensive sheep (Mecca *et al.*, 1985). Most reports demonstrated that the efficacy of ketanserin-induced decreases in BP in hypertensive animals was greater than those in normotensive controls. In our experiments, BP-lowering effects of ketanserin (0.5 and 5mg/kg, *i.v.*) in stroke-prone spontaneously hypertensive rats (SHRSP) were significantly greater than those in Wistar-Kyoto rats (Sano *et al.*, 1985). These findings indicate the ketanserin has both hypotensive and antihypertensive effects.

The ketanserin-induced hypotensive effect is thought to result predominantly from reduced vascular resistance (Wenting *et al.*, 1982; Woittiez *et al.*,1985). Whether its hypotensive effects are due to antagonistic action for 5-HT_2 receptors or $alpha_1$-adrenoceptors in vascular smooth muscles is unclear. Early studies attributed the ability of ketanserin to block 5-HT_2 receptors to its hypotensive mechanism. However, numerous investigators claim that the hypotensive effect of ketanserin is related to the antagonistic action for $alpha_1$-adrenoceptors rather than that for 5-HT_2 receptors, because more powerful and selective 5-HT_2 antagonists, ritanserin and LY53857 have no significant effect on BP (Hosie, *et al.*, 1987; Cohen *et al.*, 1983). In addition, poor correlation between antihypertensive potential and 5-HT_2 receptor antagonism has been observed in a number of drugs (Kalkman *et al.*, 1983).

Although ketanserin as well as prazosin produced a shift to the right on the phenylephrine-induced dose-response curve in normotensive subjects, the extent of shift was small compared with that of prazosin (Wenting *et al.*, 1982). In patients with essential hypertension, the pressor response to phenylephrine was not shifted by acute administration of ketanserin (Wenting *et al.*, 1982; Zabludowski *et al.*, 1984). This antagonistic action of ketanserin was also not observed in SHR (Pettersson *et al.*, 1984). Wenting *et al.* (1984) reported that intravenous administration of ketanserin induced a marked fall in BP in patients with autonomic insufficiency. In this subjects, the physiological activation of endogenous alpha-adrenoceptor is prevented by the efferent sympathetic lesions, as demonstrated by the absence of any change in BP and heart rate (HR) in response to alpha-adrenoceptor antagonist, phentolamine. These results demonstrate that $alpha_1$-adrenoceptor antagonistic action alone cannot always or entirely account for the hypotensive mechanism of ketanserin. There is a possibility that 5-HT_2 receptors antagonistic action of ketanserin may be contributed to its antihypertensive mechanism. In other words, the hypotensive mechanism of ketanserin in a hypertensive situation may differ from that in a normotensive state.

Marwood and Stokes (1984) have proposed that the vascular 5-HT_2 receptors and alpha-adrenoceptors overlap through a common binding site. This is because the 5-HT-induced vasoconstriction and amplifying effect are inhibited by phentolamine, while NE- and methoxamine-induced vasoconstriction are inhibited by methysergide or cyproheptadine. Based on their speculation, the antihypertensive effect of ketanserin may be explained by its potential to inhibit the common binding site of both 5-HT_2 receptor and $alpha_1$-adrenoceptors in the vascular smooth muscles. These hypotheses are supported by the fact that ritanserin, at doses that selectively block 5-HT_2 receptor-mediated responses but do not lower BP, potentiates the BP-lowering effect of prazosin (Gradin *et al.*, 1985).

Ketanserin has no effect on the A-II-induced pressor response (Zabludowski *et al.*, 1984; Pettersson *et al.*, 1984). Acute and chronic administration of ketanserin do not significantly alter on plasma levels of renin activity, A-II and aldosterone in hypertensive patients (Woittiez *et al.*,1986; Zabludowski, *et al.*, 1984) or SHRSP (Minami *et al.*, 1988). These findings suggest that the renin-angiotensin system is not involved in the hypotensive mechanism of ketanserin.

Central Actions

An early report suggested that ketanserin could not cross through the blood-brain barrier. However, clinical data have demonstrated that despite its being a vasodilator, ketanserin rarely produces reflex tachycardia. Furthermore, dizziness, sleepiness and sedation have been reported as side-effects of ketanserin (Milei *et al.*, 1987; Jennings

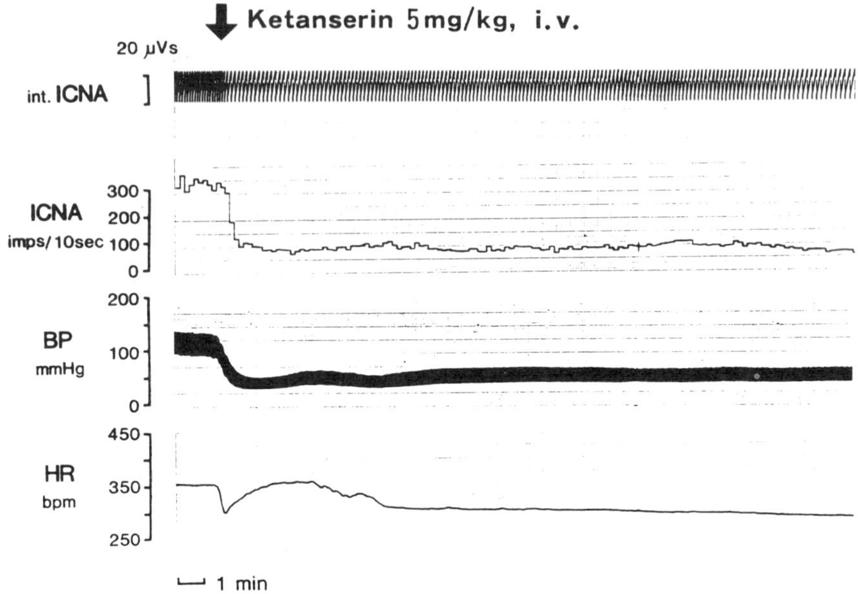

Figure 3. Effect of intravenous (*i.v.*) administration of ketanserin on inferior cardiac nerve activity (ICNA) in rats. Traces from top to bottom are integrated (int.) ICNA, discharge rate of ICNA, blood pressure (BP) and heart rate (HR).

and Opie, 1987). In most anesthetized animals, ketanserin produces significant decreases in HR during the hypotensive phase. These findings suggest the possibility that ketanserin is able to penetrate the brain and interfere with the central cardiovascular system. In order to clarify whether or not ketanserin acts on the central nervous system (CNS), we carried out experiments to elucidate the effect of ketanserin on the sympathetic nerve activity in anesthetized rats. Attention was focused on the following points: 1) the effect of ketanserin on postganglionic inferior cardiac nerve activity (ICNA), postganglionic renal nerve activity (RNA) and preganglionic adrenal nerve activity (ANA) and 2) the central site of sympathoinhibitory action of ketanserin. The following section summarizes the results of our experiments.

Materials and Methods

Normotensive male Wistar rats (9-12 weeks old) were anesthetized with alpha-chloralose and urethane (50 mg/kg and 500 mg/kg, respectively). After immobilization with gallamine triethiodide, respiration was maintained artificially. BP and HR were monitored via the femoral artery. Sympathetic nerve activity - ICNA, RNA and ANA - recorded from the central cut end of the nerve with bipolar platinum iridium wire electrodes and amplified. Amplified discharges were passed through a window discriminator and counted every 10 seconds. ICNA and RNA were also quantitated using a cumulative integrator. Baroreceptor afferent nerve activity-aortic depressor nerve activity (ADNA) were recorded from the peripheral cut end of the nerve and recorded as aforementioned. Methods of drug administration were as follows : Intravenous (*i.v.*) administration was performed through a cannula inserted into the femoral vein; Intracerebroventricular (*i.c.v.*) administration was carried stereotaxically with a microsyringe; Intrathecal (*i.t.*) administration was performed via cannula that was inserted through an incision in the atlanto-ocipital membrane to the caudal region of the thoracic enlargement. In order to deplete the 5-HT or NE content of the brain, rats were treated with monoamine neurotoxins. A selective 5-HT neurotoxin, 5,7-dihydroxytryptamine (5,7-DHT) (200 μg/rat, *i.c.v.*) and a selective catecholamine neurotoxin, 6-hydroxydopamine (6-OHDA) (200 μg/rat, *i.c.v.*) were injected after systemic administration of desipramine and pargyline, respectively.

Effect on the Sympathetic Nerve Activity

After *i.v.* administration of ketanserin (0.05, 0.5 and 5 mg/kg), a significant decrease in ICNA accompanied with decreases in both BP and HR was observed. These ketanserin-induced changes in ICNA coincided well with the changes in HR (Figs. 3 and 4). It has been generally accepted that the change of ICNA is correlated with HR. For instance, the vasodilator such as hydralazine, which decreases BP via the peripheral blood vessels, cause compensatory tachycardia with increased ICNA (Yoshioka *et al.*, 1986). Accordingly, ketanserin-induced decreases in ICNA may be responsible for its bradycardic and/or antitachycardic effects. Our suggestion is supported by McCall and Schuette (1984) who reported that ketanserin reduced the cardiac sympathetic discharge in baroreceptor-denervated cats. However, Fozard *et al.* (1987) proposed that ketanserin-induced bradycardia is due to increased vagal tone because the reduction of HR was prevented by either vagotomy or treatment with atropine. On the other hand, another group reported that ketanserin produced dose-dependent negative inotropic and chronotropic effects in isolated, blood-perfused dog atrium (Furukawa *et al.*, 1986). This suggests that ketanserin may act directly on the heart to decrease HR.

Ketanserin (0.5 and 5 mg/kg, *i.v.*) also significantly reduce RNA (Fig. 5). RNA is proposed to be mainly responsible for the constriction of renal blood vessels. Therefore, decreased RNA produced by ketanserin may contribute to its hypotensive effects. To elucidate the contribution of the decreasing potency of RNA to ketanserin-induced hypotension, we compared the effect of ketanserin on RNA with that of clonidine, which possesses the most potent sympathoinhibitory action(Togashi, 1983). The decrease in RNA induced by ketanserin (0.5 mg/kg, *i.v.*) was significantly less than that caused by clonidine (3 μg/kg, *i.v.*), at equihypotensive and equibradycardic doses (Yoshioka *et al.*, 1987). Moreover, no significant reduction of RNA was observed at dose of 0.05 mg/kg, *i.v.*, despite a significant hypotensive effect. These findings suggest that the decreased RNA may be partially, but not mainly, responsible for ketanserin-induced hypotension.

I.v. administered ketanserin (0.5 and 5 mg/kg) significantly reduced preganglionic ANA accompanied with the decreased BP and HR (Fig. 5). Therefore, the present findings that ketanserin decreases both postganglionic and preganglionic sympathetic nerve activity indicates that ketanserin possesses central sympathoinhibitory action in

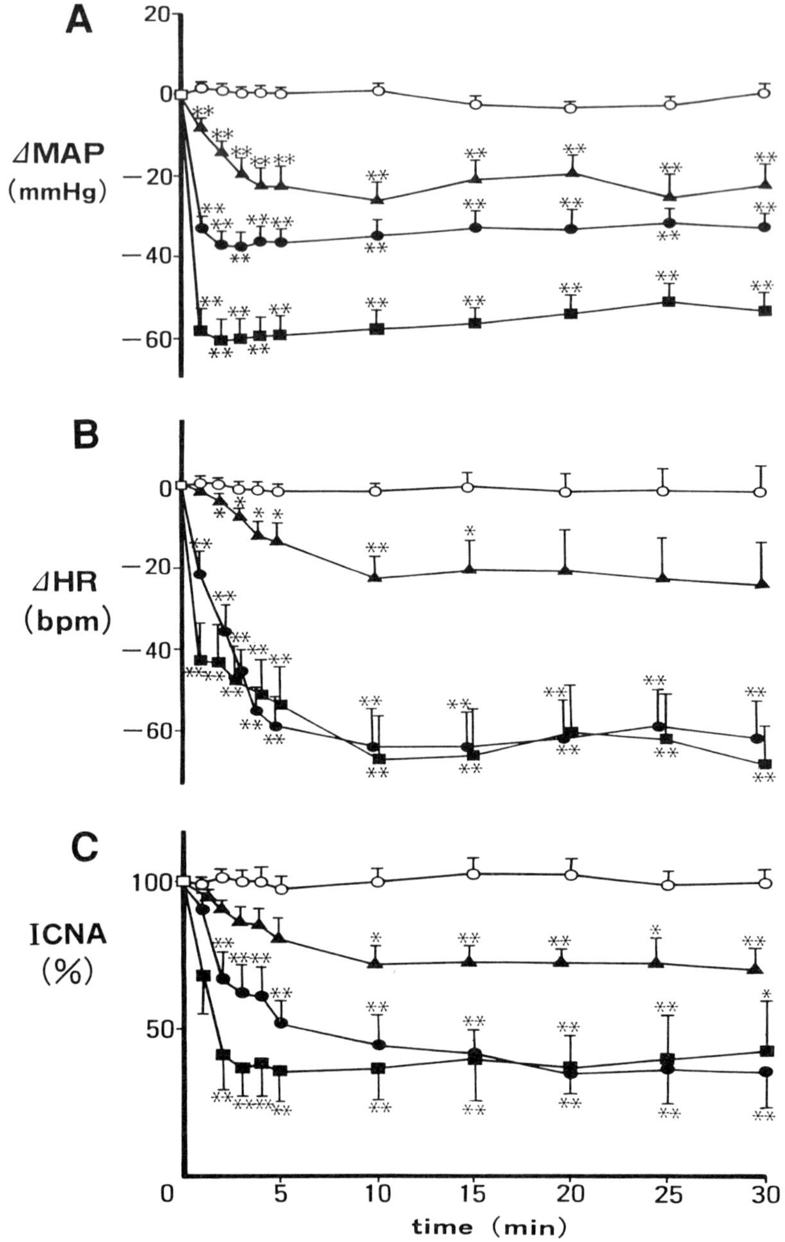

Figure 4. Effect of intravenous (*i.v.*) administration of ketanserin on mean arterial blood pressure (MAP)(A) heart rate (HR)(B) and inferior cardiac nerve activity (ICNA)(C) at the dose of 0.05. (▲ - ▲)(n = 6-10), 0.5 (● - ●)(n = 6) and 5 mg/kg (■ - ■)(n = 6-9) in rats. Each value represents the mean ± SE. * $p < 0.05$ and ** $p < 0.01$ *vs*. saline-treated control rats (○ - ○)(n = 6-10).

rats. This finding is supported by several reports that ketanserin has sympatho-inhibitory action in dogs (Phillips *et al.*, 1985) and cats (Ramage, 1985; McCall and Schuette, 1984; McCall and Harris, 1987).

In the present study we used buffer nerve intact rats. The changes in BP correlate well with the changes in baroreceptor afferent nerve activity, because peripheral information that accompanies hypotension has an input in the baroreflex center. In order to eliminate the possibility that the central sympathoinhibitory action was induced indirectly by afferent nerve modifications, we examined the effect of ketanserin on ADNA originating in the baroreceptor. After *i.v.* administration of ketanserin (0.5 and 5 mg/kg), ADNA decreased dose-dependently. This indicates that the reduction of sympathetic outflow after ketanserin administration was not due to an indirect action via baroreceptor. In conclusion, the central sympathoinhibitory action of ketanserin may play an important role in its bradycardia and may be partially responsible for its hypotensive effect in rats.

Site of Central Sympathoinhibitory Action

As shown in Fig. 6 A, intracerebroventricular (*i.c.v.*) administration of ketanserin (200 μg/rat) produced a decrease in ANA that was gradually accompanied with hypotension and bradycardia. Thirty minutes after *i.c.v.* administration, significant decreases in ANA, BP and HR were observed. Whereas, intrathecal (*i.t.*) administration of equivalent doses of ketanserin had no effect on ANA, BP and HR (Fig. 6 B). Namely, ketanserin did not act at the spinal cord level where the origin of the preganglionic nerves, the intermediolateral cell column (IML), is located. Our findings coincided with previous results that showed the decreased ANA after *i.v.*

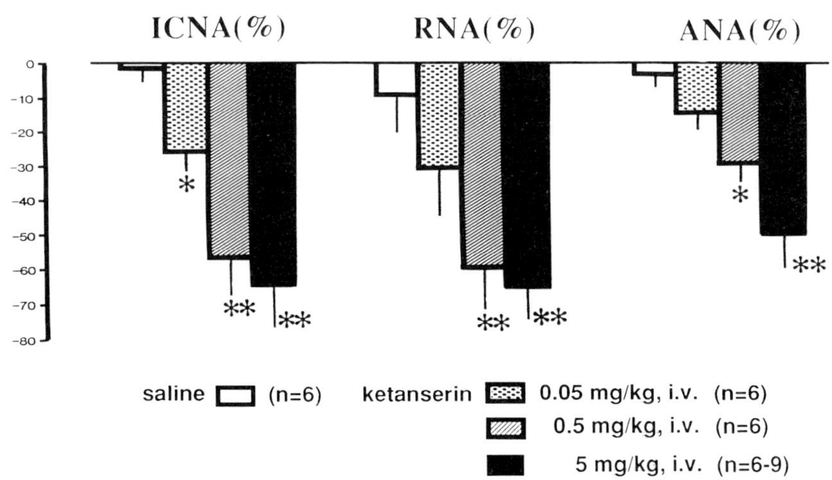

Figure 5. Effect of intravenous (*i.v.*) administration of ketanserin on postganglionic inferior cardiac nerve activity (ICNA), renal nerve activity (RNA) and preganglionic adrenal nerve activity (ANA) 10 min after administration in rats. Each value represents the mean ± SE. $*P < 0.05$ and $**p < 0.01$ *vs* saline-treated control rats.

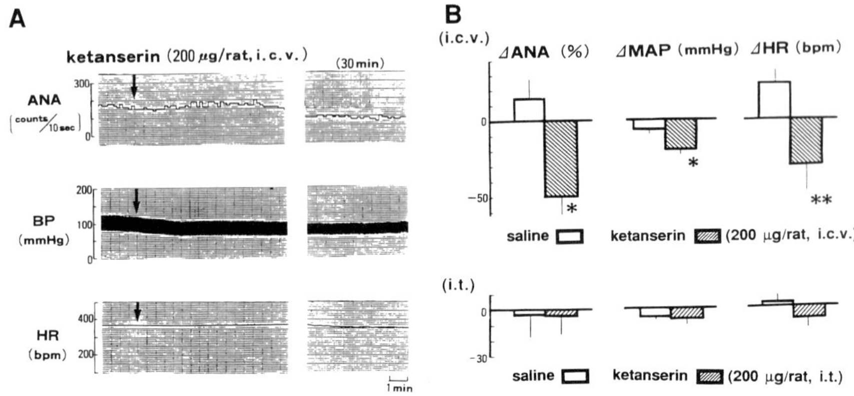

Figure 6. Effect of intracerebroventricular (i.c.v.) and (i.t.) administration of ketanserin (200 μg/rat) on adrenal nerve activity (ANA), mean arterial blood pressure (MBP) and heart rate (HR) in rats. (A) Traces from top to bottom are discharge rate of ANA, BP and HR. (B) The change of ANA, MAP and HR 30 min after i.c.v. (n = 5) and i.t. (n = 7) administration of ketanserin. Each value represents the mean ± SE. * $p < 0.05$ and ** $p < 0.01$ *vs* saline-treated control rats (n =7).

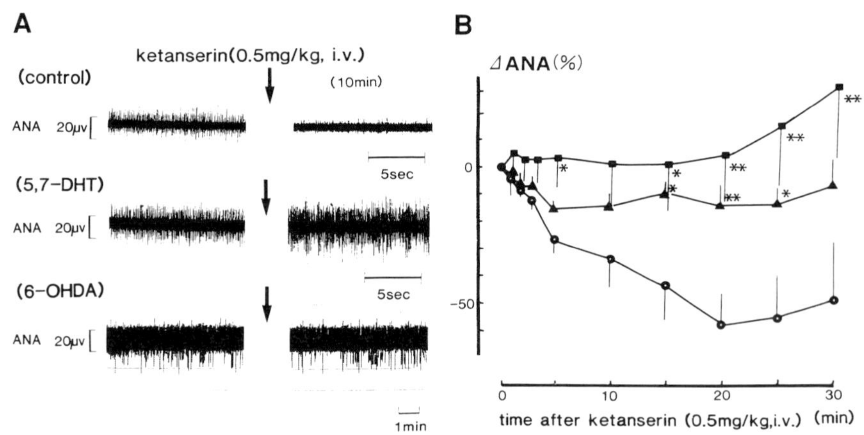

Figure 7. Effect of intravenous (*i.v.*) administration of ketanserin (0.5 mg/kg) on adrenal nerve activity(ANA) in 5,7-dihydroxytryptamine (5,7-DHT) and 6-hydroxydopamine (6-OHDA) pretreated rats. (A) Trace from actual ANA before and after *i.v.* administration of ketanserin. (B) Time course observation. Monoamine neurotoxin pretreated groups are indicate(■ - ■, 5,7-DHT pretreated rats n = 5; ▲ - ▲, 6-OHDA pretreated rats, n = 6; ○ - ○, non-treated control rats, n = 6). Each value represents the mean ± SE. * $p < 0.05$ and ** $p< 0.025$ *vs* non-treated control rats.

administration of ketanserin (5 mg/kg) disappeared after transection of the spinal cord at the C1-C2 level (Sano *et al.*, 1985). These results demonstrate that ketanserin acts at the supraspinal level to reduce ANA in rats.

Decreases in ANA produced by *i.v.* administration of ketanserin (0.5 and 5 mg/kg) were significantly attenuated by pretreatment with 5,7-DHT (Fig. 7) (Matsumoto, 1988). Ketanserin's failure to decrease the ANA in 5-HT-depleted rats indicates the possibility that the central serotonergic pathway is involved in its adrenal sympathoinhibitory action, and it may be mediated via 5-HT_2 receptors. It has been postulated that the central 5-HT activates the 5-HT_2 receptors to cause behavior like head shake response. Our studies demonstrated that the 5-HT precursor, 5-hydroxytryptophan-induced head shake response was inhibited significantly and dose-dependently by ketanserin, at doses that significantly decreased the ANA (unpublished data). These results support the possibility that decrease in ANA caused by ketanserin may be mediated by the 5-HT_2 receptors. Furthermore, ketanserin-induced decrease in ANA was attenuated by pretreatment with 6-OHDA (Fig. 7). Thus, the adrenal sympathoinhibitory action of ketanserin is mediated not only via the serotonergic pathway, but also via the noradrenergic pathway, which may be through the 5-HT_2 receptors and $alpha_1$-adrenoceptors. However, McCall and Schuette (1984) reported that ketanserin as well as prazosin reduced ICNA in baroreceptor-denervated cats. They speculated this central sympathoinhibitory action of ketanserin resulted from the ability to antagonize $alpha_1$-adrenoceptors rather than 5-HT_2 receptors. In our study, however, prazosin (0.01 mg/kg, *i.v.*) which had an effect that was equihypotensive to that of ketanserin (0.5 mg/kg, *i.v.*), did not decrease RNA. This may be due to a difference in either experimental animal species or characteristics of sympathetic nerve activity. In summary, ketanserin acts on the supraspinal level to decrease ANA in a way that may be centrally mediated via both serotonergic and noradrenergic pathways in the rat.

ROLE OF 5-HT IN BLOOD PRESSURE REGULATION

This section focuses the role of 5-HT in the regulation of BP, based on the relationship between 5-HT-induced pharmacological properties and its blockade by ketanserin. As previously described, ketanserin inhibits 5-HT-induced various responses, *i.e.*, vasoconstriction in most blood vessels and aggregation of platelets. Many investigators have claimed that the antihypertensive effect of ketanserin may be associated with the blocking action for 5-HT_2 receptors. These findings could help to establish an important pathogenetic role for peripheral 5-HT including platelet 5-HT in the development of hypertension.

Several studies concerning the responsiveness of vascular constriction to 5-HT in hypertensive animals have been published. For instance, both the pressor response to 5-HT and the amplifying effect of 5-HT to NE or AII in SHR is more pronounced than those in normotensive rats (Collis and Vanhoutte, 1977; Collis *et al.*, 1980). The tachyphylaxis, the responsiveness to repeated doses of 5-HT decreases with time, is reduced in SHR (Collis and Vanhoutte, 1977). These findings suggest the possibility that the responsiveness to 5-HT is enhanced in hypertensive states, that is, the sensitivity of the vascular smooth muscle to 5-HT is augmented by hypertension.

Few reports are available concerning the levels of free 5-HT in plasma of either hypertensive patients or hypertensive animals. Most of the plasma 5-HT is stored in the platelets and the remainder is free in plasma. The physiological significance of free

Table 1. Effects of Serotonin (5-HT) Agonists and Antagonists on Sympathetic Nerve Activity (SNA), Blood Pressure (BP) and Heart Rate (HR).

	Doses (mg/kg,*i.v.*)	Species	SNA	BP	HR	Ref.
non-selective 5-HT antagonists						
cyproheptadine	1.0-3.0	cat	↓(T)	↓	↓	a)
	0.05-1.6	cat**	↓(S) ↓(E)	↓	↓	b)
metergoline	1.0-3.0	cat	↓(T)	↓	↓	a)
	0.01-0.3	cat**	↓(S) ↓(E)	↓	→	b)
methiothepin	1.0-3.0	cat	↓(T)	↓	↓	a)
methysergide	2.0*	cat	↑↓ (T)	↓	↓	c)
	0.025-1.6	cat**	↓(S) ↓(E)	↓	↓	b)
spiperone	0.3-1.0	cat	↓(I)	↓	↓	d)
5-HT_2 antagonists						
ketanserin	0.5-5.0	rat	↓(A) ↓(I) ↓(R)	↓	↓	e) f)
	0.2-0.8	cat**	↓(S) ↓(I)	↓	↓	g)
	0.5-1.0*	cat	↓(T)	↓	↓	c)
LY53857	0.1-3.0	cat	→(I)	→	→	h)
	10.0*	cat	↑(T)	→	→	c)
ritanserin	3.0	cat	(↓)(T)	↓	↓	a)
5-HT_2 agonists						
DOI	0.1-1.0	cat	↑(I)	↑	→	d)
MK212	1.0	cat	↑(I)	↑	↓	d)
5-HT_{1A} agonists						
8-OH-DPAT	0.032-0.128	cat	↓(T)	↓	↓	i)
	0.1-1.0	cat	↓(S) ↓(I)	↓	↓	d)
PAPP	0.3-1.0	cat	↓(S) ↓(I)	↓	↓	d)
5-HT_{1B} agonists						
TFMPP	1.0	cat	↑(I)	→	↓	d)
mCPP	1.0	cat	↑(I)	→	→	d)

* infused over 1 hour, ** baroreceptor-denervated cats, (↓) decreased but not significant. 8-OH-DPAT, 8-hydroxy-2-(di-n-propylamino) tetralin; DOI, 1-(2,5-dimethoxy-4-iodophenyl)-2-aminopropane; PAPP, p-aminophenyl-ethyl-*m*-trifluoromethylphenyl piperazine; TFMPP, 1-(3 trifluoromethyl-phenyl) piperazine; mCPP, 1-(3-chlorophenyl) piperazine. SNA: Preganglionic adrenal (A), splanchnic (S) and thoracic (T) sympathetic nerve activity. Postganglionic external carotid (E), inferior cardiac (I) and renal (R) sympathetic nerve activity. For references, see a) Ramage, 1988; b) McCall and Humphrey, 1982; c) Ramage, 1985; d) McCall *et al.*, 1987; e) Yoshioka *et al.*, 1987; f) Matsumoto, 1988; g) McCall and Schuette, 1984; h) McCall and Harris, 1987; i) Ramage and Fozard, 1987

5-HT in the plasma is uncertain because of difficulty of excluding 5-HT in the platelets from the measurement. Platelet 5-HT is taken up from the circulating 5-HT which escapes inactivation by the liver or lung. It has been reported that both the ability to take up 5-HT and the 5-HT content of platelets are much reduced in hypertensive patients (Kamal *et al.*, 1984; Bhargava *et al.*, 1979). In SHR, a reduced ability of platelets to take up 5-HT and a reduced capacity of the lung to inactive 5-HT have been reported (Prina *et al.*, 1981). From these results it can be speculated that the amounts of free 5-HT in the plasma might be increased in hypertension, and furthermore, this increased 5-HT could possibly produce pronounced vasoconstriction in hypertensive blood vessels. In other words, it is conceivable that peripheral 5-HT may contribute to the initiation and maintenance of augmented vascular resistance and may be one pathogenic cause of hypertension.

On the other hand, the physiological role of central 5-HT should be considered apart from its peripheral one since 5-HT does not readily cross the blood-brain barrier due to its high polarity. Immunohistochemical studies that the location of 5-HT-containing cell bodies and their projections in the CNS coincides with the distribution of regions involved in BP regulation suggests that the serotonergic nervous system may play an important role in cardiovascular regulation. The role of central 5-HT, however, is still poorly understood because it has sometimes opposite effect on BP depending on the route of administration, the experimental conditions and animal species.

As previously discussed, ketanserin-induced adrenal sympathoinhibitory action is mediated with both serotonergic and noradrenergic pathway at the supraspinal level in rats. Namely, the interaction of serotonergic and noradrenergic pathways may be involved in the role of 5-HT in cardiovascular regulation. Anatomical studies demonstrate the existence of a dense serotonergic innervation to the locus ceruleus or anterior hypothalamus/preoptic area (AH/PO), which are the most important regions with respect to cardiovascular functions. In addition, the noradrenergic pathway was densely innervated to the B7 (nucleus raphe dorsalis) area. Benarroch *et al.* (1983) reported that the hypertension produced by 6-OHDA administered into AH/PO was blocked by the pretreatment of either 5,7-DHT or ketanserin. Baraban and Aghajanian (1980) reported that the firing activity of 5-HT cells in the B7 area was suppressed by microiontophoretically applied alpha-adrenoceptor antagonists, while it was enhanced by alpha-adrenoceptor agonists. Experiments to elucidate the effect of 6-OHDA and 5,6-dihydroxytryptamine (5,6-DHT) on the BP in the pontine rabbits suggest that the suprapontine pressor pathway includes ascending serotonergic neurons in series with ascending noradrenergic neurons, whereas the descending neurons have antagonistic control of the activity of bulbospinal preganglionec neurons (Kornel, 1987). These findings suggest the existence of a functional interaction between serotonergic and noradrenergic pathway in cardiovascular regulation.

Although central serotonergic neurons is generally thought to facilitate the sympathetic nerve activity (McCall and Humphrey, 1982), very little is known about the nature of 5-HT receptors that mediate this excitatory effect. Recently McCall *et al.* (1987) reported that 5-HT_2 receptor agonist, 1-(2,5-dimethoxy-4-iodophenyl)-2-aminopropane (DOI) and MK 212, increased the ICNA in cats. DOI-induced increases in ICNA were inhibited by pretreatment with ketanserin and LY53857

(McCall and Harris, 1988). Selective central 5-HT_{1A} receptor agonist, 8-hydroxy-2-(di-n-propylamino) tetralin (8-OH-DPAT) and p-aminophenyl-ethyl-*m*-trifluoromethylphenyl piperazine (PAPP) produced decreases in BP and sympathetic nerve activity in cats (McCall *et al.*, 1987; Ramage and Fozard, 1987). Systemic administration of 8-OH-DPAT also caused hypotension and bradycardia in conscious SHR and anesthetized normotensive rats (Fozard *et al.*, 1987). Details are shown in Table 1. These results lead to the speculation that the cardiovascular regulation of sympathetic nerve activity via central 5-HT may differ from that via 5-HT receptor subtypes. That is, central 5-HT may cause decreases in both BP and sympathetic nerve activity via 5-HT_{1A} receptors. Central 5-HT_2 receptors, on the other hand, may be concerned with increases in sympathetic nerve activity. Furthermore, noradrenergic nervous systems may play an important role in serotonergic pathway-facilitated sympathetic nerve activity.

CONCLUSION

There are reliable evidences that ketanserin reduces total vascular resistance to decrease BP. However, the hypotensive mechanism of ketanserin requires more clarification, *i.e.*, its mechanism cannot be attributed to 5-HT_2 receptor blockade alone. $Alpha_1$-adrenoceptor blockade is not an adequate explanation for the antihypertensive effect. Combined antagonistic action of both 5-HT_2 receptors and $alpha_1$-adrenoceptors or blockade of common binding sites of these receptors should be considered. Central sympathoinhibitory action through the 5-HT_2 receptors and/or the $alpha_1$-adrenoceptors cannot be excluded. Ketanserin-induced bradycardia may be responsible for the sympathoinhibitory action, however, vagally mediated reflex bradycardia or a negative chronotropic effect by direct action on the heart cannot be ignored.

Numerous clinical studies have claimed that ketanserin is an effective therapeutic drug for the treatment of hypertension. For instance, ketanserin is less apt to produce reflex tachycardia, whereas the markedly lowered HR caused by most beta-blockers has not been observed with its use. Ketanserin does not cause the orthostatic hypotension that follows the first dose of prazosin (Ferrara *et al.*, 1986). Severe adverse effects such as withdrawal hypertension that are shown by clonidine have not been reported. For detailed clinical data, we recommend the specialized review by P. Vanhoutte and co-workers (1988). On the other hand, ketanserin is suitable for monotherapy because it does not significantly change lipid metabolism, water and sodium retension (Wottiez *et al.*, 1986). Especially, ketanserin monotherapy appears to be effective in hypertensive patients aged 65 years or more. Because of advantages such as usefulness in treating elderly hypertensive patients, ketanserin is highly evaluated as an antihypertensive drug. Furthermore, ketanserin is of great interest as a pharmacological tool for the investigation of cardiovascular regulation by 5-HT.

REFERENCES

Baraban, J. M. and Aghajanian, G. K. (1980). Suppression of firing activity of 5-HT neurons in the dorsal raphe by alpha-adrenoceptor antagonists. *Neuropharmacology* **19,** 355-363.

Benarroch, E. E., Balda, M. S., Finkielman, S. and Nahmod, V. E. (1983). Neurogenic hypertension after depletion of norepinephrine in anterior hypothalamus induced by 6-hydroxydopamine administration into the ventral pons: Role of serotonin. *Neuropharmacolgy* **22,** 29-34.

Bhargava, K. P., Raina, N., Misra, N., Shanker, K. and Vrat, S. (1979). Uptake of serotonin by human platelets and its relevance to CNS involvement in hypertension. *Life Sci.* **25,** 195-200.

Blauw, G. J., van Brummelen, P., Chang, P. C., Vermeij, P. and Van Zwieten, P. A. (1987). Effects of 5-hydroxytryptamine and ketanserin on the forearm vascular resistance in man [Abstract]. *Br. J. Clin. Pharmacol.* **23,** 631p

Cohen, M. L., Fuller, R. W. and Kurz, K. D. (1983). LY53857, a selective and potent serotonergic ($5\text{-}HT_2$) receptor antagonist, does not lower blood pressure in the spontaneously hypertensive rat. *J. Pharmacol. Exp. Ther.* **227,** 327-332

Collis, M. G. and Vanhoutte, P. M. (1977). Vascular reactivity of isolated perfused kidneys from male and female spontaneously hypertensive rats. *Circ. Res.* **41,** 759-767.

Collis, M. G., DeMey, C. and Vanhoutte, P. M. (1980). Renal vascular reactivity in the young spontaneously hypertensive rat. *Hypertension* **2,** 45-52.

De Clerck, F., David, J.-L, and Janssen, P.A.J. (1982a). Inhibition of 5-hydroxytryptamine-induced and -amplified human platelet aggregation by ketanserin(R41 468), $5\text{-}HT_2$-receptor antagonist. *Agents Actions* **12,** 388-397

De Clerck, F., Jageneau, A. and Dom, J. (1982b). Haemorheological effects of ketanserin, a selective $5\text{-}HT_2$ receptor antagonist, in aged, spontaneously hypertensive dogs. *Arch. int. Pharmacodyn.* **258,** 100-115.

Ferrara, L. A., Pasanisi, F., Fasano, M.L.and Soro, S.(1986). Response to tilting in hypertensive patients receiving ketanserin. *Eur. J. Clin. Pharmacol.* **31,** 505-506.

Fozard, J.R. (1982). Mechanism of the hypotensive effect of ketanserin. *J. Cardiovasc. Pharmacol.* **4,** 829-838.

Fozard, J. R., Mir, A. K. and Middlemiss, D. N. (1987). Cardiovascular response to 8-hydroxy-2-(di-n-propylamino)tetralin (8-OH-DPAT) in the rat: Site of action and pharmacological analysis. *J. Cardiovasc. Pharmacol.* **9,** 328-347.

Frenken, M. and Kaumann, A.J. (1984). Interaction of ketanserin and its metabolite ketanserinol with $5\text{-}HT_2$ receptors in pulmonary and coronary arteries of calf. *Naunyn-Schimiedeberg's Arch. Pharmacol.* **326,** 334-339.

Furukawa, Y., Saegusa, K., Ogiwara, Y. and Chiba, S. (1986). Effects of ketanserin on the pacemaker activity and contractility in the isolated, blood-perfused dog atrium. *J. Cardiovasc. Pharmacol.* **8,** 967-972.

Gradin, K., Pettersson, A., Hedner, T. and Persson, B. (1985). Antihypertensive effects of ketanserin in the spontaneously hypertensive rat: *Mode of action. J. Cardiovasc. Pharmacol.* **7** (suppl 7), S110-S111.

Hedner, T., Persson, B. and Berglund, G. (1983). Ketanserin, a novel 5-hydroxytryptamine antagonist: monotherapy in essential hypertension. *Br. J. Clin. Pharmacol.* **16,** 121-125.

Hosie, J., Stott, D. J., Robertson, J.I.S. and Ball, S. G. (1987). Does acute serotonergic type-2 antagonism reduce blood pressure? Comparative effects of single doses of ritanserin and ketanserin in essential hypertension. *J. Cardiovasc. Pharmacol.* **10** (suppl 3), s86-s88.

Jageneau, A.H.M., Hörig, C., Loots, W. and Symoens, J. (1980). Plethysmographic registration of volume changes in a hand vein; effects of serotonin and of specific antagonist. *Angiology* **31,** 828-832

Jennings, A. A. and Opie, L. H. (1987). Effects of intravenous ketanserin on severely hypertensive patients with double-blind crossover assessment of central side-effects. *J. Cardiovasc. Pharmacol.* **9,** 120-124.

Kalkman, H. O., Harms, Y. M., van Gelderen, E. M., Batink, H. D., Timmermans, P.B.M.W.M.

and Van Zwieten, P. A. (1983). Hypotensive activity of serotonin antagonists; Correlation with α_1-adrenoceptor and serotonin receptor blockade. *Life Sci.* **32,** 1499-1505.

Kalkman, H. O., Timmermans, P.B.M.W.M. and Van Zwieten, P. A. (1982). Characterization of the antihypertensive properties of ketanserin (R 41 468) in rats. *J. Pharmacol. Exp. Ther.* **222,** 227-231

Kamal, L. A., Le Quan-Bui, K. H. and Meyer P. (1984). Decreased uptake of ^{3}H-serotonin and endogenous content of serotonin in blood platelets in hypertensive patients. *Hypertension* **6,** 568-573.

Korner, P. I. (1987). Role of central monoaminergic systems and of beta-adrenoceptros in cardiovascular regulation. In: *Alpha & Beta Adrenoceptors and the Cardiovascular System,* W. Kobinger and R.P. Ahlquist (Ed). Excerpta Medica, Princeton, pp. 72-105.

Kurowski, M. (1985). Bioavailability and pharmacokinetics of ketanserin in elderly subjects. *Eur. J. Clin. Pharmacol.* **28,** 411-417.

Leysen, J. E., Niemegeers, C.J.E., Van Nueten, J. M. and Laduron, P. M. (1981). [^{3}H]Ketanserin (R 41 468), a selective ^{3}H-ligand for serotonin2 receptor binding sites. *Mol. Pharmacology* **21,** 301-314.

Marwood, J. F. and Stokes, G. S. (1984). Serotonin (5-HT) and its antagonists: Involvement in the cardiovascular system. *Clin. Exp. Pharmacol. Physiol.* **11,** 439-456.

Matsumoto, M. (1988). Effect of ketanserin on adrenal sympathetic nerve activity in rats. *Japan. J. Pharmacol.* **48,** 57-66.

McCall, R. B. and Harris, L. T. (1987). Characterization of the central sympathoinhibitory action of ketanserin. *J. Pharmacol. Exp. Ther.* **241,** 736-740.

McCall, R. B. and Harris, L. T. (1988). 5-HT_2 receptor agonists increase spontaneous sympathetic nerve discharge. *Eur. J. Pharmacol.* **151,** 113-116.

McCall, P. B. and Humphrey, S. J. (1982). Involvement of serotonin in the central regulation of blood pressure: Evidence for a facilitating effect on sympathetic nerve activity. *J. Pharmacol. Exp. Ther.* **222,** 94-102.

McCall, R. B., Patel, B. N. and Harris, L. T. (1987). Effects of serotonin-1 and serotonin-2 receptor agonists and antagonists on blood pressure, heart rate and sympathetic nerve activity. *J. Pharmacol. Exp. Ther.* **242,** 1152-1159

McCall, R. B. and Schuette, M. R. (1984). Evidence for an alpha-1 receptor-mediated central sympathoinhibitory action of ketanserin. *J. Pharmacol. Exp. Ther.* **228,** 704-710.

Mecca, T. E., Mitchell, J., Bohr, D. F. and Webb, R. C. (1985). Effects of serotonin antagonists on blood pressure in mineralocorticoid hypertensive sheep. *J. Cardiovasc. Pharmacol.* **7,** 660-665.

Milei, J., Lemus, J., Schiavone, M. and Lucioni, M. C. (1987). Ketanserin in parenteral treatment of acute essential hypertension: A dose-response curve. *J. Cardiovasc. Pharmacol.* **10** (suppl 3), S96-S100.

Minami, M., Togashi, H., Matsumoto, M., Yoshioka, M., Hamaue, N., Amamiya, N. and Saito, H. (1988). Effects of chronic oral administration of ketanserin and trichlormethiazide on blood pressure in stroke-prone spontaneously hypertensive rats. *Hokkaido J. Med. Sci.* **63,** 232-239.

Ochs, H.R., Verburg-Ochs, B., Hoeller, M. and Greenblatt, D.J. (1985). Effect of ketanserin on the kinetics of digoxin and digitoxin. *J. Cardiovasc.Pharmacol.* **7,** 205-207.

Pettersson, A., Persson, B., Henning, M. and Hedner, T. (1984). Antihypertensive effects of chronic 5-hydroxytryptamine (5-HT_2) receptor blockade with ketanserin in the spontaneously hypertensive rat. *Naunyn-Schmiedeberg's Arch. Pharmacol.* **327,** 43-47.

Phillips, C. A., Mylecharane, E. J., Markus, J. K. and Shaw, J. (1985). Hypotensive actions of ketanserin in dogs: Involvement of a centrally mediated inhibition of sympathetic vascular tone. *Eur. J. Pharmacol.* **111,** 319-327.

Prina, R., Dolfini, E., Mennini, T., Palermo, A. and Libretti, A. (1981). Reduced serotonin uptake by spontaneously hypertensive rat platelets. *Life Sci.* **29,** 2375-2379.

Ramage, A. G. (1985). The effects of ketanserin, methysergide and LY 53857 on sympathetic nerve activity. *Eur. J. Pharmacol.* **113,** 295-303.

Ramage, A. G. (1988). Examination of the effects of some 5-HT_2 receptor antagonists on central sympathetic outflow and blood pressure in anaesthetised cats. *Naunyn-Schmiedeberg's Arch. Pharmacol.* **338,** 601-607.

Ramage, A. G. and Fozard, J. R. (1987). Evidence that the putative 5-HT_{1A} receptor agonists, 8-OH-DPAT and ipsapirone, have a central hypotensive action that differs from that of clonidine in anaethetised cats. *Eur. J. Pharmacol.* **138,** 179-191.

Sano, M., Minami, M.,Togashi, H., Yoshioka, M., Tochihara, M., Hirokami, M. and Saito, H. (1985). Effect of ketanserin on blood pressure in rats. *Folia Pharmacol. Japon.* **86,** 401-409.

Togashi, H.(1983). Central and peripheral effects of clonidine on the adrenal medullary function in spontaneously hypertensive rats. *J. Pharmacol. Exp. Ther.* **225,** 191-197

Trenk, D., Mosler, A., Kirch, W., Meinertz, T.and Jahnchen, E. (1985). Pharmacokinetics and pharmacodynamics of the 5-HT_2 receptor antagonist ketanserin in man. *J. Cardiovasc. Pharmacol.* **5,** 1034-1039.

Van Nueten, J. M. (1983). 5-Hydroxytryptamine and precapillary vessels. *Fed. Proc.* **42,** 223-227.

Van Nueten, J. M., Janssen, P.A.J., Symoens, J., Janssens, W. J., Heykants, J., De Clerck, F., Leysen, J.E., Vancauteren, H. and Vanhoutte P.M. (1987). Ketanserin. In: *New Cardiovascular Drugs*, A. Scriabine (Eds). Raven Press, New York, pp. 36-37.

Van Nueten, J. M., Janssen, P.A.J., Van Beek, J., Xhonneux, R., Verbeuren, T. J. and Vanhoutte, P.M. (1981). Vascular effects of ketanserin (R 41 468), a novel antagonist of 5-HT_2 serotonergic receptors. *J. Pharmacol. Exp. Ther.* **218,** 217-230.

Vanhoutte, P., Amery, A., Birkenhäger, W., Breckenridge, A., Buhler, F., Distler, A., Dormandy, J., Doyle, A., Frohlich, E., Hansson, L., Hedner, T., Hollenberg, N., Jensen, H-E., Johansen, P.L., Meyer, P., Opie, L., Robertson, I., Safar, M., Schlekamp, M., Symoens, J., Trap-Jensen, J. and Anchetti, A.(1988). Serotoninergic mechanisms in hypertension. *Hypertension* **11,** 111-133

Wenting, G.J., Man in't Veld, A.J., Woittiez, A.J., Boomsma, F. and Schalekamp, M.A.D.H. (1982). Haemodynamic effects of ketanserin, a selective 5-hydroxytryptamine (serotonin) receptor antagonist, in essential hypertension. *Clin. Sci.* **63** (supple 8), 435s-438s.

Wenting, G. J., Woittiez, A. J. J., Man in't Veld, A. J. and Schalekamp, M.A.D.H. (1984). Effects of ketanserin in essential hypertension and autonomic insufficiency. *Hypertension* **6,** 100-109.

Williams, F. M., Leeser, J. E. and Rawlins, M. D. (1986) Pharmacodynamics and pharmacokinetics of single doses of ketanserin and propranolol alone and in combination in healthy volunteers. *Br. J. Clin. Pharmacol.* **22,** 301-308.

Woittiez, A.J.J., Wenting, G. J., Man in't Veld, A. J., Boomsma, F. and Schalekamp, M.A.D.H. (1985). Ketanserin: A possible tool for studying the role of serotonin in hypertension. *J. Cardiovasc. Pharmacol.* **7** (suppl 7), S130-S136.

Woittiez, A.J.J., Wenting, G.J., Van den Meiracker, A. H., Ritsema, Van Eck, H.J., Man in't Veld, A.J., Zantwoort, F.A. and Schalekamp, M.A.D.H. (1986). Chronic effect of ketanserin in mild to moderate essential hypertension. *Hypertension* **8,** 167-173.

Yoshioka, M., Matsumoto, M., Togashi, H., Minami, M. and Saito, H. (1987). Central sympathoinhibitory action of ketanserin in rats. *J. Pharmacol. Exp. Ther.* **243,** 1174-1178.

Yoshioka, M., Togashi, H., Minami, M. and Satio, H. (1986). Effects of hydralazine on adrenal and cardiac sympathetic nerve activity in anesthetized rats. *Res. Commun. Chem. Pathol. Pharmacol.* **54,** 313-320.

Zabludowski, J. R., Ball, S. G. and Robertson, J.I.S. (1985). Ketanserin and α_1-adrenergic antagonism in humans. *J. Cardiovasc. Pharmacol.* **7** (suppul 7), S123-S125.

Zabludowski, J. R., Zoccali, C., Isles, C. G., Murray, G. D., Rovertson, J.I.S., Inglis, G. C., Fraser, R. and Ball, S. G. (1984). Effect of the 5-hydroxytryptamine type 2 receptor antagonist, ketanserin, on blood pressure, the renin-angiotensin system and sympatho-adrenal function on patients with essential hypertension. *Br. J. Clin.. Pharmacol.* **17,** 309-316

Progress in Hypertension, Vol. 2, pp. 207-231.
Saito *et al.* (Eds)

BASIC AND CLINICAL PHARMACOLOGY OF FENOLDOPAM AND QUINPIROLE

PATRICIA VAN DER NIEPEN and ALAIN G. DUPONT

Dept. of Pharmacology, Faculty of Medicine, Vrije Universiteit Brussel, Brussels, Belgium, AZ-VUB, Laarbeeklaan 103, B-1090 Brussels, Belgium.

INTRODUCTION

It is clearly established that stimulation of peripheral dopamine receptors results in profound cardiovascular and renal changes (Cavero *et al.*, 1982; Goldberg, 1972; 1984; Felder *et al.*, 1989b). The peripheral dopamine receptors are now pharmacologically classified into two distinct subtypes, designated as D_1 and D_2 receptors (previously called DA_1 and DA_2 receptors). This classification is based on differences in potency of series agonists and antagonists (Hilditch and Drew, 1987; Andersen *et al.*, 1990). Highly selective D_1 and D_2 receptor agonists and antagonists have been developed.

The postsynaptic dopamine receptors on vascular smooth muscle cells are clearly characterized as D_1 receptors, while the presynaptic dopamine receptors on sympathetic nerve endings are D_2 receptors (Andersen *et al.*, 1990). Stimulation of postsynaptic D_1 receptors produces active vasodilation, resulting in a blood pressure fall (Beart, 1989), which is accompanied by a baroreceptor-mediated reflex tachycardia. Stimulation of D_2 receptors results in inhibition of norepinephrine release, inducing passive vasodilation and hence a blood pressure fall. Ganglionic dopamine receptors, mediating an inhibition of ganglionic neurotransmission, have generally been shown to belong to the D_2 dopamine receptor subtype, although some studies suggest the existence of a D_1 receptor as well (Grega *et al.*, 1984; Lokhandwala and Sabouni, 1985; Lokhandwala *et al.*, 1985; 1987; Alkadhi *et al.*, 1986; Sabouni *et al.*, 1987; Van der Niepen *et al.*, 1987). In the kidney, tubular D_1 receptors (Felder and Jose, 1988; Hughes and Sever, 1989c; Huo and Healy, 1989) are present and mediate an inhibition of sodium reabsorption, but also glomerular dopamine receptors belonging to the D_2 subtype have been discovered (Felder *et al.*, 1984b). Moreover, the aldosterone secretion from the adrenal cortex appears to be under tonic dopaminergic suppression (Missale *et al.*, 1989); the subtype of adrenal dopamine receptor involved is not yet well characterized. Recent evidence also suggests the presence of a dopamine (D_1) receptor in the rat juxtaglomerular apparatus, involved in the regulation of renin secretion (Kurtz *et al.*, 1988).

Biochemically, the peripheral D_1 receptor is positively linked to adenylate cyclase, although some studies suggest the existence of a D_1-like receptor, not linked to adenylate cyclase (Felder *et al.*, 1988). From a theoretical point of view one can expect that peripheral dopamine receptor stimulation may result in a number of beneficial cardiovascular and renal effects, *e.g.* lowering blood pressure in hypertension. In this chapter, we will summarize the most relevant data obtained so far with fenoldopam, a selective D_1 receptor agonist, which has in the meantime proven to be a useful drug for the treatment of hypertensive emergencies, and of the D_2 receptor agonist quinpirole, for which only few human pharmacology data are available so far.

FENOLDOPAM

Fenoldopam, SKF 82526 or 6-chloro-7,8-2,3,4,5-tetrahydro-1-(4-hydroxyphenyl)-(1H)-3-benzazepine-7,8-diol methanesulfonate (Kaiser *et al.*, 1980; Pfeiffer *et al.*, 1982) is a potent D_1 receptor agonist, lowering blood pressure, increasing renal and mesenteric blood flow and enhancing urinary sodium excretion. It does not interact with peripheral D_2 receptors and β-adrenoceptors, but it has a weak α_2 adrenoceptor antagonistic activity.

Basic pharmacology

The D_1 receptor stimulant activity of fenoldopam has been demonstrated *in vitro* (Lokhandwala and Steenberg, 1984b) and *in vivo* (Dupont *et al.*, 1987) in different experimental set-ups and species. The D_1 agonistic activity was confirmed by demonstrating competitive antagonism with different dopamine receptor antagonists, including the selective D_1 receptor antagonist SCH 23390. Studies with rabbit retina cells (Hensler and Dubocovich, 1986a,b; Hensler *et al.*, 1987) and with bovine and human ciliary body preparations (De Vries *et al.*, 1986) showed that the R-enantiomer is 50-100 fold more potent than the S-enantiomer. Radioligand binding studies with rat striatal membranes reported binding constants for fenoldopam to the D_1 receptor in order of 1.6-1.8 nM (Sibley *et al.*, 1982; Flaim *et al.*, 1985; Hess *et al.*, 1986). Although most studies suggest that fenoldopam is selective for the D_1 receptor and does not interact with the D_2 receptor, some studies suggest an interaction with the D_2 receptor. *In vitro* electrophysiological studies using preparations of substantia nigra revealed that the inhibitory effect of fenoldopam was antagonized by haloperidol and not by SCH 23390, suggesting that fenoldopam produced its effect through D_2 receptor stimulation. However, the dose of SCH 23390 used was rather low (Bull, 1987). Receptor binding studies also demonstrated an interaction of phenoldopam with the D_2 receptor, but with an affinity 0.1-0.2 times that for the D_1 receptor (Hess *et al.*, 1986; Bull, 1987; Arnt *et al.*, 1988).

Several reports demonstrated an inhibitory effect of fenoldopam on ganglionic neurotransmission. This was evidenced *in vitro* by neuroinhibitory effects in the isolated rat (Alkadhi *et al.*, 1984) and dog sympathetic ganglia (Lokhandwala *et al.*, 1984; Shebuski *et al.*, 1985), and also by a potent inhibition of the response to electrical stimulation of preganglionic cardioaccelerator nerves (Sabouni and Lokhandwala, 1984b) and of the lumbar sympathetic chain (Lokhandwala *et al.*, 1984) in the dog. Fenoldopam produces neurogenic vasodilation in the canine forelimb (Grega *et al.*, 1984), the canine hindlimb (Lokhandwala and Sabouni, 1985; Lokhandwala *et al.*, 1985; 1987) and in the rat hindquarters vascular bed (Van der Niepen *et al.*, 1987) by activation of ganglionic D_1 or D_1-like receptors. In the dog, the inhibitory effect of fenoldopam was antagonized by R-sulpiride and not by SCH 23390, suggesting that the ganglionic dopamine receptor is different from the vascular D_1 receptor. Furthermore fenoldopam failed to increase cAMP in this preparation (Sabouni *et al.*, 1987). These findings are in contrast with those of Van der Niepen *et al.* (1987) who demonstrated that the neurogenic vasodilation produced by fenoldopam in the rat hindquarters is due to stimulation of ganglionic dopamine receptors, similar to the vascular D_1 receptor. Indeed, in the rat hindquarters vascular bed, the vasodilator effect of fenoldopam was antagonized by both, RS-sulpiride and SCH 23390.

Fenoldopam has been reported to interact also with postsynaptic α-adrenergic receptors. *In vitro* studies with isolated rabbit aorta (Ohlstein *et al.*, 1985; Nakamura *et al.*, 1986), splenic arteries and dog mesenteric artery preparations (Nakamura *et al.*, 1986) showed that fenoldopam weakly inhibited the vasoconstrictor response to norepinephrine. Furthermore, fenoldopam protected the α-adrenoceptors against irreversible block by phenoxybenzamine, indicating that the same receptor was occupied by these two antagonists. Postsynaptic α-adrenergic antagonistic activity was also clearly demonstrable *in vivo* in the autoperfused anesthetized rat superior mesenteric vascular bed (Dupont *et al.*, 1987) and in the renal and femoral arteries of anesthetized dogs (Kohli *et al.*, 1988). This careful analysis in dogs showed that this effect occurred only at doses 25-30 times higher than that producing renal vasodilation. Several studies also reported that fenoldopam is an antagonist at presynaptic α_2-adrenoceptors. Fenoldopam increased the vasoconstrictor response to sympathetic nerve stimulation in the autoperfused mesenteric vascular bed, an effect that was attributed to blockade of presynaptic α_2-adrenoceptors (Dupont *et al.*, 1987). Foley and Sarau (1984) showed that fenoldopam competed for [^{3}H] clonidine binding in rat brain cortical membranes; in human platelet membranes fenoldopam competed for [^{3}H] yohimbine binding. Yet, De Keyser *et al.* (1988) demonstrated the presence of D_1-like receptors (*i.e.* not coupled to the adenylate cyclase system) on blood platelets membranes. In the dog isolated saphenous vein and in the isolated field-stimulated guinea-pig ileum, fenoldopam antagonized the effects of the α_2-agonist BHT-920 (Ohlstein *et al.*, 1985). Hahn *et al.* (1982) and other investigators (Ohlstein *et al.*, 1985; Schmidt *et al.*, 1985; Lass *et al.*, 1987) were not able to demonstrate stimulant activity on α- and β-adrenoceptors. Fenoldopam did not alter tone in relaxed isolated vessels or increased contractility in isolated, paced papillary muscle. Moreover, high doses of fenoldopam had no effect on blood pressure or on heart rate in the pithed rat.

Fenoldopam does not cross the blood brain barrier (Weinstock *et al.*, 1980) and has no significant effects on central nervous system activity after systemic administration (Hahn *et al.*, 1982; Kamien and Woolverton, 1985; Arnt *et al.*, 1988). Moreover, Lokhandwala *et al.* (1985) demonstrated that intracisternal administration of fenoldopam had no effect on blood pressure, heart rate and hindlimb vascular resistance in the anesthetized dog.

The antihypertensive activity of fenoldopam is well documented in different species and in both hypertensive and normotensive animals. Increasing doses of *i.v.* bolus injections of fenoldopam produce dose-dependent decreases in blood pressure in anesthetized normotensive rats (Van der Niepen *et al.*, 1988), in anesthetized spontaneously hypertensive rats (McCoy *et al.*, 1986) and in conscious unrestrained spontaneously hypertensive rats (Lappe *et al.*, 1986). The blood pressure fall in response to fenoldopam is immediate, but short lasting (± 5 minutes). This hypotensive effect is antagonized by the non-selective dopamine receptor antagonist metoclopramide (Lappe *et al.*, 1986) and by the selective D_1 receptor antagonist SCH 23390 (Lappe *et al.*, 1986; McCoy *et al.*, 1986; Van der Niepen *et al.*, 1988), but not by the selective D_2 receptor antagonist domperidone (McCoy *et al.*, 1986; Van der Niepen *et al.*, 1988).

Some authors (Cavero and Hicks, 1985b; Lèfevre-Borg *et al.*, 1988; Montier, 1989) observed that tolerance to the hypotensive effect of fenoldopam does develop in anesthetized rats and dogs following repeated dosing or during *i.v.* infusion. Intravenous infusion of fenoldopam for 15 minutes in anesthetized normotensive rats and dogs decreased blood pressure with a peak effect within 3-5 minutes. However, this effect wained by 30-60% at the end of infusion. Further experiments (Cavero and Hicks, 1985b; Lèfevre-Borg *et al.*, 1988) led to the conclusion that the fading hypotensive response is partly due to the fact that fenoldopam exerts also some D_1 receptor antagonistic activity. A more important factor in the development of tolerance may be the activation of the renin-angiotensin system. Fenoldopam increases plasma renin activity not only by cardiovascular reflexogenic activation, hemodynamic changes in the kidney or changes in sodium metabolism, but also by increasing renin secretion. In primary cultures of rat kidney juxtaglomerular cells fenoldopam was shown to increase renin release through direct stimulation of D_1 receptors located on the juxtaglomerular apparatus (Cavero *et al.*, 1987b; Montier *et al.*, 1989). However, tolerance was only observed in anesthetized animals. Intravenous infusion of fenoldopam for up to 24 hours in non-anesthetized animals results in a sustained fall in blood pressure (Hieble *et al.*, 1987; Lèfevre-Borg *et al.*, 1988).

The hypotension induced by fenoldopam is accompanied by an increase in heart rate and plasma norepinephrine levels. The tachycardic effect is probably due to an increase in sympathetic tone due to the fall in blood pressure (Lappe *et al.*, 1986; McCoy *et al.*, 1986).

One of the most remarkable effects of fenoldopam is its capacity to increase renal blood flow as well in normotensive as in hypertensive, and anesthetized or conscious animals. The threshold dose is about 0.1 μg/kg or 0.1 μg/kg/min *i.v.*. Maximum increases, up to 80% are achieved at about 40 μg/kg (Hieble *et al.*, 1987; Lechaire and Lèfevre-Borg, 1987; Nichols *et al.*, 1987b; Rosenkranz and McClelland, 1987; Van der Niepen *et al.*, 1988). Fenoldopam appears to be 4 to 10 times more potent than dopamine in reducing renal vascular resistance (Ackerman *et al.*, 1982; 1983;

Hahn *et al.*, 1982; Nichols *et al.*, 1987; Kohli *et al.*, 1988). Quite marked reductions in renal vascular resistance can occur without a corresponding fall in blood pressure (Ackerman *et al.*, 1982; Cavero *et al.*, 1987b; Hieble *et al.*, 1987). No tolerance to the renal dilation response to fenoldopam was observed in anesthetized rats following repeated dosing or during the course of an infusion. This was attributed to a high D_1 receptor density in that vascular bed (Lèfevre-Borg *et al.*, 1988). Intraaortic infusion of fenoldopam in anesthetized normotensive rats (Pollock and Arendshorst, 1987) increased renal blood flow without affecting glomerular filtration rate and with only a slight decrease in blood pressure. Fenoldopam had no effect on the glomerular stop flow pressure (= measure of glomerular capillary pressure) or on the maximum decrease in stop flow pressure, indicating that fenoldopam dilates the preglomerular and efferent arteriolar vasculature without affecting glomerular capillary pressure and glomerular filtration rate. This effect is in agreement with the findings of Edwards (1986) who demonstrated *in vitro* the existence of D_1 receptors on the afferent and efferent renal arterioles.

The effect of fenoldopam on non-renal vascular beds depends on the distribution and accessibility of D_1 receptors. In the anesthetized normotensive rat, *i.v.* fenoldopam increased mesenteric but not hindquarters blood flow (Lechaire and Lèfevre-Borg, 1987; Lèfevre-Borg *et al.*, 1988; Van der Niepen *et al.*, 1988). Hilditch and Drew (1987) reported fenoldopam to be more potent than dopamine in the dog mesenteric vascular bed, equipotent in the rat mesenteric vascular bed, and less potent in the rabbit splenic artery. In contrast, fenoldopam produced only a small increase in splenic, duodenal and intestinal blood flow in the rat (Hieble *et al.*, 1987). Intravenous administration of fenoldopam also decreased blood pressure with an increase in heart rate and lateral cecal arterial blood flow in conscious healthy horses (Clark and Moore, 1989). In conscious SHR (Lappe *et al.*, 1986) *i.v.* fenoldopam (1-1000 μg/kg) increased mesenteric blood flow dose-dependently. Hindquarters vascular resistance was increased with the lowest doses (1-30 μg/kg, *i.v.*), but was decreased with the higher doses. After ganglionic blockade with hexamethonium, hindquarters vasodilation was observed with all doses, indicating that at the lower doses the response to fenoldopam was counteracted by neurally mediated vasoconstriction. Intracoronary administration of fenoldopam in anesthetized dogs increased coronary blood flow via D_1 receptor stimulation (Kopia and Valocik, 1989). Intravenous infusion of fenoldopam in the canine limb also had little effect. In contrast, intravenous bolus injections dilated the femoral arteries dose-dependently by stimulation of peripheral D_1 receptors (Lokhandwala *et al.*, 1985). However, intra-arterial administration of high doses to a denervated limb also had no effect. Close-arterial injections to the canine forelimb produced vasodilation, probably due to postsynaptic α-adrenoceptor antagonism (Grega *et al.*, 1984). The vasodilator effect of fenoldopam in the extremities has, however, mainly been ascribed to inhibition of ganglionic neurotransmission (Lokhandwala and Sabouni, 1985; Lokhandwala *et al.*, 1985; Shebuski *et al.*,1985).

Bugelski *et al.* (1989) reported that *i.v.* fenoldopam induced an arterial lesion in rats, but not in dogs or monkeys. These lesions were seen in the renal and splanchnic arteries, but not in the subserosal gastric and interlobular pancreatic arteries. The primary sites of damage were the medial smooth muscle cells. The fibrinoid lesions of small arteries appear to be α-adrenoceptor-mediated and morphologically repaired after two weeks, whereas the hemorrhagic lesions of large arteries appear to be D_1 receptor-mediated and still exist after 28 days (Morgan *et al.*, 1983; Yuhas *et al.*, 1985; Bugelski *et al.*, 1989; Kerns *et al.*, 1989a,b).

Fenoldopam also has significant effects on renal function in anesthetized and conscious rats and dogs. Intravenous infusion of fenoldopam not only increased renal blood flow and PAH and inulin clearance, but also urine flow, sodium and potassium excretion, even without changes in blood pressure (Bhat *et al.*, 1986; Watkins *et al.*, 1988). Although renal hemodynamic changes may be partly involved, the natriuretic effect of fenoldopam is mainly due to a direct stimulation of proximal tubular D_1 receptors. Indeed, in some experiments, where renal flow was kept constant, fenoldopam still induced a natriuretic effect (Jose *et al.*, 1987). Moreover, Schoors and Dupont (1990) recently showed that blockade of renal dopamine production with carbidopa results in an enhanced natriuretic response (suggesting upregulating of tubular D_1 receptors) to fenoldopam, but does not modify the renal hemodynamic response. Recent studies show that the natriuretic effect of fenoldopam appears to be due to both, hemodynamic and tubular cells mechanisms (Jose *et al.*, 1987; Watkins *et al.*, 1988). From a theoretical point of view, one could hypothesize that the natriuretic and diuretic effects of fenoldopam could be entirely due to renal vasodilation: the increased renal plasma flow and decreased filtration fraction can decrease proximal tubular sodium reabsorption by increasing peritubular capillary hydrostatic pressure and decreasing peritubular capillary colloidal osmotic pressure (Brenner *et al.*, 1973). Secondly, redistribution of renal blood flow from superficial cortical to juxtamedullary nephrons, which are believed to be more efficient in reabsorbing sodium and water, resulting in a "washout" of renal medulla, could also enhance sodium excretion (Fidem *et al.*, 1982). An increased glomerular filtration rate, which increases the solute load entering the proximal tubule could also explain natriuresis. However, most studies show an unchanged glomerular filtration rate, suggesting a balanced effect of fenoldopam on both afferent and efferent arterioles which corroborates with the finding that the sensitivity of the pre- and post-glomerular vasculature to D_1 agonists is identical (Edwards, 1986). Finally, it is possible that fenoldopam exerts a direct effect on the renal proximal tubules to prevent sodium reabsorption (Aperia *et al.*, 1987). Microperfusion studies (Bello-Reuss, 1982) showed that dopamine inhibits sodium transport in the rabbit proximal tubule. Hughes *et al.* (1986a, 1988a,b) showed that fenoldopam in man increased distal sodium delivery (proximal tubule) and decreased sodium-potassium exchange in the distal tubule without affecting systemic blood pressure, glomerular filtration rate and with a dissociation of renal blood flow from urinary sodium excretion and fractional sodium excretion. Moreover, *in vitro* studies demonstrated dopamine-dependent adenylate cyclase activity in rabbit proximal tubules (Felder *et al.*, 1984a) and in human renal cortex tissue (Baldi *et al.*, 1988) suggesting that the receptor subtype belongs to the D_1 subtype. This was confirmed by Felder and Jose (1988) and Hedge *et al.* (1989) who demonstrated the presence of D_1 receptors in the cortex of the rat kidney with autoradiographic studies. Sodium transport studies in frog skin, using short-circuit techniques suggest that the natriuretic effect of fenoldopam is due to inhibition of sodium reabsorption at the tubular level (Mann *et al.*, 1982).

In anesthetized normotensive rats, cardiac output was not affected by *i.v.* fenoldopam (Lechaire and Lèfevre-Borg, 1987). In contrast, in conscious rabbits fenoldopam increased cardiac output (Minson *et al.*, 1987). In rabbits with low output congestive heart failure due to adriamycin toxicity (Jover and McGrath, 1988) *i.v.* fenoldopam markedly increased cardiac output despite a blood pressure fall.

In *in vitro* studies with isolated saline-perfused rat lung, fenoldopam dilated prostaglandin $F_2\alpha$-vasoconstricted pulmonary arteries dose-dependently (Polak *et al.*, 1989). In *in vivo* studies with anesthetized normoxic neonatal lambs fenoldopam had no effect on pulmonary artery resistance, but increased arterial compliance (Goetzman *et al.*, 1987). In another study with conscious neonatal lambs (Drummond *et al.*, 1983b) and with hypoxic lambs (Drummond *et al.*, 1983a) fenoldopam induced pulmonary vasoconstriction. However, the doses used were about 300-fold higher than those required to lower blood pressure.

Kurtz *et al.* (1988) recently demonstrated the presence of D_1 receptors in rat renal juxtaglomerular cells. Antonipillai *et al.* (1989) showed that fenoldopam dose-dependently stimulated renin secretion in static incubations and perifusions of rat renal cortical slices. Nor α-blockade nor β-blockade interfered, whereas the D_1 receptor antagonist SCH 23390 blocked fenoldopam-induced renin release. In contrast, quinpirole, a D_2 receptor agonist, was completely ineffective. These results indicate that fenoldopam is a direct renin secretagogue due to specific D_1 receptor stimulation, and corroborates the results of Mizoguchi *et al.* (1983) who demonstrated that intrarenal administration of dopamine stimulates renin release directly, without any effect at α and β adrenergic receptors. However, the cellular mechanisms by which fenoldopam induces its stimulatory effect on renin remain to be established: D_1 receptors have been closely linked to the adenylate-cyclase-cAMP messenger system. The secretion of renin is known to be under control of both calcium and cAMP messenger systems. In *in vivo* studies, however, the increase in renin release after fenoldopam may largely be baroceptor reflex mediated.

Clinical pharmacology

Pharmacokinetics A high performance liquid chromatographic assay for determination of fenoldopam and its metabolites in human plasma (Osborne and Randolph, 1983; Boppana *et al.*, 1989) and urine (Boppana *et al.*, 1984) has been developed and used to study pharmacokinetics and metabolism of fenoldopam in healthy subjects.

Oral administration of phenoldopam is followed by rapid absorption and undergoes an extensive first-pass metabolism in the liver (Cyronak, 1987). The latter results in a low bioavailability 2-32%, and a relatively short elimination halflife ($t_{1/2}$ = 5-10h). Consequently, relatively high and frequent doses are required by oral route. In conscious dogs also, it was previously demonstrated by Ackerman *et al.* (1983) that by the oral route, a much higher dose was needed than by intravenous administration to obtain a similar increase in renal blood flow. The inactive sulphate and glucuronide conjugates are excreted in urine (80%) and in feces (20%) (Ziemniak *et al.*, 1985).

Constant intragastric (Ziemniak *et al.*, 1988) and intra-intestinal (O'Donnell and

Beck, 1987) infusion of fenoldopam in normal volunteers showed that the action of fenoldopam can be prolonged with sustained drug administration. This suggests that sulphation of fenoldopam by hepatic and /or intestinal enzymes is a saturable process.

A recent pharmacokinetic study (Weber *et al.*, 1988) showed that*i.v.* fenoldopam has a rapid but smooth onset of action (within 30 minutes), a short elimination halflife (<10 minutes) and a linear dose-response relationship. There was no evidence of tachyphylaxis in this study.

Pharmacodynamics and clinical studies Carey *et al.* (1984) showed that fenoldopam did not appreciably change blood pressure or heart rate in normal volunteers, but lowered blood pressure in hypertensives, suggesting that reduced dopaminergic activity expressed at the D_1 receptor may contribute to the pathogenesis and/or maintenance of blood pressure in essential hypertension. In contrast, in normal man, a single oral dose (100 mg) of fenoldopam produced a rapid (within 45 min) fall in diastolic blood pressure which returned gradually to baseline values after 3 to 4 hours and an increase in heart rate (Harvey *et al.*, 1985). Administration of fenoldopam as a single oral dose or by intravenous infusion lowers blood pressure in hypertensive patients. A single oral dose of 100 mg fenoldopam decreased blood pressure in hypertensive patients with a peak effect after 1 hour (Carey *et al.*, 1984; Ventura *et al.*, 1984); the hypotensive effect was sustained for ± 2 hours. The fall in blood pressure was associated with an increase in heart rate. Fenoldopam 100 mg q.i.d. during 4 weeks decreased blood pressure acutely and the hypotensive effect was sustained also for the remainder of the study. The increase in heart rate was transient and heart rate returned to baseline within 24 hours (Carey *et al.*, 1984; Harvey *et al.*, 1986). The observations of Carey *et al.* (1984) further showed that fenoldopam did not induce orthostatic hypotension or sustained tachycardia. The decrease in blood pressure is dose-dependent both after oral (Harvey *et al.*, 1986) and after intravenous (Murphy *et al.*, 1987) administration in patients with mild essential hypertension. After a single oral dose of 100 mg fenoldopam, intra-arterial blood pressure monitoring resulted in a blood pressure fall that was maximal after 50 minutes, but no longer apparent after 130 minutes (Caruana *et al.*, 1987). After 8 weeks of treatment with 100 mg fenoldopam b.i.d. there was only a small and insignificant blood pressure reduction. Tilt testing produced a small blood pressure fall and heart rate increase, yet three patients complained of postural hypotension. The rise in blood pressure induced by dynamic cycle exercise and isometric handgrip was also attenuated by fenoldopam.

All studies performed in patients with severe hypertension (diastolic blood pressure > 120 mmHg) showed that *i.v.* fenoldopam (0.1-0.4 µg/kg/min) lowered blood pressure. Blood pressure fall was apparent after the first 20 minutes and normalized diastolic values were generally obtained after 2 hours of infusion. Blood pressure remained controlled throughout the infusion and no rebound hypertension was noted (Ruilope *et al.*, 1988). The efficacy and safety of *i.v.* fenoldopam was also compared to that of *i.v.* nitroprusside in patients with severe hypertension or in hypertensive crisis. Both antihypertensive drugs successfully controlled blood pressure with a slight increase in heart rate. Additional salutary renal effects were seen in the fenoldopam-treated group (White and Halley, 1989; Elliott *et al.*, 1990; Reisin *et al.*, 1990).

Murphy *et al.* (1988) concluded from experiments with *i.v.* fenoldopam (0.5 μg/kg/min) in man that the hemodynamic effects of this compound are entirely due to selective D_1 receptor stimulation and do not result from α-adrenoceptor blockade as some reports may suggest (Ohlstein *et al.*, 1985; Nakamura *et al.*, 1985; Szabo *et al.*, 1986; Murphy *et al.*, 1987). Indeed, the dose of fenoldopam used in the referred *in vitro* and *in vivo* animal studies far exceeded those used in man.

In vitro studies (Hughes *et al.*, 1986c; 1987; Hughes and Sever, 1987; 1989a,b) with human artery and vein preparations and *in vivo* studies (Hughes *et al.*, 1986b,c,d; 1987) in hand arteries and veins of normal volunteers demonstrated a vasodilatory effect of fenoldopam in human arteries, confirming the presence of D_1 receptors but not in veins.

The effect of fenoldopam on renal hemodynamics, observed in animals, has been confirmed in humans. Stote *et al.* (1983) reported fenoldopam to be a potent oral renal vasodilator in healthy volunteers. Oral fenoldopam (25-100 mg) produced dose-dependent increases in renal plasma flow without affecting glomerular filtration rate and therefore a reduction in filtration fraction, suggesting vasodilation of both afferent and efferent glomerular arterioles. Urinary volume, sodium and chloride excretion increased. Nasogastric infusion of fenoldopam in healthy volunteers (Ziemniak *et al.*, 1988) markedly increased renal plasma flow dose-dependently, beginning 1 hour after initiation of the infusion, reaching a peak at ± 4 hours, which remained during the infusion and returned to baseline 1 hour post-infusion. In contrast to the results of Stote *et al.* (1983) natriuresis did not increase in this study. Intravenous administration of fenoldopam in healthy man increased renal blood flow by 75% and the fractional sodium excretion by 220% (Allison *et al.*, 1987).

In hypertensive patients, single oral doses of fenoldopam (50-100 mg) increased renal blood flow up to 70% and decreased renal vascular resistance without affecting glomerular filtration rate. Urine flow increased stepwise and natriuresis was markedly (+90%) increased after 2 hours. As the fractional sodium excretion only increased 120-150 minutes after fenoldopam administration, a direct effect on tubular cells is likely. Urinary and fractional excretion of potassium remained unchanged (Carey *et al.*, 1984; Ventura *et al.*, 1984; Harvey *et al.*, 1986).

An increase in urinary and fractional excretion of uric acid has been observed in some (Stote *et al.*, 1983) but not in all studies (Harvey *et al.*, 1985). A single oral dose of fenoldopam appears to have a more marked effect on blood pressure in hypertensive patients than in healthy volunteers (Carey *et al.*, 1984). Nevertheless, fenoldopam has a similar effect on renal function in hypertensive patients as in normal controls (Harvey *et al.*, 1985 and 1986). Glück *et al.* (1987) and Bughi *et al.* (1989) reported that *i.v.* fenoldopam in hypertensive patients decreased blood pressure without affecting renal blood flow but with an increase in sodium excretion. In patients with severe hypertension, *i.v.* fenoldopam was reported to increase renal plasma flow, urine flow and natriuresis (White *et al.*, 1988; White and Halley, 1989; Elliott *et al.*, 1990; Reisin *et al.*, 1990). Plasma potassium decreased after ± 150 minutes following an increase excretion which is due to an increase in distal delivery of sodium, as plasma aldosterone did not change. In patients with advanced renal failure (creatinine clearance between 6 and 35 ml/min) *i.v.* infusion of fenoldopam enhanced renal excretion of sodium and water, while ensuring an adequate blood pressure control (Andres *et al.*, 1989).

The aldosterone secretion appears to be under a tonic inhibitory dopaminergic tone. This concept is based on the observations in healthy volunteers that the non-selective dopamine receptor antagonist metoclopramide stimulates aldosterone secretion (Carey *et al.*, 1979 and 1980), that metoclopramide-induced aldosterone secretion is blocked by dopamine (Carey *et al.*, 1979; 1980), and that dopamine inhibits angiotensin II-induced aldosterone secretion (Drake and Carey, 1984; Drake *et al.*, 1984; Carey and Drake, 1986). The subtype of the adrenal dopamine receptor has not yet been determined. Oral administration of fenoldopam (100 mg b.i.d.) for two weeks did not block the aldosterone response to metoclopramide suggesting that the adrenal dopamine receptor does not belong to the D_1 subtype (Dupont *et al.*, 1986a,b). On the other hand, *i.v.* fenoldopam was reported to attenuate the aldosterone response to metoclopramide (Hughes *et al.*, 1988a). Recent radioligand binding studies, however, suggest the presence of a D_2 receptor in the rat adrenal cortex (Missale *et al.*, 1986).

In man, most studies revealed a moderate increase in plasma renin activity in response to fenoldopam (Carey *et al.*, 1984; Ventura *et al.*, 1984; Harvey *et al.*, 1985; 1986; Murphy *et al.*, 1987) which has been attributed to its vasodilatory and blood pressure lowering properties, resulting in a baroreflex-mediated increase in sympathetic tone and hence a β-receptor-mediated increase in renin release. In some studies plasma renin activity did not change after fenoldopam.

In patients with moderate to severe hypertension, fenoldopam improved left ventricular function and diastolic function not only after a single oral dose (Ventura *et al.*, 1984; Caruana *et al.*, 1987), but also after chronic oral treatment (Caruana *et al.*, 1987) and *i.v.* administration (White *et al.*, 1988). Heart rate, cardiac index, stroke volume and left ventricular ejection fraction were increased by fenoldopam, without any change in plasma norepinephrine levels, so that a reflex stimulation of the heart is unlikely. Although fenoldopam has no positive inotropic effects (Hahn *et al.*, 1982), the velocity of circumferential fibre shortening improved. The beneficial effect of fenoldopam on cardiac function is probably due to ventricular unloading.

In patients with congestive heart failure (NYHA functional class III-IV) a single oral dose of fenoldopam acutely reduced blood pressure and increased stroke volume without heart rate changes, suggesting that the impedance to left ventricular ejection is reduced. Pulmonary capillary wedge pressure was reduced without a decrease in right atrial pressure, reflecting improved left ventricular function rather than venodilation (Young *et al.*, 1985). Left ventricular filling pressure and systemic vascular resistance were decreased (Francis *et al.*, 1988; Kukin *et al.*, 1988a).

Treatment of patient with congestive heart failure (NYHA class II-III) for 8 weeks with oral fenoldopam improved exercise tolerance and subjective assessment of shortness of breath (Francis *et al.*, 1988). However, in patients with more severe congestive heart failure (NYHA class III-IV), treated for 3 months with oral fenoldopam, no hemodynamic or clinical benefits were noticed. The latter was ascribed to stimulation of neurohormones (*i.e.* PRA and PNE), particularly in those patients whose neurohumoral systems were already activated (Kukin *et al.*, 1988b).

The latter corroborates with the earlier findings of MacDonald *et al.* (1990) reporting no beneficial effect of chronic oral fenoldopam in ischemic heart failure. However, *i.v.* infusion of fenoldopam in patients with severe congestive heart failure (NYHA class III-IV) increased cardiac and stroke volume indexes. Right atrial pressure, pulmonary capillary wedge pressure and pulmonary artery pressure were not changed. Systemic vascular resistance and pulmonary arteriolar resistance were both reduced (Leon *et al.*, 1986). The onset of action as well as the return to baseline values after stopping the infusion were rapid. The hemodynamic changes were sustained in one study (Young *et al.*, 1988) but abated after 22-24 hours despite continuation of the infusion in another study (Kimmelstiel *et al.*, 1988; Munger *et al.*, 1990).

The potent D_1 agonist activity of fenoldopam thus results in afterload reduction in patients with congestive heart failure, leading to improved left ventricular function.

Safety Fenoldopam appears to be well tolerated, both after oral and *i.v.* administration. The most frequently reported side-effects are related to the vasodilator action, *i.e.* headache, flushing, dizziness and palpitations (Harvey *et al.*, 1986; Caruana *et al.*, 1987; Murphy *et al.*, 1987; Weber *et al.*, 1988). Nausea, anxiety and nasal congestion are also mentioned (Caruana *et al.*, 1987). Hematological and biochemical parameters are not influenced. The electrocardiogram occasionally shows T-wave abnormalities and unifocal premature ventricular beats.

Conclusion

In summary, fenoldopam, a highly selective D_1 receptor agonist, fulfills the criteria for a useful parenteral antihypertensive drug. Although its efficacy in patients with mild to moderate hypertension is proven, its major role may be in the treatment of severe and malignant hypertension. Its relatively rapid onset and short duration of action makes it possible to easily titrate the dose required for the desired hypotensive effect. The renal vasodilating, diuretic and natriuretic actions are desirable in severe hypertensive patients who commonly exhibit renal dysfunction and in whom acute reductions in blood pressure often lead to transitory impairment of renal function and sodium excretion. Fenoldopam can be used to enhance renal perfusion and to promote diuresis in oliguric patients whose conditions are refractory to diuretics. Finally, fenoldopam may have a role in the short term treatment of congestive heart failure by decreasing afterload and by improving renal blood flow.

QUINPIROLE

Quinpirole or LY 171555 is trans-(±)-4aR-4,4a,5,6,7,8,8a,9-octahydro-5-propyl-1H(or 2H)-pyrazolo[3,4-g]quinoline, monohydrochloride (Bach *et al.*, 1980; Titus *et al.*, 1983). This compound is the levorotatory enantiomer of LY 141865, a potent and selective centrally acting D_2 receptor agonist (Tsuruta *et al.*, 1981).

Basic pharmacology

Quinpirole has been characterized as a potent and selective agonist at presynaptic D_2 receptors. In *in vitro* preparations and in intact rats quinpirole inhibits the neuronal release of norepinephrine (Hahn *et al.*, 1983; Lokhandwala and Steenberg, 1984a,b; Cavero *et al.*, 1986; Nagahama *et al.*, 1986b; Lèfevre-Borg *et al.*, 1987). Confirmation that the effect of quinpirole results from selective stimulation of presynaptic D_2 receptors and not of α_2-adrenoceptors or D_1 receptors, was obtained using appropriate antagonists (Hahn and MacDonald, 1984; Kebabian *et al.*, 1984; Sengupta and Lokhandwala, 1985; Lechaire and Lèfevre-Borg, 1987; Lèfevre-Borg *et al.*, 1987; Nagahama *et al.*, 1986b; Van der Niepen *et al.*, 1988).

Fuller *et al.* (1983) have shown that quinpirole exerts no serotoninergic action at antihypertensive doses in anesthetized animals. Furthermore, studies in rats and dogs (Sabouni and Lokhandwala, 1984a) demonstrated that quinpirole does not stimulate postsynaptic muscarinic and histaminic receptors. Ruffolo and Shaar (1983), and Armstrong *et al.* (1983) showed that quinpirole, but only at a high dose and under certain experimental conditions, can act as a histamine-2-receptor agonist. Indeed, the doses of quinpirole which have been reported to interact with the H_2 receptor and also with the α_2 adrenoceptor (Cohen *et al.*, 1984) were at least 2000-fold higher than those required for activation of the pituitary D_2-receptor involved in prolactin release.

The cardiovascular effects following quinpirole administration in rats are inconsistent, depending on whether the animal is conscious or anesthetized, on the anesthetic used and whether the animal is normotensive or hypertensive.

Quinpirole, given intravenously, reduces blood pressure dose-dependently in the anesthetized normotensive rat (Cavero, 1983; Cavero *et al.*, 1987; Lechaire and Lèfevre-Borg, 1987; Lèfevre-Borg *et al.*, 1987; Van der Niepen *et al.*, 1988), in the anesthetized normotensive dog (Hahn *et al.*, 1983), and in the anesthetized normotensive rhesus monkey (Hahn and MacDonald, 1984). The threshold dose is 1 μg/kg *i.v.* The hypotensive effect reaches a maximum within the first 10 min after administration and is long lasting in pentobarbital anesthetized rats but only transient under urethane anesthesia. The hypotensive effect of quinpirole can be antagonized by the D_2 receptor antagonists S-sulpiride, haloperidol and domperidone. Blockade of the ganglionic transmission with chlorisondamine or hexamethonium prevents the hypotensive action (Cavero *et al.*, 1987; Lèfevre-Borg *et al.*, 1987; Van der Niepen *et al.*, 1988), indicating that this action necessitates the presence of an intact sympathetic nervous system. The site of action of quinpirole is probably located outside the central nervous system, as the hypotensive effect of either *i.v.* or *i.c.v.* administration is of a similar magnitude. Moreover, the effects of centrally given quinpirole can be blocked by *i.v.* domperidone, a D_2-receptor antagonist that does not readily cross the blood-brain barrier (Lèfevre-Borg *et al.*, 1987). Blood pressure also falls in conscious spontaneously hypertensive rats (Lèfevre-Borg *et al.*, 1987).

In contrast to the results obtained in anesthetized normotensive rats and conscious spontaneously hypertensive rats, *i.v.* bolus injections of quinpirole produced dose-dependent blood pressure increases in conscious normotensive rats (Nagahama *et al.*, 1986a; Lèfevre-Borg *et al.*, 1987). The discrepancy between the results obtained in anesthetized and conscious normotensive rats may be due to the lower level of sympathetic tone and the more efficient reflexogenic regulation present in conscious normotensive rats than in anesthetized normotensive rats. On the other hand, the much higher dose used by Nagahama *et al.* (1986a) and the different way of administration may also in part explain the different results. In pithed rats (Hahn *et al.*, 1983; Sengupta and Lokhandwala, 1985) quinpirole lacks pressor activity. After ganglionic blockade (Hahn *et al.*, 1983; Cavero *et al.*, 1987) quinpirole induces a small blood pressure increase; however, it is unlikely to be due to a direct vasoconstrictor action since quinpirole does not alter blood flow in the autoperfused superior mesenteric vascular bed of the rat (Dupont *et al.*, 1987). Moreover other investigators (Lèfevre-Borg *et al.*,1987; Van der Niepen *et al.*, 1988) did not observe a pressor action of quinpirole in ganglion blocked rats.

Nagahama *et al.* (1986a) reported that *i.v.* quinpirole, given in rather high doses to conscious normotensive rats stimulated vasopressin release from the neurohypophysis; this effect was attributed to activation of central D_2-receptors, as *i.v.* metoclopramide blocked the vasopressin release, whereas domperidone did not. Pretreatment with a vasopressin antagonist attenuated the enhanced pressor response to quinpirole in hexamethonium-treated rats and converted it to a rapid short-acting depressor response in phenoxybenzamine-pretreated rats. This depressor action was blocked by domperidone, suggesting that peripheral D_2-receptors may be involved in this effect. The second mechanism which is believed to be involved in the pressor effect of quinpirole is activation of the sympathetic outflow with elevated plasma catecholamines (Nagahama *et al.*, 1986a; Damase *et al.*, 1990). Indeed, the pressor effect of quinpirole was attenuated partly by phenoxybenzamine. Similarly, in bilateral adrenomedullectomized rats, quinpirole produced an early depressor response, followed by a pressor response which was smaller than that seen in control rats (Nagahama *et al.*, 1986a). This finding suggests that the catecholamine release from the adrenal medulla masks the initial depressor response to quinpirole.

Nagahama *et al.* (1987) further showed that intravenously given quinpirole increased heart rate and plasma catecholamines without altering blood pressure and plasma vasopressin in conscious Brattleboro rats (which are congenitally deficient in vasopressin), but increased blood pressure, heart rate, plasma catecholamines and vasopressin in conscious Long-Evans rats (*i.e.* control group). Moreover the vascular responsiveness to phenylephrine was reduced in the Brattleboro rats compared to the control group. The pressor effect of quinpirole in conscious Long-Evans rats is dependent on an increase in circulating vasopressin and on an activation by sympathetic outflow through a central D_2 dopaminergic system and not through a central vasopressinergic system. Brattleboro rats showed indeed similar increase in sympathetic outflow as indexed by increments in circulating plasma catecholamines, without a concomitant pressor response. The latter suggests a vascular hyporesponsiveness to enhanced sympathetic outflow.

Recently, D_2 receptors have also been demonstrated in adrenal medulla. Activation by quinpirole reduced the release of both catecholamines within a certain range of electrical stimulation in the anesthetized dog (Foucart *et al.*, 1988).

In conscious DOCA/NaCl hypertensive rats (Chen *et al.*, 1988) quinpirole increased blood pressure, but to a lesser degree than in normotensive control rats. Paradoxically, the increments in plasma norepinephrine, epinephrine and vasopressin were greater in DOCA/NaCl rats. In both groups, but more in the hypertensive group, quinpirole increased plasma atrial natriuretic peptide. Domperidone enhanced the maximal pressor response to quinpirole in both groups, restored the quinpirole-induced pressor response to control levels in the DOCA/NaCl rats and blocked the stimulatory effect of quinpirole on atrial natriuretic peptide release in both groups. These results suggest that the blunted pressor response to quinpirole in this model is related to enhanced atrial natriuretic peptide release and that peripheral D_2 receptors modulate ANP secretion in the rat. Recently, Yang *et al.* (1990) reported that the D_2 receptors mediating the central pressor action of quinpirole are located on neurons in the posterior region of the nucleus tractus solitarius.

Quinpirole reduced heart rate dose-dependently in anesthetized normotensive rats (Cavero, 1983; Cavero *et al.*, 1987), (Lechaire and Lèfevre-Borg, 1987; Lèfevre-Borg *et al.*, 1987; Van der Niepen *et al.*, 1988) in anesthetized normotensive dogs (Hahn *et al.*, 1983) and in anesthetized normotensive rhesus monkeys (Hahn and MacDonald, 1984). The fall in heart rate was maximal after 30 min and was antagonized by S-sulpiride, but not by the D_1 receptor antagonist SCH 23390 or the α_2 adrenocepter antagonist idazoxan (Hahn and MacDonald, 1984; Sengupta and Lokhandwala, 1985; Cavero *et al.*, 1987; Lèfevre-Borg *et al.*, 1987). The heart rate decrease occurred also in bilaterally vagotomized and atenolol pretreated rats (Hahn *et al.*, 1983; Sengupta and Lokhandwala, 1985; Cavero *et al.*, 1987). These findings indicate that the main pharmacological mechanism responsible for the reduction in heart rate is stimulation of D_2 receptors, leading to an increase in parasympathetic activity and an inhibition of sympathetic tone. Moreover, quinpirole failed to lower heart rate in chlorisondamine-pretreated and pithed rats, even when heart rate was increased by an isoprenaline infusion (Cavero *et al.*, 1987; Lèfevre-Borg *et al.*, 1987), indicating that the presence of an intact automatic nerve drive to the heart is necessary. However, these findings are difficult to reconcile with the belief that the rat lacks presynaptic dopamine receptors on sympathetic neurons supplying the cardiac pacemaker (Wilfert *et al.*, 1984). Nagahama *et al.* (1986b) suggested a central action as metoclopramide, a D_2-receptor antagonist that crosses the blood-brain barrier, prevented the bradycardia, whereas the peripherally acting D_2 receptor antagonist domperidone, did not. The heart rate decrease was also antagonized by sulpiride. On the other hand, quinpirole inhibits ganglionic neurotransmission and reduces cAMP in isolated stellate ganglia of the dog, an effect which can be inhibited by S-sulpiride (Sabouni *et al.*, 1986;1987). However, the dose of quinpirole necessary to produce these effects was very high, so it is unlikely that this could explain the bradycardia. In contrast, quinpirole increased heart rate in conscious rats (Nagahama *et al.*, 1987). This difference between quinpirole induced heart rate changes in conscious and anesthetized rats, may be due to anesthetic-provoked alterations in *e.g.* dopaminergic activity. The apparent inconsistency between the data of Nagahama *et al.* (1987) and Hahn *et al.* (1983) may be related to differences between strains in participation of the dopaminergic system in cardiovascular control (Kurz *et al.*, 1986). The tachycardic effect of quinpirole can also be related to an increase in plasma epinephrine (Damase *et al.*, 1990).

In anesthetized rats, quinpirole increased hindquarters blood flow (Lèfevre-Borg *et al.*, 1987; Van der Niepen *et al.*, 1988) and decreased superior mesenteric blood flow (Cavero *et al.*, 1987; Lèfevre-Borg *et al.*, 1987; Van der Niepen *et al.*, 1988). Renal blood flow was increased in one study (Lèfevre-Borg *et al.*, 1987) and decreased in another (Van der Niepen *et al.*, 1988). But in each study quinpirole reduced the vascular resistances. Hexamethonium abolished the vasodilatory effects in these vascular beds, suggesting that these effects are dependent on an intact sympathetic nervous system. Domperidone (Van der Niepen *et al.*, 1988) and S-sulpiride (Cavero *et al.*, 1987; Lèfevre-Borg., 1987) antagonized the vasodilation suggesting that the vasodilator response to quinpirole is mediated by activation of neuronal D_2 receptors. SCH 23390 had no effect in two different studies (Cavero *et al.*, 1987; Van der Niepen *et al.*, 1988), but was reported to inhibit the effects of quinpirole in the renal and hindquarters vascular beds (Lèfevre-Borg *et al.*, 1987), which was interpreted by these authors as a lack of selectivity at the level of the dopamine receptors located in these vascular beds of either the antagonists used or of quinpirole itself.

Quinpirole does not significantly change cardiac output (Hahn and MacDonald, 1984; Cavero *et al.*, 1987; Lèfevre-Berg *et al.*, 1987). In anesthetized normotensive rhesus monkeys quinpirole maintained stroke volume (Hahn and MacDonald, 1984). Left ventricular work was reduced at higher doses as a consequence of arterial hypotension.

Baseline plasma renin activity was reduced by quinpirole (Cavero *et al.*, 1987; Lèfevre-Borg, 1987) probably by a reduction in sympathetic tone.

Clinical pharmacology

Only one report is available on the effects of quinpirole in man. McNay *et al.* (1987) showed that quinpirole (350 µg/kg qid) given to two mildly hypertensive patients produced a rise in blood pressure and an increase in urinary norepinephrine levels.

Conclusion

In summary, the cardiovascular effects of quinpirole in anesthetized animals are, at least in part, mediated by inhibition of norepinephrine release from sympathetic nerve endings via a peripheral action at presynaptic D_2 receptors. On the other hand, quinpirole produces a pressor effect in conscious normotensive rats through stimulation of central D_2 receptors inducing vasopressin release and activating sympathetic outflow. This central pressor effect of quinpirole appears to predominate over the peripheral depressor effect in conscious rats, and possibly also in man.

CONCLUSION

Both D_1- and D_2-receptor stimulation lower blood pressure by reducing vascular resistance. Clinical investigations with fenoldopam showed that the drug can successfully be used in the treatment of arterial hypertension, cardiac and renal failure. Two relative disadvantages of fenoldopam are the associated tachycardia and the low bioavailability when given orally. From a theoretical point of view, the "ideal" compound would be a drug with good oral bioavailability, that acts on D_1 receptors with just enough D_2 receptor activity to counteract the reflex increase in heart rate, vasoconstrictor tone and renin secretion without inducing nausea or vomiting.

REFERENCES

Ackerman, D. M., Weinstock, J., Wiebelhaus, V. D. and Berkowitz, B. (1982). Renal vasodilators and hypertension. *Drug Dev. Rev.* **2,** 283-297.

Ackerman, D. M., Blumberg, A. L., McCafferty, J. P., Sherman, S. S., Weinstock, J., Kaiser, C. and Berkowitz, B. (1983). Potential usefulness of renal vasodilators in hypertension and renal disease: SK&F 82526. *Fed. Proc.* **42,** 186-192.

Alkadhi, K. A., Sabouni, M. H. and Lokhandwala, M. F. (1984). Characterization of dopamine receptors in a mammalian sympathetic ganglion. *Fed. Proc.* **43,** 1094, A4731.

Alkadhi, K. A., Sabouni, M. H., Ansari, A. F. and Lokhandwala, M. F. (1986). Activation of DA_1 receptors by dopamine or fenoldopam increases cyclic AMP levels in the renal artery but not in the superior cervical ganglion of the rat. *J. Pharmacol. Exp. Ther.* **238,** 547-553.

Allison, N. L., Dubb, J. W., Ziemniak, J. A., Alexander, F. and Stote, R. M. (1987). The effect of fenoldopam, a dopaminergic agonist, on renal hemodynamics. *Clin. Pharmacol. Ther.* **41,** 282-288.

Andersen, P. H., Gingrich, J. A., Bates, M. D., Dearry, A., Falardeau, P., Senogles, S. E. and Caron, M. G. (1990). Dopamine receptor subtypes: beyond the D_1/D_2 classification. *TIPS* **11,** 231-236.

Andres, A., Garcia-Robles, R., Alcazar, J. M., Alvarez, C., Martinez, J., Oliet, A., Sancho, J., Rodicio, J. L. and Ruilope, L. M. (1989). Diuretic and natriuretic properties of fenoldopam in chronic renal failure. *J. Hypertens.* **7** (Suppl. 6), S326-S327.

Antonipillai, I., Broers, M. I. and Lang, D. (1989). Evidence that specific dopamine-1 receptor activation is involved in dopamine-induced renin release. *Hypertension* **13,** 463-468.

Aperia, A., Bertorello, A. and Seri, I. (1987). Dopamine causes inhibition of Na^+-K^+-ATPase activity in rat proximal convoluted tubule segments. *Am. J. Physiol.* **252,** (1, part 2), F39-F45.

Armstrong, J. M., Duval, N. and Langer, S. Z. (1983). LY 141865 stimulates histamine-2 (H_2) receptors and dopamine-2 (DA_2) receptors in the anesthetized dog. *Eur. J. Pharmacol.* **87,** 165-166.

Arnt, J., Bogeso, K. P., Hyttel, J. and Meier, E. (1988). Relative dopamine D_1 and D_2 receptor affinity and efficacy determine whether dopamine agonists induce hyperactivity or oral stereotypy in rats. *Pharmacol. Toxicol.* **62,** 121-130.

Bach, N. J., Kornfeld, E. C., Jones, N. D. Chaney, M. O., Dorman, D. E., Paschal, J. W., Clemens, J. A. and Smalsting, E. B. (1980). Bicyclic and tricyclic ergoline partial structures. Rigid 3-(2-aminoethyl) pyrroles and 3- and 4-(2-aminoethyl) pyrazoles as dopamine agonists. *J. Med. Chem.* **23,** 481-471.

Baldi, E., Pupilli C., Amenta, F. and Mannelli, M. (1988). Presence of dopamine-dependent adenylate cyclase activity in human remal cortex. *Eur. J. Pharmacol.* **149,** 351-356.

Beart, P. M. (1989). Dopamine receptors: Classification, properties and drug development. *Clin. Exp. Pharmacol. Physiol.* **16,** 511-515.

Bello-Reuss, E., Higashi, Y. and Kaneda, Y. (1982). Dopamine decreases fluid reabsorption in straight portions of rabbit proximal tubule. *Am. J. Physiol.* **242,** F634-F640.

Bhat, S., Churchill, M., Churchill, P. and McDonald, F. (1984). Renal effects of SK&F 82526 in anesthetized rats. *Life Sci.* **38,** 1565-1571.

Boppana, V. K., Heineman, F. C., Lynn, R. K., Randolph, W. C. and Ziemniak, J. A. (1984). Determination of fenoldopam (SK&F 82526) and its metabolites in human plasma and urine by high-performance liquid chromatography with electrochemical detection. *J. Chromatogr.* **317,** 463-474.

Boppana, V. K., Dolce, K. M., Cyronak, M. J. and Ziemniak, J. A. (1989). Simplified procedures for the determination of fenoldopam and its metabolites in human plasma by high-performance liquid chromatography with electrochemical detection: comparison of manual and robotic sample preparation methods. *J. Chromatogr.* **487,** 385-399.

Brenner, B. M., Troy, J. L., Daugharty, T. M. and MacInnes, R. M. (1973). Quantitative importance of changes in post glomerular colloid osmotic pressure in mediating glomerulotubular balance in the rat. *J. Clin. Invest.* **52,** 190.

Bugelski, P. J., Walsh Vockley, C. M., Sowinski, J. M. Arena, E., Berkowitz, B. A. and Morgan, D. G. (1989). Ultrastructure of an arterial lesion induced in rats by fenoldopam mesylate, a dopaminergic vasodilator. *Br. J. Exp. Pathol.* **70,** 153-165.

Bughi, S., Horton, R., Antonipillai, I., Manoogian, C., Ehrlich, L. and Nadler, J. (1989). Comparison of dopamine and fenoldopam effects on renal blood flow and prostacyclin excretion in normal and essential hypertensive subjects. *J. Clin. Endocrinol. Metab.* **69,** 1116-1121.

Bull, D. R. (1987). Electrophysiological evidence for a D_2 receptor agonist effect of SKF 82526 in substantia nigra in vitro. *Br. J. Pharmacol.* **92**, (suppl.), 687P.

Carey, R. M., Thorner, M. O. and Ortt, E. M. (1979). Effects of metoclopramide and bromocriptine on the renin-angiotensin-aldosterone system in man. *J. Clin. Invest.* **63,** 727-735.

Carey, R. M., Thorner, M. O. and Ortt, E. M. (1980). Dopaminergic inhibition of metoclopramide-induced aldosterone secretion in man. *J. Clin. Invest.* **66,** 10-17.

Carey, R. M., Stote, R. M., Dubb, J. W., Townsend, L. H., Rose, C. E. Jr. and Kaiser, D. L. (1984). Selective peripheral dopamine-1 receptor stimulation with fenoldopam in human essential hypertension. *J. Clin. Invest.* **74,** 2198-2207.

Carey, R. M. and Drake, C. R. Jr. (1986). Dopamine selectively inhibits aldosterone responses to angiotensin II in humans. *Hypertension* **8,** 399-406.

Caruana, M. P., Heber, M., Brigden, G. and Raftery, E. B. (1987). Effects of fenoldopam, a specific dopamine recepter agonist, on blood pressure and left ventricular function in systemic hypertension. *Br. J. Clin. Parmacol.* **24,** 721-727.

Cavero, I. (1983). Cardiovascular effects of LY141865, a new dopamine recepter agonist, in rats. *Br.J.Pharmacol.* **78** (suppl.), 139P.

Cavero, I. and Hicks, P. E. (1985a). Fenoldopam, a partial agonist at DA_1 receptors mediating hypotension and an alpha 2-adrenoceptor blocker in rats. *Pharmacologist* **27,** 260, A808.

Cavero, I. and Hicks, P. E. (1985b). Studies on the transient hypotensive effects of fenoldopam, a DA1 dopamine receptor agonist, in rats. *Br. J. Pharmacol.* **86** (suppl.), 436P.

Cavero, I., Massingham, R. and Lèfevre-Borg, F. (1982). Peripheral dopamine receptors, potential targets for a new class of antihypertensive agents. *Life Sci.* **31,** 939-948.

Cavero, I., Hicks, P. E. and Lèfevre-Borg, F. (1986). Cardiovascular profile of quinpirole, a selective and specific DA_2-dopamine receptor agonist, in rats. *Brit. J. Pharmacol.* **86,** 618P.

Cavero, I., Thiry, C., Pratz, J. and Lawson, K. (1987). Cardiovascular characterization of DA-1 and DA-2 dopamine receptor agonists in anesthetized rats. *Clin. Exp. Theor. Pract.* **931,** A9.

Chen, Y. F., Jin, H., Paul, R. and Nagahama, S. (1988). Blunted pressor responsiveness to quinpirole, a specific dopamine D2 receptor agonist in conscious deoxycorticosterone acetate/NaCl hypertensive rats is related to atrial natriuretic peptide release. *J. Pharmacol. Exp. Ther.* **246,** 485-492.

Clark, E. S. and Moore, J. N. (1989). Effects of fenoldopam on cecal blood flow and mechanical activity in horses. *Am. J. Vet. Res.* **50,** 1926-1930.

Cohen, M. L., Shaar, C. J. and Colbert, W. E. (1984). α_2-agonist activity of the dopamine DA_2-agonist LY171555. *J. Cardiovasc. Pharmacol.* **6,** 1245-1248.

Cyronak, M., Boppana, V., Clancy, A., Dolce, K., Heineman, F., Cregreen, R. and Ziemniak, J. (1987). Dose-dependant first-pass metabolism of the DA_1 agonist, fenoldopam. *Clin. Pharmacol. Ther.* **41,** 236.

Damase, C. M., Montastruc, J. L., Gharib, C.,Geelen, G., De Saint Blanquat, G. and Tran, M. A. (1990). Effect of quinpirole, a specific dopamine DA_2 receptor agonist on the sympathoadrenal system in dogs. *J. Pharmacol. Exp. Ther.* **252,** 770-777.

De Keyser, J., De Waele, M., Convents, A., Ebinger, G. and Vauquelin, G. (1988). Identification of D_1-like dopamine receptors on human blood platelets. *Life Sci.* **42,** 1797-1806.

Drake, C. R. Jr. and Carey, R. M. (1984). Dopamine modulates sodium-dependent aldosterone responses to angiotensin II in humans. *Hypertension* **6** (suppl. 1), I-119-I-123.

Drake, C. R. Jr. Ragsdale, N. V., Kaiser, D. L. and Carey, R. M. (1984). Dopamine suppression of angiotensin II-induced aldosterone secretion in man: differential responses during sodium loading and depletion. *Metab. Clin. Exp.* **33,** 696-702.

Drummond, W. H., Schrager, H. H. and Dailey, W. A. (1983a). Potentiation of hypoxic pulmonary vasoconstriction by SKF-82526, a new specific dopamine receptor agonist, in unanesthetized neonatal lambs. *Pediatr. Res.* **17** (4, part 2), 148A.

Drummond, W. H., Williams, B. J. and Kelley, K. C. (1983b). Pulmonary and systemic vascular effects of SKF-82526-J, a new specific peripheral dopamine receptor agonist in unanesthetized neonatal lambs. *Dev. Pharmacol. Ther.* **6,** 1-8.

Dupont, A. G., Van der Niepen, P., Volckaert, A., Smitz J., Van Steirteghem, A. C. and Six, R. O. (1986a). Effect of fenoldopam on the aldosterone response to metoclopramide in man. *Clin. Endocrinol.* **24,** 203-208.

Dupont, A. G., Van der Niepen, P., Volckaert, A., Smitz J., Van Steirteghem, A. C. and Six, R. O. (1986b). Fenoldopam: effect on aldosterone secretion. *Br. J. Clin. Pharmacol.* **21,** 243-244.

Dupont, A. G., Lefebvre, R. A. and Van der Niepen, P. (1987). Influence of the dopamine receptor agonist fenoldopam and quinpirole in the rat superior mesenteric vascular bed. *Br. J. Pharmacol.* **91,** 493-501.

Dupont, A. G. and Schoors, D. F. (1990). Carbidopa enhances the natriuretic response to fenoldopam in normotensive, but not in spontaneously hypertensive rats. *J. Auton. Pharmacol.* (in press)

Edwards, R. M. (1986). Comparison of the effects of fenoldopam, SK&F R-87516 and dopamine on renal arterioles, *in vitro*. *Eur. J. Parmacol.* **126,** 167-170.

Elliot, W. J., Weber, R. R., Nelson, K. S., Oliner, C. M., Fumo, M. T., Gretler, D. D., McCray, G. R.and Murphy, M. B. (1990). Renal and hemodynamic effects intravenous fenoldopam versus nitroprusside in severe hypertension. *Circulation* **81,** 970-977.

Felder, C. C., Blecher, M. and Jose, P. A. (1989a). Dopamine-1-mediated stimulation of phospholipase C activity in rat renal cortical membranes. *J. Biol. Chem.* **264,** 8739-8745.

Felder, C. C., Jose, P. A. and Axelrod, J. (1989c). The dopamine-1 agonist, SKF 82526, stimulates phospholipase-C activity independent of adenylate cyclase. *J. Pharmacol. Exp. Ther.* **248,** 171-175.

Felder R. A. and Jose P. A. (1988). Dopamine 1 receptors in rat kidneys identified with ^{125}I-SCH 23982. *Am. J. Physiol.* **255,** F970-F976.

Felder, R. A., Blecher, M., Calcagno, P. L. and Jose, P. A. (1984a). Dopamine receptors in the proximal tubule of the rabbit. *Am. J. Physiol.* **247,** F499.

Felder, R. A., Blecher, M., Eisner, G. M. and Jose, P. A. (1984b). Cortical tubular and glomerular dopamine receptors in the rat kidney. *Am. J. Physiol.* **246,** F557-F568.

Felder, R. A., Felder, C. C., Eisner, G. M. and Jose, P. A. (1988). Renal dopamine receptors. In *Actions of dopamine*, C. Bell and B. McGrath (Ed.), MacMillan Press, Houndmilss, 124-140.

Felder, R. A., Felder, C. C., Eisner, G. M. and Jose, P. A. (1989b). The dopamine receptor in adult and maturing kidney. *Am. J. Physiol.* F315-F327.

Fidem, S. Z., Hernandez-Llamas, G., Patak, R. V., Rosenblatt, S. G., Lifschitz, M. D. and Stern, J. M. (1982). Studies on the mechanism of sodium excretion during drug-induced vasodilation in the dog. *J. Clin. Invest.* **69,** 604.

Flaim, K. E., Gessner, G. W., Crooke, S. T., Sarau, H. M. and Weinstock, J. (1985). Binding of novel dopaminergic agonist radioligand ^{3}H fenoldopam (SKF-82526) to D_1 receptors in rat striatum. *Life Sci.* **36,** 1427-1436.

Foley, J. J. and Sarau, H. M. (1984). Adrenergic binding studies of SKF-82526 and its enantiomers. *Fed. Proc.* **43,** 555, A1580

Foucart, S., Lacaille-Belanger, P., Kimura, T., Nadeau, R. and De Champlain, J. (1988). Modulation of adrenal catecholamine release by DA_2 dopamine receptors in the anaesthetized dog. *Clin. Exp. Pharmacol. Physiol.* **15,** 601-611.

Francis, G. S., Wilson, B. C. and Rector, T. S. (1988). Hemodynamic, renal, and neurohumoral effects of a selective oral DA_1 receptor agonist (fenoldopam) in patients with congestive heart failure. *Am. Heart J.* **116,** 473-479.

Fuller, R. W., Snoddy, H. D., Mason, N. R., Clemens, J. A. and Bemis, K. G. (1983). Elevation of serum corticosterone in rats by dopamine agonists related in structure to pergolide. *Neuroendocrinology* **36,** 285-290.

Glück, Z., Jossen, L., Weidmann, P., Gnädinger, M. P. and Peheim, E. (1987). Cardiovascular and renal profile of acute peripheral dopamine$_1$-receptor agonism with fenoldopam. *Hypertension* **10,** 43-54.

Goetzman B. W., Bennett S. H. and Ezzedeen F. (1987). Evidence for the dopaminergic effect in the pulmonary circulation. *Clin. Res.* **35,** 234A.

Goldberg, L. I. (1972). Cardiovascular and renal actions of dopamine: potential clinical applications. *Pharmacol. Rev.* **24,** 1.

Goldberg, L. I. (1984). Dopamine receptors and hypertension. Physiologic and pharmacologic implications. *Am. J. Med.*, Oct. **5**, 37-44.

Goldberg, L. I. (1989). The role of dopamine receptors in the treatment of congestive heart failure. *J. Cardiovasc. Pharmacol.* **14** (suppl. 5), S19-S27.

Grega, G. J., Barrett, R. J., Adamski, S. W. and Lokhandwala, M. F. (1984). Effects of dopamine (DA) and SKF 82526, a selective DA$_1$-receptor agonist, on vascular resistances in the canine forelimb. *J. Pharmacol. Exp. Ther.* **299,** 756-762.

Hahn, R. A and MacDonald, B. R. (1984). Primate cardiovascular responses mediated by dopamine receptors: effects of N,N-di-n-propyldopamine and LY171555. *J. Pharmacol. Exp. Ther.* **299,** 132-138.

Hahn, R. A., Wardell, J. R. Jr., Sarau, H. M. and Ridley, P. T. (1982). Characterization of the peripheral and central effects of SK&F 82526, a novel dopamine receptor agonist. *J. Pharmacol. Exp. Ther.* **233,** 305-313.

Hahn, R. A., MacDonald, B. R. and Martin, M. A. (1983). Antihypertensive activity of LY141865, a selective presynaptic dopamine receptor agonist. *J. Pharmacol. Exp. Ther.* **224,** 206-214.

Harvey, J. N., Worth, D. P., Brown, J. and Lee, M. R. (1985). The effect of oral fenoldopam (SKF 82526-J), a peripheral dopamine receptor agonist, on blood pressure and renal function in normal man. *Br. J. Clin. Pharmacol.* **19,** 21-27.

Harvey, J. N., Worth, D. P., Brown, J. and Lee, M. R. (1986). Studies with fenoldopam, a dopamine receptor DA$_1$ agonist, in essential hypertension. *Br. J. Clin. Pharmacol.* **21,** 53-61.

Hedge, S. S. and Lokhandwala, M. F. (1988). Involvement of cyclic-AMP in the hypotensive response to fenoldopam but not to quinpirole in the anesthetized rat. *J. Auton. Pharmacol.* **8,** 63-68.

Hedge, S. S., Ricci, A., Amenta, F. and Lokhandwala, M. F. (1989). Evidence from functional and autoradiographic studies for the presence of tubular dopamine-1 receptors and their involvement in the renal effects of fenoldopam. *J. Pharmacol. Exp. Ther.* **251,** 1237-1245.

Hensler, J. G. and Dubocovich, M. L. (1986a). Functional role for D$_1$ dopamine (DA) receptors in rabbit retina. *Invest. Ophthalmol. Vis. Sci.* **27** (suppl.), 231, A14.

Hensler, J. G. and Dubocovich, M. L. (1986b). Dopamine receptor activation mediates [^{3}H] acetylcholine release from rabbit retina. *Brain Res.* **398,** 407-412.

Hensler, J. G. Cotterell, D. J. and Dubocovich, M. L. (1987). Pharmacological and biochemical characterization of the D$_1$ dopamine receptor mediating acetylcholine release in rabbit retina. *J. Pharmacol. Exp. Ther.* **243,** 857-867.

Hess, E. J., Battaglia, G., Norman, A. B., Iorio, L. C. and Creese L. (1986). Guanine nucleotide regulation of agonist interactions at [^{3}H] SCH 23390-labeled D$_1$ dopamine receptors in rat striatum. *Eur. J. Pharmacol.* **121,** 31-38.

Hieble, J. P., Owen, D. A. A., Harvey, C. A., Blumberg, A. L., Valocik R. E. and DeMarinis R. M. (1987). Hemodynamic effects of selective dopamine receptor agonists in the rat and dog. *Clin. Exp. Ther.* **A9,** 889-912.

Hilditch, A. and Drew, G. M. (1987). Subclassification of peripheral dopamine receptors. *Clin. Exp. Hypertens.* **A9,** 853-872.

Hughes, A. and Sever, P. (1987). Demonstration of vascular DA receptors on human resistance vasculature. *Br. J. Pharmacol.* **92** (suppl.), 641P.

Hughes, A. D. and Sever, P. S. (1989a). Action of fenoldopam, a selective dopamine (DA_1) receptor agonist, on isolated human arteries. *Blood Vessels* **26,** 119-127.

Hughes, A. D. and Sever, P. S. (1989b). The action of dopamine and vascular dopamine (DA_1) receptor agonist on human isolated subcutaneous and omental small arteries. *Br. J. Pharmacol.* **97,** 950-956.

Hughes, A. and Sever, P. (1989c). Specific binding of ^{125}I SCH 23982, a selective dopamine (D_1 receptor ligand to plasma membranes derived from human kidney cortex. *Biochem. Pharmacol.* **38,** 781-785.

Hughes, J. M., Beck, T. R., Rose, C. E. Jr. and Carey, R. M. (1986a). Selective dopamine-1 stimulation produces natriuresis by a direct tubular action. *J. Hypertens.* **4** (suppl.6), S106-S108.

Hughes, A. Redman, D., Sever, P. and Thom, S. (1986b). Fenoldopam vasodilated forearm vasculature by a dopaminergic mechanism. *Br. J. Pharmacol.* 89, 708P.

Hughes, A., Thom, S., Martin, G., Redman, D., Hasan, S. and Sever, P. (1986c). The action of a DA-receptor agonist, fenoldopam, in human vasculature in vivo and in vitro. *Br. J. Clin. Pharmacol.* **22,** 535-540.

Hughes, A., Thom, S., Reman, D., and Sever, P .(1986d). Fenoldopam, a dopamine agonist, induces forearm vasodilatation by a mechanism independent of sympathetic activity. *Blood Vessels* **23,** 77, A62.

Hughes, A., Goldberg, P. and Sever, P. (1987). Absence of vascular dopamine receptors in human veins in vivo and in vitro. *Clin. Sci.* **73** (suppl. 17), A317.

Hughes, J. M., Beck, T. R., Rose, C. E. Jr. and Carey, R. M. (1988a). The effect of selective dopamine-1 receptor stimulation on renal and adrenal function in man. *J. Clin. Endocrinol. Metab.* **66,** 518-525.

Hughes, J. M., Ragsdale, N. V., Felder, R. A. Chevalier, R. L., King B. and Carey R. M. (1988b). Diuresis and natriuresis during continuous dopamine-1 receptor stimulation. *Hypertension* **11,** (suppl. 1), 169-174.

Huo, T. and Healy, D. P. (1989). Autoradiographic localization of dopamine DA_1 receptors in rat kidney with (^{3}H)SCH 23390. *Am. J. Physiol.* **257,** F414-F423.

Jose, P. A., Felder, R. A., Robillard, J. E., Felder, C. C., Felder, G. M. and Eisner, G. M. (1986). Dopamine-2-receptor in the canine kidney. *Kidney Int.* **29,** 386.

Jose, P. A., Eisner, G. M. and Robillard, J. E. (1987). Renal hemodynamics and natriuresis induced by the dopamine-1 agonist, SKF 82526. *Am. J. Med Sci.* **294,** 181-186.

Jover, B. J. and McGrath, B. P. (1988). Beneficial effects of fenoldopam on systemic and regional hemodynamics in rabbits with congestive heart failure. *J. cardiovasc. Pharmacol.* **11,** 483-488.

Kaiser, C., Ali, F. E., Bondinell, W. E., Brenner, M., Holden, K. G., Ku, T. W., Oh, H. J., Ross, S. T., Yim, N. C. F., Zirkle, C. L., Hahn, R. A., Sarau, H. M., Setler, P. E. and Wardell, J. R. Jr. (1980). Separation of potent central and renal dopamine agonist activity in substituted 6-chloro-2,3,4,5-tetrahydro-7,8,-dihydroxy-1-phenyl-1H-3-benzazepines. *J. Med. Chem.* **23,** 973-976

Kamien, J. B. and Woolverton, W. L. (1985). The D_1 dopamine agonist SKF 38393 functions as a discriminative stimulus in rats. *Psychopharmacology* **87,** 368-370.

Kebabian, J. W., Beaulieu, M. and Itoh, Y. (1984). Pharmacological and biochemical evidence for the existence of two categories of dopamine receptors. *Can. J. Neurol. Sci.* **11,** 114-117.

Kerns, W. D., Arena, E., Macia, R. A., Bugelski, P. J., Matthews, W. D. and Morgan, D. G. (1989a). Pathogenesis of arterial lesions induced by dopaminergic compounds in the rat. *Toxicol. Pathol.* **17,** 203-213.

Kerns, W. D., Arena, E. and Morgan, D. G. (1989b). Role of dopaminergic and adrenergic receptors in the pathogenesis of the arterial lesions induced by fenoldopam mesylate and dopamine in the rat. *Am. J. Pathol.* **135,** 339-349.

Kimmelstiel, C. D., Benotti, J. R. and McCue, J. E. (1988). Acute hemodynamic effects of fenoldopam in patients with advanced congestive heart failure. *Circulation.* Abstract of the 61th Sci. Sessions of the Am. Heart Assoc. Washington, Nov. 14-17.

King, B. D., Beck, T. R., Snyder, M. M. and Musser, T. K. (1987). Efficacy of fenoldopam (SK&F 82526) in chronic congestive heart failure. *Circulation* **76** (suppl. IV), A282.

Kinoshita, S., Sidhu, A. and Felder, R. A. (1989). Defective dopamine-1 receptor adenylate cyclase coupling in the proximal convoluted tubule from the spontaneously hypertensive rat. *J. Clin. Invest.* **84,** 1849-1856.

Kohli, J. D., Glock, D. and Goldberg, L. I. (1988). Relative DA_1-dopamine-receptor agonist and α-adrenoceptor antagonist activity of fenoldopam in the anesthetized dog. *J. Cardiovasc. Pharmacol.* **11,** 123-126.

Kohli, J. D., Metra, M., Satoh, Y. and Goldberg, L. I. (1989). Dopamine receptors in the stellate ganglion of the dog. *Eur. J. Pharmacol.* **164,** 265-272.

Kopia, G. A. and Valocik, R. E. (1989). Demonstration of specific dopamine-1 receptor mediated coronary vasodilation in the anesthetized dog. *J. Pharmacol. Exp. Ther.* **248,** 215-221.

Kukin, M. L., Gottlieb, S. S., Medina, N., Yushak, M. and Packer, M. (1988a). Oral fenoldopam produces favorable but short-lived hemodynamic effects in heart failure. *Circulation,* Abstract of the 61st Sci. Sessions of the Am. Heart Assoc., Washington, Nov. 14-17.

Kukin, M. L., Gottlieb, S. S., Medina, N., Yushak, M., Rochelle, L., Goldsmith, L. and Packer, M. (1988b). Neurohormoral activation limits the long-term response to dopamine-1 receptor agonist, fenoldopam in chronic heart failure. *Circulation,* Abstract of the 38th Ann. Meeting of the Am. Coll. Cardiol. Anaheim, CA, March 19-23.

Kurtz, A., Della Bruna, R., Pratz, J. and Cavero, I. (1988). Rat juxtaglomerular cells are endowed with DA-1 dopamine receptors mediating renin release. *J. Cardiovasc. Pharmacol.* **12,** 658-663.

Kurz, K., Moore, R. and Smith, T. (1986). Differential effects of the selective dopamine (DA) agonist, quinpirole, on blood pressure in conscious normotensive and hypertensive rats. *Fed. Proc.* **45,** 1071.

Lappe, R. W., Todt, J. A. and Wendt, R. L. (1986). Effects of fenoldopam on regional vascular resistance in conscious spontaneously hypertensive rats. *J. Pharmacol. Exp. Ther.* **236,** 187-191.

Lass, N. A., Bakris, G. L., Glock, D. and Goldberg, L. I. (1987). Alpha-adrenoceptor contribution to renal vascular effects of low-dose dopamine. *Clin Res.* **35,** 644A.

Lass, N. A., Glock, D. and Goldberg, L. I. (1988). Cardiovascular and renal hemodynamic effects of intravenous infusions of the selective DA_1 agonist, fenoldopam, used alone or in combination with dopamine and dobutamine. *Circulation* **78,** 1310-1315.

Lechaire, J. and Lèfevre-Borg, F. (1987). Comparison of hemodynamic effects of fenoldopam and quinpirole in anaesthetized rats. *Br. J. Pharmacol.* **90** (suppl.), 209P.

Lèfevre-Borg, F., Lorrain, J., Lechaire, J., Thiry, C., Lhoste, F., Hicks, P. E. and Cavero, I. (1987). Cardiovascular characterization of the DA_2 dopamine receptor agonist quinpirole in rats. *Fundamen. Clin. Pharmacol.* **1,** 179-200.

Lèfevre-Borg, F., Lorrain, J., Lechaire, J., Thiry, C., Hicks, P. E. and Cavero, I. (1988). Studies on the mechanisms of the development of tolerance to the hypotensive effects of fenoldopam in rats. *J. Cardiovasc. Pharmacol.* **11,** 444-455.

Lokhandwala, M. F. and Steenberg, M. L. (1984a). Selective activation by LY141865 and apomorphine of presynaptic dopamine receptors in the rat kidney and influence of stimulation parameters in the action of dopamine. *J. Pharmacol. Exp. Ther.* **228,** 161-167.

Lokhandwala, M. F. and Steenberg, M. L. (1984b). Evaluation of the effects of SKF 82526 and LY171555 on presynaptic (DA_2) and postsynaptic (DA_1) dopamine receptors in rat kidney. *J. Auton. Pharmacol.* **4,** 272-277.

Lokhandwala, M. F., Watkins, H. and Alkadi, K. A. (1984). Involvement of DA_1-receptors in the neurogenic vasodilation produced by SKF 82526 in the canine hindlimb. *Fed. Proc.* **43,** 1094, A4733.

Lokhandwala, M. F., Watkins, H. O., Sabouni, M. H and Alkadhi, K. A. (1985). Pharmacological analysis of the actions of SKF 82526 on cardiovascular dopamine receptors. *J. Pharmacol. Exp. Ther.* **234,** 337-344.

Lokhandwala, M. F. and Sabouni, M. H. (1985). Ganglionic dopamine receptors as mediators of the inhibition of neurogenic vasoconstriction produced by fenoldopam. *J. Auton. Pharmacol.* **5,** 301-305.

Lokhandwala, M. F., Watkins, H. O. and Dlewati, A. (1987). Different dopamine receptor subtypes mediate the neurogenic vasodilation produced by fenoldopam and SK&F 85174 in the dog hindlimb. *Eur. J. Pharmacol.* **143,** 383-390.

MacDonald, T. M., Jeffrey, R. F., Muir, A. L. and Lee, M. R. (1990). A preliminary study of the dopamine DA_1 agonist fenoldopam in the treatment of chronic left ventricular failure due to ischemic heart disease. *Eur. J. Clin. Pharmacol.* **38,** 199-201.

Mann, W., Gross, M., Brennan, F. and Wiebelhaus, V. (1982). The effect of the renal vasodilators dopamine, SK&F 38393-C and SK&F 82526-C on short-circuited frog skin. *Fed. Proc.* **41,** 1547.

McCoy, C. E., Douglas, F. L. and Goldberg, L. I. (1986). Selective antagonism of the hypotensive effects of dopamine agonists in spontaneously hypertensive rats. *Hypertension* **8,** 298-302.

McNay, J. L., Henry, D.P. and DeLong, A.F. (1987). LY171555, a selective (DA_2) agonist, increases the blood pressure of hypertensive patients. *Clin. Exp. Hypertens.* **A9,** 1098.

Minson, R., McRitchie, R., Chalmers, J., Goetzman, B.W., Bennet, S. H. and Ezzedeen, F.(1987). Evidence for a dopaminergic effect in the pulmonary circulation. *Clin. Res.* **35,** 234A.

Missale, C., Liberini, P., Memo M., Carruba, M. O. and Spano, P. (1986). Characterization of dopamine receptors associated with aldosterone secretion in rat adrenal glomerulosa. *Endocrinology* **119,** 2227-2230.

Missale, C., Lombardi, C., De Cotiis, R., Memo, M., Carruba, M. O. and Spano, P. F. (1989). Dopaminergic receptor mechanisms modulating the renin-angiotensin system and aldosterone secretion: An overview. *J. Cardiovasc. Pharmacol.* **14** (suppl. 8), S29-S39.

Mizoguchi, H., Dzau, V. J., Siwek, L. G. and Barger, A. C. (1983). Effect of intrarenal administration of dopamine on renin release in conscious dogs. *Am. J. Physiol.* **244,** H39-H45.

Montier, F., Katchadourian, P., Pratz, J. and Cavero, I. (1989). Increase in plasma renin activity evoked by fenoldopam in dogs is directly mediated by dopamine 1 receptor stimulation. *J.Cardiovasc. Pharmacol.* **13,** 739-747.

Morgan, D. G., Bugelski, P. J., Berkowitz, B. A., Valocik, R. E. and Arena, E. (1983). Medial necrosis and hemorrhage in arteries of rats infused with dopaminergic agents. *Lab. Invest.* **48,** 60A.

Munger, M. A., Benotti, J. R., Green, J. A., Jarvis, R. C., Nara, A. R., McCue, J. E., Pospisil, R. A. and Kasmer, R. J. (1990). Assessment of hemodynamic tolerance from a 24 hour intravenous infusion of fenoldopam mesylate in congestive heart failure. *Am. J. Cardiol.* **15,** 206-210.

Murphy, M. B., McCoy, C. E., Weber, R. R., Frederickson, E. D., Douglas, F. L. and Goldberg, L. I. (1987). Augmentation of renal blood flow and sodium excretion in hypertensive patients during blood pressure reduction by intravenous administration of the $dopamine_1$ agonist fenoldopam. *Circulation* **76,** 1312-1318.

Murphy, M. B., Weber, R. R., Nelson, K. and Goldberg, L. I. (1988). The role of α-adrenoceptor blockade in the antihypertensive effects of fenoldopam in humans. *Clin. Pharmacol. Ther.* **44,** 49-55.

Nagahama, S., Chen, Y. F., Lindheimer, M. D. and Oparil, S. (1986a). Mechanism of the pressor action of LY171555, a specific dopamine D_2 receptor agonist, in the conscious rat. *J. Pharmacol. Exp. Ther.* **236,** 735-742.

Nagahama, S., Chen, Y. F., Lindheimer, M. D. and Oparil, S. (1986b). Mechanisms of the depressor action of LY171555, a selective dopamine D_2 receptor agonist, in the anesthetized rat. *J. Pharmacol. Exp. Ther.* **239,** 426-432.

Nagahama, S., Ann, H. S., Chen, Y. F., Lindheimer, M. D. and Oparil, S. (1987). Role of vasopressin in the cardiovascular effects of LY171555, a selective dopamine D2 receptor agonist: studies in conscious Brattleboro and Long-Evans rats. *J. Pharmacol. Exp. Ther.* **242,** 143-151.

Nakamura, S., Kohli, J. D. and Rajfer, S. I. (1986). α-Adrenoceptor blocking activity of fenoldopam (SK&F 82526), a selective DA_1 agonist. *J. Pharm. Pharmacol.* **38,** 113-117.

Nichols, A. J., Smith, J. M. Jr., Shebuski, R. J. and Ruffulo, R. R. Jr. (1987a). Pharmacological analysis of the actions of SK&F 82526 on cardiovascular dopamine receptors. *J. Pharmacol. Exp. Ther.* **234,** 337.

Nichols, A. J., Smith, J. M. Jr., Shebuski, R. J. and Ruffolo, R. R. Jr. (1987b). Comparison of the effects of the novel inotropic agent, ibopamine, with epinine, dopamine and fenoldopam on renal vascular dopamine receptors in the anesthetized dog. *J. Pharmacol. Exp. Ther.* **242,** 573-578.

O'Donnell, D. M. and Beck, T. R. (1987). Assessing absorption characteristics in the small intestine during constant rate infusion. *J. Clin. Pharmacol.* **27**, 714.

Ohlstein, E. H., Zabko-Potapovich, B. and Berkowitz, B. A. (1985). The DA_1 receptor agonist fenoldopam (SK&F 82526) is also an α_2-adrenoceptor antagonist. *Eur. J. Pharmacol.* 118, 321-329.

Osborne, V. L. and Randolph, W. C. (1983). Analysis of SK&F 82526 in plasma by high-performance liquid chromatography with electrochemical detection. *J. Chromatogr.* **255,** 491-486.

Pfeiffer, F. R., Wilson, J. W., Weinstock, J., Kuo, G. Y., Chambers, P. A., Holden, K. G., Hahn, R. A., Wardell, J. R. Jr., Tobia, A. J., Setler, P. E. and Sarau, H. M. (1982). Dopaminergic activity of substituted 6-chloro-1-phenyl-2,3,4,5-tetrahydro-1H-3-benzazepines. *J. Med. Chem.* **25,** 352-358.

Polak, M. J., Knight, M. E., Gause, G. E., Bucciarelli, R. L. and Drummond, W. (1989). Effect of fenoldopam on preconstricted isolated salt-perfused rat lungs. *J. Appl. Physiol.* **67,** 1076-1080.

Pollock, D. M. and Arendshorst, W. J. (1987). Maintenance of tubuloglomerular feedback activity during DA_1-induced renal vasodilation in the rat. *Fed. Proc.* **46,** 636, A1867.

Reisin, E., Huth, M. M., Nguyen, B. P., Weed, S.G. and Gonzalez, F. M. (1990). Intravenous fenoldopam versus sodium nitroprusside in patients with severe hypertension. *Hypertension* **15** (suppl. 1), I59-I62.

Rosenkranz, R. P. and McClelland, D. L. (1987). Selective antagonism of renal hemodynamic changes induced by dopamine, DPDA, fenoldopam and quinpirole in the rat. *Proc. West. Pharmacol. Soc.* **30,** 25-36.

Ruffolo, R. R. Jr. and Shaar, C. J. (1983). Relative potency of LY141865 at the dopamine DA_2 and histamine H2 receptors. *Eur. J. Pharmacol.* **42,** 295-296.

Ruilope, L. M., Garcia Robles, R. G., Miranda, B., Tovar, J., Alcazar, J. M., Sancho, J., Rodicio, J. L., Martinez, A., Astorga, A. and Beck, T. (1988). Renal effects of fenoldopam in refractory hypertension. *J. Hypertens.* **6,** 665-669.

Sabouni, M. H and Lokhandwala, M. F. (1984a). No evidence for the involvement of (H_1, or H_2) receptors in the hypotension and presynaptic inhibition of sympathetic nerve function caused by LY141865 in dogs. *Eur. J. Pharmacol.* **97**, 145-149.

Sabouni, M. H. and Lokhandwala, M. F. (1984b). Effects of selective DA_1 and DA_2 receptor agonists on responses to pre- and postganglionic cardiac sympathetic nerve stimulation. *Fed. Proc.* **43,** 4732.

Sabouni, M. H., Alkadhi, K. A. and Lokhandwala, M. F. (1986). Pharmacological characterization of dopamine receptors in the stellate ganglia with selective DA_1 and DA_2 receptor agonists and antagonists. *J. Pharmacol. Exp. Ther.* **236,** 65-76.

Sabouni, M. H., Alkadhi, K. A. and Lokhandwala, M. F. (1987). Effect of dopamine receptor activation on ganglionic transmission and cyclic AMP levels in the stellate ganglia and renal arteries of the dog. *J. Pharmacol. Exp. Ther.* **240**, 93-97.

Satoh, J., Kohli, J. D., Glock, D. and Goldberg, L. I. (1988). DA_1 dopamine receptor agonist potency of fenoldopam in the canine intra-arterial renal assay. *FASEB J.* **2,** A284.

Schmidt, M., Imbs, J. L., Giesen-Crouse, E. M. and Schwartz, J. (1985). Effects of SK&F 82526 and SK&F 83742 on the renal vascular dopamine receptor. *J. Pharmacol.* **16,** 15-22.

Schoors, D. F. and Dupont, A. G. (1990). Carbidopa enhances the natriuretic but not the vasodilator response to fenoldopam in rats. *Eur. J. Clin. Invest.* **20,** A32.

Sengupta, S. and Lokhandwala, M. F. (1985). Characterization of the hypotensive action of dopamine receptor agonists fenoldopam and quinpirole in anaesthetized rats. *J. Auton. Pharmacol.* **5,** 289-294.

Shebuski, R. J., Fujita, T., Smith, J. M. Jr., Kopaciewicz, L. J., Blumberg, A. L. and Hieble, J. P. (1985). Suppression of sympathetic ganglionic neurotransmission by the selective dopamine-1 receptor agonist fenoldopam (SK&F 82526) in the anesthetized dog. *J. Pharmacol. Exp. Ther.* **235,** 735-740.

Shebuski, R. J., Smith, J. M. Jr., Ruffolo, R. R. Jr. (1988). Comparison of the renal and pulmonary hemodynamic effects of fenoldopam, dobutamine, dopamine and norepinephrine in the anesthetized dog. *Pharmacology* **36,** 35-43.

Sibley, D. R., Leff, S. E. and Creese, L. (1982). Interaction of novel dopaminergic ligands with D_1 and D_2 dopamine receptors. *Life Sci.* **31,** 637-645.

Soares Da Silva, P. (1990). The effects of quinpirole and fenoldopam on the potassium-evoked overflow of endogenous dopamine and noradrenaline in dog mesenteric arteries. *Naunyn-Schmiedeberg's Arch. Pharmacol.* **341,** 37-42.

Stote, R. M., Dubb, J. W., Familiar, R. G., Erb, B. B. and Alexander, F. (1983). A new oral renal vasodilator, fenoldopam. *Clin. Pharmacol. Ther.* **34,** 309-315.

Szabo, B., Hedler, L. and Starke, K. (1986). Dopamine-receptor agonist and alpha-2 adrenoceptor antagonist effects of fenoldopam in rabbits. *J. Pharmacol. Exp. Ther.* **239,** 881-886.

Titus, R.D., Kornfeld, E.C., Jones, N.D., Clemens, J.A., Smalstig, E. and Foreman, M.M. (1983). The resolution and absolute configuration of an ergoline-related dopamine agonist, trans-4, 4a, 5, 6, 7, 8, 8a, 9-octahydro-5-propyl-2H-pyrazole[3,4-g]quinoline. *J. Med. Chem.* **26,** 1112-1116.

Tsuruta, K., Frey, E. A., Grewe, C.W., Cote, T. E., Eskay, R. L. and Kebabian, J. W. (1981). Evidence that LY-141865 specifically stimulates the D_2 dopamine receptor. *Nature* **292,** 463-465.

Van der Niepen, P., Lefebvre, R. A. and Dupont, A. G. (1987). Neurogenic vasodilatation produced by fenoldopam in the rat hindquarters vascular bed. *J. Auton. Pharmacol.* **7,** 331-338.

Van der Niepen, P., Dupont, A. G., Finne, E. and Six, R. O. (1988). Hypotensive and regional haemodynamic effects of fenoldopam and quinpirole in the anesthetized rat. *J. Hypertens.* **6,** (suppl. 4), S687-S689.

Ventura, H. O., Messerli, F. H., Frohlich, E. D., Kobrin, I., Oigman, W., Dunn, F. G. and Carey, R. M. (1984). Immediate hemodynamic effects of a dopamine-receptor agonist (fenoldopam) in patients with essential hypertension. *Circulation* **69,** 1142-1145.

Vries, G. W., De Mobasser, A. and Wheeler, L. A. (1986). Stimulation of endogenous cyclic AMP levels in ciliary body by SK&F 82526, a novel dopamine receptor agonist. *Curr. Eye Res.* **5,** 449-455.

Watkins, H. O., De Feo, M. L., Dlewati, A. and Lokhandwala, M. F. (1988). Effects of DA_1 receptor stimulation with fenoldopam on renal sodium excretion, plasma renin activity (PRA), plasma aldosterone (PA) and atrial natriuretic factor (ANF) in anesthetized dogs. *FASEBJ.* **2,** A605.

Weber, R. R., McCoy, C. E., Ziemniak, J. A., Frederickson, E. D., Goldberg, L. I. and Murphy, M. B. (1988). Pharmacokinetic and pharmacodynamic properties of intravenous fenoldopam, a $dopamine_1$-receptor agonist, in hypertensive patients. *Br. J. Clin. Pharmacol.* **25,** 17-21.

Weinstock, j., Wilson, J. W., Ladd, D. L., Brush, C. K., Pfeiffer, F. R., Kuo, G. Y., Holden, K. G., Yim, N.C.F., Hahn, R. A., Wardell, J. R. Jr., Tobia, A. J., Setler, P. E., Sarau, H. M. and Ridley, P. T. (1980). Separation of potent central and renal dopamine agonist activity of substituted 6-chloro-2, 3, 4, 5-tetrahydro-7, 8-dihydroxyl-1-phenyl-1H-3-benzazepines. *J. Med. Chem.* **23,** 973-975.

White, W. B. and Halley, S. E. (1989). Comparative renal effects of intraveous fenoldopam mesylate and sodium nitroprusside in patients with severe hypertension. *Arch. Intern. Med.* **149,** 870-874.

White, W. B., Radford, M. J., Gonzalez, F. M., Weed, S. G., McCabe, E. J. and Katz, A. M. (1988). Selective dopamine-1 agonist therapy in sereve hypertension: effects of intravenous fenoldopam. *J. Am. Coll. Cardiol.* **11,** 1118-1123.

Wilfert, G., Smit, G., De Jonge, A., Thoolen, M. J. M. C., Timmermans, P. B. M. W. M. and Van Zwieten, P. A. (1984). Inhibitory dopamine receptors on sympathetic neurons innervating the cardiovascular system of the pithed rat. *Naunyn-Schmiedeberg's Arch. Pharmacol.* **326,** 91-99.

Yang, R. H., Igarashi, Y., Wyss, J. M. and Chen, Y. F. (1990). Dopamine D_2 receptors in the posterior region of the nucleus tractus solitarius mediate the central pressor action of quinpirole (LY171555). *Brain Res. Bull.* **24,** 97-103.

Young, J. B., Leon, C. A., Pratt, C. M., Suarez, J. M., Aronoff, R. D. and Roberts, R. (1985). Hemodynamic effects of an oral dopamine receptor agonist (fenoldopam) in patients with congestive heart failure. *J. Am. Coll. Cardiol.* **6,** 792-796.

Young, J. B., Leon, C. A., Pratt, C. M., Kingry, C., Taylor, A. A. and Roberts, R. (1988). Intravenous fenoldopam in heart failure: comparing the hemodynamic effects of dopamine 1 receptor agonism with nitroprusside. *Am. Heart J.* **115,** 378-384.

Yuhas, E. M., Morgan, D. G., Arena, E., Saunders, L. Z. and Lewis, H. B. (1985). Arterial medial necrosis and hemorrhage induced in rats by intravenous infusion of fenoldopam mesylate, a dopaminergic vasodilator. *Am. J. Pathol.* **119,** 83-91.

Ziemniak, J. A., Allison, N., Boppana, V. K., Dubb, J. and Stote R. M. (1987). The effect of acetaminophen on the disposition of fenoldopam: competition for sulfation. *Clin. Pharmacol. Ther.* **41,** 275-281.

Ziemniak, J. A., Boppana, V. K., Cyronak, M. J., Beck, T. R., Familiar, R. G., Dubb, J. W., Allison, N. L. and Stote, R. M. (1988). Continuous intragastric delivery of fenoldopam: relationship between plasma concentration and effects on renal function. *Br. J. Clin. Pharmacol.* **25,** 367-373.

Progress in Hypertension, Vol. 2, pp. 233-256.
Saito *et al.* (Eds)

NTS NEURONAL MECHANISMS AND CLONIDINE ACTION

TOSHIO SUGA and MASAHIKO SUZUKI

Department of Pharmacology, Saitama Medical School, Moroyama, Iruma-gun, Saitama 350-04, Japan

INTRODUCTION

The nucleus tractus solitarii (NTS) is a first sensory relay nucleus of circulatory, respiratory and other visceral reflexes, and, moreover, functions as the primary gustatory center (Palkovits and Zaborszky, 1977; Spyer, 1982, Jordan and Spyer, 1986), too. This nucleus has been defined as A2 and C2 area, a region of norepinephrine (NE)- and epinephrine (EPI)-containing neuron group, by Dahlström and Fuxe (1964; 1965), but now it is estimated to contain several neurotransmitters in addition to catecholamines (Maley and Elde, 1982; Philippu, 1988). Thus, its neuronal organization including these synaptic mechanisms is considerably complex and not enough elucidated. The NTS seems to play an important regulatory role in these reflexes, in addition to their synaptic function. In this treatise, findings concerning the neuronal mechanism of these reflexes in the NTS until today will be described, principally noticing circulatory reflex, and then, actions of clonidine, α_2-adrenergic agonist, will be mentioned in relation to these neuronal mechanism.

FUNCTIONAL STRUCTURES OF THE NTS

Neuronal mechanisms as a first synapse of circulatory reflexes in the NTS have been studied for many years, but it is only in the last few years that definitive results have been recognized. This has been due to the development of more sensitive and precise histological and physiological techniques.

Histological studies

Early histological studies using degeneration techniques after intracranial roots sectioning revealed that, in the cat, the majority of myelinated IXth and Xth afferents entered the tractus solitarius (TS) and descend in the tractus sending off collaterals in the NTS, where they terminated. Some Xth afferents, however, entered the spinal trigeminal tract, and others decussated at the level of the obex and terminated in the contralateral commissural nucleus (Com) (Ingram and Dawkins, 1945).

Using Nauta technique, which visualize also degenerating unmyelinated fibers,

Cottle (1964) investigated in detail degenerating fibers after cutting Xth nerve rootlet in the cat. TS coursed caudally, sending off small branches of fibers medially and dorsomedially, where they terminated. Further caudally, some fibers entered the lateral (Slt) and ventrolateral (Svl) regions of the NTS, and yet others crossed via Com to terminate in the contralateral dorsomedial NTS (Sdm). The pattern of degeneration indicated that the fibers terminating in the rostral third of the NTS were exclusively from IXth nerve afferents whereas those in the caudal third were of Xth nerve origin. And the maximal density of degeneration was found in the intermediate third of the NTS which is an area extending rostrally 2.5mm from the obex. The fibers in this region originated from both IXth and Xth rootlets. Cottle (1964) suggested that this was the most important area for the cardiovascular control.

The pattern of projection to this intermediate region of the NTS in the histology was shown to be similar in the rat (Torvik, 1956), cat (Kerr, 1962) , guinea-pig (Kimmel and Kimmel, 1964), monkey (Rhoton *et al.*, 1966), opossum (Culberson and Kimmel, 1972) and mallard duck (Dubbeldam *et al.*, 1979). Furthermore, the degeneration study of Gwyn and Leslie (1979) showed a heavy vagal projection to their dorsal and dorsocaudal parts of the cat NTS, a region termed the area subpostrema or subnucleus gelatinosus (SNG), and their electronmicroscopical investigation (Leslie *et al.*, 1982) revealed that the degenerating Xth terminals did make synaptic contact with neurons in this region.

Although these anterograde degeneration studies have revealed the major projection areas of the IXth and Xth nerves, it is insufficient to provide a specific information on the projection from particular organs.

Autoradiographic localization of transported radioactively labelled amino acid such as proline and leucine was induced in this study. Beckstead and Norgren (1979) injected labelled amino acid into the petrosal ganglion (IXth nerve) of monkeys, and showed the fibers, running in the NTS, dispersed mainly into the lateral part of the NTS, both rostral and caudal to the site of entry, only sparse labelling being noted in the medial parts of the nucleus. Following amino acid injections into the nodose ganglion (Xth nerve), labelled fibers were seen entering the medulla in a similar fashion to IXth afferents and the labelled terminals were concentrated in the lateral region at the level of the Xth nerve entry and more caudally most dense in the medial regions, including Com. The contralateral Sm and Com were also labelled. Outside the NTS, dense bilateral labelling of fibers was noted in the area postrema (AP) and sparse labelling was noted in the dorsomedial portion of the spinal trigeminal (V) nucleus. Similar results obtained concerning the IXth and Xth afferents of the cat (Gwyn *et al.*, 1979; Kalia and Mesulam, 1980a) and rat (Contreras *et al.*, 1982).

Technique using transganglionic retrograde transport of the enzyme horseradish peroxidase (HRP) provided more specific information regarding the central projection of visceral sensory afferents. HRP is taken up by cut nerve or after injection into sensory ganglia, whole nerves or organs of innervation. Many studies have been done on both IXth and Xth nerves or their various branches and organs of innervation. While the NTS is shown to be the prime site of termination of these afferents, there appears to be a different termination for afferents from different organs (Kalia and Mesulam, 1980b; Katz and Karten, 1983).

HRP studies following its application to the central cut ends of IXth and Xth nerve

(Ciriello *et al.*, 1981), or injection into the nodose ganglion (Gwyn *et al.*, 1979; Kalia and Mesulam, 1980a; Chernicky *et al.*, 1984) has confirmed the earlier degeneration studies and moreover found that most subnuclei received some input, of which relative density varied, high density sites are Sm (medial NTS), Com, and Spc (parvocellular NTS), middle density sites are Sdm, Slt, AP and low density sites are Is (interstital NTS) and Int (intermediate NTS).

Relations between the baroreceptor nerves (carotid sinus nerve and aortic depressor nerve) and subnuclei of the NTS as the projection sites of the nerves have been investigated with HRP method (Kalia and Mesulam, 1980b; Katz and Karten, 1983; Ciriello *et al.*, 1981b; Chernicky *et al.*, 1984). From these data it seems that there is no distinct specific region of the NTS assigned to receive afferent information from any particular visceral organ. Differences, however, in the central representation of viscera have been documented in terms of the relative importance of the different subnuclei of the NTS. A similar conclusion has been reported in the pigeon (Katz and Karten, 1983).

Physiological studies

Baroreceptor afferents Studies in rabbits and cats indicate that baroreceptor afferents project to distinct and localized regions of the NTS. Donoghue *et al.* (1982) investigated the distribution of the myelinated aortic baroreceptor afferent terminals by plotting the depth-threshold contour of the NTS stimulation for antidromic aortic nerve. In the cat, five of six afferents projected to the Slt, and these projected further to the Sm. And a single case continued to send the afferent into the Com and crossed the midline to innervate the contralateral Sm. The central projections of unmyelinated afferents of the cat have been investigated by the same method, too. No marked differences were observed in patterns of projections and termination of these afferents.

In summary, both baroreceptors in the aortic arch and carotid sinus, project to essentially similar regions of the NTS. The dorsolateral and dorsomedial regions at levels rostral to the obex appear to receive the densest innervation, and lateral and to a lesser extent ventrolateral subdivisions also receive a significant input.

Chemoreceptor afferents In the cat both the carotid sinus nerve (CNS) and aortic depressor nerve (ADN) contain chemoreceptor afferents as well as baroreceptor afferents. Central projection of 13 afferents with carotid body chemoreceptor were investigated using antidromic mapping (Donoghue *et al.*, 1984). They were unmyelinated axons and showed respiratory activity. Their central projections were fairly different from those of arterial baroreceptor. All were activated on stimulation with ipsilateral Sm, and in 12 of these there was an additional projection into the Com. Twenty-five per cent of these cases continued into Com on the contralateral side. There were also four cases of ipsilateral projection to the Slt at levels rostral to the obex.

Afferent interactions in the NTS

As mentioned in the previous section, while IXth and Xth nerve afferents have distinctive patterns of projection to the NTS, specific regions of the nucleus receive inputs from several different afferents. In particular, the Sm at levels rostral to the obex receives a variety of inputs and an extensive innervation. However, there is as yet little information available on the synaptic mechanism of this apparent convergence (Donoghue *et al.*, 1985) and only a limited appraisal of the role of presynaptic interactions has been made (Rudomin, 1967; 1968; Barillot, 1970; Jordan and Spyer, 1979; Ballantyne *et al.*, 1985). On the other hand, pre- and post-synaptic processes have been much investigated regarding interaction between cardiovascular and respiratory reflexes (Daly, 1984).

Presynaptic interaction There is a limited literature indicating that presynaptic interactions between vagal afferent fibers (Rudomin, 1967; 1968). Their interactions have been investigated by measurement of the terminal excitability of vagal afferent to conditioning stimuli delivered to other Xth branches and to stimuli delivered within the NTS (Rudomin, 1967; 1968; Barillot, 1970; Jordan and Spyer, 1979). This is due to the primary afferent depolarization (PAD) of central sensory fiber terminations. PAD occurs in central endings of superior laryngeal nerves; it is mediated by the vagus nerve, and particularly by fibers originating from pulmonary stretch receptors (PSR). Furthermore, PSR fibers themselves are also affected by conditioning stimuli delivered to the superior laryngeal nerve (SLN) and the NTS (Barillot, 1970). The phenomena related with these interactions were reported by Ballantyne *et al.* (1986). They recorded PAD of the slowly adapting lung stretch receptor afferent (SAR) in close proximity to their termination within NTS. Two waves of depolarization, the latency of 3-8 and 27-35 msec, respectively, were evoked by SLN stimulation, long latency response showing considerable sensitivity to the period of the respiratory cycle. These changes in terminal potential of SAR can be considered to be due to presynaptic interaction.

The process of these presynaptic interaction can be evoked by peripheral inputs and also by central inputs related to inspiratory activity and presumably involving the dorsal group of respiratory neurons (von Baumgarten and Kanzow, 1958).

While there have been presented convincing evidence for afferents with ending in the lung and airway receiving presynaptic influences, the same phenomenon has not been seen to occur in CNS and ADN afferent (Rudomin, 1967; 1968; Barillot, 1970; Jordan and Spyer, 1979; Ballantyne *et al.*, 1986). Stimulation of the ADN can evoke presynaptic excitability changes in SLN afferents (Rudomin, 1967) but conditioning stimuli within the NTS do not affect the response in the ADN to a test stimulus within the NTS (Rudomin, 1967; Jordan and Spyer, 1979). While both groups of afferents terminate within the Sm, SARs make monosynaptic connections with R_β neurons whose cell bodies are located in the Svl. Conversely, baroreceptor afferents appear to provide their densest innervation to the NTS in dorsomedial and dorsolateral divisions of the Slt and Sm. The K^+ activity in the vicinity of the dorsal group of respiratory neurons has been shown to change dramatically during the respiratory cycle, with a periodic increase of 0.05-0.2 mM in the Svl during inspiration and of 0.3-0.5 mM on repetitive vagal stimulation (Richter *et al.*, 1978). Such changes could account for the waves of depolarization observed in recording from SARs. However, in the dorsal regions of the NTS described above, there are no measurable changes in K^+ activity.

Electronmicroscopic studies have identified axo-axonic synapse within the Com (Chiba and Doba, 1976) but not intermediate (Chiba and Doba, 1975) regions of the

NTS, and Kalia and Richer (1985b) described SAR terminals, labelled with HRP, which form the postsynaptic element of axo-axonal synapses in the Sv and Svl.

Postsynaptic interaction The afferent inputs of circulatory reflexes, such as arterial baroreceptors, appear to be devoid of presynaptic influence, different from that of respiratory reflexes. However, many of these afferents are known to interact substantially when producing their reflex effects (Heymans and Neil, 1985; Korner, 1979; Daly, 1984). Such interactions at postsynaptic site are still a matter of argument. This problem relates with whether the NTS is simply a relay station for its wide range of afferent input or whether it can contribute to their postsynaptic interaction suitable for their physiological functions, too.

Certainly, the majority of the NTS neurons receive an input for peripheral receptors of one sort or another, e.g. , baroreceptor, chemoreceptor, receptors in the thorax or respiratory tract. However, few are known whether neurons in this region can receive inputs from more than one type of convergence of afferent in the IXth, CSN, ADN and SLN onto medullary neurons. Gabriel and Seller (1970) investigated in detail the time course of the field potential evoked by stimulation of the two sets of baroreceptor nerves. This indicated clearly postsynaptic interactions between these afferents, but, in this method, McAllen (1973) failed to observe neurons in the NTS which received a convergent input from the CSN, ADN or the cardiac branches of the Xth. Later, Ciriello and Calaresu (1980, 1981) have shown a significant population of NTS neurons (27 of 122) which could be activated from both CNS and ADN, these being located primarily in Sm, Svl and Int. In the area outside the NTS, a higher percentage of cells received convergent input (26 of 55).

The difficulty to find out convergent sensory input in many of these studies could be as one reason due to the fact that all these studies used only extracellular recording techniques which can observe only suprathreshold interactions. Donoghue *et al.* (1985) have tried to use the intracellular recording technique, in order to overcome these problems. They showed that predominant effects of stimulation of CSN, ADN or Xth was as excitatory input. These responses showed the postsynaptic, including temporal summation, summation of evoked EPSPs and an after-hyperpolarization. A significant population of neurons had a convergent input from more than one nerve in nearly 25 per cent (10 of 43) of cases.

There are several reports in the literature that activation of sensory fibers can evoke inhibitory responses within NTS. Humphrey and McCall (1984) described some neurons in the ventrolateral NTS whose spontaneous discharge was reduced by repetitive CSN stimulation. A similar inhibitory action on NTS neurons mediated by carotid sinus baroreceptors was confirmed by Lipski and Trzebski (1975). Bennet *et al.* (1982) show that NTS neurons can receive long-lasting inhibitory inputs from myelinated and unmyelinated afferents from the heart and lungs, and Sessle (1973) noted that NTS neurons with SLN or IXth nerve inputs could show depressed responses if conditioning stimuli were applied to either nerve. However, the reports of these inhibitory effects are rare. Extracellular recording can only demonstrate inhibition in spontaneously active neuron or by inhibition of other evoked responses and cannot distinguish between effects mediated pre- and postsynaptically. In the few studies where intracellular recordings have been made, reports of inhibitory postsynaptic responses are more common. Hildebrandt (1974) noted inhibitory postsynaptic potentials (IPSPs) evoked by CSN stimulation in three cells, though they were found within the medial reticular formation. Within NTS, Miura (1975) described IPSPs evoked in neurons by activation of A-fibers in the CSN, and Donoghue *et al.* (1985) observed IPSPs by stimulations of slower-conducting CSN, ADN and Xth afferents. Champagnat *et al.* (1985), using an in vitro rat medullary slice preparations, provided a variety of evidence for synaptic inhibition in NTS neurons evoked by ST stimulation. The majority of recorded neurons showed a prolonged hyperpolarization following the initial EPSP which were found in medial, ventral and ventrolateral parts of the NTS. A smaller number of neurons showed a prolonged depolarization, which were found in the dorsomedial and commissural part of the NTS. The neurons with prolonged hyperpolarization would receive a variety of input, including polysynaptic pathway, and they are supposed to play a important role in regulations of visceral reflexes. This might relate with the phenomenon that there are found very few neurons with pulsatile discharges in the NTS in spite of many obvious pulsatile input of baroreceptor inputs.

MAIN PUTATIVE NEUROTRANSMITTERS IN THE NTS

Catecholamines

The region including the NTS and DMV (dorsal motor nucleus of the vagus nerve) corresponds to the A2 NE- and C2 EPI- containing cell area (Palkovits and Zaborszky, 1977; Spyer, 1982). Further, this region has a high density of catecholaminergic nerve terminals which appear to arise from local circuitry within the Com and NTS (Armstrong, *et al.*, 1981; Sawchenko and Swanson, 1982; Swanson and Hartman, 1975) and adrenergic innervation of the DMV arising from the ventrolateral medulla (Armstrong *et al.*, 1982; Bleesing *et al.*, 1981; Sawchenko and Swanson, 1982).

Many experiments indicate the separate existence of the α_1- and α_2-noradrenergic receptors. In the periphery, it was originally proposed that α_2 receptors were presynaptically located on noradrenergic neuron to inhibit norepinephrine release (Langer, 1977; Berthelsen and Pettinger, 1977). In the brain, some of α_2 receptors may be located presynaptically (Taube *et al.*, 1974; Phillipu *et al.*, 1974). However, it has been shown that some of α_2 receptors may be located postsynaptically because, even after depletion of catecholamine by pretreatment of reserpine, α-methyl-*p*-tyrosine or 6-hydroxydopamine, clonidine produced a decrease in blood pressure and sympathetic discharge similar to that for non-pretreated animals (Haeusler, 1974; Kobinger and Pichler, 1976; Suzuki and Suga, 1980).

Localization of α_1- and α_2-receptors were investigated by microscopic autoradiography labelled by [^{3}H] WB-40101 and *p*-[^{3}H] aminoclonidine (PAC), respectively (Young and Kuhar, 1980). This study revealed a distribution of α receptors throughout the nervous system. Certain areas had elevated levels, including the NTS and locus coeruleus in the medulla oblongata. The distribution of α_1- and α_2-receptors was markedly different in certain areas. One of the examples is the NTS, where a higher density of α_2 receptors was found than α_1 receptors. The observed colocalization of α_2 receptors with opiate receptors, would provide an explanation of the observation that α_2 agonist block opiate withdrawal (Unnerstall *et al.*, 1984; Dashwood *et al.*, 1985). Unnerstall *et al.*(1984) investigated in detail the distribution of α_2 receptor using a specific [^{3}H] PAC binding site. High grain densities were found in nucleus commissuralis, subnucleus medialis and DWV. The lateral site of the TS, area postrema, hypoglossal nucleus were lower grain densities. In the other regions of medulla oblongata, high grain densities were observed in the substantia gelatinosa of the trigeminal nucleus and the posterior raphe pallidus, whereas the rostral raphe pallidus, raphe magnus and inferior olivary nucleus were seen as low grain sensitivities.

Functional studies on the NTS adrenergic neurons

The NTS has been investigated exclusively on the α_2 effect in relation with hypotensive action of clonidine. Although microinjection of α_2 agonist has been reported to produce remarkable hypotension, concerning α_1 agonist, some studies using this technique have reported to fail to elicit apparent cardiovascular responses (DeJong and Petty, 1982; Gurtu *et al.*, 1982). However, one group of investigators has reported an increase of both the blood pressure and heart rate (Kubo *et al.*,1987).

Feldman and Felder (1989) have shown clear results on the responsiveness of α_1- and α_2-receptors of the NTS neurons in slice preparation. Spontaneously active neurons in the NTS without ST input had no apparent response to methoxamine, α_1 agonist, infusion, while silent neurons with ST input were depressed dose-dependently by methoxamine infusion. Spontaneously active neurons with TS input were shown to induce increases of both spontaneous firing rate and responsiveness to ST input. On the other hand, clonidine, α_2 agonist, induced increases in responsiveness to ST input of silent neurons, but decreased both spontaneous activity and responsiveness to ST input of spontaneously active neurons. They presumed that the presence or absence of spontaneous activity in the slice preparation might eventually be used as an index of physiological role played by the individual neurons in the NTS. But it could not be well understood that both the silent and spontaneously active neurons showed similar evoked responses to the ST stimulation in spite of their opposite effects produced by methoxamine, or likewise by clonidine. Moreover, they showed in the previous report (Feldman & Felder, 1988), that all spontaneously firing neurons were depressed by both α_1- and α_2- agonists under condition of *in vivo* experiment. These discrepancies were discussed with regard to the difference in preparation, effects of anesthetics and destruction of local circuitry in slice preparation. From these experiments, however, the possibility of opposite effects between α_1- and α_2-receptor activation could be seen in the NTS neurons. This relation is consistent with the findings that microinjection of α_1 agonist into the NTS elicits a sympathoexcitation (Kubo *et al.*, 1987), while injection of α_2 agonists is associated with sympathoinhibition and parasympathoexcitation (Kubo *et al.*, 1987; Kubo and Misu, 1981).

Table 1. Responsiveness of the NTS neurons to baroreceptor and chemoreceptor stimulation in the rabbit.

	Baro-stim	Chemo-stim	No. of Neurons	Total
Baroreceptive	↑	±	14 (35.0%)	
neuron	↓	±	26 (65.0%)	40 (37.4%)
Chemoreceptive	±	↑		20 (90.9%)
neuron	±	↓	2 (9.1%)	22 (20.6%)
Converged	↑	↑	11 (33.3%)	
neuron	↓	↓	6 (18.2%)	
	↑	↓	1 (3.0%)	
	↓	↑	15 (45.5%)	33 (30.8%)
No response	±	±	12	12 (11.2%)
Total			107	107

↑ : activation, ↓ : depression, ± : no effect

Concerning the precise location of the receptors in the NTS neurons, a presynaptic location for α_2 adrenoceptors in these experiments seems unlikely, because pretreatment with 6- hydroxydopamine failed to affect the hypotensive and bradycardic effects of α_2 agonist injected into NTS (Kubo and Misu, 1981). Thus, α_2 agonist appears to exert its effect on the postsynaptic α_2 receptors. The site of activation of α_1 receptors is even less clear.

In our laboratory, direct effects of α_1- and α_2-agonist on the NTS neurons with baroreceptor and chemoreceptor inputs have been investigated using the technique of microiontophoretic application in the rabbit (Suzuki, 1990; Suzuki *et al.*, 1990). Spontaneously active NTS neurons were examined. Four types of neurons were found, that is, activated and depressed neurons to baro- and chemo-stimulation (Table 1). In 107 NTS neurons, 14 and 26 cells were activated and depressed by baro-stimulation, respectively. The convergence of baro- and chemoreceptor input was also found in 33 cells. Twelve cells did not show any change in their activities. Evoked responses to baro-stimulation, both the activated and depressed responses, were depressed by iontophoretic applications of methoxamine (Fig. 1) and potentiated by bunazosine (Fig. 3), whereas these responses were potentiated by CLO (Fig.2) and depressed by yohimbine (Figs.3, Fig.4). On the contrary, evoked responses to chemo-stimulation were potentiated by methoxamine and depressed by bunazosine, whereas these responses were depressed by clonidine and potentiated by yohimbine. But some chemoreceptive neurons were not affected by these drugs (Fig.4).

These effects of α_1- and α_2-agonists and antagonists seemed to be parallel to the cardiovascular effects produced by these agonists (Kubo and Misu, 1981). However, the neural mechanisms, by which both the activating and depressant responses to baro-stimulation were potentiated by the iontophoretic application of clonidine and,

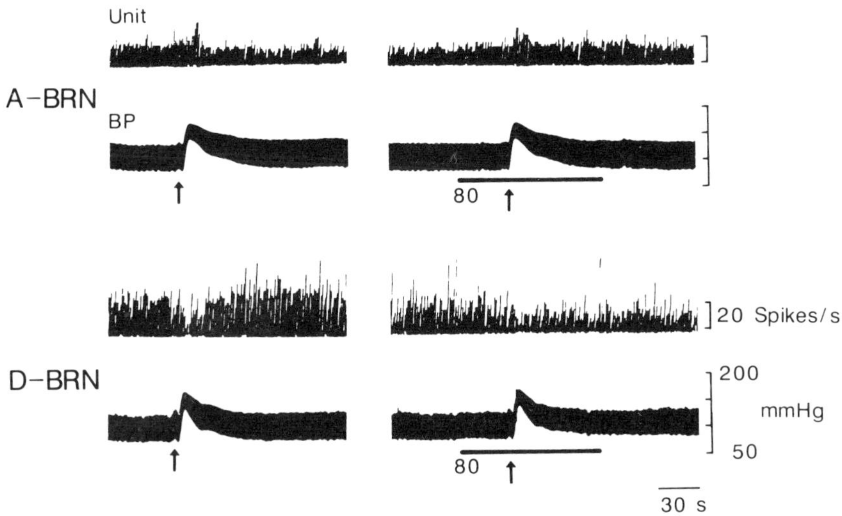

Figure 1. Two types of baroreceptive neurons, namely, activated and depressed neurons to baroreceptor stimulation (A-BRN and D-BRN, respectively) in the rabbit, and effects of methoxamine on them. Microiontophoretic application of methoxamine (80 nA) depressed both responses of A-BRN and D-BRN.

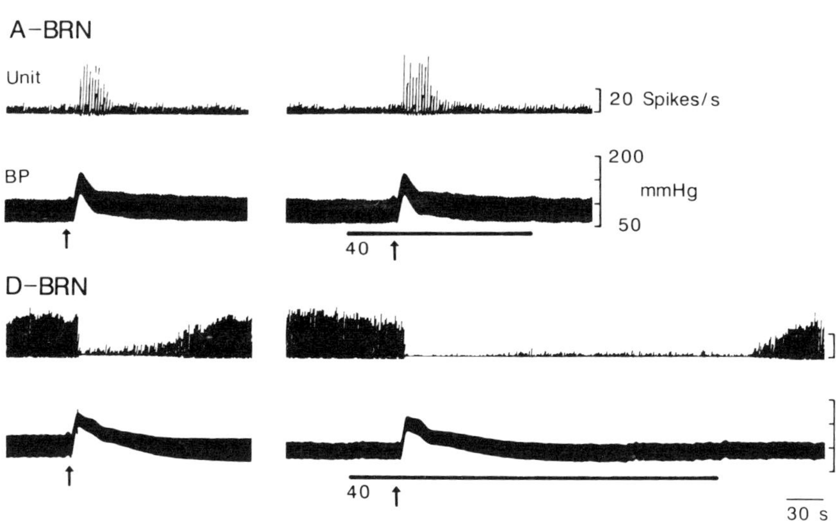

Figure 2. Effects of clonidine on baroreceptive neurons (A-BRN and D-BRN; cf. Fig.1). Microiontophoretic application of clonidine (40 nA) potentiated both responses of A-BRN and D-BRN.

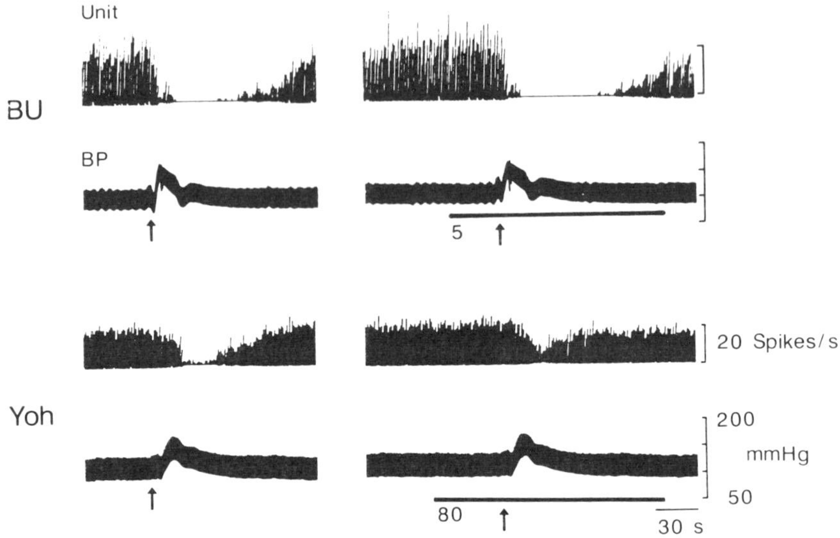

Figure 3. Effects of bunazosin (Bu) and yohimbine (yoh) on baroreceptive neurons. Microiontophoretic application of bunazosin (5 nA) potentiated the response to baroreceptor stimulation, while yohimbine (80 nA) depressed the response.

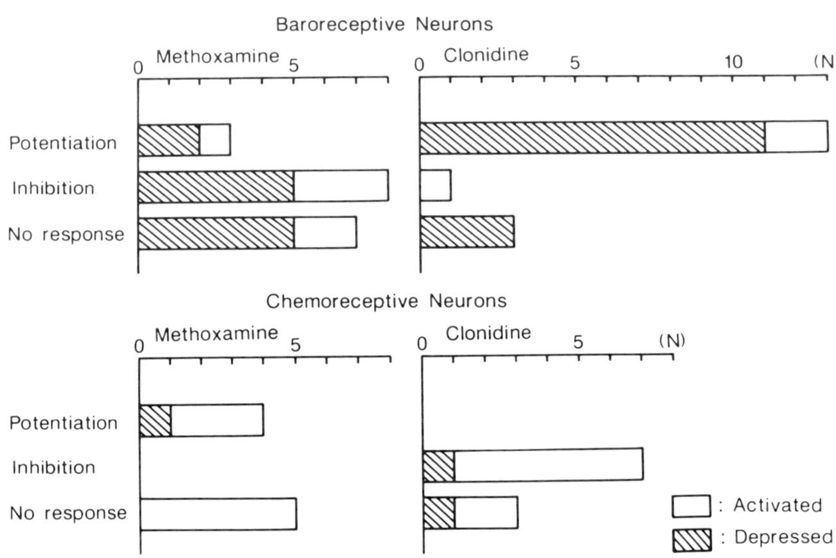

Figure 4. Totalized effects of methoxamine and clonidine on the baroreceptive and chemoreceptive neurons. Clonidine produced potentiation of the response in majority of baroreceptive neurons, while methoxamine produced more inhibition of the response than potentiation. On the contrary clonidine produced inhibition of the response in many chemoreceptive neurons, while methoxamine produced potentiation in about a half of neurons.

likewise, both responses were depressed by methoxamine, were remained to be elucidated.

Similar participation of opioid peptidergic receptors to α_2-adrenergic receptors was found in the neuronal mechanisms of baro- and chemo-receptor reflexes. It seems, however, that the opioid peptidergic system plays a more effective role in the chemoreceptor reflex than the adrenergic system (Suzuki, 1990).

Substance P

Presence of substance P in the NTS have been reported in the rat (Cuello and Kanazawa, 1978; Ljungdahl *et al.*, 1979) and in the cat (Hökfelt *et al.*, 1977). The distribution of substance P within NTS is the greatest and most varied among all the putative neurotransmitters. The highest distributions of substance P is in the medial portion of the lateral subdivision, which corresponds to a region receiving primary afferent via the pharyngeal branch of the IXth nerve (Parneton and Loewy, 1980). Lesser amounts of substance P are located in the commissural, medial, parvocellular and the remainder of the lateral subdivisions. The regions of the least amounts of substance P are the intermediate, ventrolateral and small portion of the parvocellular subdivisions (Maley and Elde, 1982). Possible source of substance P are presumed to be visceral afferent, via IXth and Xth nerves, central afferents to the NTS and neurons containing substance P within the NTS. Several studies have demonstrated neurons containing substance P immunoreactivity within the sensory ganglia of Xth nerve in rat (Pelletier and Dubé, 1978), cat (Hökfelt *et al.*, 1977; Maley and Elde, 1982), pigeon (Lundberg *et al.*, 1978) and guinea-pig (Brimijoin *et al.*, 1980). Further, centrally projecting fibers from the vagal ganglion of rat have been shown to contain substance P (Gillis *et al.*, 1980).

Sensory input from baro- and chemoreceptors terminate in the medial, commissural and a portion of the lateral subdivision (Panneton and Loewy, 1980). These areas have moderate to dense amount of substance P immunroreactivity. This data along with recent physiological evidence (Haeusler and Osterwalder, 1980) suggests that substance P may be a neurotransmitter in the afferent limb of circuits responsible for the modulation or control of cardiovascular functions. In contrast, the ventrolateral and intermediate subdivisions, areas known to receive afferents from the respiratory system (von Euler *et al.*, 1973a), contain the lowest levels of substance P immunreactivity within the NTS. This may indicate that substance P does not play an important role as a neurotransmitter for respiratory afferents (Maley and Elde, 1982).

Substance P immunoreactivity-containing neurons are found in a number of subdivisions including the commissural, intermediate, lateral, medial and parvocellular subdivisions (Armstrong *et al.*, 1981; Ljungdahl *et al.*, 1978). Most of these neurons are small (10-15 μm in diameter) with several primary dendrites arising from the cell body. These neurons may contribute as interneurons within the NTS and projection neurons from the nucleus (Maley and Elde, 1982).

Enkephalin

Existence of enkephalin immunoreactivity in the NTS have been reported in the rat (Finley *et al* ., 1981; Hökfelt *et al.*, 1977), monkey (Haber and Elde, 1981) and the cat (Malery and Elde, 1982). In the cat moderate of enkephalin immunoreactivity are found in the commissural, lateral and the dorsomedial band of the parvocellular subdivisions, while sparse accumulation are present in the medial and the majority of the parvocellular subdivisions. Occasional fibers are seen in the ventrolateral, intermediate and small area of the parvocellular subdivisions. The pattern of enkephalin immunoreactivity is similar to that of substance P immunoreactivity in the NTS, although its density is not as great. A similar relationship between substance P and enkephalin can be seen also in the dorsal horn of the spinal cord. Interaction between substance P and enkephalin is supposed.

The presence of enkephalin immunoreactivity in the areas known to receive cardiovascular afferents (the medial, commissural and parts of the lateral subdivision) suggest a role for enkephalin in central regulation or modulation of cardiovascular function.

The level of enkephalin immunoreactivity are lowest in subdivisions (intermediate and ventrolateral) related to central regulation of respiratory functions (Byritrzycka, 1980; von Euler, 1973a and b). Several investigations have reported that endogenous opioids have an inhibitory effect on the respiratory reflex system (Florez and Mediavilla, 1977; Kumazawa *et al.*, 1980). Indeed, a number of neurons containing enkephalin immuno-reactivity are present in the intermediate and ventrolateral subdivision and these enkephalinergic cells may play a direct role in the control of respiratory function. Regions containing large numbers of enkephalin neurons (the medial and commissural subdivisions) may play an indirect role. These regions have been shown to project to other respiratory areas including the retrofacial, retroambiguous and the ambigous nuclei, which then send axons to spinal cord areas known to be involved in respiratory functions.

Excitatory amino acids

Although numerous putative amino acid as neurotransmitters are estimated in the NTS, the excitatory amino acid L-glutamate has been studied most extensively with respect to its possible involvement in cardiovascular reflex control (Granata *et al.*, 1984; Perrone, 1981; Talman *et al.*, 1984; Miller and Felder, 1988). L-Glutamate microinjection into the NTS in rat causes hypotension and bradycardia, mimicking a baroreceptor response (Florez and Mediavilla, 1977). Antagonists to L-glutamate block the effects of microinjected L-glutamate (Talman *et al.*, 1981) and also block the baroreflex (Guyenet *et al.*, 1987; Humphery and McCall, 1984; Talman *et al.*, 1981). Vagal afferent stimulation causes localized release of L-glutamate in the NTS, and L-glutamate content of NTS is substantially reduced by ablating the nodose ganglia, which contain the cell body of baroreceptor afferent fibers (Perrone, 1988).

Urbanski and Sapru (1988) reported putative neurotransmitters involved in the medullary cardiovascular areas including the NTS, caudal and rostral ventrolateral medulla (cVLM and rVLM, respectively). Microinjection of L-glutamate into the NTS elicits a fall in blood pressure and heart rate in the rat. This depressor response was attenuated dose-dependently by microinjection of kynurate into the cVLM or by microinjection of bicuculline, GABA antagonist, into the rVLM. Following microinjection of kynurate into cVLM, microinjection of L-glutamate elicited conversely a presser response, which was attenuated dose-dependently by microinjection of kynurate into rVLM. These results indicated that: (1) The NTS sends excitatory glutamate inputs to both the rVLM and cVLM; the projection from the NTS to the rVLM mediates presser responses while that from the NTS to the cVLM represents one component of the depressor pathway mediating the depressor responses elicited from the NTS. (2) The pathway from the cVLM to the rVLM is GABAergic and represents second component of the depressor pathway evoked from the NTS. (3) The bradycardia evoked from the NTS may involve a pathway from the NTS to the cVLM.

In recent years, 3 major specific subtypes of excitatory amino acid receptors have been identified throughout the central nervous system, that is, N-methyl-D-asparatate (NMDA)-quisquarate- and kainate-receptors (Fagg, 1985; Krogsgaard-Larsen and Hornore, 1983; McLennan, 1983; Watkins, 1981). These 3 receptors were demonstrated in the NTS using 3 receptor antagonists, that is 2-amino-5 phosphonovalerisc acid (APV), kynreic acid (KYN) and glutamate diethylester (GDEE) (Miller and Felder, 1988). APV only have specificity for the NMDA receptor. KYN and GDEE are only relatively specific receptor antagonist, and GDEE has some anesthetic properties (Mayer and Westbrook, 1987). Miller and Felder (1988) investigated these three excitatory amino acid receptors in the synaptic transmission of the NTS on the rat's brain slice preparation, by examining effects of above-mentioned 3 antagonists on the evoked responses of single NTS neurons to ST stimulation. They obtained the following results: KYN, NMDA- and Kainate-preferring receptor blocker, had blocking activity comparable to those of APV, some of these inputs were also blocked by GDEE, a relatively selective quisqualate receptor antagonist. The results indicate that all 3 excitatory amino acid receptors can selectively participate in electrically evoked afferent input to the NTS and suggest a role in mediating input to these neurons over different pathways.

γ-Aminobutylic acid (GABA)

There is a large amount of GABA and specific GABA receptor in the NTS (Gale *et al.*, 1980), in the intermediate region in which baroreceptor afferents mainly terminate. Local injection of GABA or GABAergic agonist, muscimol, in that region of the cat produced hypertension and bradycardia, whereas the GABAergic antagonist, bicuculline, had opposite effects and prevented those of muscimol. Thus it seems that a GABAergic neuronal system modulates the activity of the baroreceptor reflex pathway within the NTS, probably by inhibiting sympatho-inhibitory and vago-excitatory neurons (Bousquet *et al.* , 1982). In the vasopressor response to GABA injection into NTS, the release of vasopressin is involved, too (Catelli *et al.* , 1987). Injection of GABA uptake inhibitor, nipecotic acid into the NTS increases artrial pressure. This effect of nipecotic acid was not antagonized by the selective $GABA_A$ receptor antagonist, bicuculline, but totally reversed by newly described $GABA_B$ antagonist, phaclofen injected into the NTS. From these results it is suggested that GABA tonically released in the NTS acts specially on $GABA_B$ receptors to increase blood pressure (Sved and Sved, 1990). It has been known that stimulation of the hypothalamic defense area (HDA) in cats suppresses the baroreceptor reflex (Coote *et al.*, 1979). Stimulation within the HDA produces inhibition of tonic and responsive activity of NTS neurons receiving a baroreceptor input, and the transmitter for this inhibition is estimated to be GABA, because GABAergic antagonists antagonize this inhibitory response elicited by HDA stimulation (Mifflin *et al.*, 1988; Jordan *et al.*, 1988).

Serotonin
Serotonergic fibers have been reported in rat and cat (Dahlström and Fuxe, 1964; Palkovits *et al.*, 1974; Steinbusch, 1981; Maley and Elde, 1982). The origins of these serotonin immunroreactive fibers in the NTS are the raphe nuclei (Bobillier *et al.*, 1976) and nodose ganglia.

The greatest amounts of serotonin immunoactivity are located in portions of the commissural and dorsomedial band of the parvocellular subdivision. Moderate accumulations are found in the remainder of the commissural, lateral and medial subdivisions, while in the ventrolateral and patches of the parvocellular subdivisions only occasional fibers are present (Maley and Elde, 1982).

Although the physiologic roles of the serotonergic system remains to be investigated, the hypotension was elicited by administration of serotonergic agonists into the NTS (Laguzzi *et al.*, 1984; Shvaloff and Laguzzi, 1986) or into the cisterna magna (Kristic and Djurkovic, 1980) and the hypertension observed after the microinjection of 5,7-dihydroxytryptamin into the NTS or in the nodose ganglion (NG) (Nosjean *et al.*, 1987) suggest that the NG-NTS serotonergic system plays a depressor role in blood pressure regulation (Nosjean *et al.*, 1990). Serotonergic projection from the NG to the NTS were evidenced by serotonin-immunohistochemical method (Nosjean *et al.*, 1990).

CLONIDINE ACTION

Clonidine (CLO) is an effective antihypertensive drug that elicits its effects by acting at distinct α_2-adrenergic receptors in the central nervous system (Anden *et al.*, 1970; Haeusler, 1974; Kobinger, 1967; Sattler and Van Zweiten, 1967; Schmitt *et al.*, 1967; Toda *et al.*, 1969; Suga and Suzuki, 1979). In addition to this effect, CLO produces other several pharmacological effects, such as anxiolytic (Svensson and Strombom, 1975; Redmond and Huang, 1979), anticonvulsant (Papanicolou *et al.*, 1982), endocrine (Eriksson *et al.* , 1981; Rudoleph *et al.* , 1980), analgesic (Schmitt *et al.*, 1974; Paalzow, 1974) and opiate withdrawal-attenuating effects (Tseng *et al.*, 1975; Aghajanian *et al.*, 1978). Thus, CLO not only affect autonomic function, but also somatosensory and affective functions. And it is due to the anatomical relationships among these functions and the participation of adrenergic system (Dahlström and Fuxe, 1964; Levitt and Moore 1979; Palkovits *et al.*, 1977; Sawchenko and Swanson, 1982; Ungelstedt, 1971), especially of α_2 adrenergic receptors in these functional structures, where CLO produces these varieties of pharmacological actions.

Cardiovascular action of clonidine
One site where CLO exerts its major antihypertensive effect is the region including the NTS and DWN. The NTS receive baro- and chemo- receptor information via the IXth and Xth cranial nerves and forms the first synapse in the reflex loop (Palkovits and Zaborszky, 1977; Spyer, 1982). This region corresponds with A2 norepinephrine-

and C2 epinephrine cell group area (Armstrong *et al.*, 1982; Levitt and Moore, 1979; Swanson and Hartman, 1975). Further, this region has a high density of catecholaminergic nerve terminals which appear to arise from the NTS, and adrenergic innervation of the DMV arising from ventrolateral medulla (Blessing *et al.*, 1981; Sawchenko and Swanson, 1982; Ungelstedt, 1971). The correlation between the density of adrenergic innervation and the presence of α_2 receptors in this region was confirmed (Pearson *et al.*, 1983).

Adrenergic neurons in the NTS are not an integral part of the primary reflex loop; rather they seem to modulate neural transmission in the reflex pathway (Haeusler, 1977; Kobinger and Walland, 1973). Microinjection of CLO into the NTS produce hypotension by acting α_2 receptors (DeJong, 1974; Kubo and Misu, 1981; Nijikamp and DeJong, 1975), whereas the primary α_2 effect within the medial NTS and DMV is the induction of bradycardia via facilitation of transmission through the baroreceptor reflex arc (Kobinger and Walland, 1971; Haeusler, 1973; Kubo and Misu, 1981). Electrolytic lesions of the NTS in the normal rat, dog and cat which abolish reflex bradycardia and produce lethal fulminating hypertension in the rat and chronic labile hypertension in cats, generally attenuate the bradycardia but not hypotensive effects of systemic CLO (Laubie *et al.*, 1976; Antonaccio and Halley, 1977). Thus, it would appear that the dorsal motor complex may be a primary site of action of CLO for the parasympathetic activation, while for attenuation of sympathetic outflow other brain regions seems to be involved additionally.

Cell groups in the ventrolateral medulla (VLM) form the second relay area in the main baroreceptor reflex pathway. Electrical or chemical stimulation of the rVLM, which includes the C1 cell group, increases sympathetic outflow, while lesion of this area produces hypotension (Dampney and Moon, 1980; Ross *et al.*, 1983). On the other hand, lesions of the caudal VLM, which includes the A1 noradrenergic cell groups and corresponds to the cardiovascular inhibitory area, produces fulminating hypertension (Blessing *et al.*, 1981; Blessing and Reis, 1982). CLO injected into the rVLM produces hypotension (Bousquet *et al.*, 1981). Further, chemical stimulation of the ventrolateral area at the pontinemedullary junction, a region which includes the A5 noradrenergic cell group, decreases blood pressure and heart rate, a response that can be abolished by pretreatment which 6-hydroxydopamine (Neil and Loewy, 1982). Intophoresis of CLO onto baroreceptive neurons in this region can depress the firing rate of these neurons (Andrade and Aghajanian, 1982). A diffuse band of [^{3}H] PAS binding can be seen to extend from the dorsomedial medulla over the lateral reticular formation to the ventrolateral medullary cell groups (Unnerstall *et al.*, 1984). This band is contiguous with the known pathways between the dorsal and ventrolateral medulla, including the parasympathetic motor nuclei (Palkovits and Zaborskzy, 1977; Pearson *et al.*, 1983; Sawchenko and Swanson, 1982). Thus, the distribution of α_2 binding sites in the ventrolateral medulla correlates with the known action of CLO in this region and appears to be associated with parasympathetic motor neurons that form this part of the baroreceptor reflex loop. On the other hand, NE and EPI neurons of A1 and C1 cell groups have been shown to innervate the A5 regions (Sawchenko and Swanson, 1982), where α_2 binding site density is higher.

α_2-Agonist can exert a sympathoinhibitory effect mainly by acting on neurons within the VLM. However, the spinal cord also may be a primary site for α_2 mediated sympathoinhibition. The medullary C1 EPI- and A5 NE-neurons provide the dense adrenergic innervation to the sympathetic preganglionic neurons located in the intermediolateral nucleus (IML) column of the thoracic spinal cord (Loewy *et al.*, 1981; Westland *et al.*, 1983). Electrical or chemical stimulation of the ventrolateral medulla in the area of catecholamine cell bodies usually increases sympathetic activity, while CLO into the IML reduces the firing rate of sympathetic preganglionic nerves via action at α_2 receptors (Coote and Macleod, 1974; Haeusler, 1974; Guynet and Cabot, 1981; Haeusler, 1974). However, in addition to the increase in sympathetic activity following stimulation of EPI-containing neuron area, a support for a sympathetic excitatory role of EPI comes from in vitro studies. Yoshimura *et al.* (1986), using intracellular recording technique, have shown an increase in neuronal excitability and a potential for repetitive discharge from sympathetic preganglionic neurons after application of EPI. Therefore, concerning the observed excitatory or inhibitory effects of EPI, it might be due to the different sites of release of EPI onto the sympathetic preganglionic neuron, because EPI-containing nerve terminals have been demonstrated to make synapses with the preganglionic neurons both on perikarya and dendrites (Milner *et al.*, 1988). Minson *et al.* (1990), using a combination of immunohistochemistry of phenylethanolamine-N-methyltransferase and of retrograde transport method, demonstrated, in the rat, the spinal projection of the EPI-containing neurons located in three medullary areas, namely, C1, C2 and C3 groups, the last of which corresponded to the area of the fasciculus longitudialis medialis and its vicinity. Since the projection of these neurons are directed toward the intermediolateral cell column, it seems likely that all three groups of medullary EPI-containing neurons contribute to the regulation of sympathetic outflow.

The higher brain centers exert a powerful influence on the circulatory reflex loop in the brainstem (Palkovits *et al.*, 1977), as it is known that full expression of CLO's hypotensive action needs intact forebrain-brainstem connection (Tadepalli and Mills, 1978). Several forebrain regions have been shown to participate in regulation of cardiovascular function. These regions include the locus coeruleus (Ward and Gunn, 1976), posterior hypothalamic area (Enoch and Kerr, 1967) paraventricular and periventricular hypothalamic nuclei (Ciriello and Calaresu, 1980; Harris, 1979), anterior hypothalamic area (Brody, 1981; Henke, 1982), bed nucleus of the stria terminalis (Struyer-Boudier *et al.*, 1974), central nucleus of the amygdala (Kapp *et al.*, 1981), septum and hippocampus (Lofring, 1961), and cingulate (Korner, 1971) and insular cortex (Saper, 1982). And there exists a strong correlation between the distribution of α_2 binding sites in the forebrain and the terminal fields of ascending NTS, COM and VLM projections (Loewy *et al.*, 1981; Otterson, 1981). These pathways via α_2 adrenergic neurons involving the transmission of visceral information to the forebrain could be modulated by α_2 agonists.

Endocrine action of clonidine

Endocrine functions can be stimulated by α_2 agonists (Eriksson *et al.* , 1981; Schaub *et al.*, 1980). Intraventricular administration of CLO decreases plasma levels of

ACTH (Ganong, 1977). Systemic CLO administration can also inhibit vasopressin secretion (Schmitt, 1977). In an isolated median eminence preparation CLO released luteinizing-hormone releasing hormone, acting at α_2 receptor (Negro-Vilar, 1982).

Anxiolytic action of clonidine

Anxiolytic action of CLO are concerned with its cardiovascular effect in the forebrain since the organic response to threating stimuli involves the simultaneous expression of both autonomic and behavioral functions (Folkow, 1982; Henke, 1982). The integration of such adaptive responses is mediated by the simultaneous involvement of several neural sites within the cortex, limbic system, hypothalamus and brain stem that are interconnected by complex bidirectional circuitry (Swanson and Mogenson, 1981; Nauta and Domesick, 1982) and contain NE- and EPI-mediated mechanisms (Mogenson *et al.*, 1980; Weiss *et al.*, 1982). CLO could exert its anxiolytic action at several sites; nucleus of the stria terminalis and insular cortex could be primary sites for the effects (Redmond and Huang, 1979; Saper, 1982).

REFERENCES

Aghajanian, G. K. (1987). Tolerance of locus coeruleus neurons to morphine and suppression of withdrawal response by clonidine. *Nature* **276,** 186-188.

Anden, N. E., Dahlström, A., Fuxe, K., Larsson, K., Hökfelt, B., Hökfelt, T., Riden, C. and Svensson, T. (1970). Evidence for a central noradrenaline receptor stimulation by clonidine. *Life Sciences* **9,** 513-523.

Andrade, R. and Aghajanian, G. K. (1982). Single cell activity in the noradrenergic A5 region: Responses to drugs and peripheral manipulation of blood pressure. *Brain Res.* **242,** 125-135.

Antonaccio, M. J. and Halley, J. (1977). Clonidine hypotension: lack of effects of bilateral lesions of nucleus solitary tract in anesthetized cats. *Neuropharmacology* **16,** 431-433.

Armstrong, D. M., Pickel, V. M., Joh, T. H., Reis, D. J. and Miller R. J. (1981). Immunocytochemical localization of catecholamine synthesizing enzymes and neuropeptides in area postrema and medical nucleus tractus solitarius of rat brain. *J. Comp. Neurol.* **196,** 505-517.

Ballantyne, D., Jordan, D., Messmann, M., Richter, D. W. and Spyer, K. M. (1986). Presynaptic depolarization in myelinated vagal afferent fibers terminating in the nucleus of the tractus solitarius in the cat. *Pflügers Arch.* **406,** 12-19.

Barillot, J. C. (1970). Dépolarization présynaptique des fibers sensitives vagales et laryngées. *J. Physiol. (Paris).* **62,** 273-294.

Beckstead, R. M. and Norgren, R. (1979). An autoradiographic examination of the central distribution of the trigeminal, facial, glossopharyngeal and vagal nerves in the monkey. *J. Comp. Neurol.* **184,** 455-472.

Bennet, J. A., Goodchild, C. S., Kidd, C. and McWilliam, P. N. (1982). Inhibition of evoked and spontaneous activity of neurons in the nucleus of the tractus solitarius and dorsal motor vagal nucleus by cardiac and pulmonary afferent fibers in the cat. *J. Physiol.* **332,** 77P.

Berthelsen, S. and Pettinger, W. A. (1977). A functional basis for classification of α–adrenergic receptors. *Life Sciences* **21,** 595-606.

Bleesing, W. W., Furness, J. B., Costa, M., West, M. J. and Chalmers, J. P. (1981). Projection of ventrolateral medullary (A1) catecholamine neurons toward nucleus tractus solitarii. *Cell Tissue Res.* 220, 270-340

Bleesing, W. W. and Reis, D. J. (1982). Inhibitory cardiovascular function of neurons in the caudal ventrolateral medulla of the rabbit: relationship to the area containing noradrenergic neurons. *Brain Res.* **253,** 161-171.

Bobillier, P., Seguin, S., Petitjean, F., Salvert, D., Touret, M. and Jouvet, M. (1976). The raphe nuclei of the cat brainstem: a topographical atlas of their efferent projections as revealed by autoradiography. *Brain Res.* **113,** 449-486.

Bousquet, P., Feldman, J., Block, R and Schwartz, J. (1981). The nucleus reticularis lateralis: a region highly sensitive to clonidine. Europ. *J. Pharmacol.* **69,** 389-392.

Bousquet, P., Feldman, J., Bloch, R. and Schwartz, J. (1982). Evidence for a neuromodulatory role of GABA at the first synapse of the baroreceptor reflex pathway. Effects of GABA derivatives injected into the NTS. *Naunyn-Schmiedeberg's Arch. Pharmacol.* **319,** 167-171.

Brimijoin, S., Lundberg, J. M., Hökfel, T., Brodin, E. and Nisson, G. (1980). Axonal transport of substance P in the vagus and sciatic nerves of the guinea-pig. *Brain Res.* **191,** 443-457.

Byritrzycka, E.K. (1980). Afferent projections to the dorsal and ventral respiratory nuclei in the medulla oblongata of the cat studied by the horseradish peroxidase technique. *Brain Res.* **185**, 59-66.

Catelli, J. M., Giakas, W. J. and Sved, A. F. (1987). GABAergic mechanisms in the NTS alter blood pressure and vasopressin release. *Brain Res.* **463,** 279-289.

Champagnat, J., Siggins, G.R., Koda, L.Y. and Denavit-Saubié, M. (1985). Synaptic responses of neurons of nucleus tractus solitarius in vivo. *Brain Res.* **325,** 49-56.

Chernickey, C. L., Barnes, K. L., Ferrario, C. M. and Conomy, J. P. (1984). Afferent projections of the cervical vagus and nodose ganglion in the dog. *Brain Res.Bull.* **13,** 401-411.

Chiba, T. and Doba, N. (1975). The synaptic structure of catecholaminergic axon varicosities in the dorso-medial portion of the nucleus tractus solitarius of the cat: Possible roles in the regulation of cardiovascular reflexes. *Brain Res.* **84,** 31-46.

Chiba, T. and Doba, N. (1976). Catecholaminergic axo-axonic synapses in the nucleus of the tractus solitarius (pars commissuralis) of the cat: possible relation to presynaptic regulation of baroreceptor reflexes. *Brain Res.* **102,** 255-265.

Ciriello, J. and Calares, F. R. (1980). Monosynaptic pathway from cardiovascular neurons in the nucleus tractus solitarii to the paraventricular nucleus in the cat. *Brain Res.* **193,** 529-533.

Ciriello, J. and Calaresu, F. R. (1981). Projection from buffer nerves to the nucleus of the solitary tract: an anatomical and electrophysiological study in the cat. *J. Auton. Nerv. Systems* **3,** 299-310.

Ciriello, J., Hrycyshyn, A. W. and Calaresu, F. R. (1981). Glossopharyngeal afferent projections to the brainstem of the cat : a horseradish peroxidase study. *J. Auton. Nerv. System.* **4,** 63-79.

Contreras, R. J., Beckstead, R. M. and Norgren, R. (1982). The central projection of the trigeminal, facial, glossopharyngeal and vagus nerves : an autoradiographic study in the rat. *J. Auton. Nerv. System.* **6,** 303-322.

Coote, J. H., Hilton, S. M. and Peres-Gonzalez, J. F. (1979). Inhibition of the baroreceptor reflex in stimulation in the brainstem defense area. *J. Physiol.* **288,** 549-560.

Cottle, M. K. (1964). Degeneration studies of primary afferents of IXth and Xth cranial nerves in the cat. *J. Comp. Neurol.* **178,** 129-156.

Cuallo, A. C. and Knazaw, I. (1978). The distribution of substance P immunoreactive fibers in the rat central nervous system. *J. Comp. Neurol.* **179,** 129-156.

Culberson, J. L. and Kimmel, D. L. (1972). Central distribution of primary afferent fibers of the glossopharyngeal and vagal nerves in the opossum *Didelphis Virginiana. Brain Res.* **44,** 325-335.

Daly, M.De.B. (1984). Breath-hold diving : Mechanisms of cardiovascular adjustments in the mammal, Baker P. F. (Ed), Recent Advances in Physiology, Vol 10, Churchill Livingstone, Edinburgh, pp 201-245.

Dampney, R.A.L. and Moon, E. A. (1980). Role of ventrolateral medulla response to cerebral ischemia. *Am. J. Physiology.* **239,** H349-H358.

Dashwood, M. R., Gilbey, M. P. and Spyer, K. M. (1985). The localization of adrenoceptors and opiate receptors in regions of the cat central neurons system involved in cardiovascular control. *Neuroscience.* **15,** 537-551.

DeJong, W. (1974). Noradrenaline: Central inhibitory control of blood pressure and heart rate.

Europ. J. Pharmacol. 29, 179-180.
DeJong, W. and Petty, M. (1982). Chemical stimulation of the nucleus of the solitary tract and the resulting blood pressure response. *J. Cardiovasc. Pharmacol.* **4** (Suppl 1), S77-S80.
Donoghue, S., Garcia, M., Jordan, D. and Spyer, K. M. (1982). Identification and brainstem projection of aortic baroreceptor afferent neurones in nodose ganglia of cats and rabbits. *J. Physiol.* **322,** 337-352.
Donoghue, S., Felder, R.B., Jordan, D. and Spyer, K.M. (1984). The central projection of carotid baroreceptors and chemoreceptors in the cats: a neurophysiological study. *J. Physiol.* **347**, 397-409.
Donoghue, S., Felder, R. B., Gilbey, M. P., Jordan, D. and Spyer, K. M. (1985). Postsynaptic activity evoked in the nucleus tractus solitarius by carotid sinus and aortic nerve afferents in the cat. *J. Physiol.* **360,** 261-273.
Dubbeldam, J. L., Brus, E. R., Menken, S.R.J. and Zeilstra, S. (1979). The central projection of the glossopharyngeal and vagus ganglia in the mallard (*Anas platyrhynchos*). *J. Comp. Neurol.* **183,** 149-168.
Enoch, D.M. and Kerr, F.W. (1967). Hypothalamic vasopressor and vasodepressor pathways. I. Functional studies. *Arch. Neurol.* **16**, 290-306.
Eriksson, E., Eden, S. and Modigh, K. (1981). Importance of norepinephrine alpha-2 receptor activation for morphine-induced rat growth hormone secretion. *Neuropharmacology* **33**, 91-96.
Fagg, G. E. (1985). L-Glutamate, excitatory amino acid receptors and brain function. *Trends Neuroscience* **8,** 207-210.
Feldman, P. D. and Felder, R. B. (1988). Electrophysiological evidence for α_1- and α_2-adrenoceptors in solitary tract nucleus. *Am. J. Pysiol.* **254,** H756-H762.
Feldman, P. D. and Felder, R. B. (1989). α-Adrenergic influences on neuronal responses to visceral afferent input in the nucleus tractus solitarius. *Neuropharmacology* **10,** 1081-1087.
Finley, J. C., Maderdrut, J. L. and Petrusz, P. (1981). The immunocytochemical localization of enkephalin in the central nervous system of the cat. *J. Comp. Neurol.* **198,** 541-565.
Florez, J. and Mediavilla, A. (1977) Respiratory and cardiovascular effects of met-enkephalin applied to the ventral surface of the brain stem. *Brain Res.* **138,** 585-590.
Folkow, B. (1982). Physiological aspects of primary hypertension. *Physiol. Rev.* **62**, 347-504.
Gabriel, M. and Seller, H. (1970). Interaction of baroreceptor afferents from carotid sinus and aorta at the nucleus tractus solitarii. *Pflügers Arch.* **356,** 65-78.
Gale, K., Hamilton, B. L., Brown, S. C., Norman, W. P., Dias Souza, J. and Gills, R. A. (1980). GABA and specific GABA binding sites in brain nuclei associated with vagal outflow. *Brain Res. Bull.* **5**(Suppl 2), 325-328.
Ganong, W. F. (1977). The renin-angiotensin system and the central nervous system. *Fed. Proc.* **36,**1711-1755.
Gillis, R. A., Helke, C. J., Hamilton, B. L., Norman, W. P. and Jacobowitz, D. M. (1980). Evidence that substance P is a neurotransmitter of baro- and chemoreceptor afferents in nucleus tractus solitarius. *Brain Res.* **181,** 476-481.
Granata, A. R., Sved, A. F and Reis, D. J. (1984). In vivo release by vagal stimulation of L-[^{3}H] gultamine. *Brain Res. Bull.* **12,** 5-9.
Gurtu, S., Sinha, J. N. and Bhargava, R. P. (1982). Involvement of α 2-adrenoceptors of nucleus tractus solitarius in baroreflex mediated bradycardia. *Naunyn-Schmiedeberg's Arch. Pharmacol.* **321,** 38-43.
Guyenet, P. G., Filtz, T. M. and Donaldson, S. R. (1987). Role of excitatory amino acids in rat vagal and sympathetic baroreflexes. *Brain Res.* **407,** 272-284.
Gwyn, D. G. and Leslie, R. A. (1979). A projection of vagus nerve to the area subpostrema in the cat. *Brain Res.* **161,** 335-341.
Gwyn, D. G., Leslie, R. A. and Hopkins, D. A. (1979). Gastric afferents to the nucleus of the solitary in the cat. *Neurosci. Lett.* **14,** 13-17.
Haeusler, G. (1973). Activation of the central pathway of the baroreceptor reflex, a possible mechanism of the hypotensive action of clonidine. *Naunyn-Schmiedeberg's Arch. Pharmacol.* **278,** 231-246.

Haeusler, G. (1974). Further similarities between the action of clonidine and a central activation of the depressor baroreceptor reflex. *Naunyn-Schmiedeberg's Arch. Parmacol.* **285,** 1-14.
Haeusler, G. and Osterwalder, R. (1980). Evidence suggesting a transmitter or neuromoduratory role for substance P at the first synapse of the baroreceptor reflex. *Naunyn-Schmiedeberg's Arch. Pharmacol.* **314,** 111-121.
Harris, M. C. (1979). Effect of chemoreceptor and baroreceptor stimulation on the discharge of hypothalamic supraoptic neurones in rats. *J. Endocrinol.* **82,** 115-125.
Helke, C. ., O'Donohue, T. C. and Jacobowitz, D. M. (1982). Substance P as a baro- and chemoreceptor afferent neurotransmitter : immunocytochemical and neurochemical evidence in the rat. *Peptides* **1,** 1-9.
Henke, P. G. (1982). The telencephalic limbic system and experimental gastric pathology : a review. *Neurosci. Biobehav. Rev.* **6,** 381-390.
Heymans, C. and Neil, E. (1985). Reflexogenic area of the cardiovascular system. Churchill, London.
Hildebrandt, J. R. (1974). Central connection of the aortic depressor and carotid sinus nerve. *Exp. Neurol.* **45,** 590-605.
Hökfelt, T., Kellerth, J. O., Nisson, G. and Pernow, B. (1975). Experimental immunohistochemical studies on the localization and distribution substance P in cat primary sensory neurons. *Brain Res.* **100,** 235-25.
Hökfelt, T., Ljungdahl, A., Terenius, L., Elde, R. and Nilsson, G. (1977). The distribution of enkepharin-immunoreactive cell bodies in the rat central nervous system. *Neurosci. Lett.* **14,** 55-60.
Humphrey, S. J. and McCall, R. B. (1984). Evidence that L-glutamic acid mediates baroreceptor function in the rat. *Clin. Exp. Hypertens.* **A6,** 1311-1329.
Ingram, W.R. and Dawkins, E.A. (1945). The intramedullary course of afferent fibers of the vagus in the cat. *J.Comp. Neurol.* **82,** 157-168.
Jordan, D. and Spyer, K. M. (1979). Studies on the excitability of sinus nerve afferent terminals. *J. Physiol.* **279,** 123-134.
Jordan, D., Mifflin, S. W. and Spyer, K. M. (1988). Hypothalamic inhibition of neurones in the nucleus tractus solitarius of the cat in GABA mediated. *J. Physiol.* 389-404.
Jordan, D. and Spyer , K. M. (1986). Brainstem integration of cardiovascular and pulmonary afferent activity. *Prog. Brain Res.* **67,** 295-301.
Kalia, M. and Mesulam, M. M. (1980a). Brainstem projections of sensory and motor components of the vagus complex in the cat : I. the cervical vagus and nodose ganglion. *J. Comp. Neurol.* **193,**435-465.
Kalia, M. and Mesulam, M. M. (1980b). Brainstem projection of sensory and motor components of the vagus complex in the cat : III. Laryngeal, tracheobronchial, pulmonary, cardiac and gastrointestinal branches. *J. Comp. Neurol.* **193,** 467-568.
Kalia, M. and Richter, D. M. (1985b). Morphology of physiologically identified slowly adapting lung stretch receptor afferents stained with intra-axonal horseradish peroxidase in the nucleus of the tractus solitarius of the cat. II. An ultrastructural analysis. *J. Comp. Neurol.* **241,** 521-535.
Kapp, B. C., Gallagher, M., Frysinger, R. C. and Applegate, C. D. (1981). The amygdala, emotion and cardiovascular conditioning. In Ben-Avii Y. (Eds), The Amygdaloid Complex, Elsevier North-Holland, Amsterdam ; pp. 355-366.
Katz, D. M. and Karten, H. J. (1983). Visceral presentation within the nucleus of the tractus solitarius in the pigeon (Columba Livia). *J. Comp. Neurol.* **218,** 42-73.
Kerr, F.W.L. (1962). Facial, vagal and glossopharyngeal nerves in the cat. *Arch Neurol.* **6,** 264-281.
Kimmel, D. L. and Kimmel, C. B. (1964). Spinal distribution of glossogharyngeal and vagal afferent fibers in the cat, rat and guinea-pig. *Anat. Rec.* **148,** 299-300.
Kobinger, W. (1967). Über den Wirkungsmechnismus einer neuen antihypertensiven substanz mit imidazolinstruktur. *Naunyn-Schmiedeberg's Acth. Pharmacol.* **258,** 48-58.
Kobinger, W. and Walland, A. (1971). Involvement of adrenergic receptors on central vagus activity. *Europ. J. Pharmacol.* **16,** 155-162.

Kobinger, W. and Walland, A. (1973). Modulation effects of central adrenergic neurones on a vagally mediated cardioinhibitory reflex. *Europ. J. Pharmacol.* **22,** 344-350.

Krogsgaard-Larsen, P. and Honre, T. (1983). Gultamate receptors and new gultamate agonists. *Trends Pharmacol.* **4,** 31-33.

Korner, P. I. (1971). Integrative neural cardiovascular control. *Physiol. Rev.* **51,** 312-36.

Korner, P. I. (1979). Central nervous control of autonomic cardiovascular function. In Berne R. M. Sperelakis N. and Geiger SR. (Eds) : *The cardiovascular system*, vol. 1, American Physiological Society, Washington, DC ; pp. 691-739.

Kristric, M. and Djurkovic, D. (1980). Analysis of cardiovascular responses to central administration of 5-hydroxytryptamine in rats. *Neuropharmacol.* **19,** 455-463.

Kubo, T. and Misu, Y. (1981). Pharmacological characterization of the alpha-adrenoreceptors responsible for a decrease of blood pressure in the nucleus tractus solitarii of the rat. *Naunyn-Schmiedebergs Arch. Pharmacol.* **317,** 120-125.

Kubo, T., Kihara, M., Hata, H. and Misu, Y. (1987). Cardiovascular effects of $alpha_1$- and $alpha_2$-adrenergic agents injected into the nucleus tractus solitarii in rats. *Naunyn-Schmiedeberg's Arch. Pharmacol.* **335,** 274-277.

Kumazawa, T., Tadak, E. and Kim, K. (1980). A possible participation of endogenous opiates in respiratory reflexes induced by thin-fiber muscular efferents. *Brain Res.* **190,** 244-248.

Laguzzi, R., Reis, D. J. and Talman, W. T. (1984). Modulation of cardiovascular and electrical activity through serotonergic mechanisms in the nucleus tractus solitarius of the rat. *Brain Res.* **304,** 321-328.

Langer, S. Z. (1977). Presynaptic receptors and their role in the regulation of transmitter release. *Br. J. Pharmacol.* **60,** 481-497.

Laubie, M., Schmitt, H. and Drouillat, M. (1976). Action of clonidine on the baroreceptor pathway and medullary sites mediating vagal bradycardia. *Europ. J. Pharmacol.* **38,** 293-303.

Lesllie, R. A., Gwyn, D. G. and Hopkins, D. A. (1982). The ultrastructure of the subnucleus geratinosus of the nucleus of the tractus solitarius in the cat. *J. Comp.Neurol.* **206,** 109-118.

Levitt, P. and Moore, R. Y. (1979). Origin and organization of brainstem catecholamine innervation in the rat. *J. Comp. Neurol.* **186,** 505-528.

Lipski, J. and Trzebski, A. (1975). Bulbospinal neurons activated by baroreceptor afferents and their possible role in inhibition of preganglionic sympathetic neurons. *Pflügers Arch.* **356,** 181-192.

Ljungdal, A., Hökfelt, T., Nilsson, G., Rehnfet, J., Elde, R. and Said, S. (1978). Distribution of substance P-like immunoreactivity in the central nervous system of the rat. I. Cell bodies and nerve terminals. *Neuroscience* **3,** 861-943.

Loewy, A. D. and Neil, J. J. (1981). The role of descending monoaminegeric systems in central control of blood pressure. *Fed. Proc.* **40,** 63-80.

Loewy, A. D., Wallach, J. and McKeller, S. (1981). Efferent connections of the ventral medulla oblongata in the cat. *Brain Res.* **40,** 2778-2785.

Lofring, B. (1961). Cardiovascular adjustments induced from the rostral cingulate gyrus with special reference to sympathoinhibitory mechanisms. *Acta Physiol. Scand.* **53** (Suppl), 1-82.

Lundberg, J. M., Hökfelt, T. and Nilsson, G. (1978). Peptide neurons in the vagus, splanchnic nerves. *Acta Physiol. Scand.* **104,** 421-450.

MacDermott, A. B. and Dale, N. (1987). Receptor, ion channels and synaptoc potentials underlying the integrative actions of excitatory amino acids. *Trends Neuroscience* **10,** 280-284.

Maley, B. and Elde, R. (1982). Immunohistochemical localization of putative neurotransmitters within the feline nucleus tractus solitarii. *Neuroscience* **7,** 2469-2490.

Mayer, M. L. and Westbrook, G. L. (1987). The physiology of excitatory amino acid in the vertebrate central nervous system. *Progr. Neuro biol.* **28,** 197-276.

McAllen, R. M. (1973). Projections of the carotid sinus baroreceptors to the medulla of the cat. Ph.D. Thesis. University of Birmingham,

McLennan, H. (1983). Receptors for the excitatory amino acids in the mammalian central nervous system. *Progr. Neurobiol.* 20, 251-271.

Mifflin, S. W., Spyer, K. M. and Withington-Wray D J. (1988). Baroreceptor input to the nucleus of the tractus solitarius in the cat : postsynaptic actions and the influence of respiration. *J. Physiol.*

399, 349-367.

Miller, B. D. and Felder, R. B. (1988). Excitatory amino acid receptors intrinsic to synaptic transmission in nucleus tractus solitarii. *Brain Res.* **456,** 333-343.

Milner, T. A., Morrison, S. F., Abate, C. and Reis, D. J. (1988). Phenylethanolamine -N-methyltransferase-containing terminals synapse directly on sympathetic preganglionic neurons in the rat. *Brain Res.* **448,** 205-222.

Minson, J., Llewellyn-Smith, I., Neville, A., Somogyi, P. and Chalmer, J. (1990). Quantitative analysis of spinally projecting adrenaline-synthesising neurons of C1, C2 and C3 groups in the rat medulla oblongata. *J. Auton. Nerv. System.* **30,** 209-230.

Miura, M. (1975). Postsynaptic potentials recorded from the nucleus of the solitary tract and its subadjacent reticular formation elicited by stimulation of the carotid sinus nerve. *Brain Res.* **100,**437-440.

Mogeson, G. J., Jones, D. L. and Yim, C. Y. (1980). From motivation to action : functional interface between the limbic system and the motor system. *Progr. Neurobiol.* **14,** 69-97.

Nauta, W.J.H. and Domesick, V. B. (1982). Neural associations of the limbic system. In: *The Neural Basis of Behavior,* Beckman A. L. (Ed), Spectrum, New York ; pp. 175-206.

Neil, J. J. and Loewy, A. D. (1982). Decreases in blood pressure in response to L-glutamate microinjections into the A5 catechoamine cell group. *Brain Res.* **241,** 271-278.

Nijikamp, F. P. and DeJong, W. (1975). Alpha-methylnoradrenaline induced hypotension and bradycardia after administration in the area of the nucleus tractus solitarii. *Eur. J. Pharmacol.* **32,** 361-364.

Nosaka, S., Kamaike, T. and Yasunaga, K. (1978). Central vagal organization in rats : An electrophysiological study. *Exp. Neurol.* **60,** 405-419.

Nosjean, A., Arluison, M. and Laguzzi, R. F. (1987). Increase in paradoxical sleep after destruction of serotonegic innervation in the nucleus tractus solitarius of the rat. *Neurosci.* **83,** 391-395.

Nosjean, A., Compoint, C., Buisseret-Delmas, C., Orer, H. S., Merahi, M., Puizillout, J. J. and Lagizzi, R. (1990). Serotonergic projections from the nodose ganglia to the nucleus tractus soltarius: an immunohistochemical and double labeling study in the rat. *Neurosci. Lett.* **114,** 22-26.

Otterson, O. P. (1981). Afferent connections to the amygdaloid complex of the rat with some observations in the cat. III. Afferents from the lower brain stem. *J. Comp. Neurol.* **202,** 335-356.

Paalzow, G. and Paalzow, L. (1982). Separate noradrenergic receptors could mediate clonidine-induced antinociception. *J. Pharmacol. Exp. Ther.* **223,** 795-800.

Palkovits, M., Brownstein, M. and Saabedra, J. M. (1974). Serotonin content of the brain stem nuclei in the rat. *Brain Res.* **80,** 237-249.

Palkovits, M. and Zaborszky, L. (1977). Neuroanatomy of central cardiovascular control. Nucleus tractus solitarii : an efferent neuronal connections in relation to the baroreceptor reflex arc. Progr. *Brain Res.* **47,** 9-34.

Panneton, W. M. and Loewy, A. D. (1980). Projection of the carotid sinus nerve to the nucleus of the solitary tract in the cat. *Brain Res.* **191**, 239-244.

Papanicolou, J., Summers, R.J ., Vajada, F. Je. and Louis, W. J. (1982). The relationship between alpha-2 adrenoreceptor serectivity and anticonvulsant effect in a series of clonidine-like drugs. *Brain Res.* **241,** 393-397.

Pelletier, G. and Dubé, D. (1978). Electronmicroscopic localization of α-MSH in the rat brain. *Am. J Anat.* **150,** 201-205.

Perrone, M. H. (1981). Biochemical evidence that L-glutamate is a neurotransmitter of primary afferent nerve fibers. *Brain Res.* **230,** 283-293.

Phillipu, A., Demmeler, R., Rosenberg, G. (1974). Effects of centrally applied drugs on pressor responses to hypothalamic stimulation. *Naunyn-Schmiedeberg's Arch. Pharmacol.* **282,** 389-400.

Phillipu, A. (1988). Regulation of blood pressure by central neurotransmitters and neuropeptides. *Rev. Physiol. Biochem. Pharmacol.* **111,** 1915.

Porter, R. (1963). Unit responses evoked in the medulla oblongata by vagus nerve stimulation.

J.Physiol. **168,** 717-735.

Redmond, D. E. Jr. and Huang, Y. H. (1979). New evidence for a locus coeruleus norepinephrine connection with anxiety. *Life Sciences* **25,** 2149-2162.

Rhoton, A. L., O'Leavy, J. L. and Ferguson, J. P. (1966). The trigeminal, facial, vagal and glossopharyngeal nerves in the monkey. *Arch. Neurol.* **14,** 530-540.

Richter, D. W., Camerer, H. and Sonnhof, V. (1978). Changes in extracellular potassium during the spontaneous activity of medullary respiratory neurones. *Pflüger Arch.* **376,** 139-149.

Ross, C. A., Armstrong, D. M., Rggiro, D. A., Pickel, V. M., Joh, T. H. and Reis, D. (1983). Adrenaline neurons in the rostral ventrolateral medulla : a possible role in tonic vasomotor control. *Brain Res.* **273,** 356-361.

Rudoleph, C. D., Kaplan, S. L. and Ganong, W. F. (1980). Sites at which clonidine act to affect blood pressure and the secretion of renin, growth hormone and ACTH. *Neuroendocrinology* **31,** 121-128.

Rudomin, P. (1967). Presynaptic inhibition induced by vagal afferent volleys. *J. Neurophysiol.* **30,**964-981.

Rudomin, P. (1968). Excitability changes of superior laryngeal, vagal and depressor afferent terminals produced by stimulation of the solitary tract nucleus. *Exp. Brain Res.* **6,** 156-170.

Saper, C. B. (1982). Convergence of autonomic and limbic connections in the insuler cortex of the rat. *J. Comp. Neurol.* **210,** 163-173.

Sattler, R. W. and Van Zweiten, P. A. (1982). Acute hypotensive action of 2-(2, 6 dichlorophenylamino)-2-imidazoline hydrochloride(st155) after infusion into the cat's vertebral artery. *Europ. Pharmacol.* **4,** 275-325.

Sawchenko, P. E. and Swanson, L. W. (1982). The organization of noradrenergic pathways from the brainstem to the paraventricular and supraoptic nuclei in the rat. *Brain Res.Rev.* **4,** 275-325.

Schaub, C., Bluet-Pajot, M. T., Mounier, F., Segalen, A. and Duhault, J. (1980). Effects of noradrenergic agonists and growth hormone secretion under gamma-hydroxybutyrate narco-analgesia in the rat. *Psychoneuro-endocrinology.* **5,** 139-146.

Schmitt, H., Schmitt, H., Boisser, J. R. and Guidicelli, J. F. (1967). Centrally mediate decrease in synaptic tone induced by 2-(2,6-dichlorophenylamino)-imidazoline(st155, Catapreson). *Europ. J. Pharmacol.* **2,** 147-148.

Schmitt, H., Le douarec, L. D. and Petillot, N. (1974). Antinociceptive effects of some alpha-sympathomimetic agents. *Neuropharmacology.* **13,** 289-293.

Schmitt, H. (1977). The pharmacology of clonidine and related products. In: *Antihypertensive Agents,* Gross F. (Ed.), Springer. New York. pp.299-396.

Shvaloff, A. and Laguzzi, R. (1986). Serotonin receptors in the rat nucleus tractus solitarii and cardiovascular regulation. *Europ. J. Pharmacol.* **13,** 557-618.

Spyer, K. M. (1982). Central nervous integration of cardiovascular control. *J. Exp Biol.* **100,** 109-128.

Steinbusch, H. W. M. (1981). Distribution of serotonin immunoreactivity in the central nervous system of the rat. Cell bodies and terminals. *Neuroscience* **6,** 558- 618.

Stryker-Boudier, H. A. J., Smeets, G. W., Brouwer, G. M. and Van Ressum, J. M. (1974). Hypothalamic alpha adrenergic receptors in cardiovascular regulation. *Neuropharmacology* **13,** 837-846.

Suga, T. and Suzuki, M. (1979). A study on the mechanism of central hypotensive action of clonidine : Effects on the responses of neurons in the nucleus tractus solitarii to stimulation of the aortic nerve. *Japan J. Pharmacol.* **29** (Suppl), 181P.

Suzuki, M. (1990). Investigation on the neuronal mechanisms of baro- and chemoreceptor reflexes in the nucleus tractus solitarii : Participation of adrenergic and opioid peptidergic system. *J. Saitama Med. School* **17,** 311-43.

Suzuki, M., Tanaka, M. and Suga, T. (1990). Participation of α_1–adrenoceptor in the neuronal mechanisms of baro- and chemoreceptor reflexes in the nucleus tractus solitarii. *Europ. J. Pharmacol.* **183**/2, 419.

Sved, A. A. F. and Sved, J. C. (1990). Endogenous GABA act on $GABA_B$ receptors in nucleus tractus solitarius to increase blood pressure. *Brain Res.* **526,** 235-240.

Svensson, T. H. and Strombom, U. (1977). Discontinuation of chronic clonidine treatment : Evidence for facilitated brain noradrenergic neurotransmission. *Naunyn-Schmiedeberg's Arch. Pharmacol.* **299,** 83-87.

Swanson, L. W. and Hartman, B. K. (1975). The central adrenergic system. An immunofluorescence study of the location of cell bodies and their efferent connection in the rat utilizing dopamine-β–hydroxylase as a marker. *J. Comp. Neurol.* **163,** 467-505.

Swanson, L. W. and Mogenson, G. J. (1981). Neural mechanism for the functional coupling of autonomic, endocrine and somatomotor responses in adaptive behavior. *Brain Res.* **3,** 1-34.

Tadepalli, A. S. and Mills, E. (1978). Contributioin of supracollicular structures of the brain to the central depression of cardiovascular function by clonidine. *J. Pharmacol. Exp. Ther.* **205,** 693-701.

Talman, W. T., Perrone, M. H., Scher, P., Kwo, S. and Reis, D. J. (1981). Antagonism of the baroreceptor reflex by glutamate diethyl ester, an antagonist to L-glutamate. *Brain Res.* **217,** 186-191.

Taube, H. D., Starke, K. and Borowski, E. (1974). *Naunyn-Schmiedebergs Arch. Pharmacol.* **290,** 123-141.

Toda, N., Fukuda, N. and Shimamoto, K. (1969). The mode of hypotensive actions of 2-(2,6-dichloro-L-amino)-imidazoline in the rabbit. *Japan J. Pharmacol.* **19,** 199-210.

Torvik, A. (1956). Afferent connections to the sensory trigeminal nuclei, the nucleus of the solitary tract and adjacent structures. *J. Comp. Neurol.* **106,** 51-141.

Tseng, L. F., Loh, H. H. and Wei, E. T. (1975). Effects of clonidine on morphine withdrawal signs in the rat. *Europ. J. Pharmacol.* **30,** 93-99.

Ungelstedt, U. (1971). Stereotaxic mapping of the monoamine pathways in the rat brain. *Acta physiol. Scand.* Suppl. **367,** 1-48.

Unnerstall, J. R., Kopajtic, T. A. and Kuhar, M. J. (1984). Distribution of α_2 agonist binding sites in the rat and human central nervous system : Analysis of some functional, anatomic correlates of the pharmacolgic effects of clonidine and related adrenergic agents. *Brain Res. Rew.* **7,** 69-101.

Urbanski, R. W. and Sapru, H. N. (1988). Putative neurotransmitters involved in medullary cardiovascular regulation. *J. Auton. Nerv. System* 181-193.

Von Baumgarten, R. and Kanzow, E. (1958). The termination of two types of inspiratory neurones in the region of the tractus solitarius of the cat. *Arch. Ital. Biol.* **96,** 361- 373.

Von Euler, C., Hayward, J. N., Marttila, I. and Wyman, R. J. (1973a). Respiratory neurones of the ventrolateral nucleus of the solitary tract of the cat : vagal input, spinal connections and morphorogical identification. *Brain Res.* **61,** 1-22.

Von Euler, C., Hyword, J. N., Marttila, I. and Wyman, R. J. (1973b). The spinal connections of the inspiratory neurones of the ventrolateral nucleus of the cats tractus solitarius. *Brain Res.* **61,** 23-33.

Watkins, J. C. and Evans, R. H. (1981). Excitatory amino acid transmitters. *Annu. Rev. Pharmacol. Toxicol.* **21,** 165-204.

Ward, D. G. and Gunn, C. G. (1976). Locus coeruleus complex : elicitation of a pressor response and a brain stem region necessary for its occurrence. *Brain Res.* **107,** 401-406.

Weiss, J. M., Bailey, W. H., Goodman, P. A., Hoffman, L. L. J., Ambrose, M. J., Salman, S. and Charry, J. H. (1982). A model for neurochemical study of depression. In Levy A. and Spiegelstein MY. (Eds), Elsevier / North Holland, New York. pp. 195-223

Yoshimura, M., Polosa, C. and Nishi, S. (1986). Noradrenaline modifies sympathetic preganglinonic neuron spike and afterpotendtial. *Brain Res.* **362,** 370-374.

Young, W. S. and Kuhar, M. J. (1980). Noradrenergic α_1 and α_2 receptors : light microscopic autoradiographic localization. *Proc. Nat. Acad. Sci. USA* **77,** 1696-1700.

Progress in Hypertension, Vol. 2, pp. 257-268.
Saito *et al.* (Eds)

CENTRALLY ACTING ANTIHYPERTENSIVE ALPHA-2 ADRENOCEPTOR AGONISTS

GÜNTER SCHOLTYSIK

University of Berne, Institute of Veterinary Pharmacology, Laenggass-Strasse 124, CH-3012 Berne, Switzerland.

INTRODUCTION

Within the armamentarium of clinically used antihypertensives centrally acting drugs play an important role in the management of hypertension. The introduction of alpha-2 adrenoceptor agonists into clinical practice in the 60 th have meanwhile not only proved to be a step forward in adequate treatment of hypertension. These drugs of which clonidine may be considered as the prototype have also been used as important pharma-cological tools to aid in the understanding of the role of the central nervous system in blood pressure regulation. In view of the importance the number of known drugs with central alpha-2 adrenoceptor stimulant properties increased considerably until now. This chapter will describe these agents as a separate group of antihypertensive drugs.

FUNCTIONAL SIGNIFICANCE OF CENTRAL ALPHA-2 ADRENO-CEPTORS IN CARDIOVASCULAR CONTROL

The nomenclature of alpha-adrenoceptors has been a matter of confusion in the past. The observation that noradrenaline regulates its own release through a negative feed-back mechanism mediated by presynaptic alpha-adrenoceptors independent from postsynaptic adrenoceptors was the basis for the now established concept of pre- and post-synaptic alpha-adrenoceptors (Starke, 1972; Rand *et al.*, 1973; Langer, 1974).

The basis for the now established classification of alpha-adrenoceptors into the major subtypes alpha-1 and alpha-2 adrenoceptors was the availability of subtype selective ligands. Consequently the identification of subtypes in terms of the selective binding of radioligands and their displacement from binding sites by a range of agonists and antagonists has become possible (see later). Furthermore, there is good evidence, that alpha-2 adrenoceptors are very heterogeneous, being divided into alpha-2A and alpha-2B subtypes and existing in different affinity states (Bylund, 1985; Nahorski *et al.*, 1985; Brown *et al.*, 1990).

In addition to their different affinities for ligands alpha-1 and alpha-2 adrenoceptors possess different mechanisms for signal transduction (Exton, 1985; Homcy and

Table 1. Centrally acting antihypertensive alpha-2 adrenoceptor agonists

Chemical structure	Name	Some references for clinical antihypertensive effectiveness
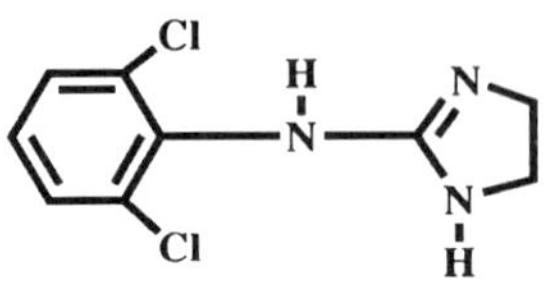	Clonidine	Weber *et al.*, 1978 Houston, 1981
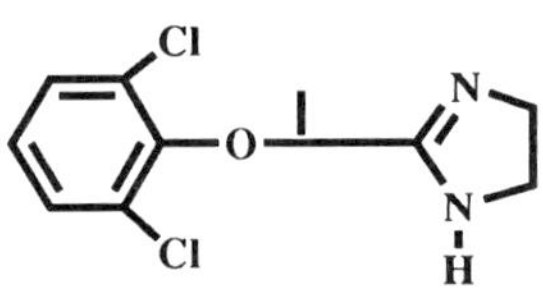	Lofexidine	Wilkins *et al.*, 1981 Fagan *et al.*, 1982
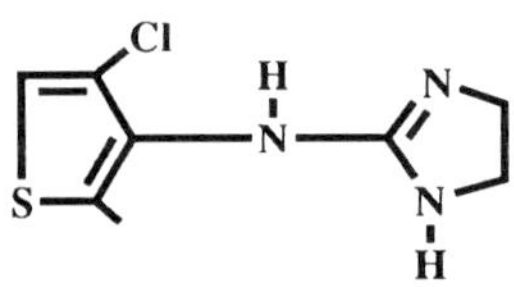	Tiamenidine	Campell *et al.*, 1980
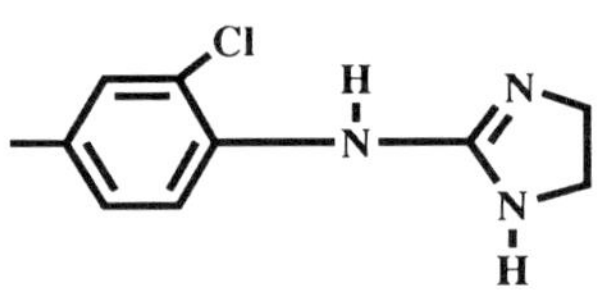	Tolonidine	Girtanner *et al.*, 1983
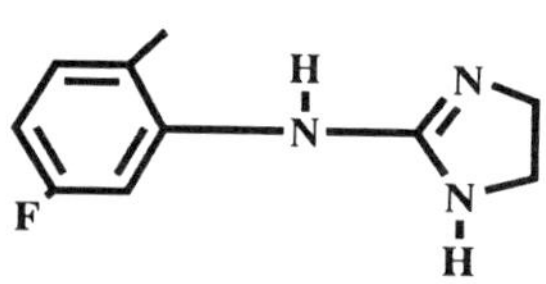	Flutonidine	Nicrosini and Pasotti, 1983
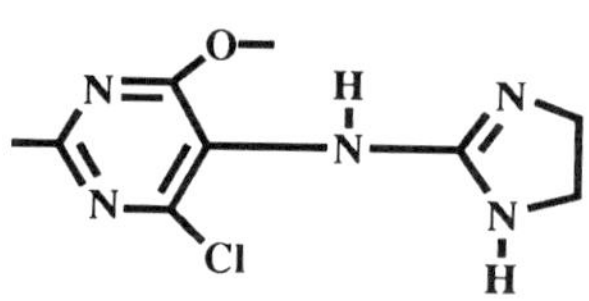	Moxonidine	Frisk-Holmberg and Plänitz, 1987; Plänitz, 1987
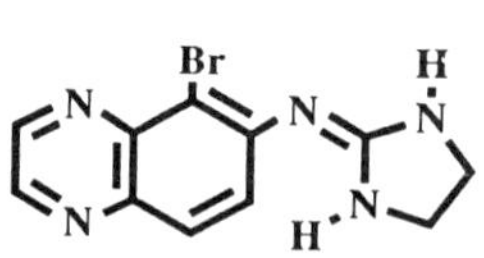	UK-14,304	Ashton and Rawlins, 1978
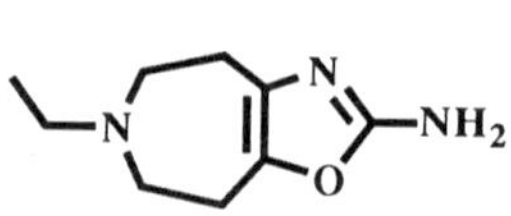	Azepexole	Pedrinelli *et al.*, 1987

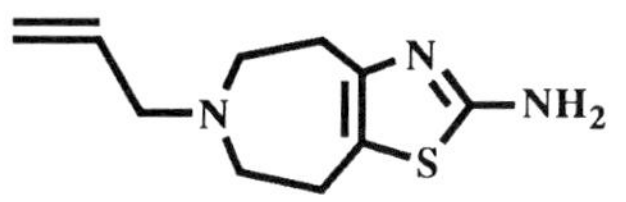	Talipexole	Wiedemann *et al.*, 1989 (trial in schizophrenia)
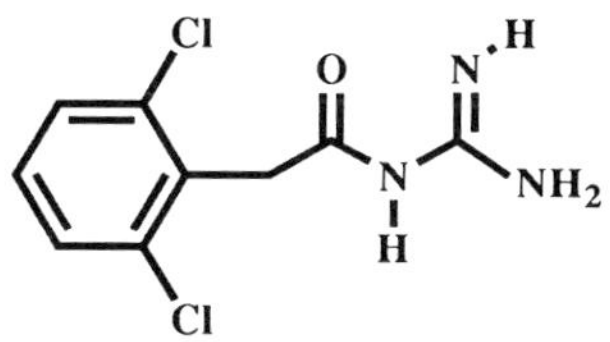	Guanfacine	Sorkin and Heel, 1986 Board *et al.*, 1988
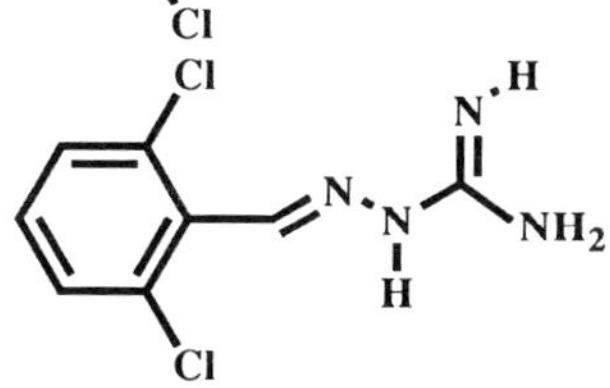	Guanabenz	Shah *et al.*, 1976 Walker *et al.*, 1980
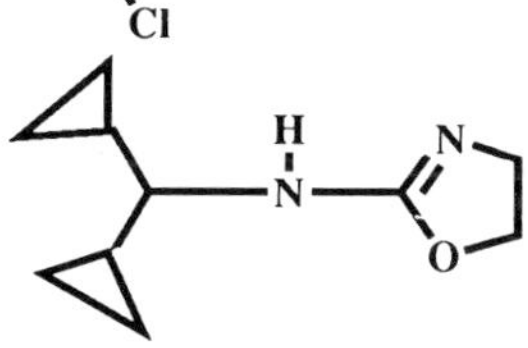	Rilmenidine	Fillastre *et al.*, 1988
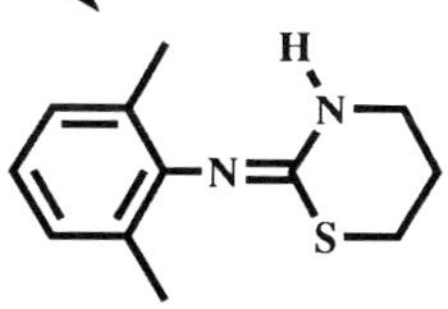	Xylazine	Kroneberg *et al.*, 1967
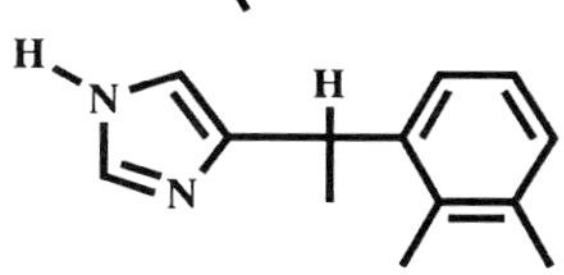	Medetomidine	Scheinin *et al.*, 1989
	L-alpha-methyl-noradrenaline	Weidmann *et al.*, 1974

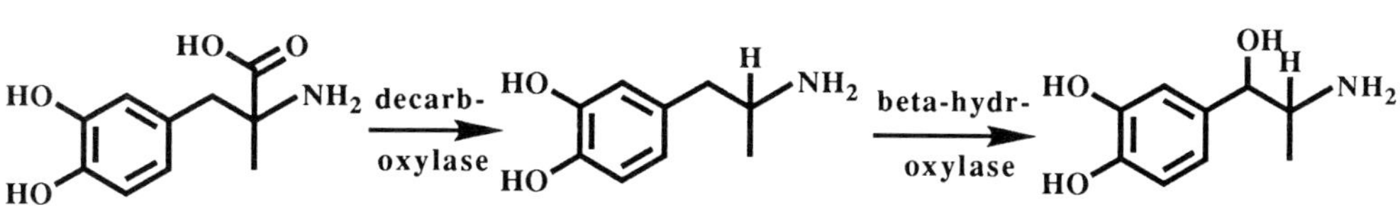

L-alpha-Methyldopa L-alpha-Methyldopamine L-alpha-Methylnoradrenaline

Graham, 1985; Michell, 1985). Alpha-agonists, like many neurotransmitters and hormones, exert their actions by altering the concentration of Ca^{2+} or cyclic adenosine 3', 5'-monophosphate (cAMP) in the cytosol of their target cells. There is evidence that alpha-1 adrenoceptors can be distinguished from alpha-2 adrenoceptors by their different effects on Ca^{2+} and cAMP. In contrast to beta-adrenoceptors and alpha-2 adrenoceptors, which are coupled to the membrane bound enzyme adenylate cyclase by guanine nucleotide regulatory proteins (Jacobs, 1985), activation of alpha-1 adrenoceptors does not alter intracellular levels of cAMP (Homcy and Graham, 1985; Exton, 1985) but may be linked to the intracellular messengers inositol-1, 4,5-triphosphate (IP_3) and 1,2,-diacylglycerol (DAG) (Berridge, 1984; Michell, 1985).

As far as functional effects are concerned, mediated by adrenoceptors in peripheral tissues, alpha-adrenoceptors located on vascular smooth muscle cells mediate vasoconstriction, alpha-1 and alpha-2 subtypes being represented in various proportions in different vascular beds (Ruffolo and Nichols, 1988; Timmermans, 1988). Presynaptically located alpha-2 adrenoceptors regulate the release of the sympathetic transmitter substance, noradrenaline, from nerve endings (Langer, 1974; 1981). In the central nervous system (CNS) alpha-2 adrenoceptors regulate the neuronal release of noradrenaline and several other transmitter substances, and are intimately involved in the modulation of sympathetic outflow, cardiovascular and endocrine functions, vigilance, emotions, cognition, and nociception (see Dubocovich, 1984; Solomon *et al.,* 1989). Alpha-2 adrenoceptor activation postsynaptically in the rat brain leads to hyperpolarization in locus coeruleus cells (Aghajanian *et al.,* 1982) and hence depression of neural activity.

As far as localization of central alpha-2 adrenoceptors is concerned most information is derived from brain of rodents. In *in vitro* autoradiographic techniques alpha-2 agonists such as clonidine and antagonists such as idazoxan as ligands have been used to map the distribution of alpha-2 adrenoceptors. Only a small proportion of these are located presynaptically on noradrenergic neurones, since destruction of these with 6-hydroxydopamine does not markedly reduce the number of alpha-2 adrenoceptors in the CNS (U'Prichard *et al.*, 1977). Dense ^{3}H-idazoxan binding appears over the anterior olfactory nuclei, fundus striatum, septum, thalamus, hypothalamus, amygdala, entorhinal cortex, central gray, inferior colliculus, dorsal parabrachial nucleus, locus coeruleus and nucleus tractus solitarius. There is a growing body of autoradiographic evidence for the anatomical and pharmacological heterogeneity of alpha-2 adrenoceptors in the rat brain (Bylund, 1988).

Postmortem autoradiographic studies in the human CNS with ^{3}H-para-aminoclonidine show high densities of alpha-2 adrenoceptors in the locus coeruleus, dorsal motor nucleus of the vagus nerve, and the intermediolateral column and substantia gelatinosa of the spinal cord (Unnerstall *et al.*, 1984; Probst *et al.*, 1985). In the human forebrain, a high density of alpha-2 adrenoceptors labelled by para-aminoclonidine exists at least in the primary visual cortex (Probst *et al.*, 1985).

CHEMISTRY

Fortunately an impressive number of synthetic chemicals is available which lower blood pressure by an action primarily mediated via the CNS. Each of them is appreciated, offering some differences in pharmacological or advantages in clinical effects in one or the other aspect. All compounds listed in Table 1 lower the blood pressure due to central alpha-2 adrenoceptor activation and have been investigated in man. The prototype of these chemicals is the imidazoline clonidine. There have been thousands of derivatives of clonidine synthesized and used as experimental tools as described in the comprehensive monograph by Timmermans *et al.* (1980). Other clinically effective antihypertensive imidazoline derivatives are lofexidine, tiamenidine, tolonidine, flutonidine, moxonidine and UK-14,304. Azepexole and talipexole which are oxazolo- and thiazolo-azepines, respectively, are highly selective alpha-2 adrenoceptor ligands. The aminoguanidine derivatives guanfacine and guanabenz have in common some preferential affinity to alpha-2 adrenoceptors. The oxazoline derivative rilmenidine differs from all the other alpha-2 adrenoceptor ligands by its absence of sedative effects and its affinity to a central "imidazoline-receptor". The thiazine xylazine and the imidazole medetomidine are extensively used sedative analgesics in veterinary medicine (Räihä *et al.*, 1989). Finally, alpha-methyldopa, which belongs to the class of centrally acting antihypertensives, is an exception in so far as it can be called a "prodrug". It has to be converted enzymatically into alpha-methyl-noradrenaline which is the active alpha-2 agonistic principle (van Zwieten, 1976; Henning, 1984).

PHARMACOLOGY

Different methods have been employed in acute animal experiments in order to detect the central site of the hypotensive drug activity. Basically the idea behind these methods is to achieve drug concentrations within the (CNS) which are markedly higher than in peripheral tissues. This can be attained by administration directly into the CNS so that drugs can penetrate the regions which contains the most important nuclei responsible for the modulation of sympathetic nerve activity. If drugs are more pronounced hypoten-sives after central than after peripheral administration, they can be considered to have a central site of action. Some techniques described are: Stereotactic micro-injections into distinct brain regions (Struyker Boudier and van Rossum, 1972), local application to the exposed medulla oblongata (Bousquet and Guertzenstein, 1973), injections into the cerebrospinal fluid either intracisternally (Kobinger, 1978) or intracerebroventricularly (Feldberg and Sherwood, 1954; Schmitt and Schmitt, 1969), intravertebral arterial administration (van Zwieten, 1975; van Zwieten and Timmermans, 1980) or crossperfusion experiments with a circulatory isolated skull (Heijmans *et al.*, 1961).

Further effects which are indicative of a central site of alpha-adrenoceptor stimulation were described as follows: Depression of preganglionic sympathetic nerve activity (Schmitt and Schmitt, 1969), facilitation of vagally mediated reflex bradycardia (Pichler *et al.*, 1980), inhibition of the central noradrenaline turnover (Scholtysik *et al.*, 1975) and sedation in young chicken (Roach *et al.*, 1982).

The mode of action of the centrally acting hypotensive drugs listed in Table 1 is clearly ascribed to their alpha-2 adrenoceptor stimulation. The early concept of Schmitt and Schmitt (1970) was based on the observation that the hypotensive effect of clonidine could be antagonized by the alpha-adrenoceptor blocking agent piperoxane. It was found later that rauwolscine, the most selective alpha-2 adrenoceptor antagonist, is the most effective antagonist of the effects of clonidine (Kobinger, 1983). Systematic pharmacological investigations with a variety of different agonists and antagonists led to the subclassification of alpha-1 and alpha-2 adrenoceptors (Starke, 1972; 1981). They are meanwhile best defined in terms of the selective binding of radioligands and their displacement from the binding sites by a

Table 2 Affinity to alpha-adrenoceptors in the CNS of centrally acting antihypertensives. The data were simply collected for rough assessment of relative potencies. For brain regions and species employd see references.

Drug	Inhibition of binding (pK_i) to alpha 1-adrenoceptors radioligand	pK_i	reference	alpha 2-adrenoceptors radioligand	pK_i	reference
Clonidine	Prazosin	6.2	[2]	Idazoxan	8.0	[1]
Lofexidine	Prazosin	6.7	[2]	Clonidine	8.6	[2]
Tiamenidine	Prazosin	5.6	[2]	Clonidine	8.6	[2]
Flutonidine	Prazosin	5.5	[2]	Clonidine	8.3	[2]
Moxonidine	Prazosin	4.3	[4]	Clonidine	8.0	[4]
UK 14,304	Prazosin	5.6	[2]	Clonidine	9.0	[2]
Azepexole	Prazosin	3.0	[6]	Clonidine	6.7	[7]
Talipexole	Prazosin	3.3	[6]	Clonidine	7.6	[7]
Guanfacine	Prazosin	5.2	[2]	Clonidine	8.7	[2]
Guanabenz	Prazosin	6.6	[2]	Idazoxan	7.7	[1]
Rilmenidine	Prazosin	5.4	[3]	Clonidine	7.8	[3]
Xylazine	Prazosin	4.5	[5]	Clonidine	6.7	[5]
Medetomidine	Prazosin	5.7	[5]	Clonidine	8.9	[5]
L-alpha-methyl-noradrenaline	Prazosin	4.8	[3]	Clonidine	8.0	[3]

[1] Brown *et al.*, 1990 [2] Summers *et al.*, 1980 [3] Louis *et al.*, 1988
[4] Armah, 1988 [5] Virtanen *et al.*, 1988 [6] Timmermans *et al.*, 1984
[7] Hammer *et al.*, 1980

range of agonists and antagonists. In receptor binding studies, using rat cortex membrane preparations the order of high selectivity for the alpha-2 adrenoceptors is as follows: guanfacine > tiamenidine > clonidine > guanabenz > lofexidine (Louis *et al.*, 1982). Other alpha-2 selective ligands listed in Table 1 were not included in that particular study but affinity data are given in Table 2 which demonstrates the preferential affinity to alpha-2 adrenoceptors for all drugs investigated.

Interestingly a central non-alpha-2 adrenoceptor binding site for clonidine has recently been proposed. This imidazoline preferring binding site with affinity to tritiated clonidine, para-aminoclonidine and idazoxan was identified in bovine, rat and human brainstem (Ernsberger *et al.*, 1987; Bricca *et al.*, 1988; 1989), and has been reported to exist in a number of peripheral tissues (Yablonsky *et al.*, 1989; Langin and Lafontan, 1989; Michel *et al.*, 1989; MacKinnon *et al.*, 1989). Moreover, an endogenous imidazoline-site ligand was described, a clonidine displacing substance (Meeley *et al.*, 1986; Atlas *et al.*, 1987; Ernsberger *et al.*, 1988). Subsequently, the existence of such receptors which are directly involved in the central blood pressure regulation gave access to a new pharmacological concept (Bousquet *et al.*, 1989). Rilmenidine, an alpha-2 adrenoceptor ligand structurally close to the imidazolines, exhibited a relative selectivity, higher than that of clonidine for the medullary imidazoline receptors as compared to the cortical alpha-2 adrenoceptors (Bricca *et al.*, 1989). The central site of action as well as the hypotensive mechanism of action of rilmenidine (antagonism by idazoxan) in the rabbit was shown by Feldman *et al.* (1990). The most interesting clinical feature (see later) of rilmenidine, initially developed as hypotensive centrally acting alpha-2 adrenoceptor agonist, was its dissociation between the antihypertensive efficacy and alpha-2 related side effects like sedation and dry mouth. Indeed, rilmenidine may represent a promising new lead structure in the search of side effect restriction of centrally acting antihypertensives.

Adrenoceptors located presynaptically in peripheral sympathetic nerves are of the alpha-2 subtype and their activation reduces stimulation-evoked release of noradrenaline in various tissues (Starke, 1977; 1987; Starke *et al.*, 1989; Scholtysik and Brugger, 1989; Brock *et al.*, 1990). Whether the presynaptic peripheral effects of clonidine and pharmacologically related compounds contribute to the therapeutic antihypertensive effect is still a matter of discussion (Brown and Harland, 1984).

CLINICAL CHARACTERISTICS OF ALPHA-2 ADRENOCEPTOR AGONISTS

In the frame of the present pharmacologically oriented survey a very brief statement on the clinical use shall round off the contribution. This class of centrally acting drugs continue to play an important role in the management of hypertension. They can be used both as monotherapy or in combination therapy. As reviewed by Weber *et al.* (1990), their advantage is that they can be administered to patients with complex clinical forms of hypertension, in cases of renal insufficiency, diabetes mellitus, ischemic heart disease, chronic obstructive airway disease and to elderly patients. There is evidence that the centrally acting drugs may produce regression of left ventricular hypertrophy, have no or negligible effects on glucose metabolism, slightly improve the plasma lipid profile and reduce circulating noradrenaline and renin plasma levels. Since there is good evidence for the fundamental role of the sympathetic nervous system in mediating hypertension (Tuck, 1986), the central suppression of sympathetic outflow by modulating (resetting) the physiologic control mechanisms offers in theory an appreciated principle (Dollery, 1988). Unfortunately the most common adverse effects like drowsiness and dry mouth appear to be linked to the basic mechanism of action of alpha-2 agonists. However, low dose therapy seems to solve in part those problems (see Weber *et al.*, 1990). On the other hand there is some hope to avoid adverse effects by chemical modification of alpha-2 adrenoceptor ligands, as first experimental and clinical experience with rilmenidine indicated, a drug which preferentially binds to imidazoline receptors (Mahieux, 1989).

Clinical indications other than hypertension (i.e. opiate withdrawal, antinociception, migraine, menopausal flush) are known for the present group of therapeutics. Some key references concerning the clinical features of the individual drugs are given in Table 1.

Acknowledgements
The author gratefully acknowledges the superb secretarial assistance of Mrs. S. Schaad and the computer drawing of chemical structures by Dr. R. Giger. This work was supported by a generous gift of a Sharp PC-5541 from the Stiftung zur Förderung der Wissenschaftlichen Forschung an der Universität Bern.

REFERENCES

Aghajanian, G. K. and Van der Maelan, C. P. (1982). Alpha 2-adrenoceptor-mediated hyperpolarization of locus coeruleus neurons: intracellurar studies in vivo. *Science* **215,** 1394-1396.

Armah, B. I. (1988). Unique presynaptic alpha-2 receptor selector selectivity and specificity of the antihypertensive agent moxonidine. *Arzneim. -Forsch./Drug Res.* **38**(II), 1435-1442.

Ashton, H. and Rawlins, M. D. (1978). Central nervous system depressant actions of clonidine and UK-14,304: partial dissociation of EEG and behavioural effects. *Brit. J. Clin. Pharmac.* **5,** 135-140.

Atlas, D., Diamant, S., Fales, H. M. and Panel, L. (1987). The brain's own clonidine : Purification and characterization of endogenous clonidine displacing substance from brain. *J. Cardiovasc. Pharmacol.* **10,** 122-127.

Berridge, M. J. (1984). Inositol trisphosphate and diacylglycerol as second messengers. *Biochem. J.* **220,** 345-360.

Board, A. W., Perry, V. P., Shepperson, B. E., Nyman, C. E. and Carchman, S. H. (1988). A post-marketing evaluation of guanfacine hydrochloride in mild to moderate hypertension. *Clin. Therapeutics.* **10,** 761-775.

Bousquet, P. and Guertzenstein, P. G. (1973). Localization of the central cardiovascular action of clonidine. *Br. J. Pharmac.* **49,** 573-579.

Bousquet, P., Feldman, J., Tribirica, E., Bricca, G., Molines, A., Dontenwill, M. and Belcourt, A. (1989). New concept on the central regulation of blood pressure. *Am. J. Med.* **87,** 10S-13S.

Bricca, G., Dontenwill, M., Molines, A., Feldman, J., Belcourt, A. and Bousquet, P. (1988). Evidence for the existence of a homogeneous population of imidazoline receptors in the human brainstem. *Eur. J. Pharmacol.* **150,** 401-402.

Bricca, G., Dontenwill, M., Molines, A., Feldman, J., Belcourt, A. and Bousquet, P. (1989). The imidazoline preferring receptor: binding studies in bovine, rat and human brainstem. *Eur. J. Pharmacol.* **162,** 1-9.

Brock, J. A., Cunnane, T. C., Starke, K. and Wardell, C. F. (1990). Alpha-2-Adrenoceptor-mediated autoinhibition of sympathetic transmitter release in guinea-pig vas deferens studied by intracellular and focal extracellular recording of junction potentials and currents. *Naunyn-Schmiedeberg's Arch. Pharmacol.* **342,** 45-52.

Brown, C. M., MacKinnon, A. C., McGrath, J. C., Spedding, M. and Kilpatrick, A. T. (1990). Alpha-2 adrenoceptor subtypes and imidazoline-like binding sites in the rat brain. *Br. J. Pharmacol.* **99,** 803-809.

Brown, M. J. and Harland, D. (1984). Evidence for a peripheral component in the sympatholytic effect of clonidine in rats. *Br. J. Pharmac.* **83,** 657-665.

Bylund, D. (1985). Heterogeneity of alpha-2 adrenergic receptors. *Pharmacol. Biochem. Behav.* **22,** 835-843.

Bylund, D. B. (1988). Subtypes of alpha-2 adrenoceptors in human and rat brain. In *Epinephrine in the central nervous systems*, (Eds) Stolk, J. M., U'Prichard, D. C. and Fuxe, K., Oxford University Press, New York, pp. 181-188.

Campbell, B. C., Elliott, H. L., Hamilton, C. A. and Reid, J. L. (1980). Changes in blood pressure, heart rate and sympathetic activity on abrupt withdrawal of tiamenidine (HOE 440) in essential hypertension. *Eur. J. Clin. Pharmacol.* **18,** 449-454.

Dollery, C. T. (1988). Advantages and disadvantages of alpha-2 adrenoceptor agonists for systemic hypertension. *Am. J. Cardiol.* **61,** 1D-5D.

Dubocovich, M. (1984). Presynaptic alpha-adrenoceptors in the central nervous system. *Ann. N.Y. Acad. Sci.* **430,** 7-25.

Ernsberger, P., Meeley, M. P., Mann, J. J. and Reis, D. J. (1987). Clonidine binds to imidazole binding sites as well as $alpha_2$-adrenoceptors in the ventrolateral medulla. *Eur. J. Pharmacol.* **134,** 1-13.

Ernsberger, P., Meeley, M. P. and Reis, D. J. (1988). An endogenous substance with colnidine like properties: selective binding to imidazole sites in the ventrolateral medulla. *Brain Res.* **441,** 309-318.

Exton, J. H. (1985). Mechanisms involved in alpha-adrenergic phenomena. *Amer. J. Physiol.* **248,** E633-E647.

Fagan, T.C. *et al.* (12 authors) (1982). Antihypertensive effects of lofexidine in patients with essential hypertension. *Br. J. Clin. Pharmacol.* **13,** 405.

Feldberg, W. and Sherwood, S.L. (1954). Injections of drugs into the lateral ventricle of the cat. *J. Physiol. London.* **123**, 148.

Feldman, J., Tibirica, E., Bricca, G., Dontenwill, M., Belcourt, A. and Bousquet, P. (1990). Evidence for the involvement of imidazoline receptors in the central hypotensive effect of rilmenidine in the rabbit. *Br. J. Pharmacol.* **100,** 600-604.

Fillastre, J. P., Letac, B., Galinier, F., Le Bihan, G. and Schwartz, J. (1988). A multicentre double-blind comparative study of rilmenidine and clonidine in 333 hypertensive patients. *Am. J. Cardiol.* **61,** 81D-85D.

Frisk-Holmberg, M. and Plänitz, V. (1987). A selective alpha-2 adrenoceptor agonist in arterial essential hypertension. Clinical Experience with Moxonidine. *Current Therap. Research.* **42,** 138-146.

Girtanner, C., Amblard, F. and Delomier, Y. (1983). Euctan trial in arterial hypertension of the aged. *J. Med. Lyon.* **64/1375,** 33-35.

Hammer, R., Kobinger, W. and Pichler, L. (1980). Binding of an imidazolidine (clonidine), an oxazoloazepine (B-HT933) and a thiazoloazepine (B-HT920) to rat brain alpha-adrenoceptors and relation to cardiovascular effects. *Eur. J. Pharmacol.* **62,** 277-285.

Heijmans, C., De Schaepdrijver, A. F., De Vleeschouwer, G. F. *et al.* (1961). Effects of reserpine on the cardio-inhibitory and vasomotor centres. *Arch. Int. Pharmacodyn. Ther.* **134,** 482.

Henning, M. (1984). Alpha-methyldopa and related compounds. In *Handbook of Hypertension vol.3: Pharmacology of Antihypertensive Drugs*, (Eds) Van Zwieten, P. A., Elsevier Science Publishers B.V.

Homcy, C. J. and Graham, R. M. (1985). Molecular characterization of adrenergic receptors. *Circ. Res.* **56,** 635-650.

Houston, M. C. (1981). Clonidine hydrochloride: review of pharmacologic and clinical aspects. *Prog. Cardiovasc. Dis.* **23**, 337-350.

Kobinger, W. (1978). Central alpha-adrenergic systems as targets for hypotensive drugs. *Rev. Physiol. Biochem. Pharmacol.* **81,** 39-100.

Kobinger, W. (1983). Central blood pressure regulation. Involvement of presynaptic or postsynaptic alpha-1- or alpha-2 adrenoceptors? *Chest* **83,** 296-299.

Kroneberg, G., Oberdorf, A., Hoffmeister, F. and Wirth, W. (1967). Zur Pharmakologie von 2-(2,6-Dimethylphenylamino)-4H-5,6-dehydro-1,3-thiazin (Bayer 1470), eines Hemmstoffes adrenergischer und cholinergischer Neurone. *Naunyn-Schmiedeberg's Arch. Pharmacol.* **256,** 257-280.

Langer, S. Z. (1974). Presynaptic regulation of catecholamine release. *Biochem. Pharmacol.* **23,** 1793-1800.

Langer, S. Z. (1981). Presynaptic regulation of the release of catecholamines. *Pharmacol. Rev.*

32, 337-362.

Langin, D. and Lafontan, M. (1989). [^{3}H] idazoxan binding at non-alpha-2 adrenoceptors in rabbit adipocyte membranes. *Eur. J. Pharmacol.* **159,** 199-203.

Louis, W. J., Summers, R. J., Dynon, M. and Jarrott, B. (1982). New developments in alpha-adrenoceptor drugs for the treatment of hypertension. *J. Cardiovasc. Pharmacol.* **4,** 5168-5171.

Louis, W. J., Jarrott, B. and Conway, E. L. (1988). Sites of actios of alpha-2 agonists in the brain and periphery. *Am. J. Cardiol.* **61,**15D-17D.

MacKinnon, A. C., Brown, C. M., Spedding, M. and Kilpatrick, A. T. (1989). [^{3}H]-idazoxan binds with high affinity to two sites on hamster adipocytes: an alpha-2 adrenoceptor and a non-adrenoceptor site. *Br. J. Pharmacol.* **98,** 1143-1150.

Mahieux, F. (1989). Rilmenidine and Vigilance. Review of Clinical Studies. *Am. J. Med.* **87,** 67S-72S.(suppl).

Meeley, M. P., Ernsberger, P. R., Granata, A. R. and Reis, D. J. (1986). An endogenous clonidine displacing substance from bovine brain: receptor binding and hypotensive actions in the ventrolateral medulla. *Life Sci.* **38,** 1119-1126.

Michel, M. C., Brodde, O-R., Schnepel, B., Behrendt, J., Tschada, R., Motulsky, H. J. and Insel, P. A. (1989). [^{3}H]idazoxan and some other alpha-2 adrenergic drugs also bind with high affinity to a nonadrenergic site. *J. Pharmacol. Exp. Ther.* **35,** 324-330.

Michell, R. H. (1985). Inositol lipid breakdown as a step in alpha-adrenergic stimulus-response coupling. *Clin. Sci.* **68**, (suppl. 10), 43s-46s.

Nahorski, S. R., Barnett, D. B. and Cheung, Y. D. (1985). Alpha-adrenoceptor-effector coupling: affinity states or heterogeneity of the alpha-2 adrenoceptors. *Clin. Sci.* **68** (suppl. 10), 39s-42s.

Nicrosini, F. and Pasotti, G. (1983). Long-term treatment with a new antihypertensive drug: flutonidine. *Minerva Med.* **74,** 1297-1300.

Pedrinelly, R., Taddei, S. and Salvetti, A. (1987). Calcium entry blockade and selective alpha-adrenergic stimulation in animals and humans. *J. Cardiovasc. Pharmacol.* **10,** S79-S86.

Pichler, L., Placheta, P. and Kobinger, W. (1980). Effect of azepexole (B-HT 933) on pre- and postsynaptic alpha-adrenoceptors at peripheral and central nervous sites. *Eur. J. Pharmacol.* **65,** 233-241.

Plänitz, V. (1987). Comparison of moxonidine and clonidine HCl in treating patients with hypertension. *J. Clin. Pharmacol.* **27,** 46-51.

Probst, A., Cortes, R. and Palacios, J. M. (1985). Distribution of alpha-2 adrenergic receptors in the human brainstem: An autoradiographic study using [^{3}H]p-aminoclonidine. *Eur. J. Pharmacol.* **106,** 477-488.

Räihä, M. P., Räihä, J. E. and Short, Ch. E. (1989). A comparison of xylazine, acepromazine meperidine and medetomidine as pre-anesthetics to halothane anesthesia in dogs. *Acta Vet. Scand.* **85,** 97-102.

Rand, M. J., Story, D. F. and Allen, S. A. (1973). In: *Frontiers in Catecholamine Research* (Eds) Usdin, E. and Snyder, S., Pergamon Press, London, pp. 579-581.

Roach, A. G., Lefevre-Borg, F., Gomeni, R. and Cavero, I. (1982). Pharmacological evidence favoring a similarity between alpha-adrenoceptor mediating sleep in chicks and cardiac presynaptic alpha-2 receptors in rats. *J. Pharmacol. Exp. Ther.* **222,** 680-687.

Ruffolo, R. R. and Nichols, A. J. (1988). Alpha-adrenoceptor coupling to vasoconstriction on the peripheral circulation. In *Epinephrine in the cental nervous system* (Eds) Stolk, J.M., U'Prichard, D.C. and Fuxe, K., Oxford University Press, New York, pp. 339-346.

Schmitt, H. and Schmitt, H. (1969). Localization of the hypotensive effect of 2-(2,6-dichlorophenylamino)-2-imidazoline hydrochloride (St 155, Catapresan). *Eur. J. Pharmacol.* **6,** 8-12.

Schmitt, H. and Schmitt, H. (1970). Interactions between 2-(2,6-dichlorophenylamino)-2-imidazoline hydrochloride (St 155, Catapresan) and alpha-adrenergic blocking drugs. *Eur. J. Pharmacol.* **9,** 7-13.

Scheinin, H., Kallio, A., Koulu, M. and Scheinin, M. (1989). Pharmacological effects of medetomidine in humans. *Acta. Vet. Scand.* **85,** 145-147.

Scholtysik, G., Lauener, H., Eichenberger, E., Bürki, H., Salzmann, R., Müller-Schweinitzer, E. and

Waite, R. (1975). Pharmcological actions of the antihypertensive agent N-amidino-2-(2,6-dichlorophenyl) acetamide hydrolchloride (BS 100-141). *Arzneim.-Forsch. Drug Res.* **25,** 1483-1491.

Scholtysik, G. and Brugger, P.(1989). In vitro cardiac effects and antiarrhythmic properties of guanfacine. *Arzneim.-Forsch. /Drug Res.* **39,** 450-454.

Shah, R. S., Walker, B. R. and Vanov, S. K.(1976). Guanabenz effects on blood pressure and noninvasive parameters cadiac performance in patients with hypertension. *Clin. Pharmcol. Ther.* **19,** 732-737.

Sorkin, E. M., Heel, R. C. (1986). Guanfacine: a review of its pharmacodynamic and pharmacokinetic properties, and therapeutic efficacy in the treatment of hypertension. *Drugs.* **31,** 301-336.

Solomon, R. E., Brody, M. J. and Gebhart, G. F. (1989). Pharmacological characterization of alpha adrenoceptors involved in the antinociceptive and cardiovascular effects of intrathecally administered clonidine. *J. Pharmacol. Exp. Ther.* **1,** 27-38.

Starke, K. (1972). Alpha-sympathomimetic inhibition of adrenergic and cholinergic transmission in the rabbit heart. *Naunyn-Schmiedeberg's Arch. Pharmacol.* **274,** 18-45.

Starke, K. (1977). Regulation of noradrenaline release by presynaptic receptor systems. *Rev. Physiol. Biochem. Pharmac.* **77**, 1-124.

Starke, K. (1981). Alpha-adrenoceptor subclassification. *Rev. Physiol. Biochem. Pharmac.* **88,** 199-236.

Starke, K. (1987). Presynaptic alpha-autoreceptors. *Rev. Physiol. Biochem. Pharmacol.* **107,** 73-146.

Starke, K., Göthert, M. and Kilbinger, H. (1989). Modulation of neurotransmitter release by presynaptic autoreceptors. *Physiol. Rev.* **69,** 864-989.

Struyker-Boudier, H.A.J. and Van Rossum, J. M. (1972). Clonidine-induced cardiovascular effects after stereotactic aplication in the hypothalamus of rats. *J. Pharm. Pharmacol.* **24,** 410-411.

Summers, R.J., Jarrott, B. and Louis, W.J. (1980). Selectivity of a series of clonidine-like drugs for alpha-1 and alpha-2 adrenoceptors in rat brain. *Neuroscience Letters.* **20,** 347-350.

Timmermans, P.B.M.W.M., Hoefke, W., Stähle, H. and Van Zwieten, P. A. (1980). Structure-activity relationships in clonidine-like imidazolidines and related compounds. *Prog. Pharmacol.* **3,** 1-104.

Timmermans, P.B.M.W.M., de Jonge, A., Thoolen, M.J.M.C., Wilffert, B., Batink, H. and Van Zwieten, P.A. (1984). Quantitative relationships between alpha-adrenergic activity and binding affinity of alpha-adrenoceptor agonists and antagonists. *J. Med. Chem.* **27,** 495-503.

Timmermans, P.B.M.W.M. (1988). Intra- and extraneuronal vascular alpha-adrenoceptors. In *Epinephrine in the central nervous system* (Eds) Stolk, J.M., U'Prichard, D.C. and Fuxe, K., Oxford University Press, New York, pp. 329-339.

Tuck, M. L. (1986). The sympathetic nervous system in essential hypertension. *Am. Heart J.* **112,** 877-886.

Unnerstall, J. R., Kopajtic, T. A. and Kuhar, M. J. (1984). Distribution of alpha-2 agonist binding sites in the rat and human central nervous system: Analysis of some functional, anatomic correlates of the pharmacologic effects of clonidine and related adrenergic agents. *Brain Res. Rev.* **7,** 69-101.

U'Prichard, D. C., Bechtel, W. D., Rouot, B. M. and Snyder, S. M. (1977). Multiple apparent alpha-noradrenergic receptor binding sites in rat brain: effect of 6-hydroxydopamine. *Mol. Pharmacol.* **16,** 47-60.

Virtanen, R., Savola, J-M., Saano, V. and Nyman, L. (1988). Characterization of the selectivity, specificity and potency or medetomidine as an alpha-2 adrenoceptor agonist. *Eur. J. Pharmacol.* **150,** 9-14.

Walker, B. R., Schneider, B. E. and Gold, J. A. (1980). A two-year evaluation of guanabenz in the treatment of hypertension. *Curr. Ther. Res.* **27,** 786-796.

Weber, M. A., Drayer, J.I.M. and Laragh, J. H. (1978). The effects of clonidine and propranolol separately and in combination on blood pressure and plasma renin activity in essential hypertension. *J. Clin. Pharmacol.* **18**, 233-240.

Weber, M. A., Graettinger, W. F. and Cheung, D. G. (1990). Centrally acting sympathetic inhibitors. In *Hypertension pathophysiology, diagnosis, and management* (Eds) Laragh, J.H. and Brenner, B. M., Ravan Press Ltd., New York, Chapter 142, pp. 2251-2261.

Weidmann, P., Hirsch, D., Maxwell, M. H. *et al.* (1974). Plasma renin and blood pressure during treatment with methyldopa. *Am. J. Cardiol.* **34,** 671-676.

Wiedemann, K., Benkert, O. and Holsboer, F. (1989). B-HT920- a novel dopamine autoreceptor agonist in the treatment of patients with schizophrenia. *Pharmacopsychiatry.* **22,** 221 Abstr.

Wilkins, L. H., Winternitz, S. R., Oparil, S. *et al.* (1981). Lofexidine and clonidine in moderate essential hypertension. *Clin. Pharmacol. Ther.* **30,** 752-757.

Yablonsky, F. and Dausse, J. P. (1989). Amiloride interacts with [^{3}H] idazoxan and [^{3}H]rauwolscine binding sites in rabbit urethra. *Eur. J. Pharmacol.* **164,** 167-170.

Van Zwieten, P. A. (1975). Antihypertensive drugs with a central action. *Prog. Pharmacol.* **1,** 1-66.

Van Zwieten, P. A. (1976). Centrally mediated action of alpha-methyl-dopa. In *Regulation of blood pressure by the central nervous system* (Eds) Onesti, G., Fernandes, M. and Kim, K.E., Grune & Stratton, New York, pp. 293.

Van Zwieten, P. A. and Timmermans, P.B.M.W.M. (1980). Centrally mediated hypotensive activity of B-HT-933 upon infusion via the cats vertebral artery. *Pharmacology* **21,** 327-332.

Progress in Hypertension, Vol. 2, pp. 269-289.
Saito *et al.* (Eds)

INDAPAMIDE AS AN ANTIHYPERTENSIVE AGENT

NOBORU TODA

Department of Pharmacology, Shiga University of Medical Sciences, Seta, Ohtsu 520-21, Japan

INTRODUCTION

Indapamide, 4-chloro-N-(2-methyl-1-indolinyl)-3-sulfamoylbenzamide hemihydrate, is an antihypertensive diuretic of the non-thiazide type developed at the Servier Research Institute in France (Fig. 1). Antihypertensive diuretics of the thiazide group, including hydrochlorothiazide (HCT), trichlormethiazide (TCM), and other analogs, have been widely used and evaluated as useful therapeutics for hypertension. During analytical studies to obtain potent and long-lasting antihypertensive drugs with a mild diuretic effect from indoline or isoindoline derivatives of sulfamoyl-benzamide, indapamide was found to be the best (Beregi, 1977). Indapamide has hypotensive and diuretic effects (Kyncl *et al.*, 1975; Morishita *et al.*,1982; Suzuki *et al.*,1977), and its vasodilator action is more evident than that of TCM (Osumi, 1982; Miyazaki *et al.*, 1985; Morishita *et al.*, 1987). The drug is more potent than TCM in lowering blood pressure in hypertensive patients given orally at doses of 2-3 mg/day (Nakashima *et al.*, 1982; Murakami *et al.*, 1980; Hashida, 1977; Ikeda *et al.*, 1983; Murakami *et al.*, 1982; Takeda, 1986).

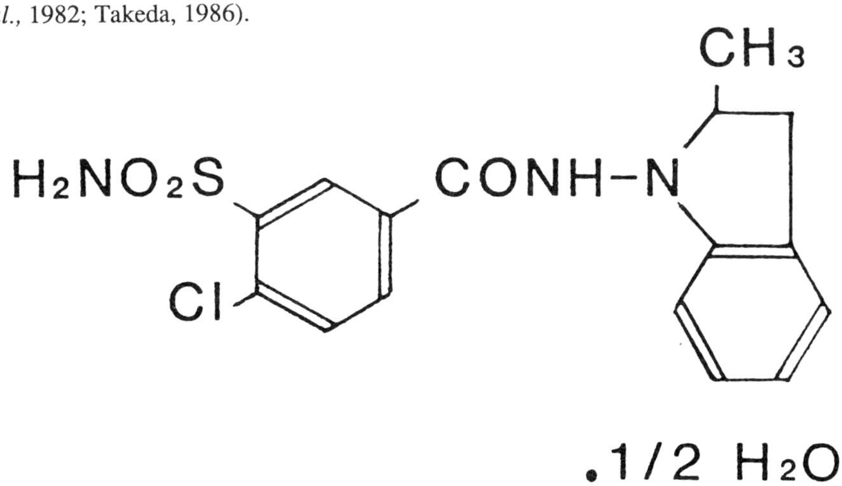

Figure 1. Chemical structure of indapamide.

These characteristics of indapamide may be explained by the following properties : Indapamide is highly soluble in lipid, easily binds to protein (Campbell *et al.*, 1977; Kitao *et al.*, 1982), is easily taken up by the vascular wall (Campbell *et al.*, 1977), is easily reabsorbed by the proximal tubule (Campbell *et al.*, 1977) and is slowly excreted in the urine (Campbell *et al.*, 1977; Kitao *et al.*, 1982).

PHARMACOLOGY

Hypotensive Effects

Indapamide showed little or no hypotensive effect in normotensive rats (30 mg/kg, single oral administration) (Morishita *et al.*, 1982) and dogs (10 mg/kg, single intravenous administration) (Kato *et al.*, 1982), but produced evident, persistent hypotension in animal models of hypertension (Kyncl *et al.*, 1975; Morishita *et al.*, 1982). In deoxycorticosterone acetate (DOCA)-salt or uni-nephrectomized-DOCA-salt hypertensive rats and spontaneously hypertensive rats (SHR), indapamide given orally at doses of 1-30 mg/kg resulted in dose-dependent and long-lasting hypotensive effects. The maximum blood pressure reduction was observed 24-29 hours after administration, and significant hypotension persisted for 72-101 hours. In dogs rendered hypertensive by bilateral renal encapsulation, indapamide at an oral dose of 5 mg/kg lowered blood pressure by 40 mmHg 48 hours later. The pressure gradually returned but was maintained at a level approximately 15 mmHg lower than that prior to the administration on the 4th day. Following the continuous oral administration of indapamide (0.3-10 mg /kg/day) for 2 weeks significantly lowered blood pressure in uni-nephrectomized-DOCA-salt hypertensive rats and SHR, and the hypotension continued during the period of administration (Fig. 2). Blood pressure slowly returned to the control level after cessation of administration, and no rebound phenomenon was observed. Furthermore, indapamide administered orally at doses of 0.5 and 10 mg/kg/day for 11 weeks prevented the development of hypertension and cardiac hypertrophy in DOCA-salt hypertensive rats (Hicks, 1979). Indapamide was three times as potent as TCM in lowering blood pressure (Kyncl *et al.*, 1975; Morishita *et al.*, 1982).

Diuretic Effects

A single oral administration of indapamide elicited significant diuresis in normal and pathological rats (Morishita *et al.*, 1982; Suzuki *et al.*, 1977). At dosages of 0.1-10 mg/kg in normal rats, 0.3-30 mg/kg in SHR, 0.05-10 mg/kg in normal rats under experimental acidosis and alkalosis, and 1.0-3.0 mg/kg in modified Masugi's nephritic rats, indapamide significantly increased the urine volume and urinary Na^+ and Cl^- excretion. Diuresis in SHR (Fig. 3) and Na^+ excretion in nephritic rats were more pronounced than those in normal rats. The effect of indapamide on K^+ excretion was less evident than that on the excretion of other electrolytes. Intravenous administration of indapamide (0.1-1.0 mg/kg) induced dose-dependent, persistent increases in the urine volume and urinary Na^+, Cl^-, and K^+ excretion in anesthetized dogs. The effect

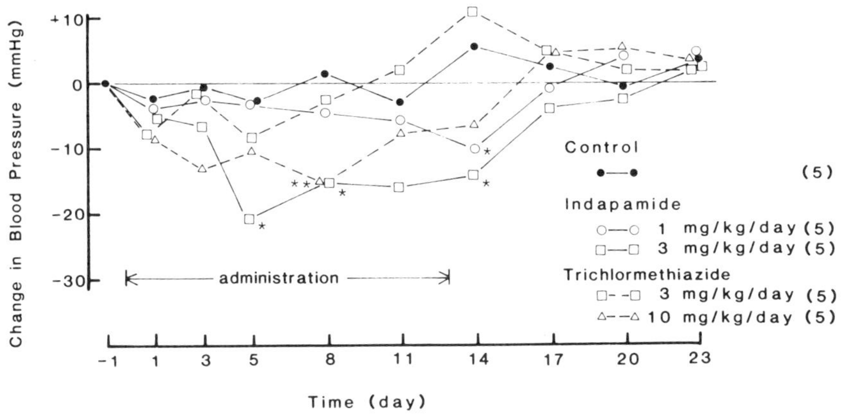

Figure 2. Changes in systolic blood pressure of spontaneously hypertensive rats during and after repetitive administration of indapamide and trichlormethiazide for 2 weeks. Significantly different from control, * $p < 0.05$, ** $p < 0.01$. Numbers in parentheses indicate the number of animals. (From Morishita *et al.*, 1982. Reproduced with permission.)

of indapamide on K^+ excretion, however, was lower than that of TCM (Suzuki *et al.*, 1977).

The mechanism and the site of diuretic action of indapamide (0.1-1.0 mg/kg, *i.v.*) were studied by the renal clearance and renal stop flow methods in anesthetized dogs (Suzuki *et al.*, 1977). Indapamide decreased the renal tubular reabsorption of water, Na^+ and K^+ without affecting renal plasma flow and glomerular filtration rates, resulting in increased urine volume and urinary Na^+ and K^+ excretion. As the result of increments in Na^+ and Cl^- excretion, osmolar clearance increased, and free water reabsorption increased in association with water diuresis. In the stop flow study, indapamide depressed Na^+ reabsorption in the distal tubules. These results indicate that the diuretic action of indapamide is not due to the increment of renal plasma flow and glomerular filtration rate, but may be associated with a suppression of reabsorption of water and Na^+ mainly in the distal tubule, which is similar to that of thiazide type diuretics. Low doses of indapamide cause lower diuretic action than TCM. This may be ascribed to the fact that indapamide is absorbed mainly at the proximal tubule (Campbell *et al.*, 1977) and acts mainly on the distal tubule for diuresis.

Effects on the Blood Vessels

Depressor effects of indapamide in hypertensive rats were greater than those of TCM, a thiazide-type diuretic, whereas diuretic effects of indapamide were less than or almost the same. Hypotensive effects of TCM are postulated to be derived from decreases in the extracellular fluid volume and peripheral vascular resistance.

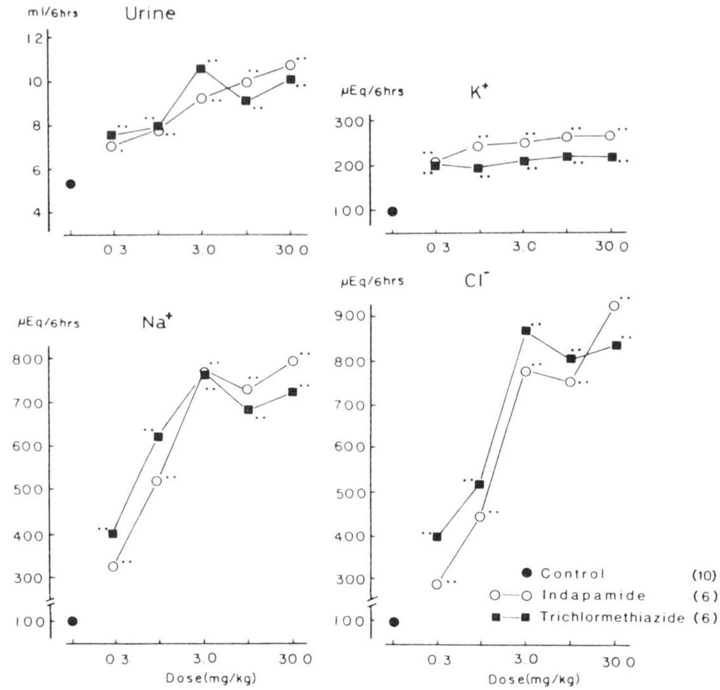

Figure 3. Dose-dependent effects of indapamide and trichlormethiazide on urine volume and electrolytes excretion in spontaneously hypertensive rats. Significant differences from controls, * $p < 0.05$, ** $p < 0.01$. Numbers in parentheses indicate the number of animals. (From Morishita *et al.*, 1982. Reproduced with permission.)

However, the hypotension caused by indapamide is due to some additional mechanisms, which are suggested by the following results. Indapamide in a dose of 10 mg/kg/day was administered orally for 2 weeks in uni-nephrectomized DOCA-salt hypertensive rats with systolic pressure of around 180 mmHg. The animals were then pithed, and the pressor responses induced by electrical stimulation of adrenergic nerves and intravenous administrations of angiotensin II, norepinephrine and tyramine were compared with those in a control group. Indapamide lowered the systolic blood pressure by 20 mmHg during the period of administration, and in the pithed animals treated with indapamide, the pressor responses to electrical nerve stimulation and angiotensin II were suppressed. Although administration of 30 mg/kg/day of TCM caused hypotension of a similar extent to that of indapamide, the pressor response to the stimuli was not suppressed in rats treated with TCM (Osumi, 1982).

In the isolated, perfused dog mesenteric vasculature, intra-arterial bolus injections of indapamide in doses of 6.3-100 μg produced a dose-dependent decrease in the perfusion pressure (Fig. 4) (Miyazaki *et al.,* 1985), and continuous infusions of 1 and 3 μg/ml persistently lowered it (Morishita *et al.,* 1987). In the isolated, perfused rat mesenteric artery, 10-100 μg/ml of indapamide dose-dependently suppressed the pressor responses to norepinephrine and KCl and to electrical stimulation of adrenergic nerves (Morishita *et al.*, 1987). TCM was evidently less potent than indapamide. In the electrophysiological study on isolated rabbit portal veins (Mironneau and Gargouil, 1979), indapamide (10^{-4} g/ml) significantly decreased the amplitude of the action potential evoked by depolarization, the influx of Ca^{2+} across cell membranes, the efflux of K^+ out of the cells, and phasic contraction. The drug also depressed the increase by angiotensin II in the transmembrane influx of Ca^{2+} and the phasic contraction. In the isolated rat portal vein, indapamide (3 x 10^{-4}M) suppressed the transient contraction evoked by norepinephrine (Mironneau *et al.*, 1986). In helical strips of dog mesenteric arteries partially contracted with prostaglandin (PG) $F_{2\alpha}$, indapamide (10^{-6}-10^{-4} M) caused a concentration-dependent relaxation, which was not influenced by pretreatment with propranolol, atropine, chlorpheniramine, cimetidine and aspirin or endothelium denudation (Miyazaki *et al.*, 1985). Similar relaxations were also observed with indapamide in concentrations of 3 x 10^{-5} M or higher in isolated rabbit cerebral, coronary, renal, mesenteric, and femoral arteries and aortae contracted with $PGF_{2\alpha}$ (Usui *et al.*, 1978). However, TCM in concentrations up to 10^{-4} M did not relax the isolated dog mesenteric arterial strips (Miyazaki *et al.*, 1985).

Effects on Prostaglandin and Thromboxane

PGI_2, PGE_2 and PGD_2 are depressor agents generated from blood vessels, and thromboxane (TX) A_2 is a potent vasopressor produced in platelets. These prostanoids are thought to participate in the control of vascular tone and blood flow by acting directly on smooth muscle and indirectly via sympathetic nerves. Because of the abnormal synthesis and metabolism of the cyclooxygenase products in SHR (Limas *et al.*, 1981; Uehara *et al.*, 1988), effects of indapamide on PGI_2 and TXA_2 generation were investigated.

Oral administration of indapamide in doses of 2 mg/kg/day for 2 weeks significantly increased the urinary excretion of 6-keto-$PGF_{1\alpha}$ in SHR (Izumi *et al.*, 1986). Indapamide in concentrations of 10^{-5} M increased the production of PGI_2 without affecting the TXA_2 production in cultured smooth muscle cells of the SHR aorta (Fig. 5) (Numabe *et al.*, 1989; Uehara *et al.*, 1990). Indapamide depressed the production of TXA_2 in the microsomal fraction of human platelets and increased the production of PGI_2 in sheep seminal vesicles (Gbeassor *et al.*, 1982). Indapamide was more active than furosemide. TCM decreased the PGI_2 production, and HCT failed to alter the production of PGI_2. Administration of indapamide (2.5 mg/day) for 6 weeks to 19 patients with essential hypertension significantly increased urinary $PGF_{2\alpha}$ excretion (LeBel *et al.*, 1983). In cultured rat vascular smooth muscle cells, indapamide (5 x10^{-5} M) significantly attenuated the free radical producing and PGI_2 synthesizing inhibitory actions of 15-hydroperoxyarachidonate (Uehara *et al.*, 1990). The scavenging effect may be involved in the enhanced PGI_2 generation by indapamide. PGI_2 vasodilates, inhibits platelet aggregation and retards vascular smooth muscle cell growth; therefore, the increased production of PGI_2 by

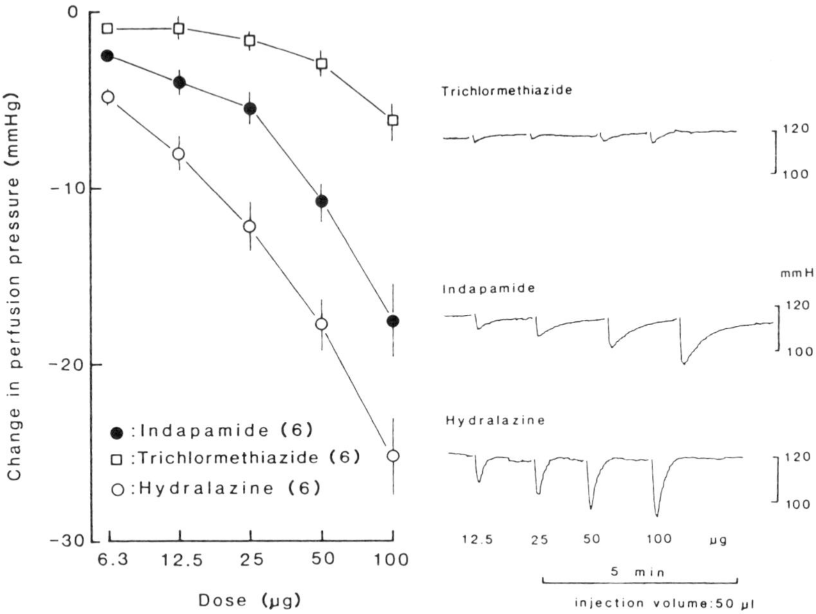

Figure 4. Effect of intra-arterial indapamide, trichlormethiazide and hydralazine (injection volume 50 µl) on isolated, perfused dog mesenteric arterial pressure. Means of 6 preparations per drug ± SEM. (From Miyazaki *et al.*, 1985. Reproduced with permission.)

indapamide would contribute to vasodilatation, improvement of circulatory disturbances and inhibition of hypertension-induced hypertrophy of blood vessels.

Peroxidation of linolenic acid by xanthine + xanthine oxidase and peroxidation of lipid in rat brain, liver and kidney homogenates were inhibited by indapamide (10^{-6} M or higher), and peroxidation of lipids in mitochondria of the rat heart by NADPH + ADP - Fe^{3+} was also inhibited (Hashizume *et al.*, 1990). TCM did not alter the lipid peroxidation (Hashizume *et al.*, 1990).

Effects on Lipid Metabolism, Platelet Aggregation, and Cell Proliferation

There are various side effects of antihypertensive diuretics of thiazide group. Clinical problems in their long-term administration are hypokalemia, hyperuricemia, and changes in blood sugar and lipid metabolism. The effects on lipid metabolism (Schoenfeld and Goldberger, 1964; Ames and Hill, 1976; Lardinois and Neuman, 1988; Weidmann *et al.*, 1983; Johnson and Danylchuk, 1989; Pollare *et al.*, 1989) and blood sugar (Pollare *et al.*, 1989; Ferguson, 1961; Hauman and Weller, 1961; Murphy *et al.*, 1982; Amery *et al.*, 1978) attenuate the ability of thiazide antihypertensives to prevent the manifestation of ischemic heart diseases (Medical Research Council Working Party, 1985; Wikstrand *et al.*, 1988; Fuller *et al.*, 1980; Jarrett *et al.*, 1982). Since these drugs were found to increase the serum cholesterol (CH) and triglyceride (TG), attention has been focused on these effects as problems in the long-term treatment of hypertension (Ames and Hill, 1976). The adverse effects of an increase in serum CH, low density lipoprotein (LDL)-CH and TG and a decrease in high density lipoprotein (HDL)-CH have been reported (Lardinois and Neuman, 1988). However, the effect of indapamide on the lipid metabolism is much lower than that of other antihypertensive diuretics (Weidmann *et al.*, 1983; Morishita *et al.*, 1990; Yoshino *et al.*, 1986; Meyer-Sabellek *et al.*, 1985; Perry, 1983; Scalabrino *et al.*, 1984; Gerber *et al.*, 1985). In addition, administration of 30 mg/kg of indapamide increased serum HDL-CH in rats fed a normal diet, and attenuated the increase of TG and the decrease of HDL-CH in rats fed with a low protein diet containing methionine, as did the increase of serum TG and LDL-CH in rats administered high lipid emulsion. The LDL-CH/HDL-CH ratio is decreased, thereby reducing the risk factor of atherogenesis. On the other hand, this beneficial effect on lipid metabolism has not been observed with TCM (Morishita *et al.*, 1990). It has been reported in clinical trials that indapamide increases HDL-CH and does not influence the level of LDL-CH in hypertensive patients (Yoshino *et al.*, 1986; Meyer-Sabellek *et al.*, 1985). HDL transports CH from peripheral tissues to the liver, thus removing CH from the body, and accelerates the transformation of PGH_2 to PGI_2 (Fleisher *et al.*, 1983). As the plasma HDL-CH level negatively correlates with the incidence of ischemic heart diseases (Miller and Miller, 1975), indapamide may be effective in preventing cardiac vascular disorders through the effect on lipid metabolism.

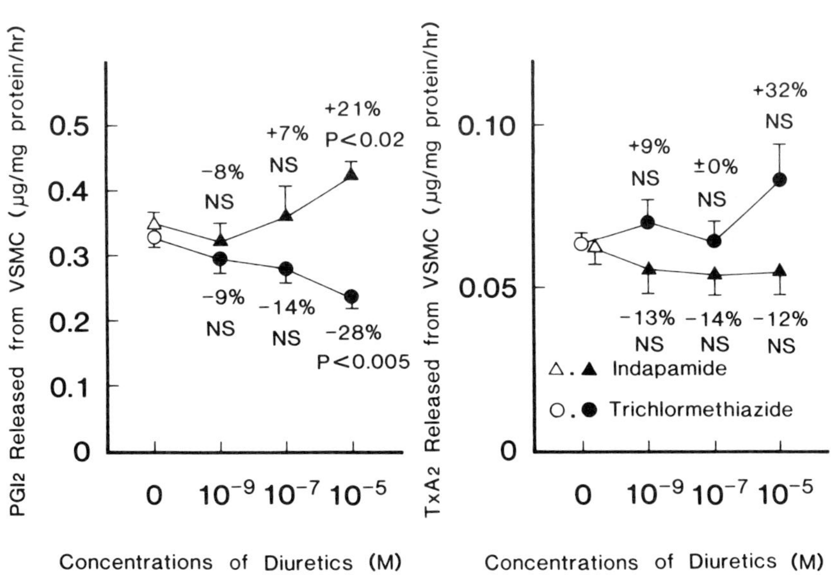

Figure 5. Effects of indapamide and trichlormethiazide on prostacyclin generation by SHR aorta smooth muscle cells (VSMC). Δ, O : indapamide- or trichlormethiazide-free control. Group difference is analyzed by one-way analysis of variance. The F-values for vascular 6-keto-$PGF_{1\alpha}$ generation are 13.9 for indapamide ($p < 0.001$) and 14.1 for trichlormethiazide ($p < 0.001$). NS: non-significant. (From Numabe *et al.*, 1989. Reproduced with permission.)

Platelet aggregation is accelerated in patients with hypertension and its complications, including angina pectoris, myocardial infarction, stroke and peripheral circulatory disorders. Platelet-derived vasoconstricting factors, such as TXA_2 and serotonin, participate in the manifestation and maintenance of hypertension. In *in vitro* experiments, indapamide in concentrations higher than 2 x 10^{-4} M inhibited the aggregation of rabbit platelets induced by collagen, arachidonate, labile aggregation stimulating substance (LASS) and U-46619, a PGH_2 analog, and in concentrations higher than 5 x10^{-5} M suppressed aggregation by thrombin, the release of serotonin,and the increment of intracellular free Ca^{2+} (by Ca^{2+} signal Fura 2) in platelets. The drug (2 x 10^{-4} M or higher) also suppressed the secondary aggregation of human platelets caused by epinephrine and ADP. However, indapamide did not affect the synthesis of LASS and malondialdehyde from arachidonic acid nor raised the cyclic AMP concentration in platelets. Decreased mobilization of Ca^{2+} in platelets may be the mechanism of the inhibitory effect of indapamide on platelet aggregation (Shirahase *et al.*, 1985; Shirahase *et al.*, 1987; Shirahase *et al.*, 1989). In a clinical study on platelet aggregation by ADP, collagen and ristocetin measured after the oral administration of indapamide (2 mg/day) for 12 weeks, the ADP-induced aggregation was significantly decreased (Shigekiyo *et al.*, 1987). Therefore, indapamide might be useful clinically in preventing circulatory complications in hypertensive patients.

It has been postulated that the intima of blood vessels injured in patients with hypertension and hyperlipemia are hypertrophied by repetitive stimulation of LDL-CH and platelet-derived growth factor (PDGF) discharged from platelets (Ross and Glomset, 1976), leading to the development of atherosclerosis. Indapamide inhibits the platelet aggregation, depresses the increment of serum LDL-CH, and decreases the atherogenic risk factor. In addition, indapamide (10^{-4} M or higher) selectively inhibits the proliferation of mouse Balb/C 3T3 cells caused by rabbit serum and human-PDGF (Shimaji *et al.*, 1989).

The inhibitory effects of indapamide on cell proliferation and platelet aggregation and its decreased adverse effects compared with thiazides on lipid metabolism may be desirable for antihypertensive therapy. In addition to an increase in PGI_2 production, these results indicate the anti-atherosclerotic effect of indapamide. Long-term follow up studies are underway.

Hypotensive Mechanism

Hypotensive effects of thiazide diuretics have been postulated to be derived from a sustained decrease in extracellular fluids caused by diuresis and a decrease in the total vascular resistance. As mentioned above, indapamide vasodilates various blood vessels *in vitro* and inhibits vasoconstriction by endogenous substances. Indapamide vasodilated *in vitro* in concentrations ranging from 0.3 to 30 μg/ml ; the low range of concentration is comparable to the maximum indapamide concentration (0.1-0.2 μg/ml) in human plasma after the administration of a therapeutic dose (Kitao *et al.*, 1982). Vasodilation observed in experimental animals would therefore contribute to lower the systemic blood pressure in hypertensive patients.

Characteristics and possible hypotensive mechanisms of indapamide action are as follows. (1) Indapamide is stable in plasma and present in an unaltered form for long periods. At an oral dose of 10 mg/kg of indapamide, approximately 1 μg/ml is observed 12 hours later in dogs (Kitao *et al.*, 1982; Sakata *et al.*, 1982), and at a dose of 4 mg of indapamide, the maximum plasma concentration of 0.2 μg/ml lasts for approximately 3 hours in volunteers (Kitao *et al.*, 1982). (2) The drug is retained for long periods in the vascular wall. In experiments with the perfused rat hindleg, the concentration of indapamide in the blood vessels is about 10 times higher than that in the perfusing fluid (Campbell *et al.*, 1977). (3) Despite the fact that the diuretic effect of indapamide is almost the same or less, than that of TCM, the hypotensive effect of indapamide is 3 times greater in SHR (Morishita *et al.*, 1982). A greater hypotensive effect of indapamide compared with TCM is observed in patients with essential hypertension (Murakami *et al.*, 1982; Takeda, 1986). (4) Indapamide increases the production of PGI_2, a potent vasodilator substance. This effect is not observed in TCM and HCT, thiazide type diuretics, that produce only a weak vasodilatation (Numabe *et al.*, 1989; Uehara *et al.*, 1990). (5) Clinical studies suggest the evident vasodilator effect of indapamide and the inhibitory effect on vasoconstriction. Pressor responses of hypertensive patients to intravenous administration of norepinephrine or angiotensin II are smaller after treatment with indapamide for 4-6 weeks than those before the treatment (Leenen *et al.*, 1988; Grimm *et al.*, 1981). After continuous administration for 4 weeks or 1 year, hypotension and decreased vascular resistance are maintained, but the cardiac output and circulating plasma volume are not altered (Leenen *et al.*, 1983; Takata *et al.*, 1982). (6) Indapamide shows hypotensive effects even in patients with renal failure and on hemodialysis (Leenen *et al.*, 1988; Acchiardo and Skoutakis, 1983; Brennan *et al.*, 1982; Raveau-Landon and Coulomb, 1984; Miach and Dawborn, 1986). (7) Species variations in the responsiveness to indapamide have been recognized. The distinct hypotensive effect is seen in humans after an oral dose of 0.03 mg/kg/day (Nakashima *et al.*, 1982; Murakami *et al.*, 1980; Hashida, 1977; Ikeda *et al.*, 1983; Murakami *et al.*, 1982; Takeda, 1986), whereas doses of 1-3 mg/kg/day are required to induce the same effect in hypertensive rats (Kyncl *et al.*, 1975; Morishita *et al.*, 1982; Suzuki *et al.*, 1977).

From the findings presented so far, indapamide-induced hypotension is probably due to the inhibitory effect on smooth muscle contractility of resistance vessels and to the release of vasodilator substances from blood vessels, together with a diuretic action. The vasodilator effect on large arteries may be beneficial in increasing vascular compliance, thus decreasing the load to the heart.

TOXICOLOGY

The acute toxicity of indapamide has been investigated after oral, subcutaneous, intraperitoneal and intravenous administration in mice, rats and rabbits. No morbidity or mortality was observed in mice and rats after oral, subcutaneous and intraperitoneal administrations (5-20 g/kg). Following intravenous administration, ataxia, respiratory depression and convulsions were observed with 400-500 mg/kg in mice and rats, and with 150-200 mg/kg in rabbits. LD_{50} values ranged from 337-1050 mg/kg (Osumi *et al.*, 1981).

Subacute toxicity in rats was studied by oral administration with 100, 300, 1000 and 3000 mg/kg/day for 3 months. Increased water intake was noted in the first half of the treatment period, together with an increase in end-of-study kidney and liver weight in males and females receiving 1000 mg/kg or more. Increased adrenal weight was seen in females receiving 3000 mg/kg. No other pathological changes were seen in the major organs (Osumi *et al.*,1981).

Chronic toxicity was studied by oral administration with 3, 30, 300 and 1500 mg/kg/day for 12 months in rats (Osumi *et al.*, 1981) and with 3, 30 and 300 mg/kg/day for 6 months in dogs (Osumi *et al.*, 1983). No changes were observed in rats. In dogs, decreased food intake and body weight were seen in those receiving 30 and 300 mg/kg. Additionally, 4/5 of the males and 2/3 of the females in the 300 mg/kg group developed exhaustion that they came to autopsy during the treatment period. Dogs exhibited the following changes at autopsy : pericarditis, myocarditis and perivascular hemorrhagic changes with associated tissue inflammation and granulation. Since concomitant KCl administration significantly attenuated the decrease in food intake and body weight and the hemorrhagic changes in the high dose group (Osumi *et al.*, 1983), a probable causal factor in these effects is hypokalemia induced at this dose level.

Reproductive and developmental toxicity was studied by administering indapamide prior to gestation and in early gestation in rats (Seg. I), during fetal organogenesis in rats and rabbits (Seg. II) and during the perinatal period and lactation in rats (Seg. III) (Seki *et al.*, 1982). In rats, no changes were seen in fertility, gestation, parturition or the post-natal period. No morphological changes or malformations were noted in fetuses or neonates, though a low birth weight due to growth retardation was observed in the 1000 mg/kg group. However, post-natal catch-up growth was such that the mature animals were normal in weight. Morphogenic and functional differentiations were not affected, while reproductive activity and gestation in mature animals were also normal. No F_1 fetal abnormalities were noted. In rabbits, the only abnormality was a transient decrease in food intake in pregnant dams in the 5 mg/kg group. There were no fetal abnormalities. It can therefore be assumed that indapamide has no adverse effects on reproduction and development.

Indapamide was tested for antigenicity in guinea-pigs using agar gel diffusion, the Schultz-Dale reaction, histamine release from extracted lung sections, cutaneous anaphylaxis over 3 hours and systemic anaphylaxis; results were negative in all cases (Nozaki *et al.*, 1982).

Carcinogenicity was tested by administering 10, 30 and 100 mg/kg indapamide daily with food for 24 months to rats and for 21 months to mice; results were negative in all cases (unpublished results).

PHARMACOKINETICS

Peak blood levels of indapamide in healthy adult volunteers after single oral dosing with 2 and 4 mg were approximately 85 and 180 ng/ml respectively, 2 hours after administration. Elimination from the blood was relatively slow, with a half-life of some 10 hours (Fig. 6). The principal metabolites are 2,3-dehydro-indapamide, which is present in low levels in blood, 4-chloro-3-sulphamoylbenzoic acid and 5-hydroxy-indapamide, which are mostly present as glucuronide conjugates in urine (Kitao *et al.,* 1982). Blood concentrations reach a plateau 4 days after starting treatment with 2 mg daily, with peak levels ranging from 110 -130 ng/ml. The half-life of elimination after cessation of a 2-week course of treatment is virtually identical to that after single dosing (Orita *et al.*, unpublished observation). Indapamide 2 and 3 mg daily was administered to patients with essential hypertension for 7.6 and 6.5 months, respectively; mean blood levels were approximately 120 and 170 ng/ml, respectively, i.e. similar to those in the 2-week study, and no drug accumulation effects were noted, despite long-term administration (Orita *et al.*, unpublished observation).

The absorption, elimination and distribution of ^{14}C-labelled indapamide have been investigated in rats, dogs and goats (Kitao *et al.,* 1982; Kitao *et al.,* 1982). Indapamide is absorbed relatively quickly after oral administration, with peak blood levels of unchanged drug being detected after 1-2 hours. Elimination from plasma is biphasic, with the half-life of the beta phase exceeding that of other antihypertensive diuretics. Studies in rats suggest that the bulk of the drug is eliminated in the feces, via the bile. Only 0.13 % of an administered dose is secreted in milk in goats.

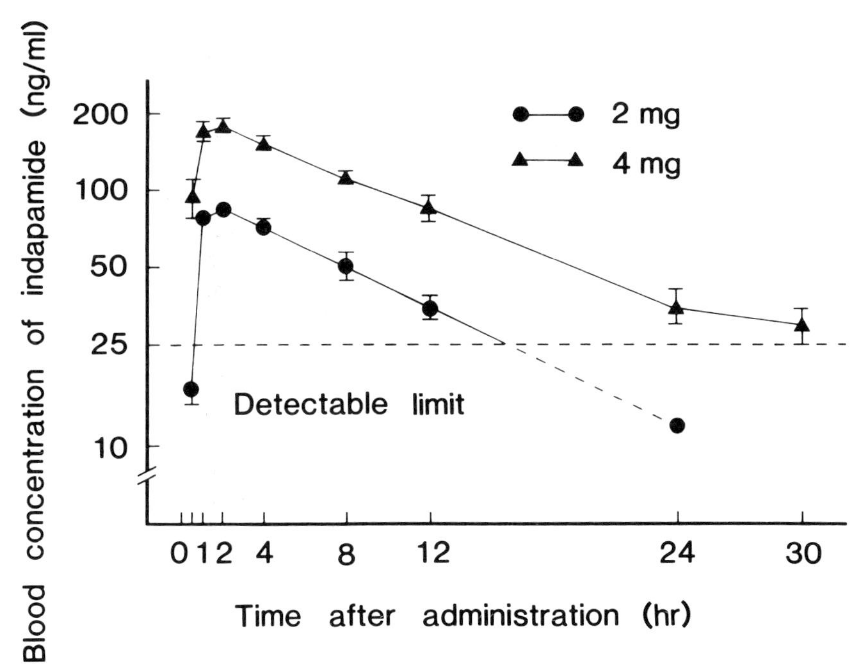

Figure 6. Blood indapamide concentrations following oral dosing in healthy volunteers. Mean ± SEM (n = 12). (From Kitao *et al.*, 1982. Reproduced with permission.)

Radioactivity is rapidly distributed into all organs in normal rats, with maximum concentrations being observed 0.5 -1 hour after administration. Concentrations are higher in the liver and kidney than in plasma, and very low in the brain. The half-life of radioactivity elimination in virtually all organs is 3 -6 hours, and 9 -10 hours in the kidney. Expired gases account for 2 -2.5 % of the excretion of the total dose. Tissue radioactivity reaches a plateau 2-3 weeks after repeated administration, with elimination from each tissue after cessation of administration being virtually identical to that after a single administration. Thus, no residual drug effects and no accumulation in specific organs are observed. The distribution pattern in pregnant rats is almost the same as in non-pregnant rats, and transplacental transfer is low.

Unlabeled and ^{14}C-labeled indapamide metabolism has been studied in rats and dogs (Sakata *et al.*, 1982). The principal metabolic pathway is hydroxylation of the indoline ring, hydrolysis of the acid-amide bond and aromatization of the indoline ring to an indole ring. The principal metabolites are 2,3-dehydro-indapamide and 5-hydroxy-indapamide in blood, 5-hydroxy-indapamide and 4-chloro-3-sulphamoyl-benzoic acid in urine, and 5-hydroxy-indapamide in faeces. The plasma concentration of 2,3-dehydro-indapamide 4-10 hours after administration is higher than the concentration of intact indapamide, and also elimination is slow. Since the metabolites have almost the same hypotensive and diuretic effects as the unchanged compound (Nishimura *et al.*, 1982), they are assumed to contribute to the pharmacological actions of indapamide. The bulk of 5-hydroxy-indapamide is excreted in bile as glucuronides or sulphates.

The serum protein binding of indapamide is relatively high : 80-85 % in rats and man (Kitao *et al.*, 1982).

Table 1. Seated blood pressure values (mmHg) before and after treatment with indapamide.

Treatment	Before trial	After 2 weeks	After 4 weeks	After 6 weeks
Indapamide 1 mg/day (n = 35)				
Systolic	175.5±3.05	159.1±3.36	159.4±3.49	150.3±2.74
Diastolic	100.7±1.80	93.2±1.79	94.2±1.82	90.9±2.13
Mean arterial	125.4±1.92	114.5±2.04	115.9±2.12	111.6±1.90
Indapamide 2 mg/day (n = 30)				
Systolic	174.4±3.07	158.1±4.66	153.1±3.17	150.9±3.72
Diastolic	101.9±2.04	93.5±2.04	89.4±2.04	88.5±1.84
Mean arterial	126.0±1.91	115.0±2.63	111.3±2.24	110.2±2.36

Note: changes from baseline levels statistically significant ($p < 0.05$) for all treatment values.
n : Number of hypertensive patients, mean ± SD. (From Hashida, 1977. Reproduced with permission.)

CLINICAL STUDIES

The safety and diuretic effect of indapamide were addressed in a Phase I trial in healthy adult volunteers using single dosing of 1,2 and 4 mg (Orita, Y. *et al.*, unpublished study) and in adult male volunteers with borderline hypertension given a daily oral dose of 2 mg for 2 weeks (Nakashima *et al.*, 1982). No significant safety problems were encountered and increases were noted in urine volume and electrolyte excretion. The diuretic effect appeared delayed compared to TCM (4 mg once daily) and was more sustained, though maximum activity was lower. After treatment for 2 weeks, blood pressure was non-significantly reduced, and cardiac and stroke indices significantly decreased.

Antihypertensive efficacy and the dose-response relationship were investigated by a double-blind trial in 101 patients with essential hypertension orally dosed with 1 or 2 mg daily (Murakami *et al.*, 1980; Hashida, 1977). Both doses of indapamide showed significant and dose-related hypotensive effects (Table 1). Side-effects were mild and virtually identical to those in controls. These findings suggest that indapamide is a clinically effective antihypertensive at doses of 1-2 mg daily. This was confirmed in a circadian rhythm study in patients with essential hypertension (Ikeda *et al.*, 1983).

The efficacy and safety of indapamide were also evaluated in double-blind clinical trials versus TCM (Murakami *et al.*, 1982; Takeda, 1986) and meticrane (MTC) (Ikeda *et al.*, 1982). Indapamide 2-3 mg once daily proved to be a more effective antihypertensive than TCM 4-6 mg once daily (294 cases). Although there were a few side-effects such as dizziness, tiredness, anorexia and slight changes in laboratory parameters such as a decreased serum K^+ and an increased uric acid in all groups, they were significantly fewer in patients receiving indapamide 2 mg daily than those receiving MTC 150 mg twice daily (219 cases).

Indapamide 1-3 mg once daily has been administered to 567 hypertensive patients in open trials, including 191 patients on a long-term basis (Goto *et al.*, 1982; Baba *et al.*, 1982) (Saitoh, T. *et al.*, unpublished studies). Indapamide was effective in 76.1 % of cases and well-tolerated. Blood pressure reduction was also sustained in the long-term. 131 of the 191 patients in the long-term trial received the drug for more than 53 weeks. Although slight changes such as a decreased serum K^+ and an increased uric acid were observed, there were no serious or irreversible side-effects. Indapamide would therefore be appropriate in the long-term management of hypertension.

In another clinical trial, indapamide 2-3 mg daily for 12 months increased plasma renin activity, decreased left ventricular internal diameter, fractional shortening and contraction time, and reduced total peripheral resistance, without any effect on cardiac output or circulatory plasma volume. Since a hypotensive effect was still observed throughout the study, it is probable that the decrease in total peripheral resistance contributes to the hypotensive action of indapamide in addition to the diuretic effect, and that the drug enhances left ventricular function by decreasing blood pressure and fluid volume (Takata *et al.*,1982).

A number of studies have addressed the effects of indapamide on blood glucose (Baba *et al.*, 1982; Beling *et al.*, 1983; Raggi *et al.*, 1985; Harrower *et al.*, 1985; Leonetti *et al.*, 1988) and lipid metabolism (Weidmann *et al.*, 1983; Yoshino *et al.*, 1986; Meyer-Sabellek *et al.*, 1985; Perry, 1983; Scalabrino *et al.*, 1984; Gerber *et al.*, 1985). Indapamide 1-3 mg once daily for 52 weeks failed to impair glucose tolerance in 44 patients with essential hypertension (Goto *et al.*, 1982; Baba *et al.*, 1982) or in 60 patients with essential hypertension complicated by diabetes (borderline diabetes = 18 cases, overt diabetes = 42 cases) (Baba *et al.*, 1982), and almost no changes in insulinogenic index were observed after administration for 26 weeks (borderline = 10, overt diabetes = 5) (Baba *et al.*, 1982). Fasting blood glucose values were virtually unchanged from the pre-treatment baseline in 173 hypertensive patients receiving indapamide 2.5 mg once daily for one year (Leonetti *et al.*, 1988), as were blood glucose and insulin levels after glucose loading in 10 diabetic hypertensives (Harrower *et al.*, 1985). Furthermore, serum HDL-CH levels in 11 patients with essential hypertension receiving indapamide 2 mg once daily for 3 months were higher than pre-treatment values, with no elevation in LDL-CH (Yoshino *et al.*, 1986). In another study, although no changes from baseline were observed in serum total CH after treatment with indapamide 2.5 mg for 6 weeks and 6 months (n = 20), HDL-CH levels were significantly increased and the atherogenic risk factor (LDL-CH/HDL-CH) significantly decreased (Fig. 7) (Meyer-Sabellek *et al.*, 1985). Serum total CH, LDL-CH, VLDL-CH, HDL-CH, TG and apoprotein A_1 and A_2 were unchanged in 29 hypertensive patients after treatment with indapamide 2.5 mg for 6-8 weeks (Gerber *et al.*, 1985). It may therefore be assumed that indapamide is an

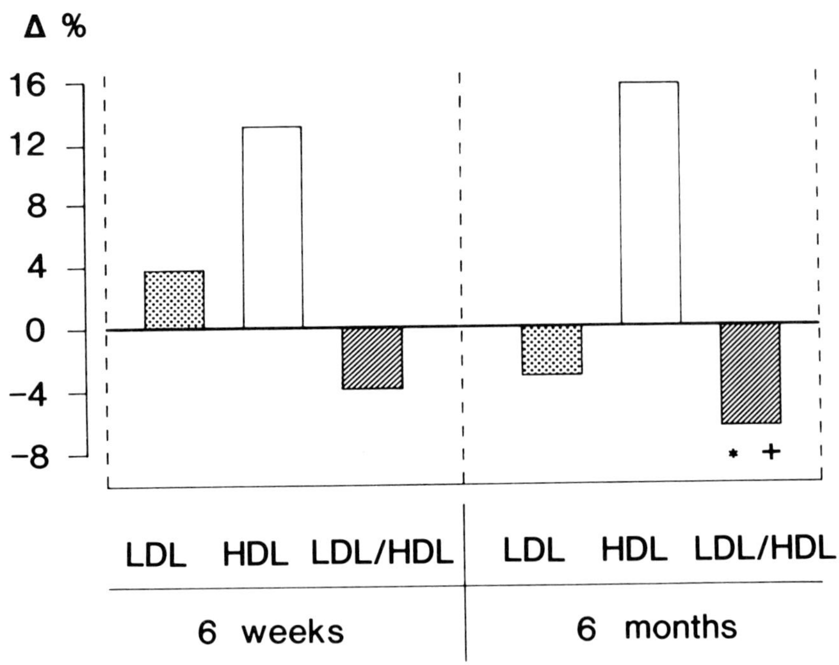

Figure 7. Differences from baseline (after placebo) in LDL cholesterol, HDL cholesterol, and LDL/HDL cholesterol ratio after treatment with indapamide 2.5 mg for 6 weeks (left columns) and 6 months (right columns). * $p < 0.05$ (6 months vs placebo), + $p < 0.05$ (6 months vs 6 weeks). (From Meyer-Sabellek *et al.*, 1985. Reproduced with permission.)

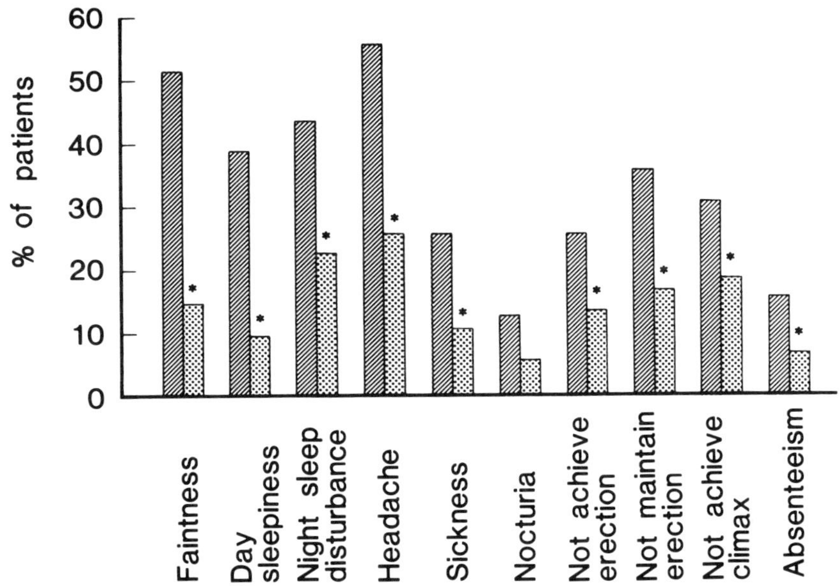

Figure 8. The percentage of patients reporting different side-effects on questionnaire at the start of the study (broken column) and 4 months after switching to indapamide 2.5 mg/day (dotted column). The order of side effects from left to right is the same as that on the questionnaire. * $p < 0.01$. (From Watters and Campbell, 1986. Reproduced with permission.)

antihypertensive diuretic with virtually no adverse effect on blood glucose and lipid metabolism.

Most antihypertensive thiazide diuretics have been reported as having no hypotensive effect in hypertensive patients with impaired renal function (glomerular filtration rate below 25 ml/min) (Acchiardo and Skoutakis, 1983; Thomas, 1985). Indapamide, on the other hand, has a distinct hypotensive effect in such patients (Leenen *et al.*, 1988; Acchiardo and Skoutakis, 1983; Brennan *et al.*, 1982; Raveau-Landon and Coulomb, 1984; Miach and Dawborn, 1986). For example, relatively good hypotensive effect was achieved in hypertensive patients with renal parenchymal disease (n = 8) and renovascular hypertension (n = 2) by administration of indapamide 1-3 mg once daily for 12 weeks (Honda, M *et al.*, unpublished study), in hypertensives with mild to moderate renal failure (creatinine clearance 36-84 ml/min, n = 13), in severely decreased renal function (creatinine clearance 8-27 ml/min, n = 4 ; 6-30 ml/min, n = 12) by 2.5 mg once daily for 6 weeks (Acchiardo and Skoutakis, 1983; Miach and Dawborn, 1986), and in hypertensives on chronic hemodialysis (n = 9) by 2.5 mg once daily for 4 weeks (Acchiardo and Skoutakis, 1983).

Given that antihypertensive treatment is likely in principle to be life-long, it is essential that it does not impair the quality of life (QOL). Several such studies have been conducted with indapamide (Watters and Campbell, 1986; Lacourciere, 1988; Guez *et al.*, 1988; Leonetti *et al.*, 1988). QOL questionnaires were presented to hypertensive patients (BP : 140/90 mmHg) 4 months after switching to monotherapy with indapamide 2.5 mg once daily ; significant improvement was observed in mood, physical symptoms, sleep disturbance, sexual activity and will to work (Fig. 8) (Watters and Campbell, 1986). Treatment of hypertensive patients with indapamide 2.5 mg once daily for 3 months had the same effect on blood pressure reduction and general well-being as captopril 12.5-50 mg twice daily (Lacourciere, 1988); significant improvement in QOL and well-being has also been reported in other studies conducted over 3 (Guez *et al.*, 1988) and 12 months (Leonetti *et al.*, 1988).

Concomitant administration of drugs with differing mechanisms of action, rather than high dose administration of a single drug, is widely used in the treatment of hypertension. Combination regimens are designed to improve hypotensive efficacy and reduce the dosage of the component drugs, and hence side-effects. Beta-blockers, nadolol (Houde *et al.*,1983), atenolol (Velasco *et al.*, 1982; De Divitiis *et al.*, 1983) and propranolol or pindolol (Takishita and Fukiyama, 1986), and ACE inhibitors (Kocijancic and Dimkovic, 1986) have been combined with indapamide, with an improved hypotensive effect in all cases, and a reduction in both dosage and adverse effects.

SUMMARY

Indapamide is a non-thiazide antihypertensive diuretic, with a more potent and sustained hypotensive action than the thiazides and a weaker diuretic effect. It has been shown in preclinical studies to be a more potent vasodilator than the thiazides and to inhibit vasoconstriction induced by pressor agents such as angiotensin II and sympathetic stimulation. For this reason, the mechanism of its antihypertensive activity is assumed to involve a direct blood vessel effect, *i.e.* vasodilatation and inhibition of vasoconstriction, together with a diuretic effect due to inhibition of Na^+ and water reabsorption in the distal tubule. Its clinical efficacy and safety have been confirmed in double-blind comparative trials in patients with essential hypertension. Treatment with 2-3 mg once daily *per os* results in better hypotensive effects than the reference antihypertensive diuretic, TCM, and fewer side-effects than MTC. Blood pressure reduction is sustained in treatment courses of 1 year or longer, acceptability is high, and adverse effects on glucose and lipid metabolism are minimal. Hypotensive activity is potentiated by combination with a beta-blocker or ACE inhibitor, and the side-effects of each drug are decreased. Quality of life, evaluated by such criteria as will to work, judgement ability and sexual activity, may actually improve in hypertensive patients on indapamide. The mechanism of its antihypertensive effect differs from that of conventional antihypertensive diuretics, and it is effective and well-tolerated, with fewer adverse effects. Indapamide therefore appears to be of value in the long-term treatment of hypertension.

Acknowledgements
The author is grateful to Mr. S. Osumi, Kyoto Pharmaceutical Industries, Ltd. for providing some of the information included in this review.

REFERENCES

Acchiardo, S. R., and Skoutakis, V. A. (1983). Clinical efficacy, safety and pharmacokinetics of indapamide in renal impairment. *Am. Heart J.* **106,** 237-244.

Amery, A., Berthaux, P., Bulpitt, C., Deruyttere, M., De Schaepdryver, A., Dollery, C., Fagard, R., Forette, F., Hellemans, J., Lund-Johansen, P., Mutsers, A. and Tuomilehto, J.(1978). Glucose intolerance during diuretic therapy. Results of trial by the European working party on hypertension in the elderly. *Lancet* **1,** 681-683.

Ames, R. P. and Hill, P. (1976). Elevation of serum lipid levels during diuretic therapy of hypertension. *Am. J. Med.* **61,** 748-757.

Baba, S., Amano, M., Iimura, M. *et al.* (1982). Study on safety of long-term administration of indapamide in patients with essential hypertension. Special study on the effects on glucose tolerance. *Geriat. Med.* **20,** 1207-1216. (in Japanese)

Beling, S., Vukovich, R. A., Neiss, E. S., Zisblatt, M., Webb, E. and Losi, M. (1983). Long-term experience with indapamide. *Am. Heart J.* **106,** 258-262.

Beregi, L. G. (1977). Antihypertensive and saluretic properties of the indoline and iso-indoline series. *Curr. Med. Res. Opin.* **5,** (Suppl. 1), 3-8.

Brennan, L., Wu, M. J. and Laquer, U. J. (1982). A multicenter study of indapamide in hypertensive patients with impaired renal function. *Clinical Therapeutics* **5,** 121-128.

Campbell, D. B., Taylor, A. R., Hopkins, Y. W. and Williams, J. R. B. (1977). Pharmacokinetics and metabolism of indapamide : a review. *Curr. Med. Res. Opin.* **5** (Suppl. 1), 13-24.

De Divitiis, O., Di Somma, S., Petitto, M., Fazio, S. and Ligouri, V. (1983). Indapamide and atenolol in the treatment of hypertension ; double-blind comparative and combination study. *Curr. Med. Res. Opin.* **8,** 493-500.

Ferguson, M. J. (1961). Saluretic drugs and diabetes mellitus. *Am. J. Cardiol.* **7,** 568-569.

Fleisher, L. N., Tall, A. R., Witte, L. D. and Cannon, P. J. (1983). Effects of high-density lipoprotein and the apoprotein of high-density lipoprotein on prostacyclin synthesis by endothelial cells. *Adv. Prost. Throm. Leukotri. Res.* **11,** 475-480.

Fuller, J. H., Shipley, M. J., Rose, G., Jarrett, R. J. and Keen, H. (1980). Coronary-heart-disease risk and impaired glucose tolerance. The Whitehall Study. *Lancet* **1,** 1373-1376.

Gbeassor, F. M., Grose, J. H. and Lebel, M. (1982). Influence of diuretics on prostaglandin and thromboxane synthesis. *Clin. Invest. Med.* **5,** 26B.

Gerber, A., Weidmann, P., Bianchetti, M. G., Ferrier, C., Laederach, K., Mordasini, R., Riesen, W. and Bachmann, C. (1985). Serum lipoprotein during treatment with the antihypertensive agent indapamide. *Hypertension* **7** (Suppl.II), 164-169.

Goto, Y., Tanabe, A., Tagawa, R. *et al.* (1982). Study on efficacy and safety of long-term administration of indapamide in patients with essential hypertension. *Geriat. Med.* **20,** 839-848. (in Japanese)

Grimm, M., Weidmann, P., Meier, A., Keusch, G., Ziegler, W., Gluck, Z. and Beretta-Piccoli, C. (1981). Correction of altered noradrenaline reactivity in essential hypertension by indapamide. *Br. Heart J.* **46,** 404-409.

Guez, D., Crocq, L., Safavian, A. and Labardens, P. (1988). Effects of indapamide on the quality of life of hypertensive patients. *Am. J. Med.* **84** (Suppl.1B), 53-58.

Harrower, A. D. B., McFarlane, G., Donnelly, T. and Gray, C. E. (1985). Effect of indapamide on blood pressure and glucose tolerance in non-insulin-dependent diabetes. *Hypertension* **7** (Suppl II), 161-163.

Hashida, J. G. (1977). A double-blind multicentre study of indapamide in the treatment of essential hypertension. *Curr. Med. Res. Opin.* **5** (Suppl. 1), 116-123.

Hashizume, Y., Shirahase, H., Kanda, M. and Osumi, S. (1990). Inhibitory effect of indapamide, an anti-hypertensive diuretic on lipidperoxidation. *Japan. J. Pharmacol.* **52** (Suppl. I), 335P.

Hauman, R. L. and Weller, J. M. (1961). Hyperglycemic effect of chlorothiazide. *Clin. Res.* **9,** 180A.

Hicks, P. E. (1979). The effects of long-term oral treatment with indapamide on the development of DOCA-salt hypertension in rats ; vascular reactivity studies. *Clin. Exp. Hypertension* **1,** 713-

731.
Houde, M., and Carrie're, S. (1983). Indapamide for out-patient treatment of hypertension ; Modifications in serum catecholamine levels. *Curr. Med. Res. Opin.* **8** (Suppl. 3), 68-76.
Ikeda, M., Abe, H., Kawai, C., Inoue, M., Kubo, S., Sakai, A., Sugitani, Y. and Kajiya, F. (1982). Clinical effect of indapamide on essential hypertension. Results of a multi-centered double blind comparative study with meticrane. *SOUGOU RINSYO* **31,** 1913-1921. (in Japanese)
Ikeda, M., Ito, K., Ishikawa, H. *et al.* (1983). Effect of indapamide on diurnal rhythm with patients in essential hypertension. *Jpn. Pharmacol. Ther.* **11,** 121-128. (in Japanese)
Izumi, Y., Honda, M. and Hatano, M. (1986). Antihypertensive mechanism of indapamide. *Jpn. Pharmacol. Ther.* **14,** 6049-6055.
Jarrett, R. J., McCartney, P. and Keen, H. (1982). The Bedford Survey : Ten year mortality rates in newly diagnosed diabetics, borderline diabetics and normoglycaemic controls and risk indices for coronary heart disease in borderline diabetics. *Diabetologia* **22,** 79-84.
Johnson, B. F. and Danylchuk, M. A. (1989). The relevance of plasma lipid changes with cardiovascular drug therapy. *Med. Clin. North Am.* **73,** 449-473.
Kato, E., Morishita, S., Saito, N., Usui, H., Shirahase, H., Kanda, M., Osumi, S., Kitao, K. and Seki, T. (1982). General pharmacology of indapamide (2). Influences on the respiratory and cardiovascular systems. *KISO TO RINSHO* **16,** 2233-2243. (Abs. in English)
Kitao, K., Fujitani, M., Kakeya, N., Honma, S., Yugi, S., Iwamura, S. and Kambegawa, A. (1982). Pharmacokinetic studies of indapamide, a new antihypertensive drugs (3). Distribution and serum protein binding in rats. *KISO TO RINSHO* **16,** 2312-2327. (Abs. in English)
Kitao, K., Sakata, T., Fujitani, M., Kakeya, N., Honma, S., Yugi, S., Iwamura, S. and Kambegawa, A. (1982). Pharmacokinetic studies of indapamide, a new antihypertensive drug (1). Absorption and elimination in rats and dogs. *KISO TO RINSHO* **16,** 2283-2293. (Abs. in English)
Kitao, K., Sakata, T., Ueshima, Y., Muro, H., Fujitani, M. and Murauchi, H. (1982). Phase I clinical study of a new non-thiazide antihypertensive drug indapamide ; Biological fate and diuretic action after a single oral dose in normal volunteers. *Jpn. Pharmacol. Ther.* **10,** 2619-2631. (Abs. in English)
Kocijancic, M. and Dimkovic, S. (1986). Antihypertensive effect of indapamide given in conjunction with captopril in severe hypertension. *Curr. Med. Res. Opin.* **10,** 313-318.
Kyncl, J., Oheim, K., Seki, T. and Solles, A. (1975). Antihypertensive action of indapamide. Comparative studies in several experimental models. *Arzneim.-Forsch.* (Drug Res.) **25,** 1491-1495.
Lacourciere, Y. (1988). Analysis of well-being and 24-hour blood pressure recording in a comparative study between indapamide and captopril. *Am. J. Med.* **84** (Suppl.1B), 47-52.
Lardinois, C. K. and Neuman, S. L. (1988). The effects of antihypertensive agents on serum lipids and lipoproteins. *Arch. Intern. Med.* **148,** 1280-1288.
LeBel, M., Grose, J. H., Belleau, L. J. and Langlois, S. (1983). Antihypertensive effect of indapamide with special emphasis on renal prostaglandin production. *Curr. Med. Res. Opin.* **8,** (Suppl.3), 81-86.
Leenen, F. H. H., Smith, D. L., Boer, W. H. and Marquez-Julio, A. (1983). Diuretic and Cardiovascular effects of indapamide in hypertensive subjects ; A dose-response curve. *Curr. Med. Res. Opin.* **8** (Suppl.3), 47-52.
Leenen, F. H. H., Smith, D. L., Farkas, R. M., Boer, W. H., Reeves, R. A. and Marquez-Julio, A. (1988). Cardiovascular effects of indapamide in hypertensive patients with or without renal failure. A dose-response curve. *Am. J. Med.* **84** (Suppl.1B), 76-85.
Leonetti, G., Rappelli, A., Salvetti, A. and Scapellato, L. (1988). Tolerability and well-being with indapamide in the treatment of mild-moderate hypertension ; An Italian multicenter study. *Am. J. Med.* **84** (Suppl.1B), 59-64.
Limas, C., Goldman, P. and Limas, C. J. (1981). Aortic phospholipid deacylation-reacylation cycle in spontaneously hypertensive rats. *Am. J. Physiol.* **240,** H33-H38.
Medical Research Council Working Party (1985). MRC trial of treatment of mild hypertension : principal results. *Br. Med. J.* **291,** 97-104.
Meyer-Sabellek, W., Gotzen, R., Heitz, J., Arntz, H. R. and Schulte, K. L. (1985). Serum

lipoprotein levels during long-term treatment of hypertension with indapamide. *Hypertension* **7** (Suppl.II), 170-174.

Miach, P. J. and Dawborn, J. K. (1986). Double-blind cross-over study of indapamide versus hydrochlorothiazide in hypertensive patient with renal impairment. In: *Indapamide and the current treatment of hypertension, Proceeding of the XIV international symposium on indapamide*, pp.67-74 Ogilvie, R. I. and Safar, M. (Eds), John Libbey, London.

Miller, G. J. and Miller, N. E. (1975). Plasma-high-density-lipoprotein concentration and development of ischemic heart disease. *Lancet* **1,** 16-19.

Mironneau, J. and Gargouil, Y-M. (1979). Action of indapamide on excitation-contraction coupling in vascular smooth muscle. *Eur. J. Pharmacol.* **57**, 57-67.

Mironneau, J., Savineau, J-P. and Mironneau, C. (1986). Effect of indapamide on the electromechanical properties of rat myometrium and rat portal vein. *J. Pharmacol. Exp. Ther.* **236,** 519-525.

Miyazaki, M., Morishita, S. and Toda, N. (1985). Vasodilator effect of indapamide in dog mesenteric vasculature. *Japan. J. Pharmacol.* **39,** 402-405.

Morishita, S., Nishimura, K., Kato, E., Shirahase, H., Osumi, S., Kitao, K. and Seki,T. (1982). Antihypertensive and diuretic effects of indapamide in normotensive and hypertensive rats. *Folia Pharmacol. Japon.* **79,** 137-146. (Abs. in English)

Morishita, S., Wada, K., Osumi, S. and Kakeya, N. (1987). Effect of indapamide on isolatated perfused mesenteric vasculature. *IGAKU TO YAKUGAKU* **18,** 1017-1020. (in Japanese)

Morishita, S., Shimaji, H., Wada, K. and Osumi, S. (1990). Effect of indapamide, a non-thiazide antihypertensive diuretic, on serum lipid levels in rats. *Folia Pharmacol. Japon.* **95,** 50P (in Japanese)

Murakami, M., Ikeda, M., Ozawa, Y. *et al.* (1980). Clinical evaluation with indapamide on patients with essential hypertension in 22 medical institutions ; Study on dose response by double-blind method. *Clin. Eval.* **8,** 437-456. (Abs. in English)

Murakami, M., Aoki, K., Arai, O. *et al.* (1982). Clinical evaluation with indapamide on patients with essential hypertension ; A double-blind multicentre study. *IGAKUNO AYUMI* **122,** 1010-1030. (in Japanese)

Murphy, M. B., Lewis, P. J., Kohner, E., Schumer, B. and Dollery, C. T. (1982). Glucose intolerance in hypertensive patients treated with diuretics ; A fourteen-year follow-up. *Lancet* **2,** 1293-1295.

Nakashima, M., Hashimoto, H., Oguchi, S., Oguro, K. and Nozue, Y. (1982). Clinico-pharmacological evaluation of a hypotensore, indapamide. Comparison with trichiormethiazide in successive administration. *RINSHOYAKURI* **13,** 383-392. (Abs. in English)

Nishimura, K., Morishita, S., Shirahase, H., Kato, E., Osumi, S., Kitao, K. and Seki, T. (1982). Antihypertensive diuretic effects and acute toxicities of optical isomers and principal metabolites of indapamide. *KISO TO RINSHO* **16,** 2275-2282. (Abs. in English)

Nozaki, Y., Morishita, S., Kato, E., Shirahase, H., Saito, N., Osumi, S., Kitao, K. and Seki, T. (1982). General pharmacology of indapamide (4) ; Influences on the motor nervous system, sensory nervous system, glucose metabolism and uric acid metabolism, and antigenicity test. *KISO TO RINSHO* **16,** 2261-2274. (Abs. in English)

Numabe, A., Uehara, Y., Hirawa, N., Takada, S. and Yagi, S. (1989). Effects of thiazide diuretic on vascular eicosanoid system of spontaneously hypertensive rats. *J. Hypertension* **7**, 493-499.

Osumi, S., Usui, H., Morishita, S., Nishimura, K., Kanda, M., Nozaki, H., Shirahase, H., Kato, E., Kitao, K. and Seki, T. (1981). Acute toxicity study in mice, rats,and rabbits, and subacute toxicity study in rats of indapamide. *IYAKUHIN KENKYU* **12,** 1110-1126. (Abs. in English)

Osumi, S., Usui, H., Morishita, S., Nishimura, K., Kanda, M., Nozaki, H., Shirahase, H., Kato, E., Kitao, K. and Seki, T. (1981). Chronic toxicity study and recovery test of indapamide in rats. *IYAKUHIN KENKYU* **12,** 1127-1157. (Abs. in English)

Osumi, S. (1982). Mechanism of antihypertensive action of indapamide. Effect on vascular response *Acta Scholae Medicinalis Universitatis in Gifu* **30,** 627-649. (Abs. in English)

Osumi, S., Seki, T., Usui, H., Kanda, M., Kitao, K., Nakayama, R., Makino, M., Horiuchi, T., Usui, T., Kusunoki, F., Masuda, S., Kuboaki, H., Ito, K. and Suzuki, R. M. (1983). Chronic toxicity study of indapamide in dogs. *KISO TO RINSHO* **17,** 79-114. (in Japanese)

Perry, H. M. (1983). Some wrong-way chemical changes during antihypertensive treatment : Comparion of indapamide and related agents. *Am. Heart J.* **106,** 251-257.

Pollare, T., Lithell, H. and Berne, C. (1989). A comparison of the effects of hydrochlorothiazide and captopril on glucose and lipid metabolism in patients with hypertension. *N. Engl. J. Med.* **321,** 868-873.

Raggi, U., Palumbo, P., Moro, B., Bevilacqua, M. and Norbiato, G. (1985). Indapamide in the treatment of hypertension in non-insulin-dependent diabetes. *Hypertension* **7** (Suppl.II), 157-160.

Raveau-Landon, C. and Coulomb, B. (1984). Indapamide in treatment of hypertension with renal failure. *Medicographia* **6** (Suppl. 1), 29-32.

Ross, R. and Glomset, J. A. (1976). The pathogenesis of atherosclerosis (Second of two parts). *N. Engl. J. Med.* **295,** 420-425.

Sakata, T., Ueshima, Y., Muro, H., Kitaura, M., Kakeya, N., Kitao, K. *et al.* (1982). Pharmacokinetic studies of indapamide, a new antihypertensive drug (2). Metabolism in rats and dogs. *KISO TO RINSHO* **16,** 2295-2311. (Abs. in English)

Scalabrino, A., Galeone, F., Giuntoli, F., Guidi, G., Birindelli, A., Natali, A. and Saba, P. (1984). Clinical investigation on long-term effects of indapamide in patients with essential hypertension. *Curr. Ther. Res.* **35,** 17-22.

Schoenfeld, M. R. and Goldberger, E. (1964). Hypercholesterolemia induced by thiazides; A pilot study. *Curr. Ther. Res.* **6,** 180-184.

Seki, T., Fujitani, M., Osumi, S., Yamamoto, T., Eguchi, K., Inoue, N., Sakka, M., Usui, T., Makino, M. and Suzuki, R. M. (1982). Reproduction studies of indapamide (1) - (4). *Jpn. Pharmacol. Ther.* **10,** 1325, 1337, 1355, 1363-1374. (Abs. in English)

Shigekiyo, T., Azuma, H., Nishiuchi, T., Takeichi, T., Saito, S. and Shirakami, A. (1987). Effect of indapamide on platelet aggregation in patients with hypertension. *KISO TO RINSHO* **21,** 6795-6799. (Abs. in English)

Shimaji, H., Morishita, S., Shirahase, H., Hashizume, Y., Osumi, S. and Kakeya, N. (1989). Inhibitory effect of indapamide on proliferation of Balb/C 3T3 cells. *Japan. J. Pharmacol.* **49,** (Suppl.), 300P.

Shirahase, H., Suzuki, Y., Kunitomo, K., Osumi, S. and Kakeya, N. (1985). In vitro anti-platelet effect of indapamide. *IGAKU TO YAKUGAKU* **14,** 1614-1618. (in Japanese)

Shirahase, H., Suzuki, Y., Kunitomo, K., Osumi, S. and Kakeya, N. (1987). In vitro anti-platelet effect of indapamide (2). *IGAKU TO YAKUGAKU* **18,** 1021-1024. (in Japanese)

Shirahase, H., Hashizume, Y., Kanda, M., Shimaji, H., Osumi, S. and Kakeya, N. (1989). Anti-platelet action of indapamide, an anti-hypertensive diuretic. *Japan. J. Pharmacol.* **49** (Suppl.), 255P.

Suzuki, Y., Hamaguti, Y. and Yamagami, I. (1977). Pharmacological studies on diuretics (8). Diuretic activity and mechanism of action of a new hypotensive diuretic, SE-1520. *Folia Pharmacol. Japon.* **73,** 321-335. (Abs. in English)

Takata, M., Iuchi, K., Yoshida, K., Mikawa, M., Iida, H., Mizumura, Y. and Sugimoto, T. (1982). The mechanism of antihypertensive effect of indapamide (SE-1520) ; Observation on the changes in hemodynamics and the vascular reactivity during cold or exercise stress test. *Jpn. Pharmacol. Ther.* **10,** 4025-4033. (in Japanese)

Takeda, T. (1986). Comparison of indapamide and trichlormethiazide in the treatment of essential hypertension. In : *Indapamide and the current treatment of hypertension. Proceeding of the XIV international symposium on indapamide,* pp51-65. Ogilvie, R. I. and Safar, M. (Eds). John Libbey, London.

Takishita, S. and Fukiyama, K. (1986). Antihypertensive effects of indapamide in combination with beta-blockers ; A comparison with thiazide diuretics. *Curr. Ther. Res.* **40,** 849-854.

Thomas, J. R. (1985). A review of 10 years experience with indapamide as an antihypertensive agent. *Hypertension* **7,** (Suppl.II), 152-156.

Uehara, Y., Ishii, M., Ishimitsu, T. and Sugimoto, T. (1988). Enhanced phospholipase C activity in the vascular wall of spontaneously hypertensive rats. *Hypertension* **11,** 28-33.

Uehara, Y., Shirahase, H., Nagata, T., Ishimitsu, T., Morishita, S., Osumi, S., Matsuoka, H. and Sugimoto, T. (1990). Radical scavengers of indapamide in prostacyclin synthesis in rat smooth muscle cell. *Hypertension* **15,** 216-224.

Usui, H., Kanda, M., Osumi, S., Seki, T. and Toda, N. (1978). Influence of indapamide on isolated rabbit arteries. *Folia Pharmacol. Japon.* **74,** 389-396. (Abs. in English)

Velasco, M., Urbina-Quintana, A. and Hernandez-Pieretti, O. (1982). Cardiovascular hemodynamic effects of indapamide and atenolol in hypertensive patients. *Curr. Ther. Res.* **31,** 1007-1017.

Watters, K. and Campbell, B. (1986). Can an antihypertensive agent be both effective and improve the quality of life ? A multicentre study with indapamide. *Br. J. Clin. Pract.* **40,** 239-244.

Weidmann, P., Gerber, A. and Mordasini, R. (1983). Effects of antihypertensive therapy on serum lipoproteins. *Hypertension* **5** (Suppl.III), 120-131.

Wikstrand, J., Warnold, I., Olsson, G. *et al.* (1988). Primary prevention with metoprolol in patients with hypertension. Mortality results from the MAPHY study. *JAMA* **259,** 1976-1982.

Yoshino, G., Kasama, T., Iwatani, I., Iwai, M., Kazumi, T. and Baba, S. (1986). Comparison of the effects of indapamide and trichlormethiazide on lipoprotein and apolipoprotein. *Curr. Ther. Res.* **39,** 1027-1032.

Progress in Hypertension, Vol. 2, pp. 291-307.
Saito *et al.* (Eds)

PHARMACOLOGY OF A NEW DIURETIC S-8666

MASASHI IMAI, TOSHIKATSU SHIMIZU* and MASUHISA NAKAMURA*

Department of Pharmacology, Jichi Medical School, Minamikawachi, Tochigi 329-04; *Shionogi Research Laboratories. Shionogi & Co., Ltd., Fukushima-ku, Osaka 553, Japan.

INTRODUCTION

Hyperuricemia is one of the major side effect of diuretic treatment. There are two factors which account for the hyperuricemia associated with the use of diuretics. First, the blood volume contraction caused by diuretic action may enhance urate reabsorption in the proximal tubule (Weinman *et al.*, 1975). Second, the most diuretics are usually excreted from the proximal tubule by the organic acid transport system. Since the excretion of urate shares the same transport system, it is inhibited by the competition with diuretics (Kahn *et al.*, 1983). Thus, the hyperuricemia or reduction of urinary urate excretion is an unavoidable side effect which is inherent in the most diuretics.

Hyperuricemic action of diuretics is unfavorable especially for the patients with gout. Although it is controversial whether hyperuricemia is a serious risk factor for cardio-vascular diseases (Mejia and Steele, 1986; Kahn and Weinman, 1986), the development of uricosuric diuretics is undoubtedly favorable because of the wide use of diuretics for the treatment of patients with hypertension, who frequently have cardiovascular complications. In addition, it has been reported that incidence of hyperuricemia is high in patients with essential hypertension (Steele, 1977).

Trials to develop an ideal uricosuric diuretic have been so far not fully successful. Although tienilic acid has been developed to serve a uricosuric diuretic, it was withdrawn from the market because of possible hepatic involvement (Zimmerman *et al.*, 1984). S-8666, 6, 7-dichloro-5-N, N-dimethyl-sulfamoyl-2,3-dihydro-benzofuran-2-carboxylic acid, is a novel uricosuric diuretic recently developed in Shionogi Research Laboratories (Harada *et al.*, 1987a ; 1987b; Itazaki *et al.*, 1988 a; 1988b; Ohsugi *et al.*, 1989). The aim of this communication is to review pharmacology of S-8666 by focusing mechanisms of the diuretic and uricosuric actions.

CHEMISTRY, PHARMACOKINETICS and TOXICITY

Chemistry

A series of dihydrobenzofurancarboxylic acid derivatives has been reported to exert uricosuric as well as diuretics actions (Hoffman *et al.*, 1981). Among many related compounds, S-8666 was found to be a favorable drug as a uricosuric diuretic. For the screening of this agent the rats treated with potassium oxonate, an inhibitor of urate oxidase, provided a useful model (Yonetani and Iwaki, 1983; Iwaki and Yonetani, 1984; Yonetani *et al.*, 1987a). S-8666 is 6,7-dichloro-5-N, N-dimethylsulfamoyl-2, 3 dihydro-2-benzofurancarboxylic acid (Fig.1), with molecular weight of 340.18 ($C_{11}H_{11}C_{12}NO_5S$), melting point of 155°C and pKa of 2.53. S-8666 is the mixture of R-(+)-and S-(-)-enantiomers, having an asymmetric carbon at position 2. Optical resolution for these enantiomers has been performed (Harada *et al.*, 1987a; Yodo *et al.*, 1988). Enantioselective synthesis of both enantiomers has been succeeded (Yodo *et al.*, 1988). Quantitative analysis of each compound is also possible (Narita and Kitagawa, 1989)

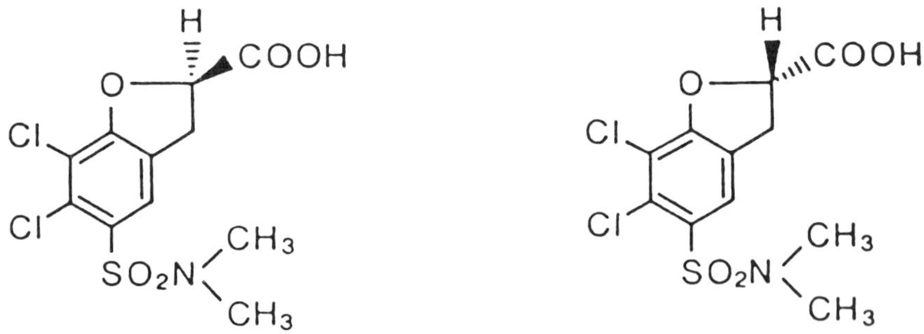

R-(+)-enantiomer S-(-)-enantiomer

Figure 1. Chemical structures of enantiomers of S-8666.

Protein binding and urinary excretion

Plasma protein binding of S-8666 was more than 95% in rabbits, and 73% in beagle dogs (Nakamura *et al.*, 1990). The most fraction of S-8666 administered intravenously or orally to rats is excreted into urine. About 50% of S-8666 excreted into urine was a demethylated metabolite, (6, 7-dichloro-5- (N-methyl-sulfamoyl)- 2, 3-dihydro- 2-benzo- furan carboxylic acid, which retains diuretic activity (Harada and Yonetani, 1988). In the dog, unchanged S-8666 was excreted in the urine when the drug was administered orally (Harada and Yonetani, 1988). In chimpanzees, major fraction of S-8666 was excreted as unchanged form and only a small fraction was monomethyl sulfamoyl metabolite (Yonetani *et al.*, 1987a). Urinary excretion of the drug or its metabolites are mediated by tubular secretion (Nakamura *et al.*, 1990).

Stop-flow studies in rabbits and beagle dogs demonstrated that major fraction of S-8666 is secreted from the proximal nephron (Nakamura *et al.*, 1990). Plasma clearance of S-8666 in rabbit was about 3 times of creatinine clearance, indicating that the tubular secretion mainly contributes to the urinary excretion of S-8666. On the other hand, the ratio of plasma clearance of S-8666 to creatinine clearance in dogs was close to unity, a value which precludes simple interpretation of tubular secretion of the drug. However, since this value decreased to 0.49 when 30 mg/kg probenecid was administered, a significant fraction of urinary excretion was mediated by the tubular secretion which is sensitive to probenecid (Nakamura *et al.*, 1990). Because of a high plasma protein binding of S-8666 in rabbits, about 99 % of the urinary excretion of S-8666 was accounted for by the tubular secretion. By contrast, in dogs about 73 % of the urinary excretion was derived from tubular secretion (Nakamura *et al.*, 1990).

Table 1. Therapeutic index of S-8666, compared with those of furosemide and tienilic acid during daily consecutive dosing for 4 weeks in DOCA-salt hypertensive rats.

Agents	Minimum Effective Dose (mg/kg/day, *p.o.*)	LD_{50} (mg/kg/day, *p.o.*)	Apparent Therapeutic Index
S-8666	60	763	12.7
Furosemide	100	232	2.3
Tienilic acid	200	606	3.0

(Cited with permission with slight modification from Matsuda *et al.* 1989.)

Toxicity

LD_{50} of S-8666 in the mouse with single oral dose ranged about 2000-3000 mg/kg, a value much higher than that for tienilic acid (1400 mg/kg), indacrinone (592 mg/kg) or furosemide (1400 mg/kg) (Harada and Yonetani, 1988). Therefore, S-8666 is the most safe among these drugs. Higaki *et al.* (1989) reported that 3mM S-8666 did not affect viability of isolated rat hepatic cell after incubation for 90min whereas the same dose of tienilic acid and indacrinone were 100% lethal at 60, and 90min, respectively.

If we take a minimum effective dose (ED_{min}) for antihypertensive effect on DOCA hypertensive rats and define an apparent therapeutic index as LD_{50}/ED_{min}, then we conclude that S-8666 has the highest therapeutic index as compared to furosemide and tienilic acid (Table 1). Although this is a very rough index, the data support the view that S-8666 is a relatively safe drug.

EFFECT ON GFR AND RPF

Yonetani *et al.* (1987b) reported that in spontaneously hypertensive rats (SHR) oral administration of 100 mg/kg S-8666 did not affect GFR assessed by inulin clearance, whereas 20 mg/kg hydrochlorothiazide or 50 mg/kg indacrinone caused significant reductions in GFR. Shimizu *et al.* (1988a) reported that in Sprague-Dawley rat i.v. administration of S-8666 at doses of 0.3, 1 and 3 mg /rat did not affect GFR but caused a small (10-20 %) but significant reduction of RPF associated with significant reduction of systemic blood pressure. However, a high dose (50 mg/kg i.v.) of S-8666 caused significant reduction of GFR in rats (Yonetani *et al.*, 1987a). Under the similar experimental condition, 20 mg furosemide (i.v.) also reduced GFR. Since under this condition, the blood volume contraction was not replaced, it is highly possible that the reduction of GFR is due to severe volume depletion caused by extensive diuretic effects with these drugs rather that direct effect of the drug on the glomerular function.

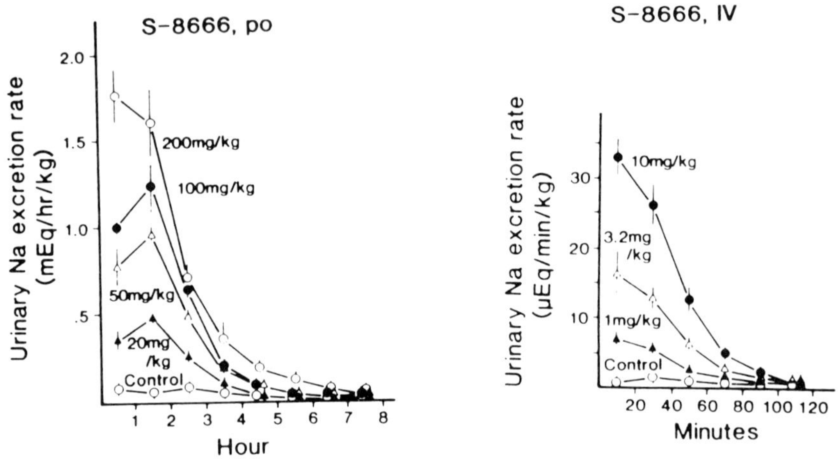

Figure 2. Time course of urinary sodium excretion after oral or intravenous administration of S-8666 to male rat. (Modified after Nakamura *et al.*, 1988.)

DIURETIC EFFECT

Urine flow and sodium excretion

S-8666 has been shown to exert diuretic effect in rats (Yonetani *et al.*, 1987a; 1987b; Nakamura *et al.*, 1988; 1990; Shimizu *et al.*, 1988a; 1989 a; 1989b; Matsuda *et al.*, 1989), mice, dogs, monkeys (Nakamura *et al.*, 1988; 1990) rabbits (Nakamura *et al.*) and chimpanzees (Yonetani *et al.*, 1987a). In rats, oral administration of S-8666 caused rapid onset of diuresis, reaching the maximum in 60-90 min (Nakamura *et al.*, 1988). The diuretic effect lasted for 4-5 hrs. (Fig. 2). The time course and magnitude of the diuretic effect were similar to those of furosemide. The duration of the effect was slightly shorter than that of thiazide diuretics. The comparison of the dose response curve among S-8666, furosemide, and trichlormethiazide suggested that S-8666 belongs to high ceiling diuretics (Nakamura *et al.*, 1988).

Diuretic responses in other animal species were qualitatively similar to those observed in rats. Nakamura *et al.* (1988) compared diuretic effects of S-8666 among rats, mice, monkeys, and dogs. Figure 3 summarizes the results of experiments in which dose-response curves for S-8666, furosemide, and trichlormethiazide were compared among these species. There are considerable species differences in the potency of furosemide as shown in the second panel of Fig. 3, with the order of potency being dogs > mice > monkeys > rats. By contrast, the species difference is less prominent in the potency of S-8666 as shown in the first panel.

S-8666 is the mixture of (+) and (-) enantiomers as already mentioned. Figure 4 shows dose-response curve for S-8666 and its enantiomers administered orally to female mice (Nakamura *et al.* 1990). The responses are compared with furosemide and trichlor-methiazide (TCM). It is clear that (-)-enantiomer showed natriuretic effect whereas (+) - enantiomer did not. Enantiospecificity of the diuretic effect of S-8666 has been also confirmed in rats (Yonetani *et al.*, 1987a; Nakamura *et al.*, 1990). Nakamura *et al.* (1990) reported that in the rat (-) S-8666 was 2 times more potent than racemate (1:1) S-8666. The enantiospecificity of diuretic effect has been reported for ozolinone (Greven *et al.*, 1980) and indacrinone (Field *et al.*, 1984). Such enantiospecificity may play some permissive roles in uricosuric action as will be discussed later.

There are considerable species differences in the potency of furosemide as shown in the second panel of Fig. 3, with the order of potency being dogs > mice > monkeys > rats. By contrast, the species difference is less prominent in the potency of S-8666 as shown in the first panel.

S-8666 is the mixture of (+) and (-) enantiomers as already mentioned. Figure 4 shows dose-response curve for S-8666 and its enantiomers administered orally to female mice (Nakamura *et al.* 1990). The responses are compared with furosemide and trichlor-methiazide (TCM). It is clear that (-)-enantiomer showed natriuretic effect whereas (+) - enantiomer did not. Enantiospecificity of the diuretic effect of S-8666 has been also confirmed in rats (Yonetani *et al.*, 1987a; Nakamura *et al.*, 1990). Nakamura *et al.* (1990) reported that in the rat (-) S-8666 was 2 times more potent than racemate (1:1) S-8666. The enantiospecificity of diuretic effect has been reported for ozolinone (Greven *et al.*, 1980) and indacrinone (Field *et al.*, 1984). Such enantiospecificity may play some permissive roles in uricosuric action as will be discussed later.

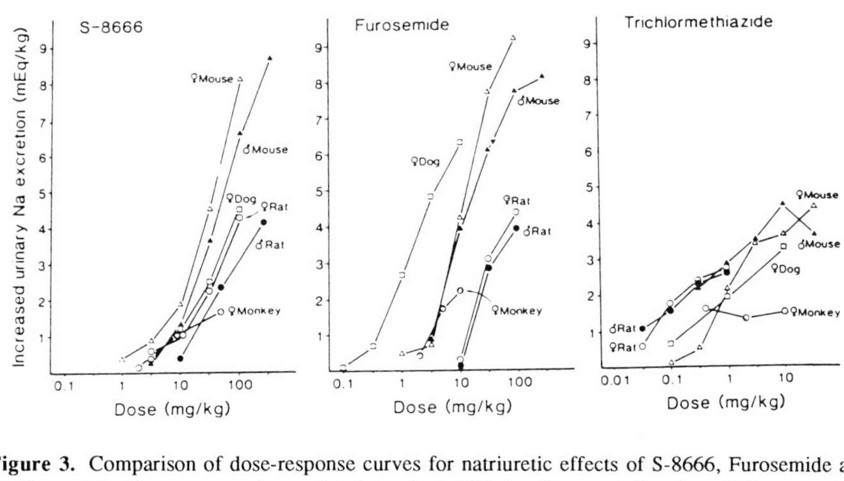

Figure 3. Comparison of dose-response curves for natriuretic effects of S-8666, Furosemide and trichlormethiazide among various animal species. (Cited with permission from Nakamura *et al*, 1988.)

Potassium excretion

The most diuretics, except for potassium sparing diuretics, increase urinary excretion of K^+. S-8666 also increases urinary excretion of K^+ in rats (Shimizu *et al.*, 1988a; Nakamura *et al.*, 1988). Nakamura *et al.* (1988) reported that the Na/K ratio in urine was similar among rats treated with S-8666, furosemide or trichlormethiazide whereas it was high in rats treated with triameterene. Although no specific study was conducted to elucidate the mechanism of increase in K^+ excretion, an increase in distal flow rate may be responsible as has been suggested for other diuretics.

Calcium excretion

Urinary excretion of Ca^{2+} usually parallels that of Na^+ (Walser, 1971). Most diuretics increase urinary Ca^{2+} excretion in parallel with Na^+. Although S-8666 also

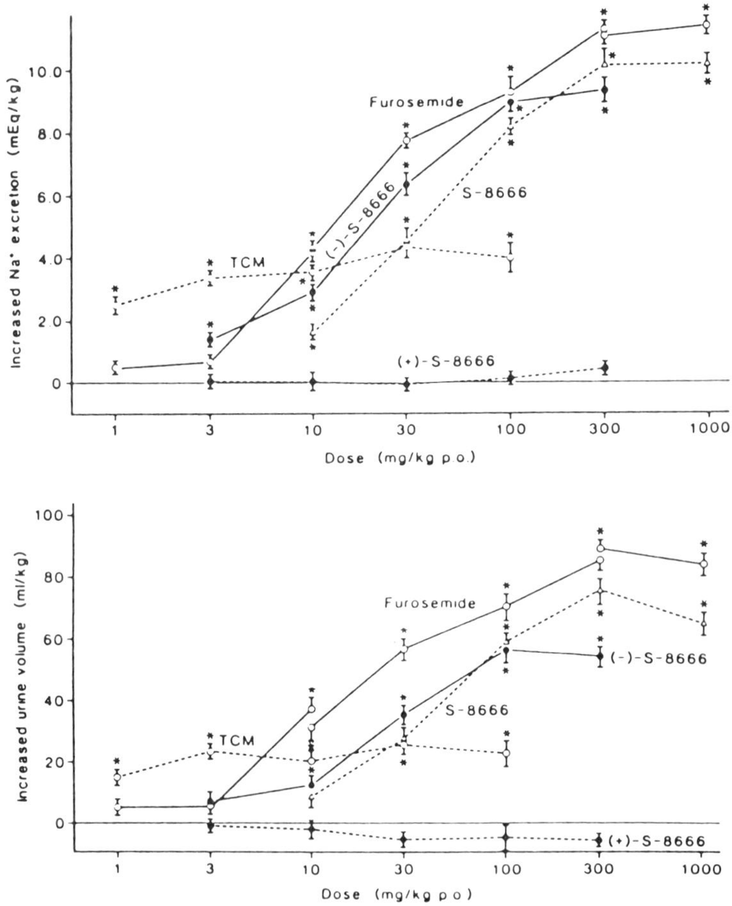

Figure 4. Comparison of dose-response curve for urine flow and urinary Na^+ excretion among (-)-S-8666, (+)-S-8666, and racimic S-8666. Dose-response curve for furosemide and trichlormethiazide (TCM) are also shown. (Cited with permission from Nakamura *et al.*, 1988.)

increases urinary excretion of Ca^{2+}, the magnitude of Ca^{2+} excretion was less than that induced by furosemide (Shimizu *et al.*, 1989a). Interpretation of this finding will be made later with respect to the sites of action of S-8666.

SITES AND MECHANISMS OF ACTION

Henle's loop

As already mentioned, the dose-response curve for S-8666 suggests that the drug belongs to a group of high ceiling diuretics. This would in turn suggest that the site of action is the loop of Henle. Shimizu et al. (1988a) conducted detailed studies on free water clearance (C_{H2O}) as well as free water reabsorption (T^{C}_{H2O}) in rats to obtain indirect information on the site of action of S-8666. The results are shown in Fig. 5 where C_{H2O} and T^{C}_{H2O} were plotted against urine flow and osmolal clearance (C_{osm}), respectively. The C_{H2O} studies were conducted in hydrated rats, which were given 12 ml water by gavage at 2 and 1 hr before the start of clearance, followed by sustained intragastric infusion of water at 0.2 ml/min. The T^{C}_{H2O} experiments were conducted in hydropenic rats. Since anesthesia is known to stimulate vasopressin in rats, no exogenous vasopressin was given. Feasibility of the experimental condition was confirmed by the fact that urine osmolality in the control period always exceeded 1500 mOsm/kg water. It is clear from Fig. 5 that both C_{H2O} and T^{C}_{H2O} were decreased by S-8666 (1.0 mg/rat) when compared with the values in the control period at comparable distal delivery. These observations are consistent with the view that the drug acts on both cortical and medullary diluting segments. Thus, the patterns are similar to those of other loop diuretics. Therefore, it was suggested that the major site of action of S-8666 is the thick ascending limb of Henle's loop.

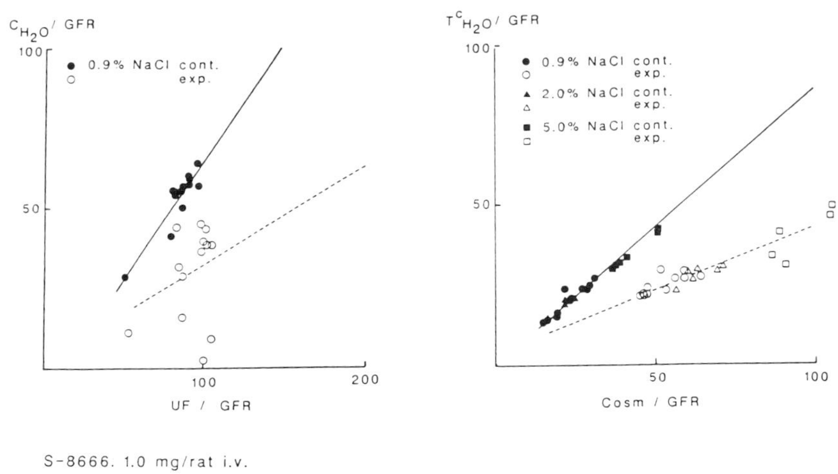

Figure 5. Free water clearance (C_{H2O}) and free water reabsorption (T^{C}_{H2O}) studied in rats. There parameters are shown as a function of urine flow (UF) and osmoral clearance (C_{osm}), respectively. Closed symbols are control experiments, whereas open symbols are the data after S-8666 (1 mg/rat, *i.v.*). (Cited with permission from Shimizu *et al.*, 1988a.)

To obtain more direct evidence for the site of action, Shimizu *et al.* (1988b) conducted in vitro microperfusion studies according to the method developed by Burg *et al.* (1966). The cortical thick ascending limbs of Henle's loop were isolated from rabbit kidneys and perfused in vitro. Transmural voltage (V_T) and lumen-to-bath ^{36}Cl flux were measured. As shown in a representative tracing (Fig. 6), the lumen positive V_T sharply decreased by intraluminal application of 10^{-4} M S-8666. This effect was reversible. The same dose of furosemide caused more profound decrease in V_T. S-8666 in reducing V_T was effective only when it was applied to the lumen but not to the bath. Administration of 10^{-3} M S-8666 in the lumen decreased $J_{Cl}(lb)$ from 375.7 ± 27.9 to 245 ± 29.5 ($n = 7, p < 0.05$) pmol mm^{-1} min^{-1}. These findings provide definite evidence that S-8666 acts on the thick ascending limb of Henle's loop. Although detailed mechanism of action has not been studied, it is reasonable to assume that S-8666 may inhibit Na^+-K^+-$2Cl^-$ co-transport in the luminal membrane as does furosemide (Greger, 1985).

By using change in V_T of the thick ascending limb as an index of the response, we compared to potency of S-8666 with that of furosemide. As shown in Fig. 7, the potency of furosemide was about 60 times higher than that of S-8666. This big difference between S-8666 and furosemide in the potency on the thick ascending limb was not expected from the potencies of both diuretics as assessed by diuretic effects in whole animals. Several possibilities to account for these differences of in vitro and in vivo potencies deserve mentioning. First, it is possible that the degradation rate differs between these diuretics. However, at the present time, no data are available to allow us such comparison. Second, since both drugs exert their diuretic effects by

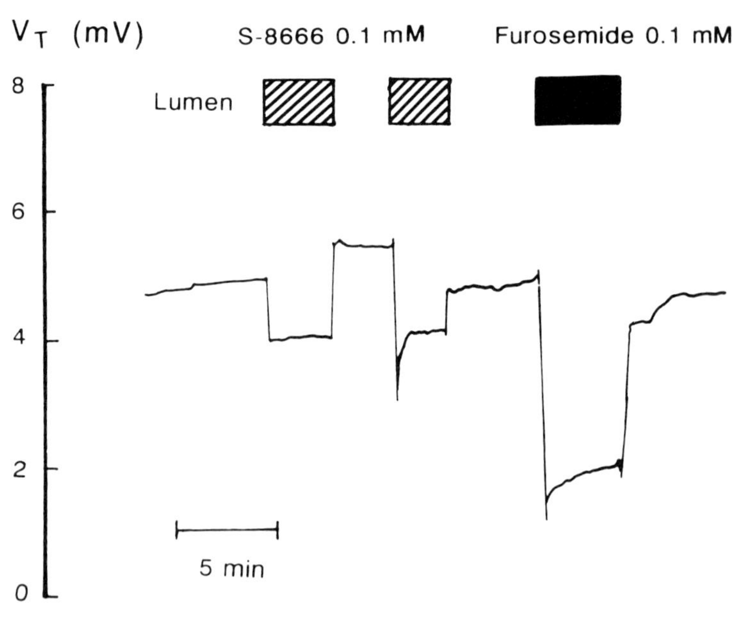

Figure 6. A representative tracing of transmural voltage (V_T) of rabbit cortical thick ascending limb perfused *in vitro*. (Cited from Shimizu *et al.*, 1988b.)

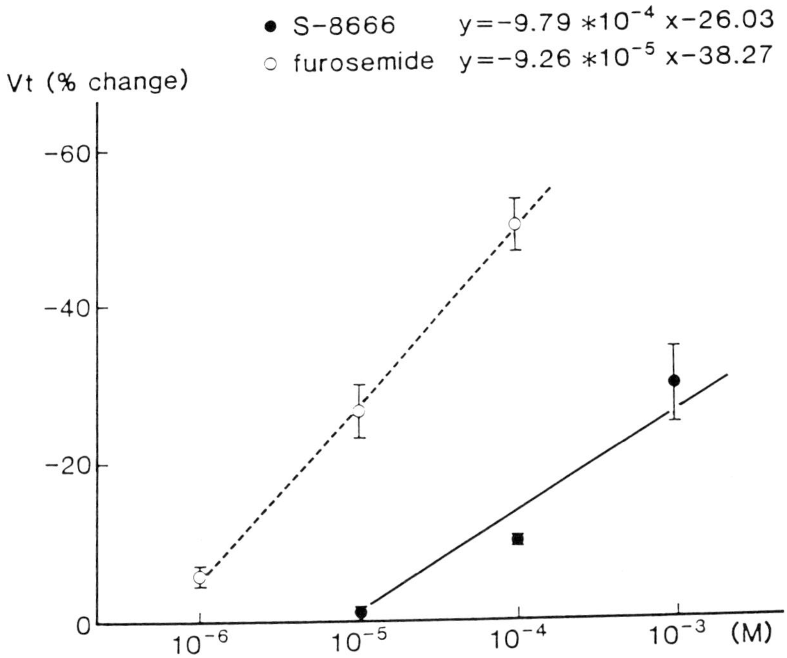

Figure 7. Comparison of dose-response curve for V_T of the cortical thick ascending limb between S-8666 and furosemide. (Cited with permission from Shimizu *et al.*, 1988b.)

acting on the luminal membrane after intratubular secretion in the proximal tubule, possible difference in secretory rate may account for the discrepancy. However, we have no definite data to make this comparison. Finally, it is possible that S-8666 has other sites of action in addition to Henle's loop. This possibility will be discussed in the following section.

Connecting tubule

In order to examine the possibility that S-8666 has an additional site of action, Shimizu et al. (1989b) explored the additive diuretic effect of S-8666 during furosemide induced diuresis in rats. While furosemide was infused at a rate of 10 mg kg^{-1} hr^{-1} following to bolus *i.v.* injection of 10 mg kg^{-1}, second bolus *i.v.* injection of 5 mg/kg furosemide did not change urine flow or urinary Na^{+} excretion, indicating that the receptive site for furosemide action is saturated under this experimental condition. When bolus injection of TCM was made during continuous *i.v.* infusion of 10 mg/kg furosemide, further increases in urine flow and Na^{+} excretion occurred. These effects were dose-dependent. Thus, this experimental protocol is useful to detect additive effects of diuretics, of which site of action is different from that of furosemide. When S-8666 was bolusly injected during furosemide infusion, further

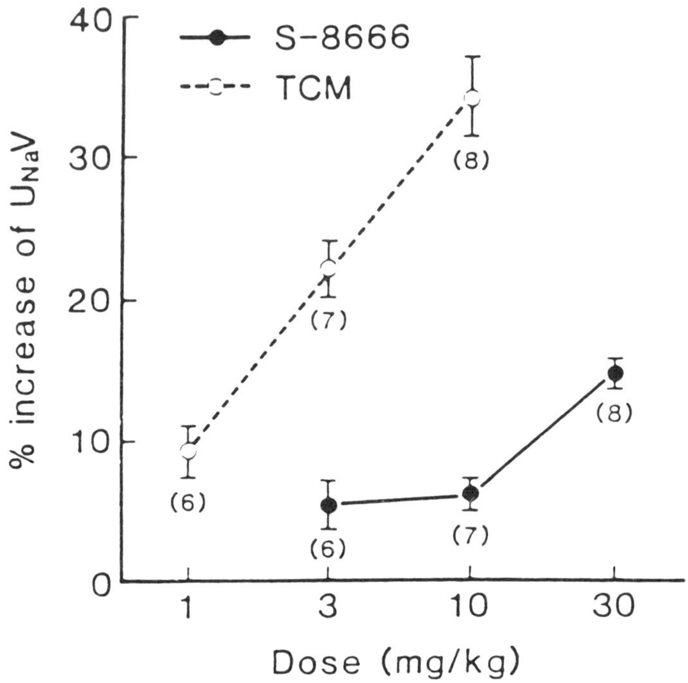

Figure 8. Dose-response curve for additive natriuretic effect of S-8666 or trichlormethiazide (TCM) to furosemide. Studies were conducted in rats. (Cited with permission from Shimizu *et al.*, 1989a.)

increases in urine flow and Na^+ excretion were observed in a dose-dependent manner but they were not as much as those observed in TCM group (Fig. 8).

Although the above data merely suggest that S-8666 exerts diuretic effect by acting on both the thick ascending limb and other additional segments, they do not identify the additional segments. However, Shimizu *et al.* (1989 a,b) suggested that the additional site of S-8666 action is the connecting tubule (CNT), the same target, on which thiazide diuretic acts (Shimizu *et al.*, 1988c). Two different types of studies supported this view. The first concerns the similarity of the effect S-8666 with TCM with respect to urinary excretion of Ca^{2+} (Shimizu *et al.*, 1989 a,b). Thiazides diuretics are well-known to have anticalciuric effect. Although S-8666 increased urinary Ca^{2+} excretion, the magnitude of the increment was less than that induced by furosemide when compared at the comparable urinary Na^+ excretion. In order to evaluate calciurlic or anti-calciuric effect, it is necessary to analyse urinary Ca^{2+} excretion as a function of urinary Na^+ excretion. Shimizu et al. (1989a) compared effects of furosemide. TCM, and S-8666 on urinary Ca^{2+} excretion as a function of Na^+ excretion in unanesthetized rats. The data are summarized in Fig. 9. In the furosemide group, an increase in Ca^{2+} excretion was associated with an increase in Na^+ excretion, with a slope being 0.30 (mg Ca^{2+}/meq Na^+). On the other hand,

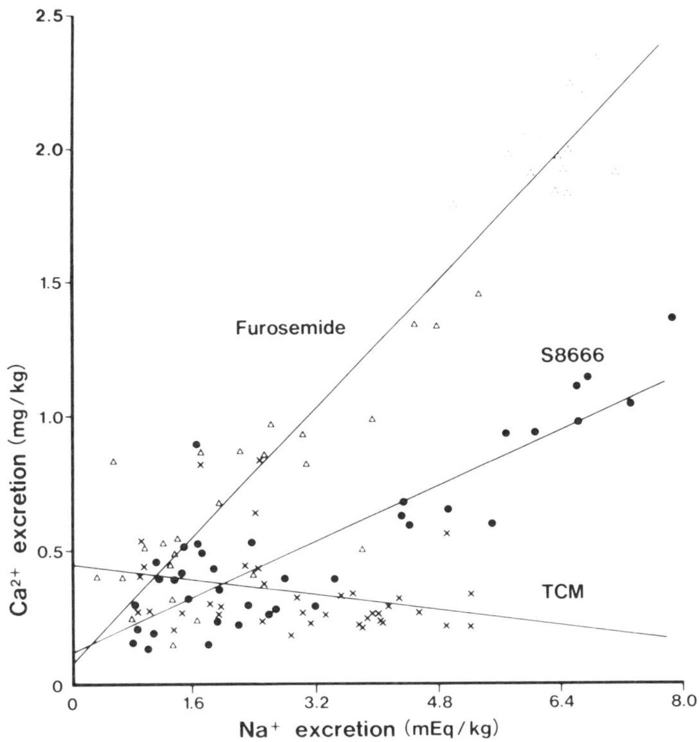

Figure 9. Relationship between urinary Na^+ and Ca^{2+} excretion. Data are compared among furosemide, S-8666 and TCM. Doses of the diuretics were varied to obtain data of wide range; 1-316 mg/kg *p.o.* for furosemide and S-8666, and 0-10 mg/kg *p.o.* for TCM. Studies were conducted in rats. (Modified after Shimizu *et al.*, 1989b.)

TCM decreased urinary Ca^{2+} excretion in spite of the increase in Na^+ excretion. In S-8666 group, there was positive correlation between these parameters. However, the slope was 0.13 (mg Ca^{2+}/meq Na^+), a value less than a half of furosemide group. These observations are compatible with the view that S-8666 shares a common site of action with TCM.

The second, and more direct evidence was provided by the observation in isolated perfused rabbit CNT (Shimizu *et al.*, unpublished). It was shown that S-8666 decreased Cl flux in the CNT. This segment has been reported to be the target of the action of TCM (Shimizu *et al.*, 1988c). Thus, we conclude that S-8666 is a unique diuretic having dual sites of action on both the thick ascending limb of Henle's loop and the CNT.

URICOSURIC ACTION

Uricosuric action of S-8666 was initially studied in rats treated with oxonate (a urate oxydase inhibitor). As shown in Fig. 10, dose-dependent increases of fractional excretion of urate (FE_{ua}) was observed when S-8666 was administered intraperitoneally. In normal rats, *i.v.* administration of 50 mg/kg S-8666 caused a small but significant increase in FE_{ua}. By contrast, administration of furosemide always caused FE_{ua} to decrease.

Extensive species difference in urate transport in the kidney is used to hamper the development of uricosuric diuretic. Although final conclusion must await studies in human, studies in chimpanzee are assumed to provide the most meaningful information. In this regard, Yonetani et al. (1987a) confirmed that oral administration of 10 mg/kg S-8666 increased FE_{ua} in chimpanzees. Although the effect was less than that of 5 mg/kg indacrinone, it was comparable to 10 mg/kg probenecid. To localize the nephron site of uricosuric action of S-8666, Shimizu et al. (1988a) conducted the microinjection study of ^{14}C-urate in rats. By using the preparation of conventional in vivo micropuncture, they observed the urinary recovery of ^{14}C-urate an ^{3}H-inulin injected into the early and late proximal tubule and distal tubule. In the control period, the recoveries of ^{14}C-urate injected in the early and late proximal tubule were 71.7 ± 1.8 and 85.1 ± 1.2 %, respectively, indicating that 13.4 % of urate is reabsorbed between two puncture sites. While the animals were treated with S-8666, the corresponding recoveries of ^{14}C-urate were 88.7 ± 2.6 and 91.5 ± 2.0 %, indicating

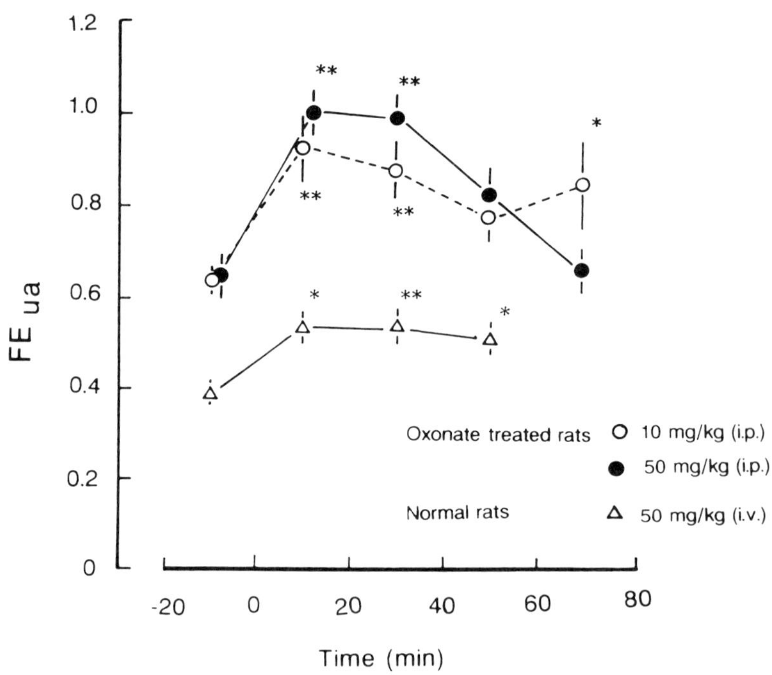

Figure 10. Time course of fractional urinary excretion of uric acid (FU_{ua}) in normal and oxonate treated rats. (Constructed from the data of Yonetani *et al.*, 1987a.)

that urate reabsorption was decreased to 2.8 %. The recovery of ^{14}C-urate injected into the distal tubule was 97.9 ± 1.5 %, a value not different from 100 %, indicating that urate reabsorption did not occur in the distal nephron segments. These observations are in accord with the view that S-8666 inhibits reabsorptive flux of urate in the proximal tubule.

More direct evidence was put forth by the *in vitro* microperfusion of the rabbit proximal convoluted (PCT) and straight (PST) tubule (Shimizu et al., 1988b). By using ^{14}C-urate, unidirectional fluxes of urate were measured in the presence or absence of 10^{-3} M S-8666 in the lumen. S-8666 decreased the lumen-to-bath urate flux (5 mol mm^{-1} min^{-1}) from 135.1 ± 18.6 to 104.5 ± 12.4 in the PCT and from 98.7± 8.4 to 66.9 ± 7.9 in the PST (Fig. 11). The rabbit is not an ideal species to test uricosuric activity, since the secretory component of urate is manifested in this species. It has been reported that probenecid (Møller, 1967; Poulsen, 1955) predominantly inhibited urate secretion in rabbits, exhibiting antiuricosuric action. In fact, Shimizu et al. (1988b) confirmed that S-8666 also inhibits the bath-to-lumen urate flux in the rabbit PST from 128.6 ± 11.0 to 73.7 ± 9.4 fmol mm^{-1} min^{-1}. Thus, it is possible that S-8666 may exhibit uricosuric effect only in those species in which reabsorptive flux of urate is dominant.

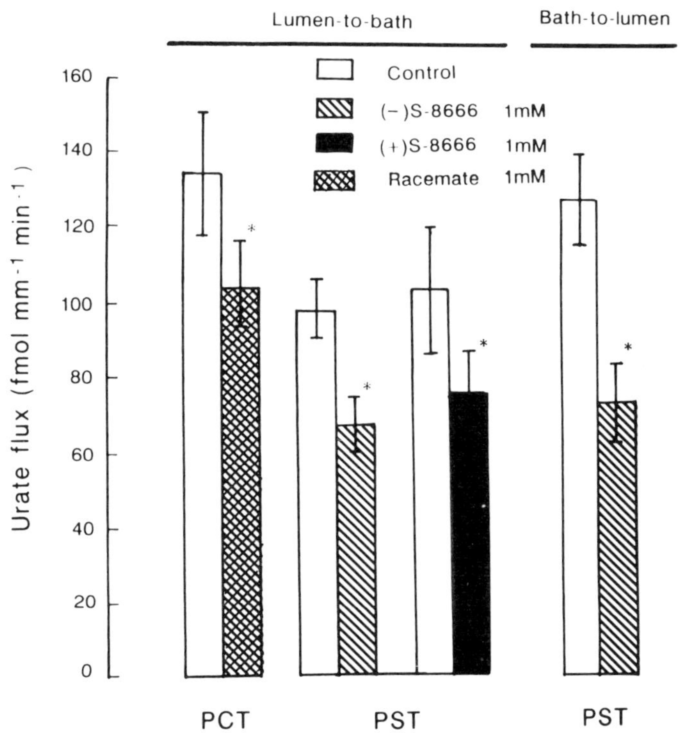

Figure 11. Effects of enantiomers of S-8666 on unidirectional fluxes of urate in the rabbit proximal convoluted (PCT) and proximal straight tubule (PST) perfused *in vitro*. (Constructed from the data of Shimizu *et al.*, 1988b.)

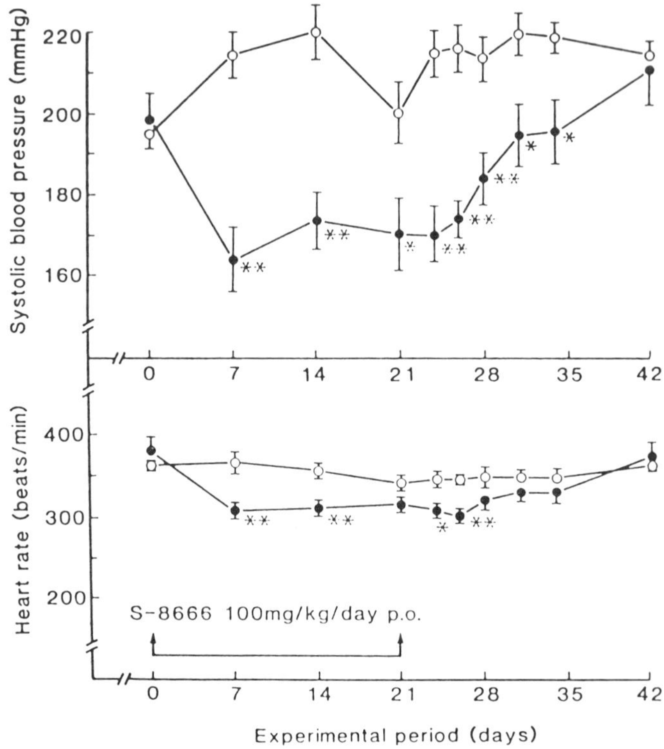

Figure 12. Antihypertensive effect of S-8666 in DOCA-salt hypertensive rats. S-8666 was given for 3 weeks (100 mg/kg/day, *p.o.*). * $p < 0.05$, ** $p < 0.01$. (Cited with permission from Matsuda *et al.*, 1989.)

Another point which deserves to be mentioned is the effect of enantiomers of S-8666 on urate transport. As already mentioned, only (S-) enantiomer has diuretic activity. However, it is important to note that both R(+) and S(-) enantiomers have uricosuric action. It was reported that R(+)-S-8666 also decreases both the lumen-to-bath and the bath-to-lumen urate fluxes in the rabbit PST to the same extents as did S (-) enantiomer (Shimizu *et al.,* 1988b).

As mentioned in the introduction, one of the most important mechanisms underlying antiuricosuric effect of diuretics is blood volume contraction. Therefore, the larger the diuretic effect, the larger the antiuricosuric effect. Thus the use of racemic compound for S-8666 is meaningful for reducing diuretic potency without affecting uricosuric activity. The similar situation also existed in case of indacrinone (Tober *et al.*, 1981).

ANTIHYPERTENSIVE ACTION

Matsuda *et al.* (1989) reported that single oral administration of 60-200 mg/kg S-8666 to DOCA-salt hypertensive rats caused a dose-dependent decrease in blood pressure and that this effect lasted for 24 hrs. Chronic oral administration of S-8666 (100mg/kg/day) to DOCA-salt hypertensive rats for 3 weeks showed antihypertensive effect after one week (Fig. 12). Antihypertensive dose of S-8666 equivalent to 3 mg/kg TCM was 60-100 mg/kg. The antihypertensive activity was shown predominantly by S(-)-enantiomer, with the R-(+)-enantiomer being only slightly active, suggesting that the antihypertensive effect is related to its diuretic effect. S-8666, like TCM, prevented the development of hypertension in DOCA-salt hypertensive rats, salt-loaded SHR, and salt-loaded Dahl-S rats (Matsuda *et al.*, 1989).

It rats with one kidney-one clip renal hypertension model, S-8666 (100 and 200 mg/kg, *p.o.*) did not show antihypertensive effect. In this model, neither TCM (3mg/kg. *p.o.*) nor indacrinone (59 mg/kg *p.o.*) showed antihypertensive effect as well. Although neither S-8666 (200 mg/kg, *p.o.*), TCM (3 mg/kg, *p.o.*) nor furosemide (100 mg/kg. *p.o.*) showed antihypertensive effect in SHR (Matsuda *et al.*,1989), S-8666 decreased blood pressure in SHR under sodium restriction as did hydrochlorothiazide, furosemide and indacrinone (Yonetani *et al.*, 1987b).

SUMMARY AND CONCLUSION

S-8666 is a novel uricosuric diuretic, having an antihypertensive effect. The diuretic effect is exerted by S(-) enantiomer. The diuretic efficacy and potency are comparable to or slightly less than furosemide. The major site responsible for the diuretic action is the thick ascending limb of Henle's loop. However, this drug is very unique in that it has an additional site of action in the CNT. Thus, S-8666 share characteristics with loop diuretics as well as thiazides. Both R(+) and S(-)-enantiomers are responsible for the uricosuric action, which is exerted by inhibition of urate reabsorption in the proximal tubules. Both agents also inhibit secretory flux of urate in the PST. The drug did not show any serious toxic effects. Because of these unique properties, S-8666 is expected to be applicable for the treatment of hypertension in human as a uricosuric diuretic.

REFERENCES

Brooks, B. A., Blair, E. M., Finch, R. and Lant, A.F. (1980). Studies on the mechanism and characteristics of action of a uricosuric diuretic, indacrinone (MK-196). *Brit. J. Clin. Pharmacol.* **10,** 249-258.

Field, M. J., Fowler, N. and Giebisch, G. (1984). Effects of enantiomers of indacrinone (MK-196) on cation transport by the loop of Henle and distal tubule studied by microperfusion *in vivo*. *J. Pharmacol. Exp. Ther.* **230,** 62-68.

Greger, R. (1985). Ion transport mechanisms in thick ascending limb of Henle's loop of mammalian

nephron. *Physiol Rev.* **65,** 760-797.
Greven, J., Defrain, W., Glaser, K., Meywold, K. and Heidenreich, O. (1980). Studies with optically active isomers of the new diuretic drug ozolinone. *Pfluegers Arch.* **3484,** 57-60.
Harada, H., Matsushita, Y., Yodo, M. and Yonetani, Y. (1987a). Studies on uricosuric diuretics. 1) 6,7-Dichloro-5-sulfamoyl-2,3-dihydrobenzofuran-2-carboxylic acids. *Chem. Pharm. Bull.* **35** (8), 3195-3214.
Harada, H., Matsushita, Y., Yodo, M., Nakamura, M. and Yonetani, Y. (1987b). Studies on uricosuric diuretics. 2) 6,7-Dichloro-4-nitro-,6,7-dichloro-4-sulfamoyl and 6,7-dichloro-4-acyl-2,3-dihydro-benzofuran-2-carboxylic acids. *Chem. Pharm. Bull.* **35**(8), 2315-3226.
Harada, H. and Yonetani, Y. (1988). 5-Sulfamoyl-2,3-dihydro-benzofuran-2-carboxylic acid derivatives. S-8666, a favorable candidate with uricosuric and diuretic activities. *Drugs of the Future* **13** (39), 257-264.
Higaki, J., Harada, H., Tonda, K. and Hirata, M. (1989). Chemical structure and toxicity of diuretics in isolated hepatocytes. *Pharmacol. Toxicol.* **65,** 21-24.
Hoffman, W. F., Woltersdorf, O. W.Jr., Novello, F. C., Cragoe, E.J.Jr., Springer, J. P., Watson, L. S. and Fanelli, G. M., Jr. (1981). (Acylaryloxy) acetic acid diuretics. 3. 2,3,-dihydro-5-acyl-2-benzofurancarboxylic acids, a new class of uricosuric diuretics. *J. Med. Chem.* **24,** 865-873.
Irvin, J. D., Vlasses, P. H., Hubert, P. B., Ferguson, R. K., Scrogie, J., J. and Davies, R. O. (1980). Comparison of oral indacrinone with furosemide. *Clin. Pharmaol. Ther.* **28,** 376-383.
Itazaki, H., Kawasaki, A., Matsuura, M., Ueda, M., Yonetani, Y. and Nakamura, M. (1988a). Synthesis of 2,3-dihydro-1,4-benzodioxin derivatives. I. 2-substituted-5(and 6)-sulfamoyl-l-2,3-dihydro-1,4-benzodioxins. *Chem. Pham. Bull.* **36,** 3387-3403.
Itazaki, H., Hayashi, K., Matsuura, M., Yonetani, Y. and Nakamura, M. (1988a). Synthesis of 2,3-dihydro-1,4-benzodioxin derivatives. I.2-substituted-5 (and 6)-sulfamoyl-2,3-dihydro-1,4-benzodioxins. *Chem. Pham. Bull.* **36,** 3387-3403.
Iwaki, K. and Yonetani, Y. (1984). Decreased renal excretion of uric acid following diuretic administration in rats. *Japan. J. Pharmacol.* **34,** 389-396.
Kahn, A. M., Bronham, S. and Weinman, E. J. (1983). Mechanism of urate and P-amino-hippurate transport in rat renal microvilus membrane vesicles. *Am. J. Physiol.* **245,** F151-F158.
Kahn, A. M. and Weinman, E. J. (1986). Effect on uric acid excretion In: *Physiological Basis of Diuretic Therapy in Clinical Medicine*, Eknoyan, G. and Martinez-Maldonado, M.(Eds). Grune and Stratton, Orlando.
Matsuda, S., Ishii, M., Kitajiri, J., Yahara, I., Harada, H. and Yonetani, Y. (1989). Antihypertensive property of a novel uricosuric diuretic, S-8666 in rats. *Drug Dev. Res.* **17,** 207-217.
Mejia, G. and Steele, T. H. (1986). Uricosuric diuretics. In: *Diuretics, Physiology, Pharmacology and Clinical Use,* Dirks, J. H., Sutton, R.A.L.(Eds), W. B. Saunders, Philadelphia.
Møller, J.V. (1967). The relation between secretion of urate and P-amino-hippurate in the rabbit kidney. *J. Physiol. (Lond.)* **192,** 505-517.
Nakamura, M., Shimizu, T., Miyata, K., Kawabata, T. and Harada, H. (1988). Diuretic effects of a novel uricosuric antihypertensive S-8666 in rat, mice, monkeys, and dogs. Comparison with furosemide and trichlormethiazide. *Drug Dev. Res.* **12,** 41-51.
Nakamura, M., Kawabata, T., Itoh, T., Miyata, K. and Harada, H. (1990). Stereoselective saluretic effect and localization of renal tubular secretion of enantiomers of S-8666, a novel uricosuric antihypertensive diuretic. *Drug Dev. Res.* **19** (1), 23-36.
Narita, S. and Kitagawa, T. (1989). S-(-)-2-(4-(1-Aminoethyl)naphtyl)-6-methoxy-N-mthyl-2H-benzotriazolyl-5-amino-dihydrochloride as a fluorescence chiral derivatization reagent for carboxylic acids in high performance liquid chromatography. *Anal. Sci.* **5,** 361-362.
Ohsugi, E., Fujioka, T., Harada, H., Nakamura, M. and Maeda, R. (1989). Synthetic studies on diuretics. 5-(3,3-N,S-substituted-2-propenoyl)-2,3-dihydro-2,2-benzo[b]furancarboxylic acids. *Chem. Pham. Bull.* **127,** 1268-1278.
Poulsen, H. (1955). Inhibition of uric acid excretion in rabbits given probenecid or salycylic acid. Acta *Pharmacol. Toxicol.* **11,** 277-286.
Shimizu, T., Nakamura, M. and Imai, M. (1988a). Renal action of a novel uricosuric diuretic, S-8666. 1) Clearance and tubular microinjection studies in rats. *J. Pharmacol. Exp. Ther.* **245,**

644-650.

Shimizu, T., Nakamura, M. and Imai, M. (1988b). Renal action of a novel uricosuric diuretic, S-8666. 2) Effect on Cl^- and urate transport in isolated perfused rabbit renal tubules. *J. Pharmacol. Exp. Ther.* **245,** 651-657.

Shimizu, T., Yoshitomi, K., Nakamura, M. and Imai, M. (1988c). Site and mechanism of action of trichlormethiazide in rabbit distal nephron segments perfused *in vitro*. *J. Clin. Invest.* **82,** 721-730.

Shimizu, T., Miyata, K. and Nakamura, M. (1989a). Effect of S-8666 on urinary Ca^{2+} excretion in rats. *J. Pharmacol. Exp. Ther.* **250,** 654-658.

Shimizu, T., Miyata, K. and Nakamura, M. (1989b). Additive diuretic effect of S-8666 during furosemide-induced diuretic in rats. *J. Pharmacol. Exp. Ther.* **250,** 659-666.

Steele, T. H. (1977). Relationship of hyperuricemia and glucose tolerance in hypertension. In: *Hypertension Physiopathology and Treatment,* Genest, J., Koiw, D., Kuchel, O. (Eds), McGraw Hill, New York.

Tober, J. A., Cirillo, V.J., Hitzenberger, G., James, I., Pryor, J., Cook, T., Buntinix, A., Holmes, I.B. and Lutterbeck, P. M. (1981). Enhancement of uricosuric properties of indacrinone by manipulation of the enantiomer ratio. *Clin. Pharmacol. Ther.* **29,** 344-350.

Walser, M. (1971). Calcium-sodium interdependence in renal transport In: *Renal Pharmacology,* Fisher, J.(Eds), Appleton-Century-Crofts, New York.

Weinman, E. J., Eknoyan, G. and Suki, W. N. (1975). The influence of the extracellular fluid volume on the tubular reabsorption of uric acid. *J. Clin. Invest.* **55,** 283-291.

Yodo, M., Matsushita, Y., Ohsugi, E. and Harada, H. (1988). Optical resolution and chiral synthesis of methyl 6,7-dichloro-2,3-secretory(b) furan-2-carboxylate. *Chem. Pharm. Bull.* **36** (3), 902-913.

Yonetani, Y., Iwaki, K., Ishii, M. and Harada, H. (1987a). A new diuretic that does not reduce renal handling of uric acid in rats, S-8666. *Jpn. J. Pharmacol.* **43,** 399-405.

Yonetani, Y., Iwaki, K., Shinosaki, T., Kawase, A., Harada, H. and Van Arther, A.E.S. (1987b). A new uricosuric diuretic, S-8666, in rats and chimpanzees. *Jpn. J. Pharmcol.* **43,** 389-398.

Zimmerman, H. J., Lewis, J. H., Ishak, K. G., and Maddrey, W. C. (1984). Ticrynafen-associated hepatic injury: Analysis of 340 cases. *Hepatology.* **4,** 315.

Progress in Hypertension, Vol. 2, pp. 309-317.
Saito *et al.* (Eds)

BUDRALAZINE: A DIRECT-ACTING VASODILATOR

MITSUHIRO YOSHIOKA, MASARU MINAMI,* HASAN S. PARVEZ**and HIDEYA SAITO

First Department of Pharmacology, Hokkaido University School of Medicine, Kita-15, Nishi-7, Kita-ku, Sapporo 060, Japan. *Department of Pharmacology, Higashi-Nippon Gakuen University Faculty of Pharmaceutical Sciences, Ishikari-Tobetsu 061-02, Japan. **Unité de Neuropharmacologie, Université de Paris XI, Centre d'Orsay, Bât. 440, 91405 Orsay cedex, France

INTRODUCTION

Recently, the development of many antihypertensive drugs possessing different mechanisms, lead us to rational and effective therapy for hypertension. Direct-acting vasodilator, like a hydralazine, is also useful not only for hypertension but congestive heart failure treatment (Chatterjee *et al.*, 1978). However, short-term and long-term therapy of vasodilator may be poorly-tolerated because of reflex cardio-stimulation (Brunner *et al.*, 1967; Saegusa *et al.*, 1986) and sodium retention (Dranov *et al.*, 1974; Dustan, 1983). This cardiac stimulation may produce awareness of angina pectoris and myocardial infarction. In addition, the immunological reactions, i.e., rheumatoid arthritis and systemic lupus erythematosus (Morrow *et al.*, 1953; Dustan *et al.*, 1954), were recognized and many people stopped using it. Lately, the combination of the β– adrenoceptor blocking drugs is becoming more effective for well-tolerated therapy since of lower doses can be used (Zacest *et al.*, 1972).

However, the pharmacokinetic problem may not disappear because of their differences in the absorption and disposition, especially biological half-life between

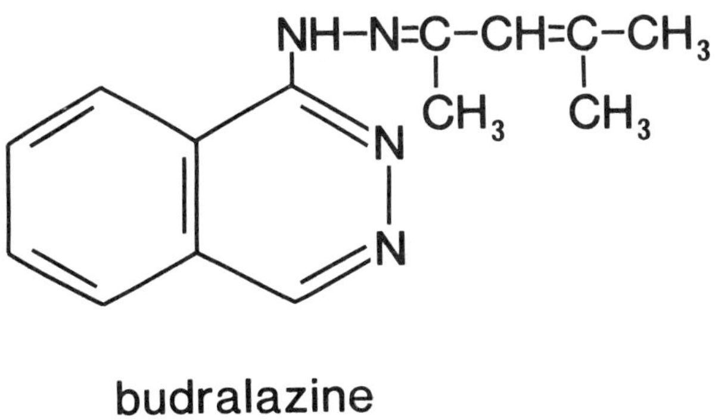

Figure 1. Chemical structure of budralazine.

hydralazine and β-blockers. These circumstances have resulted in a stimulus to search for new direct-acting antihypertensive vasodilators having no cardiac stimulation or immunological reactions. In 1974, a new phthalazine derivative, budralazine was synthetized and found to have an appreciable antihypertensive activity and almost no cardiac stimulating activity in experimental hypertensive rats (Akashi *et al.*, 1974).

In spontaneously hypertensive rats, budralazine was found to be approximately 4 times less potent than hydralazine in lowering blood pressure, but characterized by a greater separation of the antihypertensive and cardiac stimulant activities as compared with hydralazine (Akashi *et al.*, 1974; Akashi *et al.*, 1978). In addition, the effectiveness of this compound was also supported by the observation that orally administered budralazine is more potent than hydralazine in lowering blood pressure and less potent than latter in producing tachycardia in conscious renal hypertensive dogs (Konno and Ueno, 1977) and essential hypertension in humans (Takeuchi *et al.*, 1980). This review will concentrate on the pharmacodynamics and pharmacokinetics of a direct-acting vasodilator, budralazine, the possible mode of action and their clinical application, including the advantages over more conventional direct-acting vasodilators, and adverse reactions.

CHEMISTRY

Budralazine (1-[2-(1,3-dimethyl-2-butenylidene)-hydrazino]-phthalazine) is a phthalazine derivative and an antihypertensive agent synthesized by Ueno *et al.* (1976). The chemical formula is shown in Fig. 1. This compound is yellow powder, no taste, and insoluble in water but soluble in chloroform and glacial acetic acid.

PHARMACODYNAMICS

Hemodynamic actions

Blood Pressure and Heart Rate. Budralazine exerts an effective hypotensive action due to its direct dilatation of vascular smooth muscle (Akashi *et al.*, 1978; Chiba *et al.*, 1983). In vitro studies using isolated rabbit aorta demonstrated that budralazine relaxed either K^+- or norepinephrine-induced contractions and also inhibited Ca^{2+}-induced contractions of the K^+-depolarized muscle. These observations indicate that budralazine acts as vasorelaxant in a manner similar to that of hydralazine which has been reported to produce vascular relaxation by interfering with the availability of calcium (McLean *et al.*, 1978). Although hydralazine-induced depressor action is accompanied with reflex tachycardia (Stein and Hecht, 1955; Åblad, 1963; Yoshioka *et al.*, 1986), it has been reported that budralazine tends not to produce reflex tachycardia in SHR (Akashi *et al.*, 1974) and renal hypertensive dogs (Konno *et al.*, 1981) or essential hypertension in humans (Takeuchi *et al.*, 1980). It was also demonstrated that budralazine produces a significant decrease in heart rate during the hypotensive phase in normotensive rats (Fig.2) and patients with essential hypertension (Yoshinaga and Miura, 1985). These observations indicate that budralazine is characterized by a much greater separation of antihypertensive and cardiac stimulant activities.

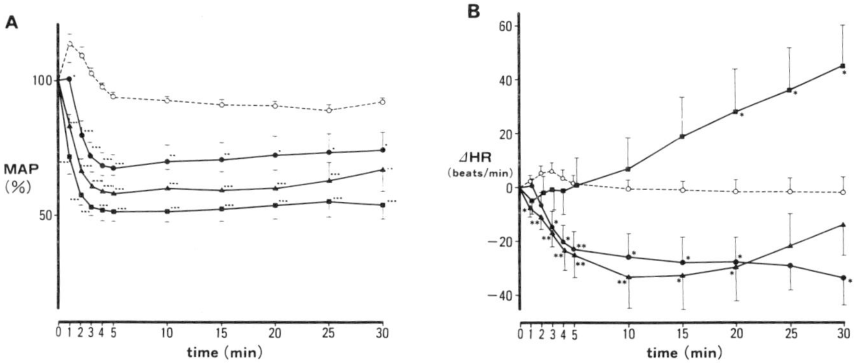

Figure 2. Effects of budralazine on mean arterial pressure (MAP) and heart rate (HR) in anesthetized rats. Budralazine: (●) 0.5mg/kg, (▲) 1.0mg/kg, and (■) 5.0mg/kg. Vehicle: (○) 0.5ml/kg. Vertical lines show standard errors of means. *, ** and *** indicate significant differences (p <0.05, 0.01 and 0.001, respectively) from the vehicle treated values.

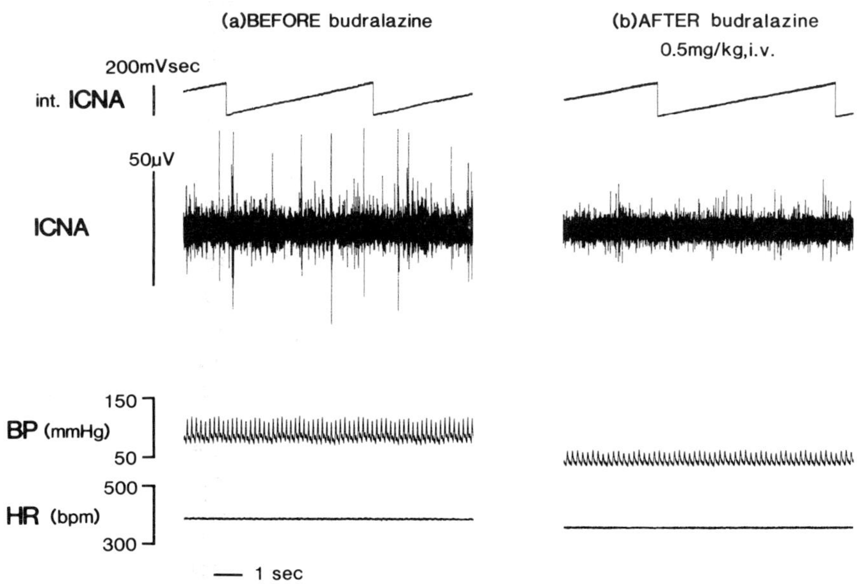

Figure 3. Effect of intravenous budralazine on inferior cardiac nerve activity (ICNA) in an anesthetized rat. Traces from top to bottom are integrated (int.) ICNA, actual ICNA, blood pressure (BP) and heart rate (HR).

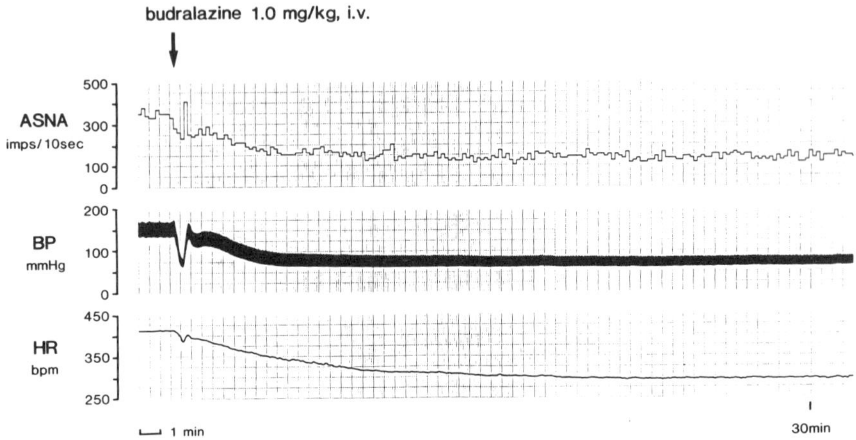

Figure 4. Effect of budralazine on adrenal sympathetic nerve activity (ASNA) in an anesthetized rat. Traces from top to bottom are discharge rate of ASNA, blood pressure (BP) and heart rate (HR).

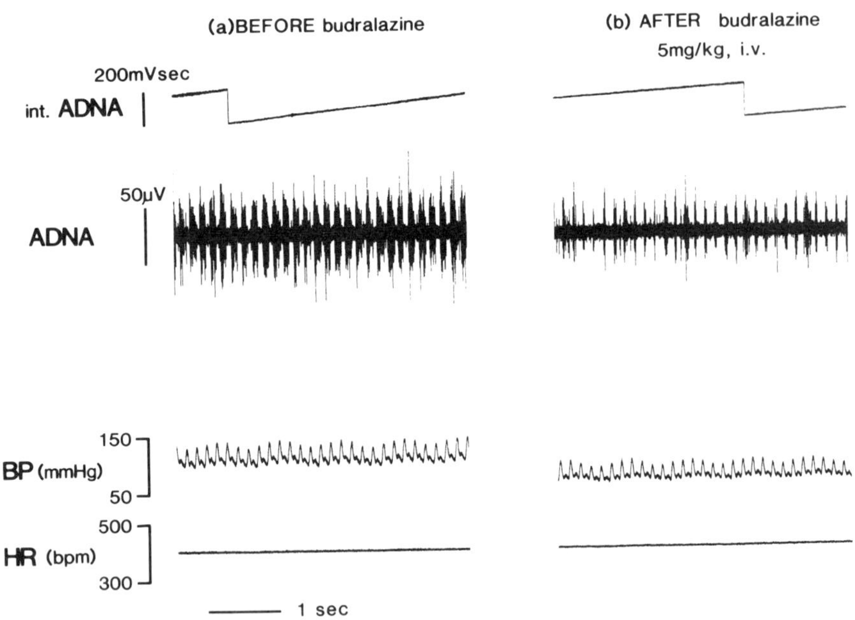

Figure 5. Effect of budralazine on aortic depressor nerve activity (ADNA) in an anesthetized rat. Traces from top to bottom are integrated (int.) ADNA, actual ADNA, blood pressure (BP) and heart rate (HR).

Regional Blood Flow. After the intravenous administration of 6 mg/kg dose of budralazine, there were increases in regional blood flow in the vascular beds, carotid, vertebral, mesenteric, renal and femoral arteries in anesthetized dogs (Chiba *et al.*, 1983). Budralazine, 1 mg/kg, was without significant effect on mean blood pressure, but produced a significant increase in renal blood flow and considerable but non-significant increase in coronary blood flow (Konno *et al.*, 1981). In anesthetized cats, budralazine (10 mg/kg, i.v.) produced an increase in blood flow in the cerebral cortex and hippocampus in spite of an effective decrease in blood pressure. In the hypothalamus and amygdala, however, the blood flow was inconsistent (Hagihara, 1981).

Other actions

Central sympathoinhibitory action. Intravenous administration of budralazine (0.5-5.0 mg/kg i.v.) produced a dose- dependent reduction of mean arterial pressure and

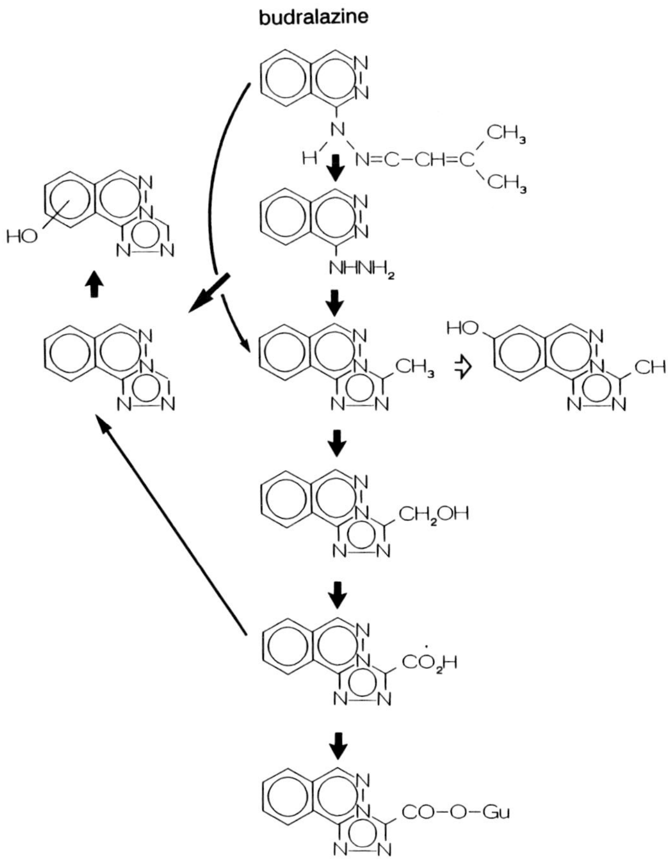

Figure 6. Metabolic pathway of budralazine.

0.5 and 1.0 mg/kg doses of budralazine produced a simultaneous decrease in both cardiac sympathetic nerve activity and heart rate in anesthetized rats (Fig.3). Moreover, intravenous administration of budralazine at a dose of 1.0 mg/kg decreased the preganglionic adrenal sympathetic nerve activity significantly (Fig.4). These findings suggest that the central sympathoinhibitory action of budralazine may be responsible for its anti-tachycardia or bradycardia. Furthermore, carotid sinus nerve activity was not changed significantly by 0.5 mg/kg of budralazine. Although a slight decrease in aortic depressor nerve activity was observed with a budralazine dose of 0.5 mg/kg, this decrease was not significant as compared with the vehicle treated group. Because the aortic depressor nerve components are mostly baroreceptor afferent fibers in rats (Sapru and Krieger, 1977), the decrement of aortic depressor nerve activity is not due to chemoreceptor inhibition. Therefore, we excluded the possibility that a 0.5 mg/kg dose of budralazine produces a decrease in peripheral sympathetic outflow via its indirect actions such as the modification of the baro- or chemo-receptor reflexes. Our results firmly suggest that doses of 0.5 and 1.0 mg/kg budralazine may act centrally to reduce the peripheral sympathetic outflow. In this experiment, the fact that the initial hypotension induced by budralazine (0.5 mg/kg i.v.) was not accompanied with the decreased sympathetic nerve activity, leads us to suggest that this initial hypotension may be associated with the direct vasodilating action of budralazine. Another important observation in the present experiment is that when a high dose of budralazine (5.0mg/kg) was administered intravenously, mean arterial pressure decreased approximately 50% and aortic depressor nerve activity was reduced significantly (Fig.5). Moreover, budralazine-induced hypotension was accompanied with significant increases in cardiac and adrenal sympathetic nerve activity as well as tachycardia. Our findings suggest that a baroreceptor-mediated reflex is associated with the increased sympathetic nerve activity and tachycardia induced by 5.0 mg/kg of budralazine.

Presynaptic α-adrenergic action. In normotensive pithed rats, budralazine (6 and 12 mg/kg, i.v.) not only inhibit transiently the sympathetic tachycardia but also enhance the clonidine bradycardia, although this drug was almost without effect on resting heart rate. These effects of budralazine on the heart rate response in the stimulated rats were partially but significantly antagonized by pretreatment with phentolamine (Shibamura *et al.*, 1981).

Effects on cyclic nucleotide levels. Budralazine, at 10, 20 and 30 mg/kg p.o. that produced a gradual decrease in systolic blood pressure with maximum of 50 mmHg at 4 hours in SHR (Akashi *et al.*, 1978), affected significantly neither cyclic AMP nor cyclic GMP levels in aorta at any times after dosing. When equihypotensive dose (4mg/kg, p.o.) of hydralazine was administered, no significant changes in the levels of cyclic nucleotide occurred, although the cyclic AMP to cyclic GMP ratio was significantly lower at 10 min after dosing as compared to the control values (Chiba *et al.*, 1983).

Effects on glucose and lipid metabolism. Repeated administration of budralazine (15-60 mg/kg, p.o.) had no effect on glucose tolerance, insulin secretion, electrolytes and lipid levels (cholesterol, triglyceride and non-esterified fatty acid) in Wistar rats, SHR, insulin-dependent diabetes SHR and non- insulin-dependent diabetes SHR. From these observations, it is possible that budralazine is useful for the treatment of hypertension associated with diabetes.

PHARMACOKINETICS

Absorption and Elimination

The concentration of [^{14}C]-budralazine in whole blood and plasma in normotensive and hypertensive rats reaches peaks of about 3.8 and 3.6 μg/ml 2 hours after oral administration of [^{14}C]-budralazine (15 mg/kg), followed by a decline with half-lives of 11.0 and 6.4 hours, respectively (Ono *et al.*, 1979).

The distribution of budralazine in normotensive and hypertensive rats were quite similar. Tissue concentration of radioactivity occurred in the following order: kidney > liver > adrenal > lung > heart > spleen. The concentration of radioactive materials in aorta wall was significantly higher than that in the blood for budralazine and declined more slowly than that in blood. The disappearance curve of radioactivity from the aorta wall paralleled the time course of the reduction in blood pressure.

Approximately 41% of the dosed radioactivity was eliminated in the urine and 40% in the faeces of normotensive and hypertensive rats within 24 hours. After oral administration to normotensive rats with cannulated bile ducts, radioactivity was excreted rapidly in bile; maximum excretion was reached at 2 hours, and during 24 hours about 17% of the dose was excreted in the bile.

Metabolism

On the basis of urinary metabolites observed in vivo studies, the scheme in figure 6 represents the possible metabolic pathways of budralazine in rats. Budralazine was

Table 1. Adverse effects of budralazine

Central nervous system adverse effects	
Headache	1.9%
Dizziness	0.3%
General or lower limbs fatigue	0.3%
Cardiac adverse effects	
Palpitation	2.4%
Chest pain	0.3%
Angina pectoris	0.4%
Arrhythmias	0.4%
Gastrointestinal adverse effects	
Nausea, vomiting	0.4%
Anorexia	0.3%
Diarrhea	0.4%
Constipation	1.0%
Flatulence	0.3%
Malaise	1.5%
Epigastric pain	0.4%
Other adverse effects	
Itching	0.4%
Flushing	0.6%

mainly metabolized via hydrazino phthalazine (hydralazine), followed by acetylation, dehydration, ring closure and conjugation reactions.

Preparations and dosage

Preparations currently used in Japan are 30-, and 60-mg tablets (Buterazine®) and also powder containing 100 mg/gram is available. The initial dose should be 90-120 mg. The maintenance dosage is usually 90-180 mg per day, given in two or three divided doses.

ADVERSE EFFECTS

No patients with essential hypertension showed positive antinuclear antibody or factor (ANF) in over a 10-month-period trial. A 5% incidence of gastrointestinal upset has been noted (Table 1).

PRESENT STATUS OF BUDRALAZINE IN THE MANAGEMENT OF HYPERTENSION

Results of comparative trials of budralazine and ecarazine in essential hypertension for 12 weeks indicate that although no significant difference in hypotensive activity was apparent among them, the total effectiveness of budralazine was greater and greater during the trial. Eventually budralazine was found to be much better than ecarazine ($p < 0.01$) in terms of effectiveness. This difference might be linked to the effective onset of budralazine more gradually than that of ecarazine.

A double-blind, multicenter trial evaluated budralazine, hydralazine and placebo for their values as a "second drug" when added to ongoing thiazide diuretic treatment. Over all of hypotensive effectiveness, adverse effects, laboratory tests and subjective symptoms, budralazine was a more suitable second drug than hydralazine as an advent therapy with a thiazide.

Acknowledgment

Budralazine hydrochloride was kindly supplied by Daiichi Seiyaku Co, Ltd., Tokyo, Japan.

REFERENCES

Åblad, B. (1963). A study of the mechanism of hemodynamic effects of hydralazine in man. *Acta Pharmacol. Toxicol.* **20** (suppl.1), 1-15.

Akashi, A., Chiba, T., Suzuki, I. and Hashimoto, H. (1978). Pharmacology of budralazine, a new antihypertensive drug. *Arch. int. Pharmacodyn.* **135,** 134-146.

Akashi, A., Chiba,T. and Kasahara, A. (1974). Antihypertensive activity of 1-[2- (1,3)-dimethyl-2-butenylidene)-phthalazine (DJ-1461), a new phthalazine derivatives. *European J. Pharmacol.* **29,** 161-164.

Brunner, H., Hedwall, P.R. and Meier, M. (1967). Influence of adrenergic β-receptor blockade on the acute cardiovascular effects of hydralazine. *Br. J. Pharmacol.* **30,** 123-133.

Chatterjee, K., Massie, B., Rubin, S., Gelberg, H., Brundage, B. H. and Ports, T. A. (1978). Long-term outpatient vasodilator therapy of congestive heart failure. *Am. J. Med.* **65,** 134-145.

Chiba, T., Hirohashi, M., Suzuki, I., Tanaka, S., Watanabe K. and Akashi, A. (1983). Characterization of the Vasodilator Action of Antihypertensive Drug Budralazine. *Arzneim.-Forsch.* **33,** 112-116.

Dranov, J., Skyler, J. S. and Gunnells, J. C. (1974). Malignant Hypertension: Current Modes of Therapy. *Arch. Intern. Med.* **133,** 791-801.

Dustan, H.P. (1983). Causes of inadequate response to antihypertensive drugs. *Hypertension* **5** (suppl. III), 26-30.

Dustan, H. P. Taylor, R. D., Corcoran, A. C. and Page I. H. (1954). Rheumatic and febrile syndrome during prolonged hydralazine treatment. *JAMA* **154** (1), 23-29.

Hagihara, Y. (1981). Effect of 4-methyl-3-penten-2-one(l-phthalazinyl)hydrazone (budralazine) on intra- and extracranial circulation in cats. *Folia pharmacol. Japon.* **77,** 41-50.

Konno, K. and Ueno, A. (1977). Cardiovascular effect of 1-[2-(1,3-dimethyl-2-butenylidene) hydrazino] phthalazine in unanesthetized normo- and hypertensive dogs. *Folia pharmacol. Japon.* **73,** 98P.

Konno, K., Nonaka, K. and Ueno, A (1981). Hypotensive and cardiovascular action of budralazine in unanesthetized and unrestrained dogs. *Arzneim.-Forsch.* **31,** 1072-1075.

McLean, A. J., D-U Souich, P., Barron, K. W., Briggs, A. H. (1978). Interaction of Hydralazine with Tension Development and Mechanisms of Calcium Accumulation in K+-Stimulated Rabbit Aortic Strips. *J. Pharmacol. Exp. Ther.* **207,** 40-48.

Morrow, J. D., Schroeder, H. A. and Perry Jr., H. M. (1953). Studies on the control of hypertension by Hyphex. II. Toxic reactions and side effects. *Circulation* **8,** 829-839.

Ono, K., Saito, T., Sasano, H., Moroi, R., Sano, M. and Akimoto, T. (1979). Metabolism of budralazine, a new antihypertensive agent, III. *Xenobiotica* **9,** 227-236.

Saegusa, K., Furukawa, Y. and Chiba, S. (1986). Pharmacological analysis of hydralazine-induced cardiac action in intact dogs and isolated, blood-perfused canine atria. *J. Cardiovasc. Pharmacol.* **8,** 614-620.

Sapru, H. N. and Krieger, A. J. (1977). Carotid and aortic chemoreceptor function in the rat. *J. Appl. Physiol.* **42,** 344-348.

Shibamura, S., Chiba, T., Suzuki, I. and Akashi, A. (1981). Pre- and Postsynaptic α-Adrenergic Effects of the Antihypertensive Drug Budralazine. *Arzneim.-Forsch.* **31,** 1076-1079.

Stein, D. H. and Hecht, H. H. (1955). Cardiovascular and renal responses to the combination of hexamethonium and 1-hydrazinophthalazine (Apresoline) in hypertensive subjects. *J. Clin. Invest.* **34,** 867-874.

Takeuchi, J., Taniguchi, K. and Iesaka, Y. (1980). Long effects of budralazine (DJ-1461) on essential Hypertension. *Shinryo to Shinyaku* **17,** 2381-2385 (in Japanese).

Yoshinaga, K. and Miura, Y. (1985). Treatment of hypertension with budralazine. *Progress in Medicine* **5,** 2223-2232 (in Japanese).

Yoshioka, M., Togashi, H., Minami, M. and Saito, H (1986). Effects of hydralazine on adrenal and cardiac sympathetic nerve activity in anesthetized rats. *Res. Comm. Chem. Pathol. Pharmacol.* **54,** 313-320.

Zacest, R., Gilmore, E. and Koch-Weser, J. (1972). Treatment of essential hypertension with combined vasodilation and β-adrenergic blockade. *N. Engl. J. Med.* **286,** 617-622.

Progress in Hypertension, Vol. 2, pp. 319-329.
Saito *et al.* (Eds)

POSSIBLE ROLE OF ENDOTHLIUM-DERIVED NITRIC OXIDE IN THE REGULATION OF BLOOD PRESSURE

ICHIRO SAKUMA, TAKESHI KOBAYASHI, MASATOSHI SATO, HISAKAZU YASUDA, MASAMICHI KATO,* HIROKO TOGASHI,** MITSUHIRO YOSHIOKA** and HIDEYA SAITO**

Department of Cardiovascular Medicine, *2nd Department of Physiology and **1st Department of Pharmacology, Hokkaido University School of Medicine, Sapporo 060, Japan.

INTRODUCTION

Endothelium-derived relaxing factor (EDRF) (Furchgott and Zawadzki, 1980) is released from endothelial cells in response to various pathophysiological stimuli; it relaxes smooth muscle and inhibits platelet aggregation and adhesion, and plays an important role in maintaining blood circulation *in vivo* (Fig. 1) (Ignarro, 1989; Furchgott and Vanhoutte, 1988; Moncada *et al.*, 1989; Bassenge and Heusch, 1990). The mode of action of EDRF results from stimulation of cytosolic soluble guanylate cyclase which is followed by the elevation of intracellular cyclic GMP (cGMP) levels (Fig. 1) (Ignarro, 1989).

Recently, one of EDRFs was identified to be nitric oxide (NO) which originates from guanidino nitrogen(s) of L-arginine (Fig. 2) (Palmer *et al.*, 1987). Since the synthesis of NO from L-arginine is selectively inhibited by N^G-monomethyl-L-arginine (NMA) (Fig. 2) and since this inhibition can be reversed by excess L-arginine (Palmer *et al.*, 1988; Sakuma *et al.*, 1988; 1990b), one can investigate the role L-arginine-derived NO in particular experimental settings by using NMA and L-arginine.

It is of interest whether EDRF/NO continuously released from vascular endothelium contributes to modify blood pressure. Indeed, Aisaka *et al.* (1989) reported that intravenous administration of NMA to guinea pig elicits a sustained increase in arterial blood pressure: L-arginine reverses this response. Furthermore, Rees *et al.* (1989), Whittle *et al.* (1989) and Kilbourn *et al.* (1990) reported similar results in rabbits, rats and dogs, respectively. These findings indicate that basal EDRF/NO release from peripheral vessels elicits a continuous vasorelaxant action and thus lowers blood pressure *in vivo* (Fig. 1).

L-arginine-derived NO is also produced in the brain; Garthwaite *et al.* (1988) found that upon activation of L-glutamate receptors rat cerebellar cells release EDRF/NO which in turn elevates intracellular cGMP. Additional investigations have demonstrated the enzymatic formation of NO or nitrogen oxides from L-arginine in rat forbrain synaptosomal cytosol (Knowles *et al.*, 1989) and in bovine brain cytosol (Schmidt, *et al.*, 1989). More recently, production of L-arginine-derived NO by serotonin, which results in intracellular increase in cGMP, was reported in NG108-15 mouse neuroblastoma x rat glioma hybrid cells (Tohda and Nomura, 1990). These

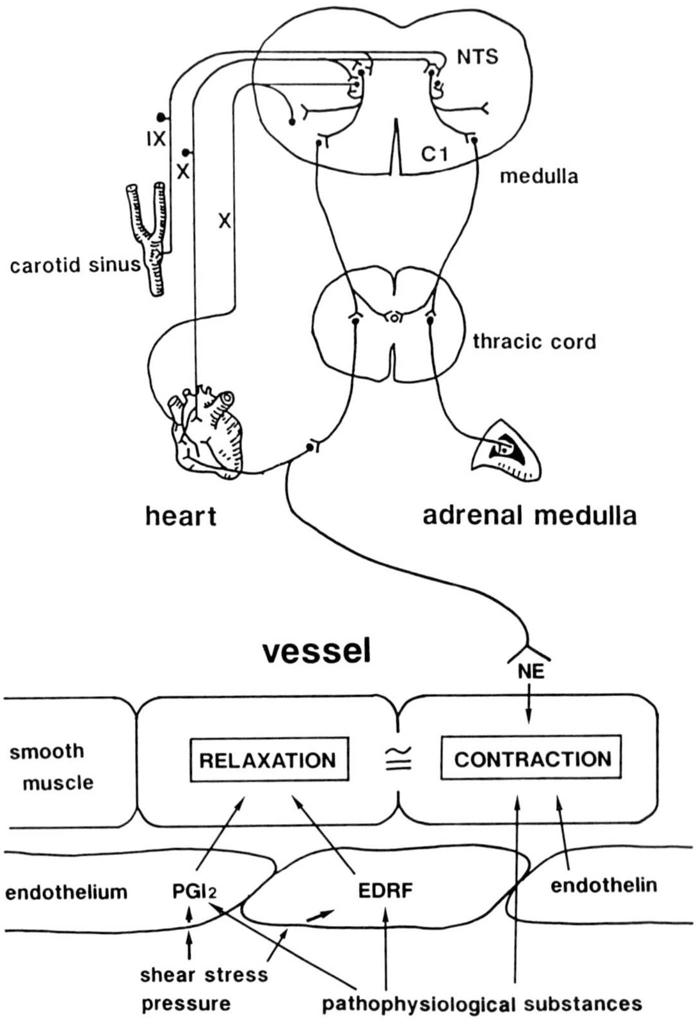

Figure 1. Peripheral and central regulation of vascural tone. In the periphery, vascular tone is set at a certain level by a balance between vasorelaxation and vasoconstriction resulting from various stimuli or substances. Vascular tone can be modulated centrally or by baroreceptor reflex. Abbereviations used: IX: afferent fiber of the ninth nerve, X: afferent and efferent fibers of the tenth nerve, NTS: nucleus tractus solitalius, C1: C1 area of medulla, NE: norepinephrine, PGI_2: prostacyclin.

observations have raised a possibility that EDRF/NO-cGMP system may function as one of important signal transduction systems in the brain (Moncada *et al.*, 1989). Indeed, new evidence by Bredt *et al.* (1990) that NO synthase is present in descrete neuronal populations further implies a neural messenger role for NO.

Since L-glutamate and serotonin receptors are involved in central autonomic reflexes (Reis *et al.*, 1984), it is conceivable that sympathetic nerve activity would be modulated by L-arginine-derived NO. If so, NMA might be expected to alter sympathetic tone by inhibiting the formation of NO in the brain and consequently elicit changes in blood pressure and heart rate (Fig. 1).

In this chapter, we show our data suggesting that L-arginine-derived NO generated in the brain modifies sympathetic outflow and thus contributes to the cardiovascular regulation (Sakuma, *et al.*, 1990a; 1991). We also refer to the other sites where L-arginine-derived NO has been reported to be generated and the released NO is presumed to contribute to the regulation of blood pressure.

MATERIALS AND METHODS

Measurement of hemodynamic parameters

Male Wistar rats (Scl: Wistar/ST rats), weighing between 350 and 400g, were anesthetized with alpha-chloralose (50 mg/kg, *i.p.*) and urethane (500 mg/kg, *i.p.*) and immobilized with gallamine triethiodide (10 mg/kg, *i.p.*). Ventilation was maintained artificially (Harvard apparatus, model 683) through a tracheal cannula. The core temperature was monitored intrarectally and maintained close to 38°C with a heating pad. The left femoral vein was cannulated for the injection of drugs. In some rats a polyethylene tube (PE-10) was inserted onto the ventral surface of medulla oblongata via cisterna magna. Blood pressure (BP) and heart rate (HR) were continuously monitored from the left femoral artery via a polyethylene cannula filled with heparinized 0.9 % NaCl solution which was connected to a pressure transducer (Nihon Kohden, TP-101T). In some rats total deafferentation was performed by bilateral surgical section of vagal and sino-aortic nerves (denervated rats).

The second set of experiments was performed on adult cats of either sex weighing between 2 and 4 kg. The cats were anesthetized with alpha-chloralose (50 mg/kg, *i.p.*) and urethane (500 mg/kg, *i.p.*). The right femoral vein was cannulated for the injection of drugs, while the right femoral artery was cannulated and connected to a pressure transducer via a polyethylene cannula filled with heparinized 0.9 % NaCl solution for the continuous measurement of BP and HR. In some cats total deafferentation was performed by bilateral surgical section of vagal and sino-aortic nerves (denervated cats).

Nerve Recordings

In the rat experiment postganglionic sympathetic nerve was approached retroperitoneally and dissected under a microscope (Togashi *et al.*, 1990). The exposed nerves were cut and covered with liquid paraffin. Efferent renal nerve activity (RNA) was recorded from the central cut end of the nerve with a bipolar platinum-iridium electrode. The impulses were amplified by a biophysical preamplifier (Nihon

Kohden, AVB-10) with high and low cut-off frequency of 1 kHz and 50 Hz, respectively. The amplified signals were separated from noise levels with a window discriminator and counted every 10 sec using a real-time data analyser (Nihon Kohden, ATAC-450). Similar recording procedure was performed in the feline experiment. Nerve activity is expressed as the percent of control value measured before drug administration.

Chemicals
N^G-monomethyl-L-arginine (NMA) was prepared from L-ornithine and thiopseudourea. The flavinate salt of NMA was converted to the hydrochloride salt by stirring with Dowex 1 (Cl^-); the concentration of pure compound was determined by amino acid analysis. All other compounds were of the highest grade available from Sigma (St. Louis, MO).

RESULTS

Intravenous administration of NMA (10 to 125 μmol/kg) to anesthetized rats elicited an immediate increase in BP which was dose-dependent in the magnitude and duration (Fig. 2 and 3). The pressor response to NMA at a dose of 50 μmol/kg started within 30 sec after NMA administration and lasted more than 20 min (Fig. 2). The pressor

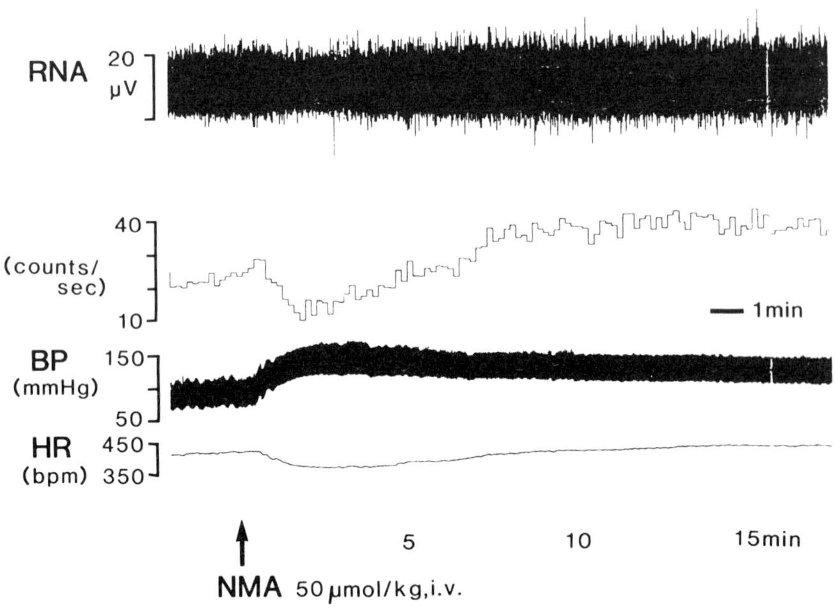

Figure 2. Effects of *NG* -monomethyl-L-arginine (NMA; 50 μmol/kg, *i.v.*) on sympathetic renal nerve activity (RNA), arterial blood pressure (BP) and heart rate (HR) in the nerve-intact α-chloralose/urethane-anesthetized rat. (modified from Sakuma *et al.*, 1991.)

response was associated with biphasic changes in HR and RNA; both of which initially decreased, until the concomitant increase in BP reached a plateau, and subsequently tended to increase, usually beyond their pre-NMA levels, while BP remained elevated (Fig. 2, 3 and 4). The increase in RNA is the opposite of what would be expected as a reflux response to elevated blood pressure.

To confirm that the initial decrease in HR and RNA observed with NMA occurs as a reflux to NMA's pressor effect, we studied whether these actions could be blocked by vagotomy and sino-aortic deafferentation. The total deafferentation did not significantly alter NMA-induced pressor response, however, abolished the initial decrease in HR and RNA, and only a monophasic increase in RNA remained (Fig. 5). Thus, the direct effect of NMA on RNA is thought to be stimulation. Since NMA intravenously administered to the deafferentated rat elicited a similar increase in preganglionic adrenal nerve activity and since spinal transection of the deafferentated

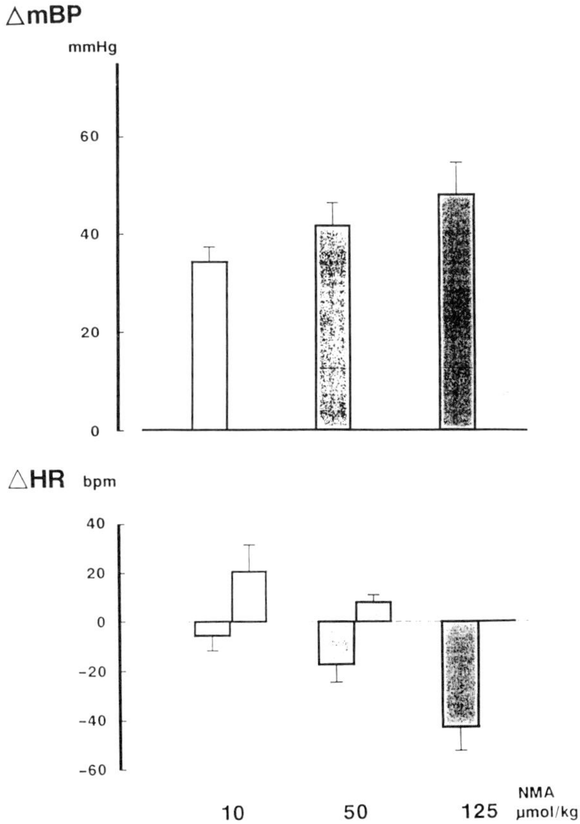

Figure 3. Effects of intravenous administration of *NG* -monomethyl-L-arginine (NMA) on mean arterial blood pressure (mBP) and heart rate (HR) in the α-chloralose/urethane-anesthetized rat. Bars depict means of changes from control values ± SE (n = 3-8). In HR, the left side bars of each dose of NMA represent maximal changes in the initial decrease and the right side bars represent those in the subsequent increase.

rat abolished the increase in RNA caused by NMA (data not shown), we infer that the site of NMA's action is likely to be the brain, rather than the periphery (Fig. 1).

We next investigated whether the effects of NMA on RNA were specifically due to inhibition of NO synthesis. If so, we would expect L-arginine to inhibit the increase in RNA caused by NMA. Indeed, the NMA induced-increase in RNA as well as BP was inhibited by the pretreatment of the deafferentated rat with L-arginine (50 to 500 μmol/kg, *i.v.*) (data not shown). Thus, we conclude that NMA increases sympathetic nerve activity by an action in the brain that results from inhibition of neuronal NO synthesis.

Because of this possibility we further investigated the effect of NMA on RNA and BP when NMA is administered directly into the brain. Intracisternal administration of NMA (1 μmol) to the deafferentated rat caused a rapid and marked increase in RNA and a slow and small increase in BP (data not shown). These changes in RNA and BP were attenuated by the intravenous administration of L-arginine. Furthermore, Intracisternally-administered L-arginine (1 μmol) abolished the increase in RNA, and attenuated the increase in BP, induced by intravenously-administered NMA (data not shown). These observations further support the view that inhibition of L-arginine-derived NO synthesis in the brain contributes to NMA's pressor effect by increasing sympathetic outflow; alternatively, L-arginine-derived NO continuously formed in the central neurons may act to reduce sympathetic outflow and consequently lower blood pressure.

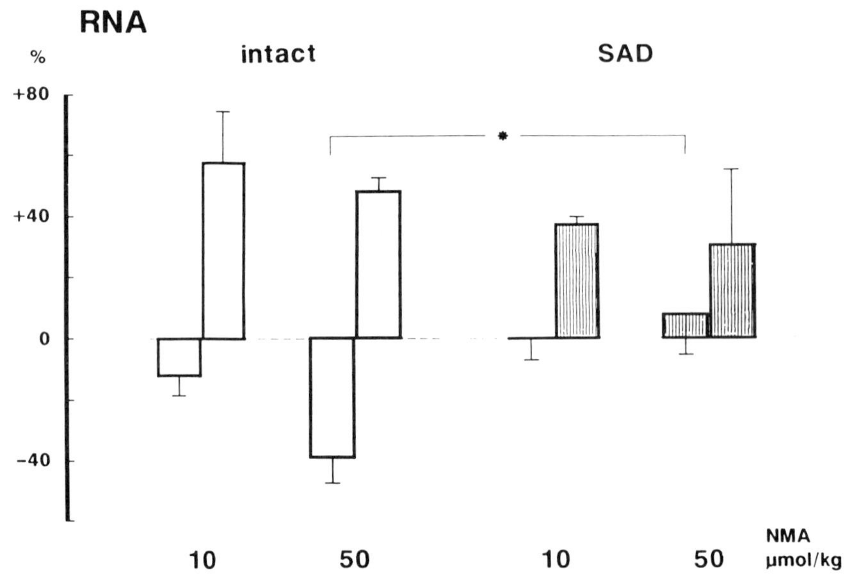

Figure 4. Effects of intravenous administration of *NG*-monomethyl-L-arginine (NMA) on sympathetic renal nerve activity (RNA) in the nerve-intact (intact) and totally-deafferentated (SAD) α-chloralose/urethane-anesthetized rat. Bars depict means of changes from control values ± SE (n = 4-8). The left side bars of each dose of NMA represent maximal changes in the initial decrease and the right side bars represent those in the subsequent increase. *; $p < 0.05$.

Basically similar results were observed in the feline experiment. Intravenous administration of NMA (3 to 10 μmol/kg) to the nerve-intact cat resulted in a rapid increase in BP which was associated with a monophasic increase in RNA and an equivocal change in HR (data not shown). Following total deafferentation, the pressor response to NMA was potentiated and the increase in RNA was enhanced; moreover, HR increased while BP remained elevated (data not shown). The increase in RNA and HR by NMA, but not BP, was abolished by spinal transection (data not shown). Furthermore, the changes induced by NMA were attenuated by pretreatment of the cat with L-arginine (50 μmol/kg, *i.v.*) (data not shown). These findings obtained from cats further strengthen our notion that L-arginine-derived NO formed in the brain may play a role in the central regulation of blood pressure by influencing (reducing) sympathetic outflow.

DISCUSSION

On the basis of the findings presented here, we propose that NO continuously formed from L-arginine in the brain may attenuate adrenergic drive and thus lower blood pressure (Fig. 5-1). We cannot determine from our present results, however, the precise site(s) in the brain that is responsible for the effects of L-arginine-derived NO.

The modulatory role of the L-arginine-derived NO for the regulation of blood pressure would be expected in various sites of the body other than the brain. These sites are drawn summarized in Fig. 5. Palacios *et al.* (1989) reported that L-arginine-derived NO is formed in adrenal cortex and medulla cytosol and stimulates soluble guanylate cyclase; of course, this raises the cGMP level. The increase in cGMP of adrenal cortex and medulla may regulate the release of steroid hormones and catecholamines, respectively, and, in turn, modulate blood pressure (Fig. 5-2).

The field stimulation of perivascular nerves had been reported to produce vasorelaxation which is not cholinergic or adrenergic in its origin (Fig. 5-3). Toda and Okamura (1990) demonstrated recently that this non-cholinergic, non-adrenergic relaxation is inhibited by NMA in an L-arginine-reversible manner. Hence, L-arginine-derived NO is thought to be released from perivascular nerves to relax vessels and thus might function to reduce blood pressure. In this regard, it should be noted that Bredt *et al.* (1990) reported the presence of NO synthase in the body of perivascular nerves immunohistochemically.

Release of norepinephrine from adrenergic nerve endings innervating vascular smooth muscles have been reported to be inhibited by EDRF/NO through activation of soluble guanylate cyclase and the resultant increase in the cGMP levels in the sympathetic neurons (Greenberg *et al.*, 1989, Fig. 5-4). Speculatively, electrical or chemical stimulation of sympathetic nerve endings may give rise to the intraneuronal generation of NO from L-arginine and result in the elevation of intraneuronal cGMP. Alternatively, especially in the case of small resistant arteries, EDRF released from vascular endothelium may be able to penetrate to perivascular adrenergic nerve endings and increase intraneuronal cGMP. Whichever the pathway, L-arginine-derived NO, if it acts on the sympathetic nerve endings, would lower blood pressure by the presynaptic inhibition of norepinephrine release.

Wood *et al.* (1990) described that vascular smooth muscle cells synthesize NO from L-arginine in their cytosol (Fig. 5-5). This smooth muscle derived-NO may relax and hence contribute to lower blood pressure.

Generation of EDRF from endothelium (Fig. 5-6) is one of the major determinants of vascular tone and this has been proven even in humans (Vallance *et al.*, 1989). Various pathophysiological stimuli including mechanical stress, neurotransmitters and hormones can liberate EDRF from endothelium. Among them, the main stimulus for the generation of EDRF in physiological condition *in vivo* may be share stress.

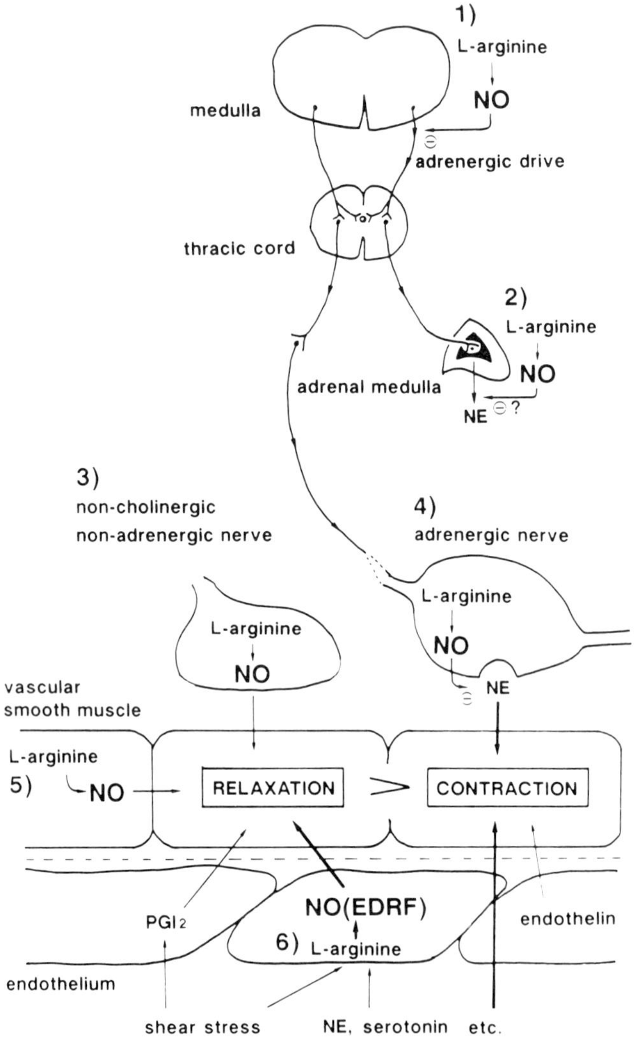

Figure 5. Possible sites where L-arginine-derived nitric oxide (NO) is synthesized and the generated NO may contribute to modulate blood pressure. NE: norepinephrine, PGI_2: prostacyclin.

Vargas *et al.* (1990) demonstrated that NMA, but not phenylephrine, failed to elevate blood pressure in the pithed rat indicating that a certain level of blood pressure is necessary for NMA to elicit its vasoconstricting effect; in other words, EDRF may not be released from vascular endothelial cells when share stress onto the endothelium is very small.

It is conceivable that, if EDRF release from vascular endothelium is reduced *in vivo* in certain pathological conditions, blood pressure would rise and, consequently, may result in hypertension. Indeed, the impairment of endothelium-dependent relaxing mechanism has been demonstrated in many animal models of hypertension (Gryglewski *et al.* 1988; Vane *et al.*, 1990). In those experiments the vessels from animals with hypertension have been shown to release less EDRF/NO than those from normal animals. Also in humans, Panza *et al.*(1990) reported that the ability of acetylcholine to dilate bronchial artery by EDRF/NO-dependent mechanism is decreased in patients with essential hypertension. Thus, the impairment of EDRF/NO release from endothelium is presumed to be a common feature of hypertension in all species. However, it is not determined whether the endothelial dysfunction is primary or secondary to the hypertensive status. Studies in animal models have shown that endothelium can be injured by short-term and long-term elevations of blood pressure; resulting in impaired EDRF-mediated vasorelaxation (Lamping *et al.*, 1987; Miller *et al.*, 1987). Moreover, impaired EDRF-mediated vasorelaxation in the hypertensive animals can be normalized after treatment with antihypertensives (Luscher *et al.*, 1987). Thus, the endothelial abnormality in hypertension may not be a cause but a consequence of the elevated blood pressure. Nevertheless, the endothelial impairment in hypertension, if not a primary phenomenon, would exacerbate the existing vascular dysfunction and render the establishment of systemic hypertension. It is tempting to speculate, however, that the impairment of a generation of L-arginine-derived NO or failure of NO to increase cGMP levels in various sites of the body mentioned above other than vascular endothelium contribute to the elevation of blood pressure and final consequences of systemic hypertension.

Furthermore, it is conceivable that, if the alteration in the generation of L-arginine-derived NO is related to the cause of hypertension, hypertension may be able to be manipulated by compounds that modulate the release of NO. In this context it is of much interest that Nakaki *et al.* (1990) reported that intravenous infusion of L-arginine to humans elicits hypotension and the hypotensive effects of L-arginine is more prominent in hypertensive patients than normal controls. The supplementary L-arginine or L-arginine analogs in food or water might thus be used as an antihypertensive treatment.

REFERENCES

Aisaka, K., Gross, S. S., Griffith, O. W. and Levi, R. (1989). N^G-methylarginine, an inhibitor of endothelium-derived nitric oxide synthesis, is a potent pressor agent in the guinea pig: does nitric oxide regulate blood pressure in vivo? *Biochem. Biophys. Res. Comm.* **160,** 881-886.

Bassenge, E. and Heusch, G. (1990). Endothelial and neuro-humoral control of coronary blood flow in health and disease. *Rev. Physiol. Biochem. Pharmacol.* **116,** 77-165.

Bredt, D., Hwang, P. M. and Snyder, S. H. (1990). Localization of nitric oxide synthase indicating a neural role for nitric oxide. *Nature (London)* **347,** 768-770.

Furchgott, R. F. and Vanhoutte, P. M. (1989). Endothelium-derived relaxing and contracting factors. *FASEB J.* **3,** 2007-2018.

Furchgott, R. F.and Zawadzki, J. V. (1980). The obligatory role of endothelial cells in the relaxation of arterial smooth muscle by acetylcholine. *Nature (London)* **228,** 373-376.

Garthwaite, J., Charles, S. L. and Chess-Williams, R. (1988). Endothelium-derived relaxing factor release on activation of NMDA receptors suggests role as intercellular messenger in the brain. *Nature (London)* **336,** 385-388.

Greenberg, S., Diecke, F. P. J., Peevy, K. and Tanaka, T. (1989). The endothelium modulates adrenergic neurotransmission to canine pulmonary arteries and veins. *Eur. J. Pharmacol.* **162,** 67-80

Gryglewski, R. J., Botting, R. M. and Vane, J. R. (1988). Mediators produced by the endothelial cell. *Hypertension* **12,** 530-548.

Ignarro, L. J. (1989). Endothelium-derived nitric oxide: actions and properties. *FASEB J.* **3,** 31-36.

Kilbourn, R. G., Gross, S. S., Jubran, A., Adams, J., Griffith, O. W., Levi, R. and Lodato, R. F. (1990). N^G -methyl-L-arginine inhibits tumor necrosis factor-induced hypotension: Implication for the involvement of nitric oxide. *Proc. Natl. Acas. Sci. USA* **87,** 3629-3632.

Knowles, R. G., Palacios, M., Palmer, R. M. G. and Moncada, S. (1989). Formation of nitric oxide from L-arginine in the central nervous system: A transduction mechanism for stimulation of the soluble guanylate cyclase. *Proc. Natl. Acad. Sci. USA* **86,** 5159-5162.

Lamping, K. G., Cannon, R. O. III. and Talbot, T. L. (1987). Acute hypertension selectively potentiates constrictor responses of large coronary arteries to serotonin by altering endothelial function *in vivo*. *Cir. Res.* **61,** 904-913.

Luscher, T. F., Vanhoutte, P. M. and Raji, L. (1987). Antihypertensive treatment normalizes decreased endothelium-dependent relaxations in rats with salt-induced hypertension. *Hypertension* **9** (Suppl III), III-193-III-197.

Miller, M. J. S., Pinto, A. and Mullane, K. M. (1987). Impaired endothelium-dependent relaxation in rabbits subjected to aortic coarctation hypertension. *Hypertension* **10,** 164-170.

Moncada, S., Palmer, R. M. J. and Higgs, E. A. (1989). Biosynthesis of nitric oxide from L-arginine. *Biochem. Pharmacol.* **11,** 1709-1715.

Nakaki, T., Hishikawa, K., Suzuki, H., Saruta, T. and Kato, R. (1990). L-arginine-induced hypotension. *Lancet* **336,** 696.

Palacios, M., Knowles, R. G., Palmer, R. M. J. and Moncada, S. (1989). Nitric oxide from L-arginine stimulates the soluble gaunylate cyclase in adrenal glands. *Biochem. Biophys. Res. Comm.* **165,** 802-809.

Palmer, R. M. J., Ferige, A. G. and Moncada, S. (1987). Nitric oxide release accounts for the biological activity of endothelium-derived relaxing factor. *Nature (London)* **327,** 524-526.

Palmer, R. M. J., Ashton, D. S. and Moncada, S. (1988). Vascular endothelial cells synthetize nitric oxide from L-arginine. *Nature (London)* **333,** 664-666.

Panza, J. A., Quyyumi, A. A., Brush, J. E. and Epstein, S. E. (1990). Abnormal endothelium-dependent vascular relaxation in patients with essential hypertension. *N. Engl. J. Med.* **323,** 22-27.

Reis, D. J., Granata, A. R., Joh, T. H., Ross, C. A., Ruggiero, D. A. and Park, D. H. (1984). Brain stem catecholamine mechanism in tonic and reflex control of blood pressure. *Hypertension* **6** (Suppl II), II-5-II-15.

Rees, D. D., Palmer, R. M. J. and Moncada, S. (1989). Role of endothelium-derived nitric oxide in the regulation of blood pressure. *Proc. Natl. Acad. Sci. USA* **86,** 3375-3378.

Sakuma, I., Stuehr, D. F., Gross, S. S., Nathan, C. and Levi, R. (1988). Identification of arginine as a precursor of endothelium-derived relaxing factor. *Proc. Natl. Acad. Sci. USA* **85,** 8664-8667.

Sakuma, I., Togashi, H., Yoshioka, M., Saito, H., Yanagida, M., Tamura, M., Kobayashi, T., Yasuda., H., Gross., S.S. and Levi, R. (1991). N^G-methyl-L-arginine, an inhibitor of L-arginine-derived nitric oxide synthesis, stimulates renal sympathetic nerve activity in vivo: a role for nitric oxide in the central regulation of sympathetic tone? *Circ. Res.* in press.

Sakuma, I., Togashi, H., Yoshioka, T., Kobayashi, T., Saito, H., Yasuda., H., Gross., S. S. and Levi, R. (1990a). Effects of intravenous and intracisternal administration of N^G-monomethyl-L-arginine on renal sympathetic nerve activity in anesthetized rats. In: *Nitic oxide from L-arginine: a bioregulatory system*, S. Moncada and E.A. Higgs (Eds). Elsevier Science Publishers, Amsterdam, pp. 493-494.

Sakuma, I., Yasuda, H., Gross., S. S. and Levi, R. (1990b). L-arginine is a precursor of endothelium-derived relaxing factor in various animal species and vascular beds. In: *Nitric oxide from L-arginine: a bioregulatory system*, S. Moncada and E.A. Higgs (Eds). Elsevier Science Publishers, Amsterdam, pp. 445-449.

Schmidt, H.H.H.W., Wilke, P, Evers, B. and Böhme, E. (1989). Enzymatic formation of nitrogen oxides from L-arginine in bovine brain cytosol. *Biochem. Biophys. Res. Comm.* **165,** 284-291.

Toda, N. and Okamura, T. (1990). Possible role of nitric oxide intransmiting information from vasodilator nerve to cerebroarterial muscle. *Biochem. Biophys. Res. Comm.* **170,** 308-313.

Togashi, H., Yoshioka, M., Tochihara, M., Matsumoto, M. and Saito, H. (1990). Differential effects of hemorrhage on adrenal and renal nerve activity in anesthetized rats. *Am. J. Physiol.* **259,** H-1134-H1141.

Tohda, M. and Nomura, Y. (1990). Serotonin stimulates both cytosolic and membrane-bound guanylate cyclase in NG-108-15 cells. *J. Neurochemistry* **55,** 1800-1805.

Vallance, P., Collier, J. and Moncada, S. (1989). Effects of endothelium-derived nitric oxide on peripheral arteriolar tone in man. *Lancet* **2,** 997-1000.

Vane, J. R., Änggård, E. E. and Botting R. M. (1990). Regulatory function of the vascular endothelium. *N. Engl. J. Med.* **323,** 27-36.

Whittle, B.J.R., Lopez-Belmonte, J. and Rees, D. D. (1989). Modulation of the vasodepressor actions of acetylcholine, bradykinin, substance P and endothelin in the rat by a specific inhibitor of nitric oxide formation. *Br. J. Pharmacol.* **98,** 646-652.

Wood, K. S., Buga., G. M., Byrns, E. and Ignarro, L. J. (1990). Vascular smooth muscle-derived relaxing factor (MDRF) and its close similarity to nitric oxide. *Biochem. Biophys. Res. Comm.* **170,** 80-88.

Vargas, H., Ignarro, L. J. and Chaudhuri, G. (1990). Physiological release of nitric oxide is dependent on the level of vascular tone. *Eur. J. Pharmacol.* **190,** 393-397.

Progress in Hypertension, Vol. 2, pp. 331-353.
Saito *et al.* (Eds)

IMMUNOLOGICAL FACTORS IN GENETIC HYPERTENSION

BATAILLARD, A., TOURAINE, J. L.,* and SASSARD, J.

Department of Physiology and Clinical Pharmacology, CNRS URA 606, School of Pharmacy, Lyon and * Department of Nephro-Urology Transplantation and Immunology, E. Herriot Hospital, Lyon, France.

INTRODUCTION

In most cases (95%) hypertension is described as "essential" since mechanisms which interfere with its onset and/or its maintenance remain unknown. To date, among the physiopathological hypothesis proposed, abnormalities of the permeability of the cellular membrane to sodium, or hyperactivity of the sympathetic nervous system and/or of the renin-angiotensin system have the favor of the investigators. In this paper we consider another hypothesis, which has aroused interest over the last ten years, according to which immunological factors may be involved in the pathogenesis of hypertension.

In humans, an increased serum level of immunoglobulins (Ebringer and Doyle 1973; Olsen, 1973; Kristensen, 1978) and a higher prevalence and amount of autoantibodies occur more often in hypertensive patients than in normotensive subjects (Mathews *et al.*,1973; Wilson *et al.*, 1987, Kristensen, 1984). Likewise, patients with essential hypertension have more frequently rheumatoid and thyroid factors than controls (Bastenie *et al.*, 1971; Ebringer and Doyle, 1973). Regarding cellular immunity, inhibition of macrophage migration by exposure to human arterial wall extracts as well as increased reactivity of T-cells against the same extracts are observed in 50% of the patients with essential hypertension (Olsen, 1974; Gudbransson *et al.*, 1981). Finally, various associations between the HLA-system and hypertension have been described. A greater frequency of B15 positivity is noticed in patients with a family history of hypertension (Kristensen *et al.* , 1970). Although interesting, this finding must be interpreted with caution since other groups have found a higher frequency of B8 (Low *et al.*, 1975) or B12(Gelsthorpe *et al.*, 1975) in essential hypertension and, more recently, of DR4 in gravidic hypertension (Simon *et al.*, 1989).

In animals, the first evidence arguing for an involvement of immunological factors in hypertension was reported by Okuda and Grollman (1967) who showed that hypertension induced by renal infarction could be transferred into normotensive recipients by lymph node cells. This was confirmed by Olsen (1980) who used splenocytes from one-kidney Goldblatt or DOCA/salt hypertensive rats. Furthermore, secondary forms of hypertension induced by angiotensin (Olsen, 1970) or DOCA/salt administration (Svendsen, 1977a) are accompanied by inflammatory reactions including mononuclear cell infiltration and oedema of the arterial walls. In secondary hypertension as well as in the genetically hypertensive rat of the Japanese strain (SHR) immunological abnormalities are observed: decreased number and *in vitro* responses to mitogens of thymic and splenic lymphocytes (Chatelain *et al.*, 1980; Takeichi *et al.*, 1980, 1981a; Ventola and Strausser, 1984), hyperplasia and hypertrophy of thymic epithelial cells (Chatelain and Ferrario, 1978; Rojo-Ortega *et al.*, 1973) and thymocytotoxic autoantibodies (Takeichi *et al.*, 1981b). More persuasive, a reduction in the degree of hypertension can be obtained in SHR by a restoration of their immunological functions either by administration of thymic hormonal factors (Ba *et al.*, 1982; Strausser *et al.*, 1983) or by thymus graft from normotensive rat donors (Norman *et al.*, 1985, 1986). Surprisingly, an improvement of hypertension is also obtained by immunosuppressive therapy against T-cells such as short treatment with anti-rat thymocyte serum (Bendish, 1981) or thymectomy (Khraibi *et al.*, 1987). Finally, an association between blood pressure and gene(s) within the major histocompatibility complex has been recently found using recombinant inbred strains obtained by mating normotensive (Brown Norway) and hypertensive (SHR) rats (Pravenec *et al.*, 1989).

Taken as a whole, these data strongly suggest that immunological factors maybe of pathogenic importance in hypertension. Nevertheless, several questions should be posed as to the relationship between immune system and hypertension. In particular, in experimental hypertension, it is necessary to distinguish secondary forms with a known etiology (renovascular, endocrinological) from primary forms which are genetically determined. Since secondary forms are experimentally induced, the immunological abnormalities observed might be the consequence but cannot contribute to the initial rise of blood pressure. Furthermore, it remains to determine whether these abnormalities are involved in the maintenance of hypertension or whether they constitute an epiphenomenon. By contrast, in essential hypertension as well as in genetic models of hypertension, one can wonder whether immunological abnormalities contribute to hypertension, namely, whether hypertension is not a complication of an immunological disease. As in secondary forms of hypertension, immunological alterations could also occur subsequently to the rise of blood pressure. A third possibility would be the presence of one of several genetic factors which predispose to both hypertension and immunological abnormalities.

In order to clarify the role of immunological factors in the pathogenesis of essential hypertension, we studied a model of genetic hypertension. Among the various existing models, the Lyon model appeared appropriate since genetically hypertensive rats (LH) can be compared with two simultaneously selected control strains of normotensive (LN) and low-blood pressure (LL) rats (Vincent, 1984). Subsequently, this model allows valid interstrain comparisons. The purpose of our studies was to determine whether the immune system and, particularly, the thymus might contribute to the genesis and/or maintenance of hypertension in LH rats. We have: (1) searched for the presence of immunological abnormalities during the initial and chronic phases of hypertension; (2) determined the effects of neonatal thymectomy as well as of two pharmacological manipulations of the immune system on blood pressure and immunological characteristics of LH rats; (3) attempted to transfer hypertension or resistance to hypertension by lymphoid cells.

MATERIALS AND METHODS

Experiment 1 : Neonatal thymectomy

Groups of 40 hypertensive (LH), 40 normotensive (LN) and 40 low-blood pressure (LL) male rats were used. Half of the rats were thymectomized at birth and the other half were sham-operated. Blood pressure was measured at 5, 7, 9, 11, 14 and 20 weeks of age. A group of animals was sacrificed at 5 weeks of age and another one at 21 weeks of age for the measurement of immunological parameters and for histological study of the thymus.

Experiment 2 : Cyclophosphamide (Cy) treatment

Groups of 20 hypertensive (LH) and 20 normotensive (LN) male rats were used. Half of the animals were treated from 3 to 11 weeks of age with cyclophosphamide (Endoxan, Lucien-Colombe, France). The first week they received a dose of 75 mg/kg (2 x 37.5 mg/kg/wk, i.p.). diluted in phosphate buffered saline (PBS, Eurobio, France). Then, the treatment was maintained at a rate of 7.5 mg/kg(2 x 3.75mg/kg/wk, i.p.). Controls received an equivalent volume (250 µl) of PBS. Three days after the last injection, animals were killed and immunological parameters were measured.

Experiment 3 : Thymostimulin (TP1) treatment

Groups of 10 thymectomized and sham-operated LH male rats received a thymus extract-Thymostimulin (TP1, Serono, Italy) (Falchetti, Bergesi, Eshkol, Cafiero, Adroni & Caprino, 1977) diluted in PBS at 2 doses: 1 and 10 mg/kg/48h, i.p., from 2 to 5 weeks of age. Controls received PBS alone. Both PBS and thymostimulin were endotoxin-free. Blood pressure was measured at 5, 7, 9, and 11 weeks of age. Animals were sacrificed at 11 weeks of age and immunological parameters were measured.

Experiment 4 : Fetal liver cell (FLC) transplantation

The experiment was conducted in LH and LN rats. At 4 weeks of age, LH and LN rats were lethally irradiated and transplanted with allogeneic or syngeneic fetal liver cells. Isotransplanted and non-irradiated animals were used as controls. Blood pressure was measured at 10, 12, 14, and 20 weeks of age. Animals were sacrificed at 21 weeks of age and immunological parameters were measured.

Animals

They were housed in controlled conditions (temperature 21±1 °C, lighting from 8 a.m. to 8 p.m., humidity 60±10%) and received standard rat chow (UAR A03 Entretien) and tap water *ad libitum*.

Measurement of systolic blood pressure

Systolic blood pressure (SBP) was measured by an indirect tail-cuff plethysmographic method (Narco Biosystem DMP 4B, USA) in the preheated (38°C, 10 min) conscious unrestrained rat.

Thymectomy

Anesthesia was induced by cooling rats for 6 minutes on melting ice. An anterior incision led to the sternum which was cut at the level of the first two ribs. The thymus was sucked out through a Pasteur pipette connected to a vacuum pump. The incision was closed with 2 silk sutures. The completeness of thymectomy was assessed at the time of sacrifice by macroscopical examination of the anterior mediastinum. Animals with incomplete thymectomy or suspect tissues were discarded.

Fetal liver cell transplantation

Pregnancies were timed and fetal liver were obtained from LH or LN fetuses, 13 -14 days postconception. Uteri were removed aseptically and placed in a cold culture medium-RPMI 1640- (Eurobio, France). Fetal livers were dissected and processed into a single cell suspension by repeated aspiration with a Pasteur pipette. After washing, nucleated cells were counted and tested for viability before *i.v.* injection. Recipient LH and LN rats (4 week-old) were given acidified water 1 week before grafting and 3 weeks thereafter. Irradiation was delivered by a Therac 20 Saturne roentgenogram unit (18 meV) at the rate of 3.0 Gy/min. The animals were placed in plexiglass containers and exposed to a dose of 7.5 and 7.0 Gy for LH and LN rats respectively. Ten hours later, irradiated rats were given 15 x 10^6 fetal liver cells in a volume of 250 µl. Rats were then housed, during 6 weeks, in groups of 2 in autoclaved cages and given autoclaved food and sterile acidified water, containing Suanovil (1g/l), *ad libitum*. Afterwards, the rats were kept under normal conditions.

Immunological parameters

Hematological analysis: blood samples (300 µl on EDTA) were drawn by cardiac puncture and processed for blood cell count analysis by an automatic method (Coulter counter, Coultronics, France).

Cell suspensions: the thymus and the spleen were aseptically removed and placed into sterile Petri dishes containing 5 ml of Hepes buffered RPMI 1640 medium. The organs were minced through a fine wire-mesh. Cell suspensions were then washed 3 times in RPMI 1640 (200 g, 10 min).

Mitogen response : Proliferative assays were performed in quadruplicate. One million lymphocytes in RPMI 1640 supplemented with 2 mmol L-Glutamine (Biomerieux, France), 100 U/ml Penicillin & 50 mg/ml Streptomycin (Biomerieux, France) and 10% inactivated human serum were cultured in flat-bottomed microtiter plates (Greiner, France). Concanavalin A (Con A, 2 µg/well, Sigma, USA), Pokeweed-Mitogen (PWM, 1 µg/well, Gibco, USA) and Lipolysaccharide W Salmonella Thyphimurium (LPS, 20 µg/well,Difco, USA) were added to the lymphocyte suspensions. Cells were incubated 72 h at 37°C in humidified atmosphere containing 5% CO_2. Tritiated thymidine (1 µCi, CEA, France) was added for the last 24 h. Cells were harvested on paper filter (Schleicher & Schull, FRG) for scintillation counting. The stimulation indices (SI) were calculated according to the following formula : SI = average of cpm of stimulated cells/average of cpm of

Table 1. Proliferative responses (SI) to Con A, PWM, PHA and LPS of thymocytes and splenocytes from low-blood pressure (LL), normotensive (LN) and hypertensive (LH) rats of the Lyon strain.

age	organ	mitogen	LLs	LLt	LNs	LNt	LHs	LHt
5 weeks	Thymus	ConA	63.5 ± 14.4	-	45.4 ± 10.0	-	51.2 ± 13.7	-
		PWM	9.9 ± 3.0	-	7.5 ± 1.7	-	17.7 ± 5.2	-
	Spleen	ConA	3.8 ± 0.7	0.9 ± 0.1	7.6 ± 1.6	1.2 ± 0.2	9.1 ± 2.1†	0.7 ±0.1
		PWM	1.4 ± 0.2	1.6 ± 0.3	2.6 ± 0.3†		3.7 ± 0.5¶#	0.7 ±0.1
21 weeks	Thymus	ConA	78.2 ± 6.6	-	75.9 ± 6.0	-	56.3 ± 5.9†#	-
		PWM	8.2 ± 0.8	-	8.7 ± 0.9	-	5.6 ± 0.9†#	-
		PHA	16.9 ± 1.6	-	18.3 ± 1.3	-	14.1 ± 1.4#	-
		LPS	1.5 ± 0.2	-	1.4 ± 0.3	-	1.3 ± 0.2	-
	Spleen	ConA	39.4 ± 2.9	1.0 ± 0.2***	42.3 ± 4.1	0.9 ± 0.1***	31.1 ± 2.6†#	1.1 ± 0.2***
		PWM	6.1 ± 0.7	1.0 ± 0.1***	5.4 ± 0.6	1.1 ± 0.1***	5.9 ± 0.5	1.1 ± 0.1***
		PHA	10.1 ± 3.2	1.0 ± 0.1***	9.6 ± 2.9	0.9 ± 0.1***	7.8 ± 2.6	0.9 ± 0.1***
		LPS	5.3 ± 1.0	6.4 ± 0.5*	5.6 ± 1.3	6.7 ± 0.4*	4.0 ± 0.4†#	5.8 ± 0.6*

SI = stimulation index ; s = sham-operated ; t = thymectomized. † $p < 0.05$, ¶ $p < 0.01$ vs sham-operated LL rats. # $p < 0.05$ vs sham-operated LN rats. * $p < 0.05$,*** $p < 0.001$ neonatally thymectomized vs respective sham-operated rats. $n = 10$ in each group.

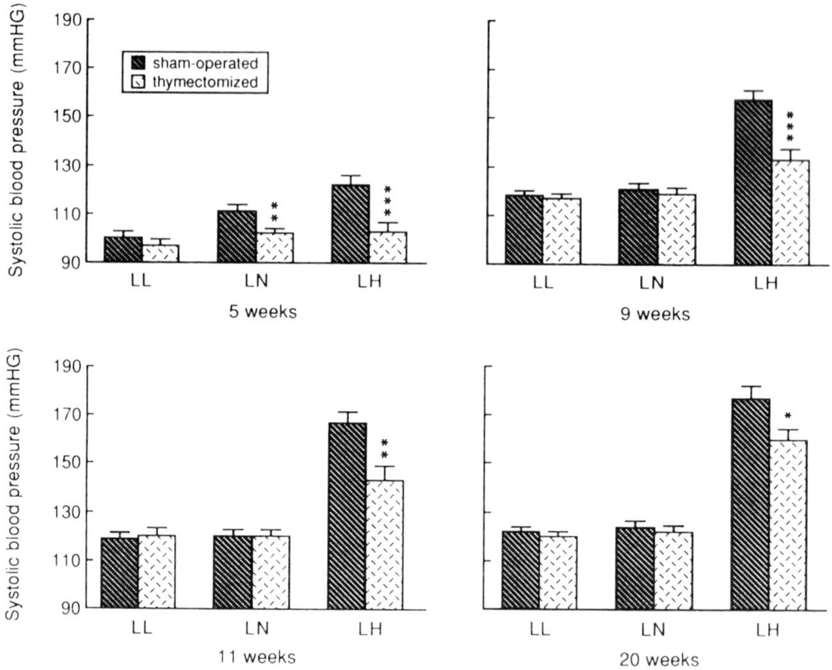

Figure 1. Effect of neonatal thymectomy on systolic blood pressure from low-blood pressure (LL), normotensive (LN) and hypertensive (LH) rats of the Lyon strain. * $p < 0.05$, ** $p < 0.01$, *** $p < 0.001$ thymectomized *vs* respective sham-operated ($n = 10$ in each group).

unstimulated cells.

Immunofluorescence study : The percentage of splenic and thymic T-lymphocytes and T-cell subsets was determined by indirect immunofluorescence. One million lymphocytes in 200 µl PBS (Biomerieux, France) containing 1% BSA (Biomerieux, France) and 0.1% NaN_3 (Merck, FRG) were incubated for 30 min with the following monoclonal antibodies: W3/13 mouse IgG anti-rat T cell, W3/25 mouse IgG anti-rat T helper cells, MRC-OX8 mouse IgG anti-rat T suppressor/cytotoxic cells (Sera-Lab, GB) diluted to 1/100 (final concentration). After incubation the cells were washed twice and a fluoresceine labeled goat IgG anti-mouse IgG serum (Bionetics, USA) diluted to 1/40 final was added. The percentage of B lymphocytes was determined by incubating for 30 min one million lymphocytes with a goat anti-rat Ig serum (Nordic, USA) diluted to 1/30 (final concentration).

Histological study

After sacrifice, pieces of thymuses and kidneys were immediately removed, splited and fixed in 2.5% glutaraldehyde. Sections of 0.5 µm were cut from tissue embedded in Resin ACM (Fluka, Suisse) and staining performed.

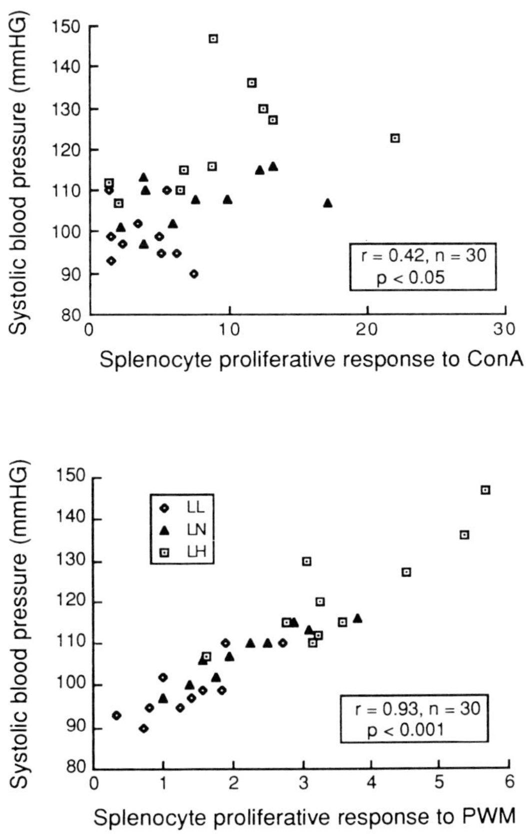

Figure 2. Relationships between systolic blood pressure and splenocyte proliferative responses to Con A and PWM among the 3 Lyon strains.

Statistical analysis

Results are expressed as mean ± s.e.m. and Student's t -test for unpaired data was used for further analysis.

RESULTS

Experiment 1: Neonatal thymectomy

As shown in Fig. 1, 5 week -old sham-operated LH rats exhibited a higher SBP (122 ± 4 mmHg) than age-matched LN (112 ± 2 mmHg) and LL (99 ± 2 mmHg) rats. The SBP of LH rats rapidly increased until the 11th weeks of age and then rose more slowly to reach 177 ± 6 mmHg at 20 weeks of age. Neonatal thymectomy markedly lowered ($p < 0.001$) SBP in 5 week-old LH rats. Afterwards, the SBP of neonatally thymectomized LH rats remained significantly decreased as compared to that of sham-

Table 2. Percentages of W3/13+, W3/25+, OX8+ and Ig+ cells in the thymus and in the spleen from low-blood pressure (LL), normotensive (LN) and hypertensive (LH) rats of the Lyon strains.

age	organ	antibody	LLs	LLt	LNs	LNt	LHs	LHt
5 weeks	Thymus	W3/13	95.3±0.7	-	91.9±0.4	-	91.9±3.5	-
		W3/25	65.8±1.9	-	59.4±1.3	-	62.9±3.5	-
		OX8	62.7±1.9	-	58.2±1.2	-	58.9±5.6	-
	Spleen	W3/13	41.8±2.9	13.6±1.5***	48.5±1.0†	16.6±1.8***	47.1±1.6†	12.2±1.1***
		W3/25	22.8±0.9	7.1±1.3***	23.6±1.5	7.9±1.2***	26.7±1.5†	6.2±0.9***
		OX8	21.1±2.4	8.6±1.8***	24.6±1.4	7.7±1.5***	22.7±1.3	5.6±1.0***
		Ig	48.2±2.5	58.6±5.5	46.7±0.2	54.8±2.5	48.1±0.9	49.9±2.2
21 weeks	Thymus	W3/13	93.3±1.4	-	94.5±1.5	-	88.4±1.3†§	-
		W3/25	79.6±1.4	-	82.4±1.6	-	76.9±1.9	-
		OX8	81.2±2.0	-	78.9±1.7	-	71.1±1.6¶#	-
	Spleen	W3/13	38.9±1.6	5.8±1.0***	41.9±1.8	4.9±0.9	34.0±1.5†#	5.5±1.2***
		W3/25	24.1±1.9	n.d.	24.9±1.4	n.d.	24.1±2.0	n.d.
		OX8	15.7±1.4	n.d.	16.7±1.9	n.d.	11.3±1.5†#	n.d.
		Ig	61.7±1.7	79.6±3.2***	59.4±2.1	81.8±3.1***	62.9±1.7	82.4±3.6***

s = sham-operated ; t = thymectomized ; n.d. = non defined, † $p < 0.05$, ¶ $p < 0.01$ vs sham-operated LL rats. # $p < 0.05$, § $p < 0.01$ vs sham-operated LN rats. *** $p < 0.001$ neonatally thymectomized vs respective sham-operated rats. $n = 10$ in each group.

operated animals until they were 20 weeks of age (160 ± 4 *vs* 177 ± 6, $p < 0.05$). By contrast, except in 5 week-old LN rats, neonatal thymectomy did not modify SBP in the two normotensive control strains.

In 5 week-old animals, the number of white blood cells as well as their differential counts were similar in the 3 stains of sham-operated animals. By contrast, at 21 weeks of age, sham-operated LH rats showed a decreased ($p < 0.05$) number of circulating lymphocytes (4717 ± 161/mm^3) as compared to LN (5082 ± 167/mm^3) and LL (5187 ± 184/mm^3) rats whereas the number of polymorphonuclear cells was increased ($p < 0.05$) (1091 ± 83/mm^3 vs 755 ± 70/mm^3 and 700 ± 65/mm3 in LH, LN and LL rats respectively). Thymectomy induced a pronounced reduction ($p < 0.001$) in the number of circulating lymphocytes in all the animals and significant difference between the 3 strains could no longer be observed.

As shown in Table 1, in 5 week-old sham-operated animals the proliferative response of thymocytes to ConA, PHA and PWM did not significantly differ between the 3 strains. In the spleen, the proliferative response to PWM was higher in LH than in LN ($p < 0.05$) and LL ($p < 0.001$). Furthermore, a positive correlation was established between the SBP level and the splenocyte proliferative response to PWM ($r = 0.93$, $n = 30$, $p < 0.001$) or, to a lesser extent to ConA ($r = 0.42$, $n = 30$, $p < 0.05$) (Fig. 2). By contrast, in 21 week-old sham-operated animals, the proliferative response of thymocytes to PWM and ConA and of splenocytes to ConA and LPS was decreased ($p < 0.05$) in LH rats as compared to LN and LL rats. Splenocytes from thymectomized animals of the 3 strains lost their ability to be stimulated by T-mitogens ($p < 0.001$) whereas their response to LPS was significantly increased ($p < 0.05$).

As shown in Table 2, the percentages of W3/13$^+$, W3/25$^+$ and OX8$^+$ thymocytes were similar in 5 week-old rats of the 3 strains. On the opposite, 21 week-old LH rats exhibited a decreased ($p < 0.05$) percentage of W3/13$^+$ and OX8$^+$ thymocytes as

Table 3. Effects of cyclophosphamide on body weight and number of white blood cells, lymphocytes, thymocytes and splenocytes from normotensive and hypertensive rats of the Lyon strain.

	Normotensive (LN)		Hypertensive (LH)	
	solvent	cyclophosphamide	solvent	cyclophopsphamide
Body-weight (g)	238 ± 4	211 ± 4.7***	320 ± 4.9	277 ± 5.7***
White blood cells (/mm3)	7750 ± 578	3383 ± 388***	8150 ± 460	3606 ± 450***
Lymphocytes (/mm3)	6375 ± 449	2276 ± 352***	6812 ± 373	2455 ± 307***
Thymocytes (x 1,000,000)	694 ± 37	236 ± 30***	861 ± 46	272 ± 36***
Splenocytes (x 1,000,000)	292 ± 14	132 ± 13***	371 ± 19	144 ± 20***

***$p < 0.001$ cyclophosphamide treated *vs* respective controls. $n = 10$ in each group.

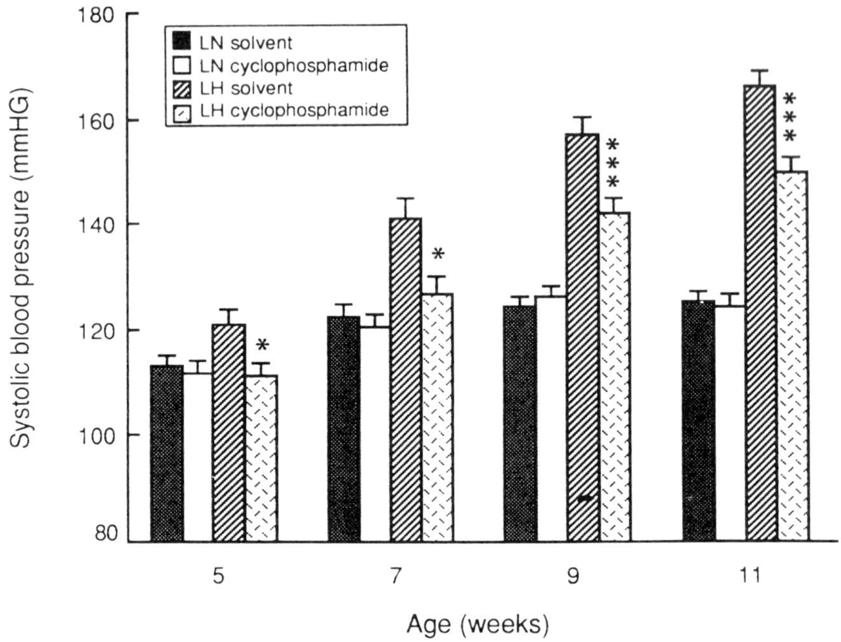

Figure 3. Effect of Cy on systolic blood pressure from normotensive (LN) and hypertensive (LH) rats of the Lyon strain. * $p < 0.05$, *** $p < 0.001$ Cy-treated *vs* control LH rats ($n = 10$ in each group).

Table 4. Effects of cyclophosphamide on the proliferative response (SI) to Con A, PWM and LPS of thymocytes and splenocytes from normotensive and hypertensive rats of the Lyon strain.

		Normotensive (LN)		Hypertensive (LH)	
organ	mitogen	solvent	cyclophosphamide	solvent	cyclophosphamide
	Con A	92.2 ± 8.6	56.2 ± 4.7**	81.6 ± 6.1	44.7 ± 5.5***
Thymus	PWM	15.7 ± 1.5	9.9 ± 0.9**	12.3 ± 1.2	7.4 ± 1.1**
	LPS	1.6 ± 0.2	1.4 ± 0.2	1.4 ± 0.12	1.5 ± 0.2
	Con A	52.3 ± 6.9	12.4 ± 2.3***	42.2 ± 5.7	10.8 ± 1.6***
Spleen	PWM	8.9 ± 1.6	3.9 ± 0.5**	8.2 ± 0.9	3.4 ± 0.4***
	LPS	5.9 ± 0.6	2.9 ± 0.6***	5.2 ± 0.7	2.6 ± 0.5**

SI = stimulation index. * $p < 0.05$, ** $p < 0.01$, *** $p < 0.001$ cyclophosphamide treated *vs* respective controls. $n = 10$ in each group.

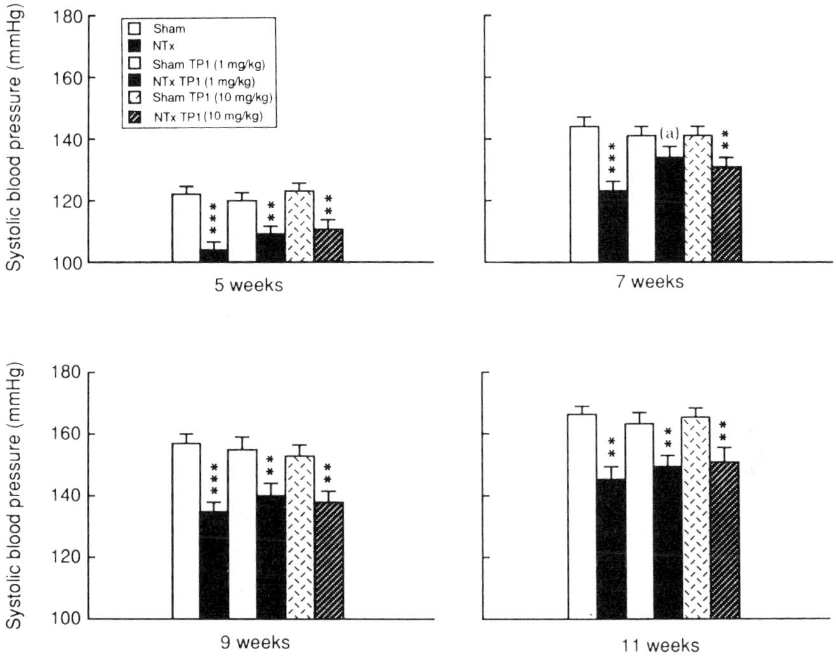

Figure 4. Effects of two doses of TP1 on systolic blood pressure from sham-operated (Sham) and thymectomized (NTx) LH rats of the Lyon strain. ** $p < 0.01$, *** $p < 0.001$ thymectomized versus respective sham-operated rats ; (a) $p < 0.05$ TP1-treated thymectomized *vs* non-treated thymectomized ($n = 10$ in each group).

compared to LN and LL rats. In the spleen of 5 week-old sham-operated animals the percentages of Ig^+ and $W3/25^+$ were similar in the 3 strains whereas the percentages of $W3/13^+$ and of $OX8^+$ cells were higher ($p < 0.05$) in LH than in LN and LL rats. At 21 weeks of age, sham-operated LH rats showed an increased percentage of Ig^+ and a decreased percentage of $W3/13^+$ as well as of $OX8^+$ as compared to LN and LL rats. Since the percentage of $W3/25^+$ splenocytes was similar in the 3 strains, the ratio $W3/25^+$/ $OX8^+$ was higher in LH (2.1 ± 0.1) than in LN (1.6 ± 0.1) and LL (1.5 ± 0.2) rats. Thymectomy strongly decreased ($p < 0.001$) the percentage of $W3/13^+$, $W3/25^+$ and $OX8^+$ splenocytes in 5 and 21 week-old animals of the 3 strains. In 5 week-old thymectomized animals the percentage of Ig^+ cells was not significantly increased whereas it was markedly enhanced ($p < 0.001$) in 21 week-old animals.

At 5 weeks of age, LH thymuses showed a normal structure, comparable to that of LN and LL thymuses. By contrast, structural alterations were observed in the thymus

At 5 weeks of age, LH thymuses showed a normal structure, comparable to that of LN and LL thymuses. By contrast, structural alterations were observed in the thymus of 21 week-old LH rats including a loss of the normal line of demarcation between the cortex and medulla and a cellular depletion in the two areas. The medulla appeared severely depleted and showed several plaques of lysis as well as abundant, hypertrophic epithelial cells. Lastly, some of these thymuses were infiltrated by mastocytes and eosinophils.

Experiment 2 : Cyclophosphamide treatment

As shown in Fig. 3, Cy lowered SBP in 5 week-old LH rats (111 ± 2.2 vs 121 ± 2.0 mmHg, $p < 0.02$). Afterwards, Cy-treated LH rats kept a significant decreased SBP as compared to untreated LH rats until the 11th week of age (149 ± 2.9 *vs* 166 ± 2.0 mmHg, $p < 0.001$). By contrast, Cy did not modify the SBP of LN rats during the 6 weeks of the study.

Cy induced a significant ($p < 0.001$) loss of body weight in both LH (-11%) and LH (-13%) rats (Table 3). Except for this loss of weight, no noticeable side-effects were observed in Cy-treated animals and their behavior was similar to that of untreated animals. The number of circulating lymphocytes as well as the number of mononuclear cells in the thymus and in the spleen were strongly decreased ($p <$ 0.001) by Cy in LH and LN rats (Table 3). As indicated by the proliferative responses to mitogens, the functional capacity of the remaining lymphocytes was altered. Splenocytes from Cy-treated LN and LH rats kept only about 20% of their mitotic activity to ConA and PWM and about 40% to LPS (Table 4). In the thymus, the proliferative response to ConA and PWM was decreased by 50%. Concerning the lymphoid populations and subpopulations in the spleen and in the thymus, no differences were observed between untreated LN and LH rats. Cy strongly decreased the percentage of $W3/13^+$ and Ig^+ splenocytes whereas the percentage of $W3/25^+$ and $OX8^+$ cells was not significantly reduced. In the thymus, the percentage of $W3/25^+$ cells was more markedly reduced by Cy in LH than in LN. On the opposite, Cy treated LN rats showed a more pronounced decrease of their $OX8^+$ thymocytes than LH rats (Table 5).

Table 5. Effects of cyclophosphamide on the percentage of W3/13+, OX8+ and Ig+ cells in the thymus and in the spleen from normotensive and hypertensive rats of the Lyon strain.

		Normotensive (LN)		Hypertensive (LH)	
		solvent	cyclophosphamide	solvent	cyclophosphamide
Thymus	W3/13	92.2 ± 1.9	88.1 ± 1.4	91.3 ± 1.9	87.6 ± 2.1
	W3/25	82.3 ± 1.8	3.9 ± 2.1**	79.1 ± 1.5	68.8 ± 1.6***
	MRC-OX8	79.9 ± 1.8	68.6 ± 2.4***	76.4 ± 1.8	64.8 ± 2.2**
Spleen	W3/13	38.0 ± 1.9	30.7 ± 1.6**	35.8 ± 1.6	28.1 ± 2.2**
	W3/25	23.5 ± 1.8	20.6 ± 1.5	23.2 ± 2.4	19.7 ± 1.8
	MRC-OX8	15.8 ± 2.2	12.5 ± 1.4	13.1 ± 1.9	10.7 ± 1.7
	Ig	53.3 ± 2.2	40.7 ± 3.1***	54.2 ± 2.7	42.1 ± 1.9***

* $p < 0.05$, ** $p < 0.01$, *** $p < 0.001$ cyclophosphamide treated versus respective controls. n = 10 in each group.

Experiment 3: Thymostimulin treatment

As indicated in Fig. 4, TP1 administration at 1 or 10 mg/kg did not significantly alter the SBP of sham-operated LH animals. The SBP values observed in thymectomized LH rats confirm the results previously reported (see experiment 1). At the dose of 1 mg/kg, TP1 had a tendency to increase SBP in thymectomized rats but the difference with untreated thymectomized was significant in 7 week-old rats only ($p < 0.05$). As this age, significant difference between 1 mg/kg treated-thymectomized and treated-sham-operated animals could no longer be observed. At the dose of 10 mg/kg, TP1 not significantly increased SBP of thymectomized rats which remained lower than that of treated and non treated sham-operated animals.

TP1 treated and non-treated LH animals showed a similar number of circulating lymphocytes. Thymectomy induced a pronounced lymphopenia in all the animals. However, TP1 administration significantly ($p < 0.05$) increased the number of circulating lymphocytes in thymectomized LH rats. This effect was not dose-dependent as a similar increase was observed with the 2 doses (Table 6).

The proliferative response of thymocytes and splenocytes to mitogens was comparable in treated and non treated sham-operated LH rats. Splenocytes from non-treated thymectomized rats could no longer be stimulated by ConA or PWM whereas the proliferative response to LPS was slightly increased as compared to sham-operated animals. TP1 significantly increased the splenocyte proliferative response to ConA ($p < 0.001$) and to PWM ($p < 0.05$) in thymectomized animals but the magnitude of the response remained markedly depressed ($p < 0.001$) as compared to sham-operated rats (Table7).

Experiment 4: fetal liver cell transplantation

The mean life span of irradiated rats that were not treated was 9.0 ± 0.4 days in LN rats and 10.1 ± 0.5 days in LH rats. Irradiated rats that had received synergic FLC showed a 83 % (LH) and 78% (LN) survival rate 17 weeks following transplantation.

Table 6. Effects of thymostimulin on the body weight and the number of white blood cells and lymphocytes from sham-operated and thymectomized LH rats.

	Vehicle		TP1: 1mg/kg		TP1: 10mg/kg	
	Sham	NTx	Sham	NTx	Sham	NTx
Body Weight (g)	322 ± 4.4	315 ± 4.7	324 ± 5.9	316 ± 4.1	311 ± 6.1	328 ± 5.8
White blood cells (/mm3)	8078±456	3464±302***	8526±399	3786±175***	8230±399	3803±181***
Lymphocytes (/mm3)	6789±393	2196±122***	6999±375	(a) 2578±113***	6642±286	(a) 2620±124***

Sham = sham-operated ; NTx, thymectomized, *** $p < 0.001$ thymectomized versus respective sham-operated, (a) $p < 0.05$ treated thymectomized *vs* non-treated thymectomized. $n = 10$ in each group.

Table 7. Effects of thymostimulin on the proliferative response (SI) to ConA, PWM and LPS of thymocytes and splenocytes from sham-operated and thymectomized LH rats.

		Vehicle		TP1: 1mg/kg		TP1: 10mg/kg	
organ	mitogen	Sham	NTx	Sham	NTx	Sham	NTx
Thymus	ConA	72.5±6.1	-	68.9±5.3	-	82.5±7.2	-
	PWM	10.2±0.6	-	11.6±0.9	-	12.6±1.3	-
	LPS	1.5±0.2	-	1.3±0.2	-	1.4±0.2	-
Spleen	ConA	39.1±4.3	1.1±0.1***	43.9±3.2	(b) 4.1±0.6***	44.2±4.7	(b) 4.9±0.5***
	PWM	6.8±0.7	1.1±0.1***	7.4±0.8	(a) 1.6±0.1***	6.0±0.8	(a) 1.7±0.3***
	LPS	5.1±0.5	6.8±0.5*	4.7±0.3	6.3±0.7*	4.9±0.5	6.4±0.8*

SI = stimulation index ; Sham = sham-operated ; NTx = thymectomized, * $p < 0.05$, *** $p < 0.001$ thymectomized *vs* respective sham-operated. (a) $p < 0.05$, (b) $p < 0.01$ treated thymectomized *vs* non-treated thymectomized. $n = 10$ in each group.

Table 8. Hematological parameters in normotensive (LN) and hypertensive (LH) rats of the Lyon strain after transplantation of syngeneic or allogeneic FLC.

	LNu $n = 10$	LN→LN $n = 11$	LH→LN $n = 11$	LHu $n = 10$	LH→LH $n = 10$	LN→LH $n = 14$
RBC (x 1,000,000/mm3)	6.5±0.4	6.4±0.2	6.1±0.5	6.7±0.3	7.0±0.2	6.5±0.3
Hb (g/dl)	13.3±0.6	13.0±0.5	12.6±0.5	14.1±0.6	13.5±0.5	12.8±0.5
Hct (%)	36.3±1.2	37.6±1.3	34.8±1.7	36.9±1.8	36.1±1.5	32.4±1.4
WBC (x 1000/mm3)	8.3±0.5	7.2±0.5	6.1±0.4**	7.9±0.5	7.4±0.6	6.2±0.4*
Lymphocytes (x 1000/mm3)	6.8±0.4	5.8±0.4	4.5±0.5**	5.6±0.3#	5.3±0.4	4.1±0.4**

u = unirradiated ; *n* = number of animals. # $p < 0.05$ LH vs LN unirradiated rats. * $p < 0.05$, ** $p < 0.01$ allotransplanted vs respective unirradiated rats.

Death in these animals occurred within 20 weeks after transplantation and was attributed to the failure of engraftment. In rats transplanted with allogeneic FLC, the number of survivors gradually decreased during the first 10 weeks after transplantation and then remained unchanged until the 17th week following transplantation. At this time, LN→LH recipients exhibited a higher survival rate (54%) than LH→LN recipients (42%) ($p < 0.05$).

Hematopoietic failure and intercurrent, opportunistic bacterial infections were the cause of death in non-survivors. Except for a loss of body-weight, no obvious signs of graft versus host disease (GvHD) were observed in animals which survived more than 10 weeks following transplantation.

As shown by Fig. 5, SBP in LH and LN rats transplanted with syngeneic FLC was found decreased as compared to that of respective unirradiated controls, the difference being significant in LH rats only ($p < 0.01$). Allotransplanted LH→LN rats showed a SBP level similar to that of LN→LN recipients and of unirradiated LN rats. Transfer of FLC from LN to LH rats tended to decrease their SBP but the difference with LH→LH recipients was significant ($p < 0.05$) only in 14 week-old animals (i.e. 10 weeks post-grafting).

The values of red blood cells, hematocrit, hemoglobin and white blood cells did not differ between LH and LN unirradiated controls (Table 8). Transfer of FLC led to a complete recovery of erythropoiesis in isotransplanted and allotransplanted rats as assessed by the normal values of red blood cells, hematocrit and hemogloblin observed in these animals. The total number of white blood cells as well as the number of circulating lymphocytes in isotransplanted rats did not differ from those of unirradiated controls. By contrast, leucocytic reconstitution remained incomplete in allotransplanted rats which exhibited a significantly lower number of circulating lymphocytes as compared to unirradiated controls ($p < 0.01$).

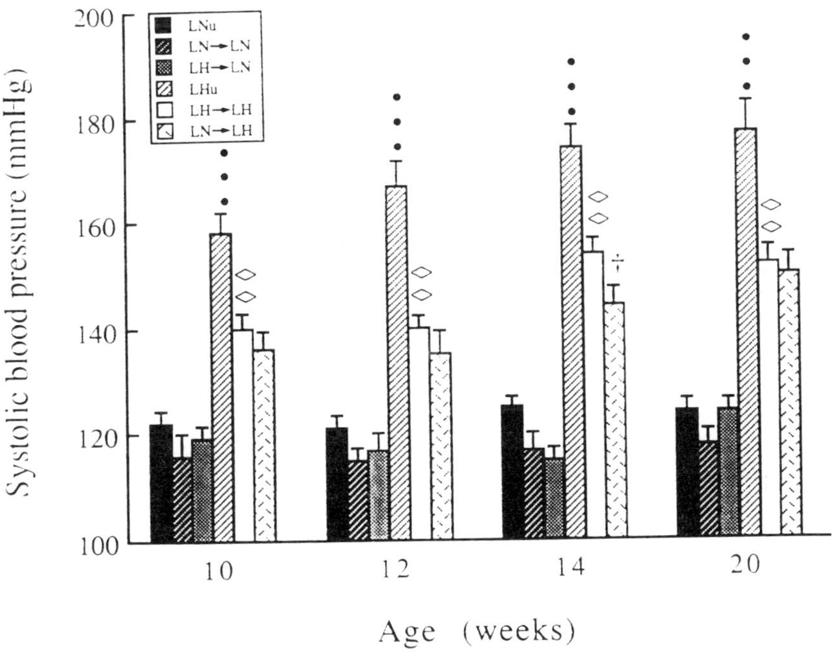

Figure 5. Systolic blood pressure after transplantation of syngeneic or allogeneic FLC in normotensive (LN) and hypertensive (LH) rats of the Lyon strain (u = unirradiated ; ••• $p < 0.001$ LH *vs* LN unirradiated rats ; ◊◊ $p < 0.01$ isotransplanted *vs* unirradiated LH rats ; † $p < 0.05$ allotransplanted vs isoplanted LH rats).

Table 9. Proliferative responses (SI) to ConA, PWM and LPS of thymocytes and splenocytes from normotensive (LN) and hypertensive (LH) rats of the Lyon strain after transplantation of syngeneic or allogeneic FLC.

organ	mitogen	LNu $n = 10$	LN→ LN $n = 11$	LH→ LN $n = 11$	LHu $n = 10$	LH → LH $n = 10$	LN → LH $n = 14$
Thymus	ConA	70.5±4.4	55.0±3.7§	*** 31.4±2.1†††	56.3±4.8#	52.4±4.5	*** 27.6±2.7††
	PWM	15.8±1.2	15.6±1.5	*** 8.9±0.9†††	12.0±1.0#	10.9±1.1	** 7.9±0.8†
	LPS	1.0±0.2	1.2±0.2	1.1±0.2	1.3±0.3	1.5±0.3	1.1±0.2
spleen	ConA	46.7±3.2	39.4±2.2	*** 20.2±1.9†††	38.7±1.9#	33.0±2.3	*** 21.6±2.6†††
	PWM	5.4±0.6	5.9±0.55	*** 3.8±0.4†††	5.9±0.6	5.0±0.51	** 4.2±0.3†
	LPS	5.6±0.7	4.9±0.4	** 4.0±0.4†	5.7±0.5	4.7±0.6	** 3.6±0.4†

SI = stimulation index ; u = unirradiated ; n = number of animals. # $p < 0.05$ LH *vs* LN unirradiated rats. § $p < 0.05$ isotransplanted *vs* unirradiated LN rats. * p < 0.05, ** p < 0.01, *** p < 0.001 allotransplanted *vs* respective unirradiated rats. † p < 0.05, †† p < 0.01, †††p < 0.001 allotransplanted *vs* respective unirradiated rats.

In unirradiated rats, the proliferative response of thymocytes to ConA and to PWM and of splenocytes to ConA was significantly decreased in LH as compared to LN rats ($p < 0.05$) (Table 9). In isotransplanted rats, mitogenic responses were slightly depressed. The decrease reached significance only for the mitogenic response of the thymocytes to ConA in LN→LN recipients ($p < 0.05$). By contrast, recovery of mitogen responsive-ness remained partial in allotransplanted rats which exhibited significantly lower mitogenic response as compared to those of isotransplanted rats or unirradiated controls. Furthermore, in allotransplanted rats, the ConA response of thymocytes and splenocytes reached an average 50% of normal values whereas PWM or LPS response was 70% of control values. On the other hand, significant differences between LH and LN rats could no longer be observed after allogeneic transplantation.

DISCUSSION

The possible involvement of the immune system in hypertension has been studied in genetically hypertensive rats used as a model of human essential hypertension. The Lyon model of genetic hypertension was chosen since it consists in three genetically defined, different strains which have been simultaneously selected from a single group of rats: one hypertensive (LH), one normotensive (LN) and one with low-blood pressure (LL). Therefore, this model allows to distinguish alterations which may be related to hypertension (LH differ similarly from both LN and LL rats) from those which cannot (LH differ from LN or LL only) (Vincent *et al.*, 1984).

The first part of this study investigated whether immunological abnormalities occurred during the initial and the chronic phases of hypertension. Young LH rats do not show immunological alterations in their thymus. By contrast, they exhibit an increased splenocyte proliferative response to PWM as compared to LN and LL controls. Furthermore, this proliferative response and, to a lesser extent, that to ConA, seem to be related to the blood pressure level since their magnitude are positively correlated with SBP using the three strains. It is noteworthy that the degree of these relationships depends on the mitogen used, probably because these two lectins do not stimulate the same lymphoid populations (Stobo, 1980). In addition, such relationships are not observed in the thymus. Since the spleen is markedly vascularized and since lymphocyte activation requires a series of metabolic events (Nel *et al.*, 1987), it can be postulated that blood pressure may modify the cellular environment and, subsequently, contribute to the variations noticed between normotensive and hypertensive animals.

Taken together, these data indicate that except for the splenocyte proliferative response to PWM, the early phase of hypertension in LH rats is not associated with major immunological abnormalities (Bataillard *et al.*, 1986a). This is in contrast with the results reported by Chatelain *et al.*(1980) who observed a thymus atrophy as well as an impairment of the number and the functional capacities of lymphocytes in the initial phase of renovascular hypertension. The rapid rise of blood pressure and the severity of hypertension exhibited by these rats may account for the difference.

By contrast, in adult LH rats, the chronic phase of hypertension is accompanied by immunological modifications: decreased number of circulating lymphocytes, depression of the proliferative response to mitogens in the thymus and in the spleen and reduced percentages of W3/13$^+$ and OX8$^+$ cells in these organs. The structural alterations observed in the thymus of LH rats might initiate these immunological abnormalities. These data indicate that, contrary to the initial phase, the chronic phase of hypertension in LH rats is associated with a moderate immune defect (Bataillard *et al.*, 1986b). In experimental conditions similar to ours, comparable results have been observed in SHR rats (Rojo-Ortega *et al.*, 1973; Takeichi *et al.*, 1981; Ventola and Strausser, 1984). It should be pointed out that, in these two models of genetic hypertension, the immuno-logical abnormalities do not precede the blood pressure rise but are posterior to it. Therefore, they may be involved in the maintenance of hypertension but do not contribute to its development.

The alterations in T-cell number and functions in adult LH rats prompted us to determine the consequences of neonatal thymectomy on blood pressure and immunological characteristics of these animals. The well-known immune defect induced by neonatal thymectomy (Jancovic *et al.*, 1962) has been confirmed here. Thymectomized animals of the three strains show a severe T-cell depletion in lymphoid organs and a virtually complete abolition of the proliferative responses to mitogens. However, the major results of this study is that neonatal thymectomy delayed the onset and attenuated the degree of hypertension in LH rats. This antihypertensive effect is specific since, except in 5 week-old LN rats, neonatal thymectomy does not alter SBP in the two normotensive control strains. The effects of thymectomy in genetically or secondary models of hypertension appear divergent. In SHR, Koga *et al.* (1984) did not observe any modification of blood pressure following thymectomy performed at 6 or 12 weeks of age. On the opposite, SHR rats thymectomized at 4 weeks of age showed a reduced SBP as compared to sham-operated animals (Khraibi *et al.*, 1987). These data suggest that, in genetically hypertensive rats, the efficacy of thymectomy depends on the age at which it is performed. Such a result should be compared with the fact that, from an immunological point of view, to be effective, thymectomy must be performed early in life, *i.e.* before the animal develops a pool of mature T-cell; otherwise its effect are delayed. In addition, the antihypertensive effect of thymectomy does not appear specific to genetic hypertension since DOCA/salt or renovascular hypertension were shown to be non persistent in nude mice whereas thymus graft in these animals allowed the maintenance of chronic hypertension (Svendsen, 1976a,b).

In summary, our findings in genetically hypertensive LH rat suggest that the thymus may be needed for the development of their hypertension. Therefore we developed experiments addressing the thymus-dependent mechanism which could contribute to the rise of blood pressure.

It is generally accepted that thymic epithelial cells secrete several hormonal factors and the thymus can thus be considered as an endocrine gland (Sztein and Goldstein, 1986). These hormonal factors exhibit two different functions. First they are involved in the differentiation process of immature T-cells (Stutman, 1978) and second they are able to influence the functional capacities of endocrine glands. Neonatal thymectomy results in various endocrinological disturbances such as decreased production of testosterone and luteotrophic hormones and increased production of corticosterone and adrenocortical hormones (Comsa *et al.*, 1982; Deschaux *et al.*, 1979). Hence, it was suggested that the thymus could be involved in the regulation of the hypophyseal-adrenocortical and hypophyseal-gonadal axis by means of these hormonal factors since such disturbances could be reversed by thymus extracts (Deschaux *et al.*, 1979) or by thymus graft (Chesnokova *et al.*, 1987) but not by isolated thymocytes (Pierpaoli, 1975). Consequently, the effect of the thymus on blood pressure elevation might be mediated by an immunological or by a so-called hormonal pathway.

As regards to the last possibility, we demonstrated that the antihypertensive effect of neonatal thymectomy in LH rats was not mediated by a decreased production of the mineralocorticoid hormones which had been found elevated in young sham-operated animals (Bataillard *et al.*, 1988; Vincent *et al.*, 1989). The possible part played by thymic hormonal secretions was also explored by investigating the effect of chronic treatment with TP1, a thymus extract, on blood pressure. Two different doses, known to be effective in vivo (Falchetti *et al.*, 1982), were given to young LH rats since the antihypertensive effect of neonatal thymectomy was greatest in young animals.

In sham-operated animals, TP1 has no effect on the immunological parameters determined and SBP of LH rats. These results differ from those reported in SHR showing that thymosin, another thymus extract, induced a transient but significant reduction of blood pressure (Ba *et al.*, 1982; Strausser *et al.*, 1983). The age of the treated animals may account for this discrepancy since at this time LH rats exhibited no immune defect. This was not the case for SHR rats and, furthermore, the thymosin-induced attenuation of the SBP level in SHR was related to an immunological restoration (Ba *et al.*, 1982). In addition, as TP1 increased SBP of thymectomized LH rats but did not restore it to the level observed in sham-operated animals, it appears that the suppression of thymic hormonal secretions cannot totally account for the antihypertensive effect of neonatal thymectomy. Nevertheless attention should be drawn to the fact that, in thymectomized LH rats, the slight increase in blood pressure induced by thymostimulin was associated with an enhancement in the number and in the proliferative response of lymphocytes. This last observation suggests that the participation of the thymus in the development of hypertension may be mediated by cellular immune reactions rather than by hormonal mechanisms.

To test this hypothesis, we studied the effects of Cy, an immuno-suppressant, on SBP in LN and LH rats. Cy was chosen since it is one of the most powerful chemical immunosuppressive agents which depresses both cellular and humoral immunity (Bach, 1987). Our results show that Cy exerted similar immunosuppressive effects in LN and LH rats. By contrast, it delayed and attenuated the full development of hypertension in LH rats but did not change the SBP of LN controls. The possibility that non-immune effects of the drug may contribute to the reduction of hypertension could not be excluded as it was accompanied by a loss of body weight. However, such an anti-hypertensive effect is unlikely to be explained by the decrease observed in body weight because a similar decrease was found in LN rats and no noticeable side-effect in behavioral patterns was observed in treated animals. Therefore, the antihypertensive effect of cyclophos-phamide may be due to a specific effect on blood pressure regulation.

It is interesting to notice that thymectomy and pharmacological immunosuppression resulted in a reduction of hypertension in LH rats which is associated with a pronounced lymphopenia and a depression of the functional capacity of the remaining lymphocytes. The fact that the spontaneous increase in SBP could be reduced either by thymectomy or Cy treatment strongly suggests that thymus-dependent immune reactions could be involved in the pathogenesis of genetic hypertension in LH rats.

This assumption agrees with previous observations showing that suppression of high blood pressure in SHR could be obtained by several types of manipulations of the immune system such as thymectomy, pharmacological immunosuppression by cyclophosphamide or short-term administration of anti-thymocytes serum (Khraibi *et al.*, 1984 and 1987; Bendish *et al.*, 1981). In addition, studies performed in NZB, a strain of mice which develops auto-immune renal disease and spontaneously elevated blood pressure, have provided evidence for an involvement of T-lymphocytes in the development of their hypertension. Athymic nude NZB mice failed to develop hypertension and cyclophosphamide treatment significantly reduced the level of blood pressure and the renal lesions in euthymic NZB mice (Svendsen, 1977b). These investigators postulated that the development of hypertension may result, at least in part, from an autoimmune process. Interestingly, the results obtained in LH rats also argue for such a process since: (1) chronically hypertensive LH rats exhibit impaired T-cell number and functions, alterations which are often associated with auto-immune disease (Eisen, 1980); (2) an improvement of hypertension is obtained by two different immuno-suppressive methods, a finding which is one of the characteristics of an auto-immune disease (Eisenbarth, 1985). Consequently, the possible immune mechanism involved in the etiology of hypertension might be mediated by delayed hypersensitivity reactions and/or by autoantibodies against components of vascular tissues. These autoantibodies might lead to vascular tissue damages inducing an increase of the peripheral resistance and, therefore, a rise in blood pressure. The fact that renal vascular lesions are observed in experimental and genetical models of hypertension (Svendsen, 1977a; Suzuki *et al.*, 1979) including the LH rat (Blanc-Brunat *et al.*, 1980), and that the significance of these lesions is reduced by neonatal thymectomy (unpublished observation) agree with such a hypothesis.

More persuasive evidence arguing for an auto-immune disease in a given disease is provided by the demonstration that it can be passively transferred into naive recipients by lymphoid cells. We tested this possibility by transferring lymphoid cells from LH to LN rats. Experiments designed to induce resistance to hypertension in LH rats by transferring lymphoid cells from LN rats were also performed. Since LH and LN rats are mismatched at the major histocompatibility complex (Moutier *et al.*, 1983), bone marrow or splenic cell transplantation could not be performed because they are responsible for severe graft versus host disease (GVHD). Using an alternative method, we transplanted cells from fetal livers which contain abundant hematopoietic cells but very few immunocomponent T-cells and have therefore a reduced capacity to induce GVHD. In these conditions, FLC transplantation from LH to irradiated LN rats fails to transfer hypertension and the same manipulation from LN to LH rats does not reduce hypertension. The failure of the passive transfer of hypertension could have two explanations. First no auto-immune process is involved in LH rats. Second, hypertension in these animals implies auto-immune process but, in our experimental conditions, SBP did not change since: (1) the number of transplanted cells was not sufficient; (2) the immunological reconstitution of allotransplanted LN rats remained partial and did not allow the expression of pathogenicity; (3) since the transplanted cells are very immature, it is possible that they did not differentiate into pathogenic cells in their new genetic environment; (4) an alternative to the last possibility would be that the manifestation of pathogenecity is MHC-restricted. Indeed, our results differ from those obtained in secondary forms of hypertension by Okuda and Grollman (1967) and later by Olsen (1980) who demonstrated that one-kidney Goldblatt hypertension or DOCA-salt hypertension could be transferred by lymphoid cells into syngeneic normotensive recipients. The use of syngeneic or allogeniec recipients, the degree of maturity of the transplanted cells (splenocytes and lymph node cells or FLC) and the type of the hypertension (secondary or genetically-determined) are likely to be responsible for such a discrepancy.

CONCLUSION

Our observations demonstrate the involvement of immunological factors and, particularly, of the thymus in the development and the maintenance of hypertension in LH rats. In addition, since reduction in the blood pressure level could be achieved by similar manipulations of the immune system in LH as well as SHR rats, these two different models of genetic hypertension may share a common immune pathogenetic mechanism. The failure to passively transfer hypertension by lymphoid cells does not support our hypothesis of an auto-immune component in the hypertension developed by LH rats. However, such a transfer did not fully restore the immune functions and thus introduced problems of interpretation. Further attempts to describe precisely the role of immunological factors in the pathophysiology of genetic hypertension and their possible interest as therapeutic targets in the treatment of hypertensive disease are necessary.

REFERENCES

Ba, D., Takeichi, N., Kodama, T. and Kobayashi, H. (1982). Restoration of T cell depression and suppression of blood pressure in spontaneously hypertensive rats (SHR) by thymus graft or thymus extracts. *J. Immunol.* **128,** 1211 - 1216.

Bach, J. F. (1987). Alkylating agents. In: *Immunopharmacology*, P. S. Sirois and M. Rola-Peszczinsky (Eds)., Amsterdam, pp.175 - 239.

Bastenie, P.A., Van Haelst, L., Bonnyns, M., Neve, P. and Sacquet, M. (1971). Preclinical hypothyroidism: a risk factor for coronary heart disease. *Lancet.* **1,** 203 - 204.

Bataillard, A., Freiche, J. C., Vincent, M., Sassard, J. and Touraine, J. L. (1986a). Antihypertensive effect of neonatal thymectomy in the genetically hypertensive LH rat. *Thymus* **8,** 321 - 330.

Bataillard, A., Touraine, J.L., Vincent, M. and Sassard, J. (1986b). Role of the thymus in genetically hypertensive rats of the Lyon strain. *J. Hypertension* **4** (Suppl 6), 87 - 89.

Bataillard, A., Gomez-Sanchez, C. E., Vincent, M.Touraine, J. L. and Sassard, J. (1988). Mineralocorticoids are not involved in the antihypertensive effect of neonatal thymectomy in the genetically hypertensive LH rat. *Clin. Exp. Pharmacol. Physiol.* **15,** 875 - 882.

Bendish, A. and Belisle, E. (1981). Immune system modulation and its effect on the blood pressure of spontaneously hypertensive male and female rats. *Bioch. Biophys. Res. Com.* **99,** 600 - 607.

Blanc-Brunat, N., Vincent, M. and Sassard, J. (1980). Evolution des lésions rénales précoces chez les trois souches de rats génétiquement hypertendus, normotendus et normotendus-bas. *Paroi Artérielle.* **6,** 145 - 154.

Chatelain, R. E. and Ferrario, C. M. (1978). Biphasic changes in the thymus structure during evolving renal hypertension. *Clin. Sci. Mol. Med.* **55,** 149 - 156.

Chatelain, R. E., Vessey, A. R. and Ferrario, C. M.(1980). Lymphoid alterations and impaired T lymphocyte reactivity in experimental renal hypertension. *J.Lab. Clin. Med.* **95,** 737 - 747.

Chesnokova, V. M., Grutenko, E. V., Ivanova, L. N. and Ignatyeva, E. V. (1987). Evidence for thymus involvement in the development of adrenal and gonadal endocrine functions in mice. *Endocrinol. Exp.* **21,** 133 - 142.

Cosma, J., Leonhardt, H. and Wekerle, H. (1982). Hormonal coordination of the immune response. *Rev Physiol. Biochem. Pharmacol.* **92,** 115 - 191.

Deschaux, P., Massengo, B. and Fontanges, R. (1979). Endocrine interactions of the thymus with the hypophysis, adrenals and testes : effect of two thymic extracts. *Thymus* **1,** 95 - 106.

Ebringer, A. and Doyle, A. E. (1970). Raised serum IgG levels in hypertension. *Br. J. Med.* **2,** 146-149.

Ebringer, A. and Doyle, A. E. (1970). Rheumatoid factor in hypertension. *Br. Heart J.* **35,** 334 - 337.

Eisen, H. N. (1980). Cell mediated hypersensitivity and immunity. In: *Immunology*. Harper and Row (Eds), New York, pp 493 - 521.

Eisenbarth, G. S. (1985). Insuffisance auto-immune des cellules béta diabéte sucré de type I. In: Maladies auto-immunes, *J. Sandoz des sciences médicales*, Triangle, **25,** 115 - 128.

Falchetti, R., Cafiero, C. and Caprino, L. (1982). Impaired T-cell functions in aged guinea-pig restored by thymostimulin(TS). *Int. J. Immunopharmac.* **4,** 181 - 186.

Gelsthorpe, K., Doughty, R. W., Bing, R.F., O'Maller, B. C., Smith, A. J. and Talbot, S. (1975). HLA antigens in essential hypertension. *Lancet* **1,** 1039 - 1040.

Gudbransson, T. H., Herlitz, H., Lindholm, L., Nilson, L.A. and Hansson, L. (1981). Immunological changes in patients with previous malignant essential hypertension. *Lancet* **1,** 406 - 408.

Jankovic, B. D., Waksman, B. H.and Arnason, B. G. (1962). Role of the thymus in immune reactions in rats. I. The immunologic response to bovine serum albumin (antibody formation, Arthus reactivity and delayed hypersensitivity) in rats thymectomized or splenectomized at various time after birth. *J. Exp. Med.* **116,** 159 - 176.

Koga, Y., Takeichi, N. and Kobayashi, H. (1984). The effect of thymectomy on the development of hypertension in spontaneously hypertensive rats. *Jpn. Heart J.* **26,** 807 - 811.

Khraibi, A.A., Norman, R.A., and Dzielak, D.J. (1984). Chronic immunosuppression attenuates hypertension in Okamoto spontaneously hypertensive rats. *Am. J. Physiol.* **247,** H722 - H726.

Khraibi, A. A., Smith, T. L., Hutchins, P. M., Lynch, C .D. and Dusseau, J. W. (1987). Thymectomy delays the development of hypertension in Okamoto spontaneously hypertensive rats. *J. Hypertension.* **5,** 537 - 541.

Kristensen, B. O., Andersen, P. L., Lamm, L. U. and Kissmeyer-Nielsen, F. (1970). HLA antigens in essential hypertension: relation to familial disposition and serum immunoglobulins. *Tissue Antigens.* **10,** 70 - 74.

Kristensen, B. O. (1978). Increased serum levels of immunoglobulins in untreated and treated essential hypertension. I. Relation to blood pressure. *Acta. Med. Scand.* **203,** 49 - 54.

Kristensen, B.O. and Andersen, P.L. (1984). Autoantibodies and vascular events in essential hypertension. A five-year longitudinal study. *J. Hypertension.* **2,** 19 - 24.

Low, B., Scharsten, B., Sanetur, G., Thulin, T. and Mittelman, F. (1975). HLA B8 and W15 in diabetes and essential hypertension. *Lancet* **1,** 695.

Mathews, J.D., Hooper, B.M., Whittingham, S. and Mackay, I. R. (1973). Association of autoantibodies with smoking, cardiovascular morbidity and death in the Busselton population. *Lancet* **6,** 745 - 758.

Nel, A. E., Wooten, M. W. and Galbraithi, R. M. (1987). Molecular signaling mechanisms in T-lympho- cytes activation pathway: a review and future prospects. *Clin. Immunol. Immunopath.* **44,** 167 - 186.

Norman, R. A., Dzielak, D. J., Bost, K. L., Kharaibi, A. A. and Galloway, P. G. (1985) Immune system dysfunction contributes to the aetiology of spontaneous hypertension. *J. Hypertension* **3,** 261 - 268.

Norman, R. A. and Dzielak, D. J. (1986). Spontaneous hypertension is the result of sympathic overactivity and immunologic dysfunction. *Proc. Soc. Exp. Biol. Med.* **182,** 448 - 453.

Okuda, T. and Grollman, A. (1967). Passive transfer of autoimmune induced hypertension in the rat by lymph node cells. *Texas Rep.Biol.* **25,** 257 - 264.

Olsen, F. (1970). The secondary and tertiary inflammatory cellular reactions in arterioles damaged by acute angiotensin-hypertension in rats. *Acta. Path. Microbiol. Scand. Sect.* A. **78,** 454 - 457.

Olsen, F., Hilden, M. and Ibsen, H. (1973). Raised levels of immunoglobulins in serum of hypertensive patients. *Acta. Path. Microbiol. Scand. Sect.* A. **81,** 498 - 501.

Olsen, F. (1974). Delayed hypersensitivity and essential hypertension. *Acta. Path. Microbiol. Scand. Sect.* B. **82,** 448 - 450.

Olsen, F. (1980). Transfer of arterial hypertension by splenic cells from DOCA-salt hypertensive and renal hypertensive rats to normotensive recipients. *Acta. Path. Microbiol. Scand. Sect.* C. **88,** 1- 5.

Pierpaoli, W. and Sorkin, E. (1975). Role of the thymus in programming of neuroendocrine functions *Clin. Exp. Immunol.* **20,** 323 - 338.

Pravenec, M., Klir, P., Kren, V., Zicha, J. and Kunes, J. (1989). An analysis of spontaneous hypertension in spontaneously hypertensive rats by means of new recombinant inbred strains. *J. Hypertension* **7,** 217 - 222.

Rojo-Ortega, J. M., Yeghijan, E. and Genest, J. (1973). The thymus of spontaneously hypertensive

rats : light and electron microscope studies. *Clin. Sci. Mol. Med.* **45,** 141 - 144.
Simon, P., Fauchet, R., Ang, K. S., Pilorge, M., Calvez, C., Fiblec, B., Gam, G., Laurent, M. C., Blanchard, J., Genetet, B. and Cloup, B. (1989). Augmentation de la fréquence de l'antigéne HLA DR4 dans l'hypertension récidivante dela grossesse. *Arch. Mal. Coeur.* **81,** 291 - 295.
Stobo, J. (1980). Mitogens. *Clin. Immunobiol.* **4,** 55 - 72.
Strausser, H. R.(1983). Immune response modulation in the spontaneously hypertensive rat. *Thymus.* **5,** 19 - 33.
Stutman, O. (1978). Intrathymic and extrathymic T cell maturation. *Immunol. Rev.* **42,** 138 - 184.
Suzuki, T., Oboshi, S. and Sato, R. (1979). Periarteritis nodosa in spontaneously hypertensive rats : incidence and distribution. *Acta. Path. Jap.* **29,** 697 - 703.
Svendsen, U. G. (1976a). The role of the thymus for the development and prognosis of hypertension and hypertensive vascular diseases in mice following renal infarction. *Acta. Path. Microbiol. Scand. Sect.* A. **84,** 235 - 243.
Svendsen, U. G. (1976b). Evidence for an initial, thymus independent and a chronic dependent phase of DOCA and salt hypertension in mice. *Acta. Path. Microbiol. Scand. Sect.* A. **84,** 523 - 528.
Svendsen, U. G. (1977a). The importance of the thymus in the pathogenesis of the chronic phase of hypertension in mice following partial infarction of the kidney. *Acta. Path. Microbiol. Scand. Sect.* A. **85,** 539 - 547.
Svendsen, U. G. (1977b). Spontaneous hypertension and hypertensive vascular disease in the NZB strain of mice. . *Acta. Path. Microbiol. Scand. Sect.* A. **85,** 548, 554.
Sztein, M. B. and Goldstein, A. L. (1986). Thymic hormones, a clinical update. Springer Semin. *Immunopathol.* **9,** 1 - 18.
Takeichi, N., Suzuki, K., Okayasu, T. and Kobayashi, H. (1980). Immunological depression in spontaneously hypertensive rats. *Clin. Exp. Immunol.* **40,** 120 - 126.
Takeichi, N., Ba, D. and Kobayashi, H. (1981a). Natural cytotoxic autoantibody against thymocytes in spontaneously hypertensive rats. *Cell. Immunol.* **60,** 181 - 190.
Takeichi, N., Suzuki, K. and Kobayashi, H. (1980). Characterization of immunological depression in spontaneously hypertensive rats. *Eur. J. Immunol.* **11,** 483 - 487.
Vincent, M., Sacquet, J. and Sassard, J. (1984). The Lyon strain of hypertensive, normotensive and low-blood pressure rats. In: *Handbook of hypertension*, DeYong, J. (Eds), Amsterdam, pp. 314 - 327.
Vincent, M., Gomez-Sanchez, C. E., Bataillard, A. and Sassard, J. (1989). Steroid during the development of genetic hypertension in rats of the Lyon strain. *Am. J. Physiol.* **257,** H506 -H 510.
Ventola, D. A. and Strausser, H. R. (1984). Evaluation of T-cell subpopulation and function in thymosin-treated spontaneously hypertensive rats. *Thymus.* **5,** 129 - 141.
Wilson, J. D., Bullock, J. Y. and Booth, R. J. (1978). Autoantibodies in essential hypertension. *Lancet* **1,** 996.

Progress in Hypertension, Vol. 2, pp. 355-366.
Saito *et al.* (Eds)

PHARMACOKINETIC PROPERTIES OF ANTI-HYPERTENSIVE DRUGS

KATSUMI MIYAZAKI, KEN ISEKI and AKIKAZU NOMURA*

Department of Pharmacy, Hokkaido University Hospital and *Department of Cardiovascular Medicine, Hokkaido University School of Medicine

INTRODUCTION

Many effective antihypertensive agents including diuretics, β-adrenoceptor antagonist, calcium channel blockers and angiotensin converting enzyme inhibitors have been developed and their pharmacokinetic properties have been studied in more detail during recent years. In the antihypertensive therapy, it is probable that many patients will receive treatment with more than one drug to obtain the accurate hypotensive effect. Moreover, for the accurate pharmacological control antihypertensive therapy often demands the long term treatment. It is important, therefore, to seize pharmacokinetic properties to increase the hypotensive action or to minimize unwanted effects.

The present chapter focuses on the fundamental pharmacokinetic properties (absorption, bioavailability, metabolism, first-pass effect, urinary and biliary excretion, elimination half-life) and the pharmacokinetics in the disease states of the antihypertensive agents which are widely used and newly developed.

DIURETICS

Hydrochlorothiazide

Absorption. The intestinal absorption of hydrochlorothiazide in fasting healthy subjects averages 65% (Beermann and Groschinsky-Grind, 1977) and increases to about 75% when it is given concomitantly with food (Beermann and Groschinsky-Grind, 1978a). Pretreatment with propantheline (60mg) increases the amount of a dose absorbed to approximately 90% (Beermann and Groschinsky-Grind, 1978b). The increased uptake resulted from the coadministration with food or propantheline appears to be caused by decreased gastric emptying rate (Beermann and Groschinsky-Grind, 1978b).

Great differences in absorption are found when this drug is given orally to patients with congestive heart failure and urinary recovery ranges from 20 to 70% during 7days (Beermann and Groschinsky-Grind, 1979).

Elimination. Plasma half-life of hydrochlorothiazide averages 10hr in subjects with normal renal function (Beermann and Groschinsky-Grind, 1980). Major elimination route is renal excretion and more than 95% is excreted as unchanged from via the kidney (Beermann *et al.*, 1976).

The high renal clearance indicates proximal tubular secretion. Plasma half-life is prolonged in renal failure (Beermann and Groschinsky-Grind, 1980).

Ethacrynic acid

Absorption and Elimination. Ethacrynic acid possesses good lipid solubility and is absorbed promptly from gastrointestinal tract (Beyer *et al.*, 1965). Active tubular secretion occurs by the probenecid-sensitive organic acid pathway in the proximal tube and true tubular excretion fraction is 35% (Odlind, 1979).

Furosemide

Absorption and Elimination. Bioavailability of oral furosemide ranges from 50 to 60% (Cutler and Blair, 1979). After oral dose, furosemide evokes a rapid saluretic response which peaks within 20 to 30 minutes. The kinetic disposition of the drug fits most closely an open two-compartment model. An elimination half-life ranges from 19 to 100 minutes. Urinary excretion and non-renal elimination contribute almost equally to plasma clearance. Proximal tubular secretion is the main route for urinary excretion and glomerular filtration appears to be of small importance in view of the high protein binding of this drug. The unbound fraction averages 2.3 to 4.1% at therapeutic concentrations (Beermann and Groschinsky-Grind, 1980).

The main urinary metabolite isolated in man has been the glucuronide (Beermann *et al.*, 1975) which is unstable under the light. In patients with acute pulmonary edema, renal and hepatic failure, the elimination half-life may be markedly prolonged (Lant, 1985).

Bumetanide and Piretanide

Absorption and Elimination. Bumetanide is rapidly and completely absorbed from the gastrointestinal tract (Pentikainen *et al.*, 1977). After oral administration of ^{14}C-bumetanide (2mg), the administered radioactivity was almost completely recovered (96%), 81% appearing in the urine and the remainder in the feces. Renal and non-renal clearance contributes equally to its elimination. Elimination half-life of bumetanide is approximately 1.2 to 1.5 hours (Beermann and Groschinsky-Grind, 1980; Halladay *et al.*, 1977) in healthy subjects and does not appear to be prolonged in renal disease (Beermann and Groschinsky-Grind, 1980).

Bioavailability of oral piretanide is approximately 80%. The pharmacokinetic profile of this drug resembles of furosemide,with an elimination half-life of the order of 30 to 50 minutes, increasing to 75 to 240 minutes in renal failure (Lant, 1985).

Spironolactone

Absorption and Elimination. According to the experiment administered ^{3}H-spironolactone orally to patients with bile drainage, the urinary and biliary recovery of radioactivity during 4 days averaged 54 and 16% respectively, which indicates a minimal uptake of 70% (Abshagen *et al.*, 1977).

Table 1. Lipid solubility (a) and Pharmacokinetic properties (b) of ß-adrenoceptor antagonists

relative lipid solubility		Distribution coefficients at pH 7.4 and 37°C	Oral bio-availability (%)	Presystemic elimination 1st pass %	t1/2(ß) (hr)	Excreted unchanged (%)
	Atenolol	0.015	50 - 60	0	6-7	100
low	Sotalol	0.039	100	0	7.5 - 15	75
	Practolol	0.060	95	0	10 - 11	100
	Nadolol	0.066	30 - 50	0	19	100
	Acebutolol	0.68	20 - 60	30	7	15 - 20
	Pindolol	0.82	100	20	2.5 - 4	40
moderate	Metoprolol	0.98	40 - 50	25 - 50	3.5	< 5
	Timolol	1.16	75	50 - 60	2.7	20
	Oxperenolol	2.28	25 - 60	25 - 80	2.2	< 5
	Labetalol	11.5	30 - 40	60 - 70	3.9	< 5
high	Propranolol	20.2	30	70	4	< 1
	Penbutolol	17.9	100	-----	26	< 10
	Alprenolol	-----	10	90	2.5	< 1

(a) (Woods and Robinson, 1981), (b) (Riddell *et al.*, 1987)

Spironolactone undergoes an enterohepatic circulation and an extensive first-pass effect. The metabolism of this drug is very extensive and complex.

Canrenone, the major pharmacologically active metabolite of spironolactone, is formed via the thiol intermediate. Canrenone is then converted to canrenoic acid which is further conjugated with glucuronic acid. Canrenone may also be hydroxylated to 15 α-hydroxy canrenone in man. After oral administration, about 80% of spironolactone is converted to canrenone (Sadee *et al.*, 1973). The elimination half-life of canrenone is 18 to 20 hours after doses of 100 to 400mg (Beermann and Groschinsky-Grind, 1980). Canrenone is eliminated as metabolites via the urine and the bile(Beermann and Groschinsky-Grind, 1980).

β-ADRENOCEPTOR ANTAGONISTS (β-BLOCKERS)

The lipid solubilities of β-blockers are extensively different among drugs (Woods and Robinson, 1981) and can be classified roughly into highly lipid soluble (alprenolol, labetalol, propranolol), low lipid-soluble (atenolol, nadolol, practolol, sotalol), and moderate lipid-soluble (others) compounds.

As shown in Table 1, their various lipid solubilities are associated with a pharmacokinetic profiles, i.e., differences in absorption, distribution and excretion properties (Riddle *et al.*, 1987).

Absorption

Hydrophilic (low lipid-soluble) β-blockers are generally less well absorbed than lipophilic β-blockers. Intestinal absorption of hydrophilic β-blockers such as atenolol and nadolol are below 50%. On the contrary, lipophilic (high and moderate lipid-soluble) β-blockers are rapidly and completely absorbed from the gastrointestinal tract.

Following absorption, however, lipophilic β-blockers are subject to presystemic elimination (first-pass effect) as shown in Table 1.

Plasma protein binding

As general properties of basic drugs, the β-blockers bind not only to albumin but also α1-acid glycoproteins. The extent of protein binding is generally increased with increased lipophilicity (alprenolol: 76%, penbutolol: > 95%, propranolol: 90-94%), but protein binding of acebutolol, labetalol and metoprolol are relatively low.

The most hydrophilic β-blockers such as atenolol, practolol and sotalol are poorly protein binding (atenolol: < 5%, sotalol: < 1%).

Elimination

The majority of β-blockers except for most hydrophilic drugs are subject to highly variable first-pass oxidative metabolism (Table 1). The hydrophilic β-blockers and the metabolites of the lipophilic drugs are excreted primarily by the kidney.

The hydrophilic drugs including atenolol, nadolol, practolol, and sotalol are excreted as unchanged form in the urine and have long elimination half-life (Table 1).

CALCIUM ANTAGONISTS

Calcium antagonists can be classified into two types, I and II according to the difference of chemical structure. Type I drug, which have not dihydropyridine structure, include verapamil, diltiazem and tiapamil. Type II classes comprise the dihydropyridines, including nifedipine and its structural analogues. Although the chemical structures of these two type drugs are different, they exhibit common pharmacokinetic properties (Echizen and Eichelbaum, 1986; Abernethy and Schwartz, 1988). These drugs are extensively metabolized and only small amounts of unchanged drugs are excreted in urine. Their bioavailabilities are low despite almost complete absorption following oral administration.

Therefore, pronounced intra- and inter-individual variations in clearance and bioavailability are observed. Moreover, during long term administration, oral clearance decreases and bioavailability increases due to saturation of first-pass metabolism. The various pharmacokinetic parameters are markedly altered in liver disease.

Type I calcuim antagonists

Verapamil

Absorption and Bioavailability. Following oral administration, verapamil is almost competely absorbed but less than 5% of the dose is excreted in the urine as unchanged drug (Echizen and Eichelbaum, 1986). The bioavailability is low and ranged from 10 to 35% because of extensive first-pass metabolism (Echizen and Eichelbaum, 1986). Compared with single-oral administration, a significant increase in bioavailability has been reported with multiple dose. Many studies indicate that steady-state plasma concentrations are 2 to 3 times higher than the predicted steady-state concentrations using pharmacokinetic data obtained from single oral dose studies. The increase of bioavailability of this drug is probably due to saturation of first-pass metabolism.

Elimination. Many metabolites accounting for 50 to 55 % of the dose administered have been isolated and identified in the urine (Eichelbaum *et al.*, 1979). N-dealkylation of the shorter side chain of the verapamil is the main metabolic step. O-demethylation of verapamil and its N-dealkylated metabolite are then occurred. Verapamil is used clinically as a racemic mixture containing equal proportion of (+)-verapamil and the more active (-)-isomer. Recent studies (Eichelbaum *et al.*, 1984; Vogelgesang *et al.*, 1984; McTavish *et al.*, 1989) have reported the stereoselective first-pass metabolism of verapamil and have shown that oral clearance of the more active (-)-verapamil was higher than that of (+)-verapamil and that bioavailability was about 50 % and 20 % for the (+)- and (-)-isomers, respectively. These properties may explain why intravenous verapamil has a more pronounced pharmacological effect compared to orally administered verapamil at a given plasma verapamil concentration.

Pharmacokinetics in disease states. The pharmacokinetics, bioavailability and pharmacodynamic response to verapamil are grossly altered in patients with liver cirrhosis (Somogyi *et al.*, 1981; Woodcock *et al.*, 1981). After intravenous administration, total plasma clearance is markedly reduced and terminal elimination half-life prolonged 4-fold. Peak plasma concentrations were much higher and occurred earlier after oral administration. The bioavailability is more than doubled. Therefore, in liver cirrhosis patients, intravenous dose of verapamil should be halved and the oral dose decreased by a factor of 5 to prevent untoward effects (Somogyi *et al.*, 1981). In patients with end-stage chronic renal failure, on the contrary, the plasma concentration-time course of verapamil and desmethylated verapamil do not differ from those in normal volunteers (Echizen and Eichelbaum, 1986).

Diltiazem

Absorption and Elimination. Several reports indicate that bioavailability of diltiazem is approximately 40 % (Echizen and Eichelbaum, 1986). During long term administration, steady-state plasma concentrations of diltiazem are higher than those predicated from single oral dose studies, indicating saturation of hepatic first-pass metabolism. Five metabolites have been identified in urine following oral administration in man (Echizen and Eichelbaum, 1986). Desacetylation occurred primarily. O- and N-desmethylation of the desacetylated form and glucuronidation of the phenolic metabolite are then occurred. The pharmacological activity and the contribution to the therapeutic effect of these metabolites have not been extensively studied in man. However, N-monodemethyldiltiazem may have about 20% of the vasodilating potency of the parent drug (Yabana *et al.*, 1985). There are few data on the pharmacokinetics of diltiazem in disease state. Renal failure seems to have no effect on the pharmacokinetic and bioavailability of diltiazem and its major metabolite, desacetyl form.

Tiapamil

Absorption and Elimination. Tiapamil is an analogue of verapamil and the pharmacokinetic profile of tiapamil is quite similar to that described for verapamil except for excretion route of metabolites. After oral administration tiapamil has absolute bioavailability of 15 to 70% with marked intersubject variability (Abernethy and Schwartz, 1988). Elimination half-life of β-phase has been reported to be 1.3 to 3.5 hours. Tiapamil is subject to more extensive biliary excretion. After extensive first-pass metabolism in the liver, tiapamil metabolites are predominantly excreted in feces (60 to 90 %), while verapamil metabolites appear principally in urine (70 %) (Abernethy and Schwartz, 1988). The excretion of tiapamil in breast milk was studied following a single oral 600 mg dose and the milk/plasma ratio of tiapamil derived from the areas under the plasma and milk concentraion-time curves was 0.44 ± 0.10 (n = 6) (Hartman *et al.*, 1988). In patients with liver cirrhosis, the elimination half-life prolonged about 2-fold (1.7 → 3.5 hours) and the absolute bioavailability tended to be greater (Hinderling *et al.*, 1986).

Type II drugs

Many dihydropyridine calcium antagonist drugs have been designed to modify or overcome both the physicochemical (photochemical stability, water solubility) and pharmacokinetic (elimination half-life) properties of nifedipine. However, the pharmacokinetic properties of the dihydropyridines are similar, with the exception of the clearance and the bioavailability of amlodipine. Amlodipine has a lower clearance, but a much larger Vd, resulting in prolonged elimination half-life (Arrowsmith *et al.*, 1986; Faulkner *et al.*, 1986).

Absorption and Bioavailability. All dihydropyridines are promptly and completely absorbed from gastrointestinal tract, but the bioavailabilities, with the exception of amlodipine, are markedly low as follows. Amlodipine: 52-88%, Nifedipine:40-50%, Felodipine: 10-25%, Isradipine: 15-20%, Nicardipine: 15-43%, Nilvadipine: 14%, Nimodipine: 7%, Nisoldipine: 8%, Nitrendipine: 10-30% (Abernethy and Schwartz, 1988).

Elimination. Each of eight dihydropyridines (nifedipine, felodipine, isradipine, nicardipine, nilvadipine, nimodipine, nisoldipine and nitrendipine), with the exception of amlodipin, have a high clearance due to the extensive first-pass metabolism, intermediate volume of distribution, and an elimination half-life less than about 10 hours. The elimination half-life of amlodipine is approximately 40 hours. In most cases studied, the presystemic clearance, first-pass metabolism, decreases with increasing oral dose and/or with multiple dose (Abernethy and Schwartz, 1988).

Pharmacokinetics in disese states. Predictably, patients with severe hepatic disease have decreased presystemic clearance as has been indicated in the patients with liver cirrhosis. Nifedipine in liver cirrhosis (Echizen and Eichelbaum, 1986): After intravenous administration, the terminal elimination half-life increase (7.0 ± 4.2 vs 1.7 ± 0.4 hours). The bioavailability almost doubles to 90%. Plasma protein binding decreases in the cirrhosis patients. Patients with renal disease also exihibit a decrease in protein binding. Nisoldipine in liver cirrhosis (Abernethy and Schwartz, 1988) : In this studies comparing intravenous and oral administration to cirrhotic and control subjects, mean (± SD) total clearance is decreased in cirrhotic patients (29.4 ± 7.2 *vs* 51.0 ± 18.6 l/h), elimination half-life markedly prolonged in cirrhotic patients (16.6 ± 4.6 *vs* 9.7 ± 5.4 hours). Moreover, absolute bioavailability is increased in cirrhotic patients (14.7 ± 10.1 vs 3.7 ± 2.1%). On the contrary, renal disease has little influence on all dihydropyridines, as has been observed with type I calcium antagonists (Echizen and Eichelbaum, 1986; Abernethy and Schwartz, 1988).

ANGIOTENSIN CONVERTING ENZYME (ACE) INHIBITORS

Captopril

Absorption. Captopril is rapidly absorbed from the gastrointestinal tract with detectable concentrations apparent as early as 15 minutes following administration, and approximately 60 to 75 % of an oral dose is absorbed (Duchin *et al.*, 1982; Kripalani *et al.*, 1980). Concurrent consumption of food has been reported to decrease absorption by 30 to 40 % (Singhvi *et al.*, 1982). In the long term study (4 to 5 weeks), however, peak plasma concentrations and the AUC of total and free captopril were only slightly reduced by food consumption and the blood pressure lowering effect was unchanged (Ohman *et al.*, 1985).

Elimination. After absorption, captopril is metabolized to several forms including a captopril-cysteine disulphide, a disulphide dimer of captopril and other mixed disulphides with endogenous thiol compounds (Komai *et al.*, 1981). Captopril and its metabolites are primarily excreted by the kidney with minor elimination in the feces. The primary mechanism of renal excretion is tubular secretion, accounting for 78% of urinary captopril. About 86% of the total administered radioactivity is recovered, with 67% of the radioactivity appearing in the urine and 18% in the feces (Kripalani *et al.*, 1980). The major metabolites excreted in the urine is a captopril-cysteine disulphide.

Pharmacokinetics in renal disease. The elimination half-life of unchanged captopril is approximately 1.7 hours in healthy volunteers, and increases with decreasing the renal function (Duchin *et al.*, 1982). A linear relationship between the elimination rate of captopril and endogeous creatinine clearance is observed, with a regression coefficient of 0.957 ($p < 0.001$) (Rommel *et al.*, 1980). In patients with creatinine clearance less than 20 ml/min, elimination half-life increases markedly to over 20 to 40 hours (Onoyama *et al.*, 1981).

Enalapril

Absorption. Enalapril is a prodrug that must be de-esterfied to the active form, enalaprilat and monoethylester of enalaprilat. Enalapril is very poorly absorbed from the gastrointestinal tract. Enalapril is rapidly absorbed and prompt inhibition of angiotensin I formation occurs. The extent of absorption of enalapril is 60 to 70 % and once absorbed, enalapril is rapidly transformed to its active form, enalaprilat (Irvin *et al.*, 1984). In contrast to captopril, concurrent consumption of food does not appear to influence the absorption of enalapril.

Elimination. The formation of enalaprilat probably occurs in the liver by hepatic esterase. There does not appear to be any other metabolite of enalapril other than enalaprilat. Like captopril, enalapril and enalaprilat are primarily excreted by the kidney. Approximately 61 and 33% of the dose is excreted in the urine and in the feces, respectively (Ulm *et al.*, 1982). Of the 61% recovered in the urine, 18% was in the form of unchanged enalapril and 43% in the form of enalaprilat (Ulm *et al.*, 1982). On the other hand, of the 33% recovered in the feces only 6% was in the form of enalapril and 27% in the form of enalaprilat (Ulm *et al.*, 1982). The slower clearance and longer half-life of enalapril might be expected to lead to a longer duration of action in patients with congestive heart failure.

Newer ACE inhibitors

Many newer ACE inhibitors have been designed to modify the rapid elimination property of captopril and to intensify the ACE inhibiting action. Most of them (benazepril, cilazapril, delapril, pentopril, perindopril, quinapril, ramipril and other several drugs) are prodrugs, with the exception of lisinopril, converted by hepatic esterolysis to active diacids of each drug. The metabolism and elimination of the prodrugs are rapid. Elimination of the diacids and lisinopril is mostly renal and therefore is affected by renal impairment in a predictable fashion (Kelly and O'Malley, 1990).

Absorption and Elimination of each drug

Benazepril. Gastrointestinal absorption is at least 37% (Kelly and O'Malley, 1990). Active metabolite (diacid form), Benazeprilat, represent 56% of urinary total recovery. Two other hydrophilic metabolites, glucuronide conjugates, are detected in the urine (Kelly and O'Malley, 1990).

Cilazapril. Cilazapril is a prodrug with a structural relationship of enalapril. Bioavailability of cilazaprilat, active form, is 77.5% after cilazapril and 29% after cilazaprilat (Kelly and O'Malley, 1990). Values of C_{max} decreases by about 30% and T_{max} increases by about 1 hour with the concurrent consumption of food (Massarella *et al.*, 1989). Plasma concentrations of cilazapril are significantly higher in patients with renal disease. AUC (0-24 hours) following multiple dosing in the renally impaired group is 5 times that in the normal patients (Shionoiri *et al.*, 1988).

Delapril. Once absorbed, delapril is converted to an active diacid derivative, which is biotransformed in turn to an active 5-hydroxy-indane diacid. Extent of absorption is more over 55% (Kelly and O'Malley, 1990). Renal impairment have no effect on the pharmacokinetics of the prodrug (delapril) but affect those of the two active metabolites (diacids). Significant positive correlation are found in the diacid metabolites between the inverse of C_{cr} and AUC (Shionoiri *et al.*, 1987).

Lisinopril. Structurally lisinopril is a diacid, similar to enalaprilat with the replacement of a methyl group by a butylamino group. Therefore, lisinopril is an active form and it does not need metabolic conversion for ACE inhibiting actions different from many other ACE inhibitors (prodrugs). Intestinal absorption is relatively low and 24-hour urinary recovery is approximately 25%. Fecal recovery is largely accounted (69%) for by unabsorbed drug (Kelly and O'Malley, 1990). The apparent elimination half-life is very long and approximately 40 hours. Severe renal impairment with values of GFR below about 30 ml/min is associated with larger changes in pharmacokinetics (Kelly and O'Malley, 1990). The AUC (0-24 hours) and the trough plasma concentrations (24 hours) are negatively related to Ccr (Jackson *et al.*, 1988).

Pentopril. Pentopril is converted *in vivo* to the active diacid with no other detectable metabolite formed subsequently. Pentopil is rapidly absorbed and its bioavailability , is 60 to 70% based upon urinary recovery, with 21% as pentopril and 40% as the active metabolite (Rakhit *et al.* 1986). Apparent half-life values of prodrug and diacid are short at about 1 and 2 hour, respectively (Rakhit *et al.*, 1986). There is no appreciable changes in pentopril pharmacokinetics with renal impairment, but clearance of diacid is sensitive to changes in renal function with the most marked and rapid changes occurring when Ccr fell below 50 ml/min (Rakhit *et.al.*, 1988).

Perindopril. Perindopril is an ester prodrug and is converted to the active diacid, perindoprilat. Bioavailability (area method) of perindopril is 66%, with 32% conversion of perindopril to perindoprilat (Kelly and O'Malley, 1990). Urinary recovery of perindoprilat up to 120 hours after intravenous administration was 57% ± 25% and 70% ± 16 % of the dose in elderly and young subjects, respectively (not significant) (Lees *et al.*, 1988). Food decreases perindoprilat bioavailability following oral perindopril administration, possibly due to the alteration of the conversion of prodrug to perindoprilat (Lecocq *et al.*, 1990). Renal clearance of perindoprilat is significantly lower in the elderly subjects (67 ± 31 ml/min) than in the young subjects (110 ± 39 ml/min) (Lees *et al.*, 1988).

Quinapril. Quinapril is rapidly converted to its active metabolite, diacid (quinaprilat). After oral administration of the ^{14}C-labeled drug, 61% and 37% of the radiolabel is recovered in the urine and in the feces, respectively (Olson *et al.*, 1989). There is a significant correlation between Ccr and plasma clearance of the diacid metabolite (Begg *et al.*, 1989).

Ramipril. Ramipril is also a prodrug and its active form is ramiprilat. Radioactive excretion data indicates that ramipril is rapidly absorbed, and that at least 60% of the administered dose is absorbed (Eckert *et al.*, 1984). About 56% of the administered radioactivity is found in the urine and about 40 % in the feces (Eckert *et al.*, 1984). Mean trough concentrations and absolute accumulation increase with worsening renal function, and the renal clearance of ramiprilat is significantly correlated with the creatinine clearance (Schunkert *et al.*, 1989). In patients with severe renal insufficiency (Ccr below 30 ml/min) smaller doses of ramipril are required than in patients with normal or borderline renal function (Schunkert *et al.*, 1989).

REFERENCES

Abernethy, D. R. and Schwartz, J. B. (1988). Pharmacokinetics of calcium antagonists under development. *Clin. Pharmacokinet.* **15,** 1-14.

Abshagen, U., Von Gridzicki, U., Hirschberger, U. and Rennekamp, H. (1977). Effect of enterohepatic circulation on the pharmacokinetics of spironolactone in man. *Naunyn-Schmiederberg's Arch. Pharmacol.* **300,** 281-287.

Arrowsmith, J. E., Campbell, S.F., Cross, P. E., Stubbs, J.K., Burges, R. A., Gardiner, D. G. and Blackburn, K. J. (1986). Long-acting dihydropyridine calcium antagonists I: 2-alkoxymethy derivatives incorporating basic substituents. *J. Med. Chem.* **29,** 1696-1702.

Beermann, B.,Dalen, E., Lindstrom, B. and Rosen, A. (1975). On the fate of furosemide in man. *Eur. J. Clin. Pharmacol.* **9,** 57-61.

Beermann, B., Groschinsky-Grind, M. and Rosen, A. (1976). Absorption, metabolism and excretion of hydrochlorothiazide. *Clin. Pharmacol. and Ther.* **19,** 531-537.

Beermann, B. and Groschinsky-Grind, M. (1977). Pharmacokinetics of hydrochlorothiazide in man. *Eur. J. Clin. Pharmacol.* **12,** 297-303.

Beermann, B. and Groschinsky-Grind, M. (1978a). Gastrointestinal absorption of hydrochlorothiazide enhanced by concomitant intake of food. *Eur. J. Clin. Pharmacol.* **13,** 125-128.

Beermann, B. and Groschinsky-Grind, M. (1978b). Enhancement of the gastrointestinal absorption of hydrochlorothiazide by propantheline. *Eur. J. Clin. Pharmacol.* **13,** 385-387.

Beermann, B. and Groschinsky-Grind, M. (1979). Pharmacokinetics of hydrochlorothiazide in patients with congestive heart failure. *Br. J. Clin. Pharmacol.* **7,** 579-583.

Beermann, B. and Groschinsky-Grind, M. (1980). Clinical pharmacokinetics of diuretics. *Clin. Pharmacokinet.* **5,** 221-245.

Begg, E. J., Bailey, R. R., Lynn, K. L., Robson, R. A., Frank, G. J. and Olson, S. C. (1989). The pharmacokinetics of angiotensin converting enzyme inhibitors in patients with renal impairment. *J. Hyper.* **7** (Suppl 5) S29-S32.

Beyer, K. H., Baer, J. E., Michaelson, J. K. and Russo, H. F. (1965). Renotropic characteristics of ethacrynic acid: A phenoxyacetic saluretic-diuretic agent. *J. Pharmacol. Exp. Ther.* **147,** 1-22.

Cutler, R. E. and Blair, A. D. (1979). Clinical pharmacokinetics of frusemide. *Clin. Pharmacokinet.* **4,** 279-296.

Duchin, K. L., Singhvi, S. M., Willard, D. A., Migdalof, B. H. and Mckinstry, D. N. (1982). Captopril kinetics. *Clin. Pharmacol. Ther.* **31,** 452-458.

Echizen, H. and Eichelbaum, M. (1986). Clinical pharmacokinetics of verapamil, nifedipine and diltiazem. *Cin. Pharmacokinet.* **11,** 425-449.

Eckert, H. G., Badian, M. J., Gantz, D., Kellner, H. M. and Volz, M. (1984). Pharmacokinetics and biotransformation of 2-(N-((S)-1-etyoxycarboxyl-3-phenylpropyl)- L-alanyl)-(1S,3S,5S)-2-azabicyclo (3.3.0) octane-3-carboxylic acid (Hoe 498) in rat, dog and man. *Arzneimittel-Forschung* **34,** 1435-1477.

Eichelbaum, M., Remberg, E. G., Schomerus, M. and Dengler, H. J. (1979). The metabolism of D,L(^{14}C) verapamil in man. *Drug Metab. Dispos.* **7,** 145-148.

Eichelbaum, M., Mikus, G. and Vogelgesang, B. (1984). Parmacokinetics of (+)-, (-)- and (±)-verapamil after intravenous administration. *Br. J. Clin. Pharmacol.* **17,**453-458.

Faulkner, J. L., McGibney, D., Chasseaud, L. F., Perry, J. L. and Taylor, I. W. (1986). The pharmacokinetics of amlodipine in healthy volunteers after single intravenous and oral doses and after 14 repeated oral doses given once daily. *Br. J. Clin. Pharmacol.* **22,** 21-25.

Halladay, S. C., Sipes, I. G. and Carter, D. E. (1977). Diuretic effect and metabolism of bumetanide in man. *Clin. Pharmacol. Ther.* **22,** 179-187.

Hartmann, D., Lunell, N.O., Friedrich, G. and Rane, A. (1988). Excretion of tiapamil in breast milk. *Br. J. Clin. Pharmacol.* **26,** 183-186.

Hinderling, P. H., Eckert, M., Gasic, S., Eichler, H G., Potzi, R. and Heizmann, P. (1986). Comparative pharmacokinetics and cardiovascular effects of tiapamil in healthy volunteers and patients with hepatic cirrhosis. *Eur. J. Clin. Pharmacol.* **31,** 397-404.

Irvin, J. D., Till, A. E., Vlasses, P. H., Hickens, M., Rotmensch, H. H., Harris, K. E., Merrill, D.

D and Ferguson, M. D. (1984). Bioavailability of enalapril maleate. *Clin. Pharmacol. Ther.* **33,** 248.

Jackson, B., Cubela, R. B., Conway, E. L. and Johnston, C. L. (1988). Lisinopril pharmacokinetics in chronic renal failure. *Br. J. Clin. Pharmacol.* **25,** 719-724.

Kelly, J.G. and O'Malley, K. (1990). Clinical pharmacokinetics of the newer ACE inhibitors: A reveiw. *Clin Pharmacokinet.* **19,** 177-196.

Komai, T., Ikeda, T., Kawai, K., Kameyama, E. and Shindo, H. (1981). In vitro studies on the metabolic pathway of SQ14225 (captopril) and mechanism of mixed disulfide formation. *J. Pharmacobiodyn.* **4,** 677-684.

Kripalani, K. J., Mckinstry, D. N., Singhvi, S. M., Willard, D. A., Vukovich, R. A. and Migdalof, B. H. (1980). Disposition of captopril in normal subjects. *Clin. Pharmacol. Ther.* **27,** 636-641.

Lant, A. (1985). Diuretics, clinical pharmacology and therapeutic use (part I). *Drugs* **29,** 57-87.

Lees, K. R., Green, S. T. and Reid, J. L. (1988). Influence of age on the pharmacokinetics and pharmacodynamics of perindopril. *Clin. Pharmacol. Ther.* **44,** 418-425.

Lecocq, B., Funck-Brentano, C., Lecocq, V., Ferry, A., Gardin, M. E., Devissaguet, M. and Jaillon, P. (1990). Influence of food on the pharmacokinetics of perindopril and the time course of angiotensin converting enzyme inhibition in serum. *Clin. Pharmacol. Ther.* **47,** 397-402.

Massarella, J. W., Defeo, T. M., Brown, A. N., Lui, A.and Willis, R. J. (1989). The influence of food on the pharmacokinetics and ACE inhibition of cilazapril. *Br. J. Clin. Pharmacol.* **27,** 205S-209S.

McTavish, D. and Sorkin, E. M. (1989). Verapamil, an updated review of its pharmacodynamic and pharmacokinetic properties, and therapeutic use in hypertension. *Drugs* **38** (1), 19-76.

Odlind, B. (1979). Relation between renal tublar secretion and effects of five loop diuretics. *J. Pharmacol. Exp. Ther.* **211,** 238-244.

Ohman, K. P., Kagedil, B., Larsson, R. and Karlberg, B. E. (1985). Pharmacokinetics of captopril and its effect on blood pressure during acute and chronic administration and in relation to food. *J. Cardiovasc. Pharmacol.* **7** (Suppl. 1), S20-S24.

Olson, S. C., Horvath, A. M., Michniewicz, B. M., Sedman, A. T., Colburn, W. A. and Welling, P. G. (1989). The clinical pharmacokinetics of quinapril. *Angiology* **40,** 351-359.

Onoyama, K., Hirakata, H., Iseki, K., Fusimi, S., Omae, T., Kobayashi, M. and Kawahara, Y. (1981). Blood concentration and urinary excretion of captopril (SQ14225) in patients with chronic renal failure. *Hypertension* **3,** 456-459.

Pentikainen, P. J., Pentilla, A., Nenveonen, P. J. and Gothoni, G. (1977). Fate of (^{14}C)-bumetanide in man. *Br. J.Clin. Pharmacol.* **4,** 39-44.

Rakhit, A., Hurley, M. E., Tipnis, V., Coleman, J., Rommel, A. and Brunner, H. R. (1986). Pharmacokinetics and pharmacodynamics of pentopril, a new angiotensin-converting-enzyme inhibitor in humans. *J. Clin. Pharmacol.* **26,** 156-164.

Rakhit, A., Radensky, P., Szerlip, H. M., Kochak, G. M., Audet, P. R., Hurley, M. E and Feldman, G. M. (1988). Effect of renal impairment of disposition of pentopril and its active metabolite. *Clin. Pharmacol. Ther.* **44,** 39-48.

Riddell, J. G., Harron, D.W.G. and Shanks, R. G. (1987). Clinical pharmacokinetics of β-adrenoceptor antagonists, An update. *Clin. Pharmacokinet.* **12,** 305-320.

Rommel, A. J., Perides, A. M. and Heald, A. (1980). Captopril elimination in chronic renal failure. *Clin. Pharmacol.Ther.* **27,** 282.

Sadee, W., Dagcioglu, M. and Schroder, R. (1973). Pharmacokinetics of spironolactone, canrenone and canrenoate-K in humans. *J.Pharmacol. Exp.Ther.* **185,** 686-695.

Schunkert, H., Kindler, J., Gassmann, M., Lahn, W., Irmisch, R., Ritz, E., Debusmann, E. R., Pujadas, J.O., Koch, K. M. and Sieberth, H.-G. (1989). Pharmacokinetics of ramipril in hypertensive patients with renal insufficiency. *Eur. J. Clin. Pharmacol.* **37,** 249-256.

Shionoiri, H., Yasuda, G., Abe, Y., Yoshimura, H., Kaneko, Y. and Shindo, Y. (1987). Pharmacokinetics and acute effect on the renin-angiotensin system of delapril in patients with chronic renal failure. *Clin. Nephrol.* **27,** 65-70.

Shionoiri, H., Gotoh, E., Takagi, N., Takeda, K., Yabana, M.and Kaneko, Y. (1988).

Antihypertensive effects and pharmacokinetics of single and consecutive doses of cilazapril in hypertensive patients with normal and impaired renal function. *J. Cardiovasc. Pharmacol.* **11,** 242-249.

Singhvi, S. M., Mckinstry, D. N., Shaw, J. M., Willard, D. A. and Migdalof, B. H. (1982). Effect of food on the bioavailability of captopril in healthy subjects. *J. Clin. Pharmacol.* **22,** 135-140.

Somogyi, A., Albrecht, M., Kliems, G., Schafer, K. and Eichelbaum, M. (1981). Pharmacokinetics, bioavailability and ECG response of verapamil in patients with liver cirrhosis. *Br. J. Clin. Pharmacol.* **12,** 51-60.

Ulm, E. H., Hichens, M., Gomez, H. J., Till, A. E., Hand, E., Vassil, T. C., Biollaz, J., Brunner, H. R. and Schelling, J. L. (1982). Enalapril maleate and a lysine analogue (MK-521): Disposition in man. *Br. J. Clin. Pharmacol.* **14,** 357-362

Vogelgesang, B., Echizen, H., Schmidt, E. and Eichelbaum, M., (1984). Stereoselective first-pass metabolism of highly cleared drugs: studies of the bioavailability of L- and D-verapamil examined with a stable isotope technique. *Br. J. Clin. Pharmacol.* **18,** 733-740.

Woodcock, B. G., Rietbrock, I., Voehringer, H. and Rietbrock, N. (1981). Verapamil disposition in liver disease and intensive-case patients: kinetics, clearance, and apparent blood flow relationships. *Clin. Pharmacol. Ther.* **29,** 27-34.

Woods, P. B. and Robinson, M. L. (1981). An investigation of the comparative liposolubilities of β-adrenocepter blocking agents. *J. Pharm. Pharmacol.* **33,** 172-173.

Yabana, H., Nagao, T. and Sato, M. (1985). Cardiovascular effects of the metabolites of diltiazem in dogs. *J. Cardiovasc. Pharmacol.* **7,** 152-157.

Author index

Bataillard, A., 331
Burszyn, M., 107

Chijiwa, T., 131

De Jong, W., 157
Dupont, A. G., 207

Gavras, H., 107

Hamaue, N., 15
Hidaka, H., 131

Iimura, O., 93
Imai, M., 291
Imai, S., 73
Iseki, K., 355
Ishibashi, T., 73
Ishikawa, T., 131

Kato, M., 319
Kobayashi, T., 319

Matsumoto, M., 15, 31, 189
Minami, M., 15, 31, 189, 309
Miyazaki, K., 355

Nakamura, M., 291
Nomura, A., 145, 355

Parvez, S. H., 15, 309

Saito, H., 15, 31, 189, 309, 319
Sakuma, I., 319
Sassard, J., 331
Sato, M., 319
Scholtysik, G., 257
Shimamoto, K., 93
Shimizu, T., 291
Suga, T., 233
Suzuki, M., 233

Tochihara, M., 31
Toda, N., 269
Togashi, H., 15, 31, 189, 319
Touraine, J. L., 331

Van den Buuse, M., 157
Van der Niepen, P., 207
Van Zwieten, P. A., 1

Wikstrand, J., 49

Yasuda, H., 145, 319
Yoshioka, M., 15, 31, 189, 309, 319

Subject index

18-502, 73
20-785, 73
5,6-DHT, 201
5,7-dihydroxytryptamine (5,7-DHT), 195
5-HT
    induced vasoconstriction, 190
    amplifying effect , 199
5-HT_1 receptors, 190
5-HT_{1A}
    agonistic potency, 1
    agonists, 200, 10
    receptors, 4, 8, 10
5-HT_{1B}
    receptors, 4
    agonists, 200
5-HT_{1C}, 4
5-HT_2
    agonists, 200
    antagonists, 189, 200
    receptors, 4, 8, 189, 200
5-HT_3 receptors, 190
5-methylurapidil, 4
6-hydroxydopamine (6-OHDA), 195
6-OHDA, 165, 201
8-OH-DPAT, 10, 202

A first dose hypotension, 11
A10, 157
A12, 157, 172
A14, 157
A2, 169
A8, 157
A9, 157
Absorption, 25, 355
Acebutolol, 95
Action potential, 75
Acute myocardial infarction, 51
Adrenal nerve activity (ANA), 20, 22, 194, 195, 197
Afterload, 23, 27, 217
    reduction, 217
Aggregation, 274
Agranulocytosis, 117
Alcepril, 108
Aldosterone levels, 108
Allogeneic fetal liver cells, 333
Alpha blockers, 148
Alpha-1 adrenoceptor, 191
    blocking drugs, 1, 21, 31, 42, 44, 148
    blocking properties, 98
Alpha-2 adrenoceptor, 257
    agonists, 257, 263
    blocking activity, 15
    ligands, 263
Alpha-2A subtypes, 257
Alpha-2B subtypes, 257
Alpha-methyl-DOPA, 8, 9, 22, 114, 148, 165
Alprenolol, 95, 357
Amlodipine, 360
Amosulalol, 94, 98, 102
    chemical structure, 98
Amount of autoantibodies, 331
Amphetamine , 173
Analgesic, 246
Angina pectoris, 114
Angioedema, 117
Angiotensin converting enzyme (ACE) inhibitors
    chronic renal failure, 112
    hemodynamics, 109
    inhibitors, 108, 113, 117, 355, 361
    newer ACE inhibitors, 362
    renovascular hypertension, 113
    side effects, 116
Angiotensin II, 108
    competitive antagonist, 113
    receptor antagonist, saralasin, 109
Anterior hypothalamus, 164
    preoptic region, 164
Anti-rat thymocyte serum, 332
Anticonvulsant, 246
Antinuclear antibody, 316
Anxiolytic, 246
    action, 249
Aortic depressor nerve activity, 314
Apomorphin, 163, 173
Aprotinin, 108
Arotinolol, 95
Arrythmias, 25
Ascending noradrenergic neurons, 201
Ascending serotonergic neurons, 201
Asthma, 94
Atenolol, 95, 101, 357
    HAPPHY study, 61
Atherosclerosis, 118
Athritis, 115
ATP, 24
Auto-immune component, 351
Azepexole, 258, 261

B-HT920, 5
B1/B3-regions, 8
B7 (mucleus raphe dorsalis), 201
    area, 201
Backcross rats, 170
Baroreceptor-denervated cats, 195, 199

Bed nucleus of the stria terminalis, 248
Benazepril, 108, 362
Benazeprilat, 362
Bendroflume-thiazide, 50, 55
Benzothiazepines, 146
Beta-receptors, 4
    blockers, 66, 110, 111, 148, 355, 357
    blocking activity, 74
Beta 1-adrenoceptor
    activity, 1
    selectivity, 64
Beta-2 adrenoceptor, 100
    stimulants, 50
Bevantolol, 95
BHT-920, 209
Biliary excretion , 355
BIO 14.6, 24
Bioavailability, 355
Blood flow, 25
Blood pressure, 78
    regulation, 199
Bopindolol, 73, 94, 99, 100
    18-502 (metabolite), 73
    20-785 (metabolite), 73
    pA_2, 85
Bradykinin, 108
    antagonists, 108
Bromocriptine, 160, 163, 172,
Budralazine, 309
    adverse effects, 315
    chemical structure, 309
    pharmacodynamics, 310
Bufetolol, 95
Buffer nerve intact rats, 197
Bulbospinal preganglionic neurons, 201
Bumetanide, 15, 18, 23, 356
Bunazosin, 16, 20, 24, 32, 34
    acute toxicity, 26
    data of clinical study, 26
    pharmacokinetics, 27
    urinary sodium excretion, 18
Bunitrolol, 95
Bupranolol, 95

C1/A1 region, 162
Ca^{2+}
    binding proteins, 131
    Ca^{2+}-PDE inhibitors, 138
    excretion, 296
Calcium antagonists, 132, 145, 146, 153
    channel blockers, 110, 355
    side effects, 152
    subclass, 146
Calmodulin, 131
    antagonists, 135
    CaM KII inhibitor, 139
    classification, 133
cAMP, 260
Canine heart, 24
Canrenone , 357
Captopril, 107, 111, 117, 361
Cardiac arrhythmias, 150
Cardiac failure, 50, 94
Cardiohemodynamic effect, 76
Cardiomyopathy, 24
Cardioprotection, 24
Cardioselectivity, 66
Cardiovascular disease, 246
    mortality, 55
Carteolol, 95
Carvedilol, 95, 97
Catecholamines, 238
Cause-specific mortality, 51, 55
Celiprolol, 94, 100, 102
    chemical structure, 99
Cell suspensions, 334
Cellular immunity, 331
Central alpha-1 adrenoceptors, 32
Central alpha-2 adrenoreceptor
    sensitivity,178
    in cardiovascular control, 257
Central dopamine
    activity, 172
    D_2 receptors, 160, 219
    D_2 receptor agonist, 217
    mechanisms, 173
    modulation, 157
    systems, 158, 165, 171, 219
Central hypotensive activity, 9
Central monoamines , 165
Central nervous system disease, 152
Central nucleus of the amygdala, 248
Central serotonergic neurons, 201
    Central 5-HT_2 receptors, 202
Central sympathoinhibitory action, 195, 313
Centrtal 6 hydroxydopamine, 175
Cerebral blood flow, 115
Cerebrovascular insufficiency, 114
Chlorisondamine, 218
Chronic Obstructive pulmonary disease, 115
Chronic renal failure, 112
Cilazapril, 108, 362
    Cilazaprilat, 362
Cingulate, 248
Cirazoline, 5
Clentiazem, 146
Clonidine, 1, 8, 33, 40, 195, 233, 257, 262
    cardiovascular action, 246
    endocrine action, 248
Coadministration with food, 355
Cold pressor test, 177
Common binding sites, 202
Competitive angiotensin II blockade, 108

Concomitant disease, 149
Conduction defect, 94
Congestive heart failure, 113, 150
Connecting tubule, 299
Conscious beagle dogs, 76
Conscious SAD-rats, 164
Coronary artery
  by-pass surgery, 61
  disease, 114, 149
  vasodilator, 145
  Coronary heart disease, 50, 58
  vasospasms, 101
Cough, 115
Creatine phosphokinase (CK), 24, 83
Cyclophosphamide treatment, 333, 342
Cyproheptadine, 189

D_1
  agonist activity, 217
  agonists, 157, 212
  antagonist, 213
  receptors, 157, 207, 211, 221
D_2 receptor,8, 207, 221
  agonist, 208, 213
  antagonists, 218
Definite cardiovascular disease events, 57
Delapril, 362
Desipramine, 175, 195
Diabetes, 315
  mellitus, 50, 111, 151
  nephropathy, 111
1, 2,-diacylglycerol (DAG), 260
Diazoxide, 23
Dihydropyridine, 146, 360
Dilevalol, 94, 102
  chemical structure, 99
Diltiazem, 145, 148, 359
Diuretic, 55, 58
  effect, 270, 294
  diuretics, 55, 61,110, 148, 355
DOCA
  hypertensive sheep, 192
  hypertensive rats, 16, 24, 163, 219, 331
  hypertention, 161, 171
Domperidone, 161, 171, 210, 221
DOPAC, 170
  levels, 170
  DOPAC/DA ration, 170
Dopamine, 158
  agonist, 163, 173
  antagonist, 165
  D_1 agonist , 174
  D_2 agonist, 157, 161, 163
  D_2 antagonist, 161, 170, 171
  D_2 receptors, 157
  turnover, 169
  receptor binding, 169
  uptake inhibitor , 173, 175
Dopamine-beta-hydroxylase, 163
  activity, 178
Dopaminergic
  drugs, 158
  hypoactivity, 173
  lesions, 175
  receptors, 4
Dorsal motor nucleus of the vagus, 163
Dorsomedial nucleus of the hypothalamus, 163
Drowsiness, 263
Drug interaction, 153
Dry mouth, 263
Dry persistent cough, 117

Effort angina, 101
Elimination half-life, 355
Enalapril, 108, 118, 361
Endocrine, 246
  action, 248
Endothelium-derived relaxing factor (EDRF), 190, 319
Enkephalin, 243
Eosinophilia, 117
Ethacrynic acid, 356
Excitatory amino acids, 244
Excretion, 25

Fatal and nonfatal coronary heart disease, 51, 56
Fatal stroke, 51
Felodipine, 146, 360
Fenoldopam, 207
  basic pharmacology, 208
  clinical pharmacology, 213
Fetal liver cell transplantation, 333, 334
First-pass effect, 355
Flutonidine, 258
Follow-up Program (HDFP), 63
Fosenopril, 108
Free water
  clearance, 297
  reabsorption, 297
Frosemide, 294, 356
Functional renal failure, 117

Ganglionic
  D_1-like receptors, 209
  dopamine receptor, 209
∂-Aminobutylic acid (GABA), 245
Gastric emptying rate, 355
GBR-12909, 173, 175
Genetic
  hypertension, 332, 350
  hypertensive animals, 192
Glomerular filtration rate, 212

Glucose
intolerance, 93
tolerance, 314
Grooming activity, 174
Guanabenz, 259, 262
Guanethidine, 22
Guanfacine, 8, 259, 262
Guinea-pigs, 74

Haloperidol, 165, 170, 208, 218
HAPPHY, 61
HDFP, 61
HDL cholesterol 102
concentrations, 83
Heart rate, 78
Henle's loop, 297
Hexamethonium, 218, 221
High polariry, 201
Hippocampus, 248
Histamine, 8
HLA-system, 331
Humoral factors, 25
Hybrid drugs, 31
Hydralazine, 22, 50, 195, 309
Hydrochlorothiazide, 11, 50, 55, 110, 355
Hyperlipidemia, 93
Hypertensive emergency, 148
Hyperuricemia, 93, 291
Hypothalamic area, 248

Idazoxan, 220, 262
^{3}H-idazoxan, 260
Immune
pathogenetic mechanism, 350
system , 350
Immunofluorescence study, 336
Immunoglobulins, 331
Immunological
factors, 331, 350
parameters, 333
In vivo
microdialysis, 165
voltammetry, 165, 170
Indapamide, 269
diuretic effect, 270
hypotensive effect in patients, 281
hypotensive mechanism, 276
metabolism, aggregation,
proliferation, 274
perfused mesenteric vasculature, 272
prostacyclin, 275
side effects, 283
Indoleamines, 157
Inferior cardiac nerve activity (ICNA),34, 35, 195
Inositol-1,4,5-triphosphate (IP3), 260
Inotropic effects, 100
Insular cortex, 248
Insulin secretion, 314

Intermediolateral column (IML), 197, 260
Intermittent clausdication, 95
Intra-NTS, 162, 164
Intracellular
Ca^{2+} antagonists, 131
Ca^{2+} release inhibitor, 134
Intrinsic
sympathetic activity, 6
sympathomimetic activity (ISA), 94, 99
IPPPSH report, 54, 61, 63
Isradipine, 146, 360
Ketanserin, 37, 189, 194
central action, 193
pharmacokinetics, 191
Kininase II , 108

L-alpha-Methyldopa, 259
L-alpha-methyl-noradrenaline, 259
L-alpha-Methyldopamine, 259
L-arginine, 319
L-arginine-derived NO, 319
L-Dopa, 173, 176
Labetalol, 94, 101, 357
LDL-cholesterol, 64
Left atria, 74
Left ventricular hypertrophy, 24, 118, 149
Leucopenia, 117
LH, 332
Lipid metabolism, 27
Lipids, 152
Lipophilicity , 64, 66
Lisinopril, 108, 115, 362
Locomotor, 173
activity, 173, 174
Locus coeruleus, 162, 201, 248, 260
Lofexidine, 258, 262
Loop diuretics, 297
Loss of taste, 118
[^{3}H]-LSD binding, 170
LY 171555, 217
LY53857, 193
Lymphocytes, 339
Lymphoid cells, 351
Lyon model , 332

M_1, 8
Macrophage, 331
MAPHY primary prevention study, 49, 61, 63
Medetomidine, 259
Medical Research Council Trial, 61
Membrane stabilizing activity (MSA), 94, 75
Mental arithmetic test, 177
Mesocortical/mesolimbic dopamine system, 178

Mesolimbic dopaminergic
  neurotransmission, 175
Metabolism, 25, 274, 355
Metabolite, 25
Methiothepine, 10
Methysergide, 9, 189
Metoclopramide, 165, 216, 219
Metoprolol, 7, 49, 61, 95, 101
  MAPHY study, 49
  cause-specific mortality, 55
Migraines, 151
Minoxidil, 23, 112
Mitochondrial dysfunction, 24
Mitogen responce, 334
MLCK inhibitors, 138
Monoamine oxidase, 159
Mortality and morbidity of CHD, 56
Mortality rate, 62
Moxonidine, 258
MRC, 61, 63
Mucocutaneous febrile reactions, 117
Muscarinic receptors, 4
Myocardial infarction, 118
Myointimal proliferation, 118

Nadolol, 95, 357
Natriuretic effects, 15, 20
Negative chronotropic effect , 100
Neonatal thymectomy, 333, 337
Neutropenia , 118
N^G-monomethyl-L-arginine, 319
Nicardipine, 146, 360
Nifedipine, 110, 145, 148, 360,
Nigrostriatal
  dopamine system, 176
  pathway, 178
Nilvadipine, 146, 360
Nimodipine, 146, 360
Nipradilol, 94, 101
Nisoldipine, 146, 360
Nitrendipine, 146, 360
Nitric oxide (NO), 319
Nitroprusside, 214
Nocturnal leg cramps, 150
Non-selective
  dopamine receptor antagonist, 216
  5-HT antagonists, 200
Noncardiovascular deaths, 58
Nonfatal
  myocardial infarctions, 51, 57
  stroke, 53
Nonstereoidal antiinflammatory drugs, 115
Noradrenergic pathway, 199
Nucleus arcuatus, 157, 172
  of the hypothalamus, 160
Nucleus tractus solitarii (NTS), 159, 162,
  233

Obstructive lung disease, 50, 94,
One-kidney goldblatt, 331
Open-field test, 173
Opiate withdrawal-attenuating, 246
Opioid-receptors, 8
Oral angiotensin converting enzyme
inhibitors, 107
Orthostatic hypotension, 11,16, 202, 214
Oxprenolol, 95

Papillary muscle, 75
PAPP, 202
Para-aminoclonidine, 262
  ^{3}H-para-aminoclonidine, 260
Pargyline, 195
Parkinson's disease, 164, 176, 178
Partial agonist activity, 77
Pathogenesis of hypertension, 331
Penbutolol, 95
Pentopril, 363
Perfused mesenteric vasculature, 272
Pergolide, 163, 170
Perindopril, 108, 110, 363
Peripheral
  dopamine receptors, 207
  vascular disease, 115
  vascular insufficiency, 95
Pharmacokinetics, 27
  in the disease states, 355
Phenoxybenzamine, 31
Phentolamine, 24, 31
Phenylalkylamines, 146
Pindolol, 10, 95, 99, 101
Piperoxane, 262
Piretanide, 356
Pithed rats, 5
Plasma concentration
  catecholamine, 19
  epinephrine, 20
  norepinephrine, 20, 177
  prolactin, 171, 172
  serotonin, 19, 20
Plasma renin activity, 23, 100, 108, 176
Platelet aggregation, 190
Post-synaptic alpha-adrenoceptors, 257
Post-transplantation, 152
Posterior
  hypothalamus, 160, 164
  NTS, 162
Postganglionic inferior cardiac nerve
  activity, 194
Postjunctional alpha1-adrenoceptors, 5
Postsynaptic
  adrenoceptors, 257
  alpha-1 adrenocepters, 2
  alpha-1 adrenoceptor antagonist, 15
Potassium excretion,296
Practolol, 357
Prazosin, 1, 6, 11, 15, 23, 31, 191, 202

Preganglionic adrenal nerve activity, 34, 40
Preload, 23, 27
Pre-synaptic
alpha-adrenoceptors, 257
alpha-2 agonistic activity, 1
alpha-2 adrenoceptors, 2
Primary aldosteronism, 116
Prolactin, 171
release, 172
Proliferation, 274
Propranolol, 93, 101, 357, 61
Prostacyclin, 275
Proteinuria, 112, 118
Pulmonary disease, 151

Quality of life, 153
Quinapril, 108, 363
Quinpirole, 161, 173, 207, 208, 217, 221
basic pharmacology, 217
clinical pharmacology, 221

Ramipril, 108, 110
Raphe nuclei , 162
Rash, 117, 118
Rat, 77
Rauwolscine, 1
Raynaud's phenomenon, 150
Receptor binding studies, 170
Reflex tachycardia, 23, 35, 202
Regional blood flow, 313
Renal
and cerebral blood flow, 149
artery stenosis, 113, 116
blood flow, 217, 212
circulation, 26
disease , 151
failure, 113
hypertensive rats, 163
nerve activity (RNA), 21, 194, 195
sympathetic nerve activity, 35, 40
angiotensin system, 107, 193, 149
Renin responses,172
Renovascular hypertension, 113, 116
Reperfusion arrythmias, 24
Reserpine , 169
Rheumatoid arthritis, 115, 309
Right atria, 74
Rilmenidine, 259, 263
Ritanserin, 193

S-8666, 291
antihypertensive action, 305
calcium excretion, 296
diuretic effect, 294
hyperuricemia, 291
potassium excretion, 296
sodium excretion, 294
uricosuric action, 301

S-sulpilide, 218, 221
SAD rats, 178, 164
SCH 23390, 208, 221
Second-and third-degree atrioventricular block, 50
Secondary hypertention, 50
Selective
D_1 receptor agonist, 208, 217
D_1 receptor antagonist, 210
D_2 receptor antagonist, 210
Septum, 248
Serotonergic
nervous system, 201
pathway, 199
Serotonin, 246
Serum lipid fractions, 83
SGB-1347, 43
SGB-1534, 36
Side effects, 283
Silent myocardial infarction, 52
Sino-aortic-denervated (SAD), 164
SK&F 102698, 163
SK&F 38393, 174
SKF 82526, 208
SM-2470, 38, 39, 43, 44
Smoking
patients, 58
status, 58
Sodium
excretion, 294
reabsorption, 23
Sotalol, 357
Spinal cord, 248
Spiperone, 10, 189
[^{3}H]-Spiroperidol, 170
Spironolactone, 50, 357
Spiroperidol (Spiperone), 170
Spiroxatrine, 10
Splenic T-lymphocytes, 336
Splenocytes, 343
Spontaneously hypertensive dogs, 192
Spontaneously hypertensive rats (SHR), 16, 24, 78, 159, 169, 171, 190, 211, 332
Stroke-prone spontaneously hypertensive rats (SHRSP), 16, 33, 192
Stable angina, 114
Stereoselective first-pass metabolism , 359
Stereotyped headwaving, 174
Stroke, 51
Substance P, 243
Substantia
gelatinosa, 260
nigra, 157, 178
Sudden cardiac death, 51
Sulphydryl group, 115

Sulpiride, 173
^{125}I-Sulpiride, 170
Suprapontine pressor pathway, 201
Supraspinal level, 199
Supraventricular tachycardia, 146
Surgery, 152
Swimming stress, 169
Sympathetic nerve
activity, 20, 33
recordings, 17
Sympathoinhibitory
action, 195
activity, 31, 42
Syngeneic fetal liver cells, 333
Syrian hamster, 24
Systemic lupus erythematosus, 309

T-cell, 331, 336
Tachyphylaxis, 199
Takayasu's disease, 115
Talipexole, 259, 261
Taste disturbance, 118
Teprotide, 107, 116
The elderly, 111
The happhy study, 67
The maphy study group, 68
Thiazide
diuretics, 50
thiazides, 110
Thick ascending limb of Henle's loop, 297
Thromboembolic complications, 64
Thymectomy, 332, 339, 341
Thymic
and splenic lymphocytes, 332
hormonal factors, 332
T-lymphocytes, 336
Thymocytes, 339, 343
Thymocytotoxic Autoantibodies, 332
Thymostimulin (TP1) treatment, 333, 343
Thymus, 350
extract-Thymostimulin, 333
Tiamenidine, 258, 262
Tiapamil, 146, 360
Tilislol (N-696), 94, 97
Timolol, 95
Tolonidine, 258
Total
cholesterol, 102
mortality, 51
peripheral resistance, 76
Tracheal smooth muscle, 74
TRH, 172
Trichlormethiazide, 294
Triglycerides, 102
Tuberoinfundibular dopamine system, 172
Tyrosine hydroxylase, 169
activity, 169

UK-14, 258, 261, 304

Urapidil, 1, 6, 11, 94, 98
binding study, 3
interaction, 6
central hypotensive activity, 7
adverse reactions, 11
Uricosuric
action, 301
diuretic, 291
Urinary
aldosterone excretion, 19
catecholamine content, 19
creatinine, 19
potassium excretion, 18
sodium excretion, 18, 26
volume excretion, 18

V_{max}, 75
Vagal C fibers, 115, 117
Valsalva's maneuver, 177
Variant angina, 101
Vascular
D_1 receptor, 209
injury, 118
Ventral tegmental area, 157
Ventrolateral medulla (VLM), 247
Verapamil, 145, 148, 358, 359
Voltage-dependent L-channel, 146
Von Bezold-Jarish reflex, 190

Water drinking behavior, 24
Wistar Kyoto rats (WKY), 17, 169, 174, 192

Xylazine, 259

YM-617, 40, 43
Yohimbine, 1

Zofenopril, 108